Handbook of Tropical Residual Soils Engineering

Handbook of Tropical Residual Soils Engineering

Editors

Bujang B.K. Huat
Department of Civil Engineering, Universiti Putra Malaysia, Serdang, Malaysia

David G. Toll
School of Engineering and Computer Sciences, University of Durham, Durham, United Kingdom

Arun Prasad
Department of Civil Engineering, Institute of Technology, Banaras Hindu University, Varanasi, India

CRC Press is an imprint of the
Taylor & Francis Group, an **informa** business
A BALKEMA BOOK

CRC Press/Balkema is an imprint of the Taylor & Francis Group, an informa business

First issued in paperback 2017

Typeset by MPS Limited, Chennai, India

Library of Congress Cataloging-in-Publication Data

Handbook of tropical residual soils engineering / editors, Bujang B.K. Huat, David G. Toll & Arun Prasad.
p. cm.
Includes bibliographical references and index.
ISBN 978-0-415-45731-6 (hardback : alk. paper)
1. Residual materials (Geology)— Handbooks, manuals, etc. 2. Soil mechanics—Handbooks, manuals, etc. 3. Soils—Tropics—Handbooks, manuals, etc.
I. Huat, Bujang B. K. II. Toll, D. G. III. Prasad, Arun.

TA709.5.H36 2012
621.4'51360913—dc23

2012002152

Published by: CRC Press/Balkema
P.O. Box 447, 2300 AK Leiden, The Netherlands
e-mail: Pub.NL@taylorandfrancis.com
www.crcpress.com – www.taylorandfrancis.com

ISBN 13: 978-1-138-07735-5 (pbk)
ISBN 13: 978-0-415-45731-6 (hbk)

Table of Contents

Preface

Residual soils are found in many parts of the world. Like other soils, they are used extensively in the construction; either to build upon or as construction materials. These soils are formed when the rate of rock weathering is more rapid than the transportation of the weathered particles e.g., water, gravity and wind, resulting in much of the soils formed remaining in place. The soils typically retain many of the characteristics of the parent rock. In a tropical region, residual soil layers can be very thick, sometimes extending to hundreds of meters before reaching un-weathered rock. Unlike the more familiar transported sediment soil, the engineering properties and behavior of tropical residual soils may differ and are more difficult to predict and model mathematically.

This handbook presents state-of-the-art knowledge of various experts, including researchers and practicing engineers, who have gained extensive experience with these soils. Despite the abundance and significance of residual soils, dedicated efforts to produce a comprehensive document is far and few in between. Compared with transported sediment soil, our knowledge and understanding of tropical residual soils is not as extensive. Meanwhile, the need for knowledge of the engineering behavior of tropical residual soils is great; thanks to the extensive construction worldwide on such soils. This handbook is specially written to serve as a guide for the engineers, researchers and students on the state-of-the-art knowledge on engineering of tropical residual soils. This unique handbook will constitute an invaluable reference and should be a standard work in the library of any engineer involved in geological, foundation and construction work in tropical residual soils.

The handbook covers almost, if not, all aspects of residual soils from genesis, classification to formation, sampling and testing, behaviour of weakly bonded and unsaturated soil, and volume change and shear strength. It also includes applications of the soil in slope, retaining walls, as road construction materials and foundations. This handbook also features regional case studies of tropical residual soils from regions/countries namely, South America, Hong Kong, Southern India and Southeast Asia.

Chapter 1 of the handbook presents the details of the residual soils, its geographical occurrence, with an emphasis on its classification. This chapter also discusses the engineering peculiarities of the residual soils.

Chapter 2 details the formation and classification of tropical residual soils. The characteristics of these soils and pedogenetic classification are also discussed. Further, the definition and classification of tropical residual soils in civil engineering practice has been given special consideration.

Sampling and testing of tropical residual soils are discussed in Chapter 3. Testing of samples in the lab are discussed including in-situ tests, such as standard penetration test, piezocone penetration test (CPTU), seismic test, pressuremeter, plate load test, dilatometer and permeability.

An overview of the important aspects of unsaturated soil behaviour that are of relevance to tropical residual soils is discussed in Chapter 4. The concepts of suction are introduced and detail the methods to measure suction. The role of suction in determining shear strength is described, together with test results. The volume change behaviour of unsaturated soils (shrinkage, swelling and collapse) is also discussed as this can be of relevance to tropical clayey soils.

The magnitude of volumetric deformations in response to changes in moisture content at given net applied stress is influenced by the mineralogy of clay fraction, soil density and moisture content, surcharge pressure and soil microstructure and these topics are dealt in detail in Chapter 5.

Chapter 6 describes the shear strength models of tropical residual soils. The real soil shear strength behaviour in saturated and partially saturated conditions is very complex. The application of the true soil shear strength behaviour will help to explain complex soil mechanical behaviour. The equipments and the test procedures required to comprehensively characterise soil shear strength behaviour is quite extensive. All these issues are detailed in this chapter.

Chapter 7 focuses on the geological factors and their effect on slope stability, landslip type, classification, effect of relict structures and rainfall, suction and finally address the stabilisation and remedial measures to soil slopes. The author's experience, covering a wide range of tropical regions including the Far East and Central and South Americas, are presented and discussed in detail. Different methods of slope stability analyses that incorporate unsaturated soil mechanics principles are described. The significance of different parameters related to slope stability assessment and slope stabilisation methods described are illustrated through results and knowledge gained from parametric studies, laboratory experiments and field observations made in residual soil slopes in Singapore.

In chapter 8, the authors, based on their combined experiences in many regions of the world where tropical residual soils exist, have tried to emphasise those aspects of foundation engineering on tropical soils which are unconventional. In many circumstances, for the bearing capacity, the ultimate limit state can be designed for without difficulty, and it is the serviceability limit state which governs. To adequately satisfy this limit state requires a reasonable knowledge of foundation performance, and particularly foundation stiffness. Much of this chapter is devoted to discussing methods of predicting foundation behaviour.

Chapter 9 details the residual soil of Hong Kong. Based on field measurements and laboratory tests on intact and recompacted specimens, the hydraulic and mechanical properties of several common Hong Kong saprolites are introduced and described. These engineering properties include the stress-dependent soil water characteristic curve, permeability function, small strain stiffness and shear strength characteristics. Measured results from the saprolites in Hong Kong are also compared with other similar decomposed materials in other places as appropriate.

The residual soils of India are described in chapter 10. The geology of the soil forming rocks is briefly considered as prelude to soil characteristics. This chapter discusses the occurrence and distribution of residual soils, its physic-chemical properties and geotechnical engineering data.

Chapter 11 describes the residual soils of South-east Asia. This chapter discusses the engineering practice in tropical residual soils in South-east Asia, drawing experiences from three South-east Asian countries namely Malaysia, Thailand and Singapore. Tropical residual soils and the factors affecting its formation in the local environment are first briefly reviewed with a discussion on the descriptive scheme used and delimitation considered as soil in engineering practice on the weathered profile. The distribution of tropical residual soils with respect to and their characteristics on different rock follows and then some of their physico-chemical properties are given. This, followed with a discussion on their engineering significance and how tropical residual soils affect slope stability, dam foundations, etc. drawing on local case histories as illustrations.

Acknowledgements

We are very grateful to every person who has contributed, directly or indirectly, to the realisation of this handbook. We would like to extend our sincere thanks to all our authors and co-authors, as well as all other contributors.

The Editors

About the Editors

Professor Bujang B.K. Huat graduated in 1983 from the Polytechnic of Central London, UK, and obtained his MSc and PhD at the Imperial College London, and the Victoria University Manchester, UK, in 1986 and 1991 respectively. He has spent his professional career as a Professor in Geotechnical Engineering, in the Department of Civil Engineering, Universiti Putra Malaysia, one of Malaysia's five research universities. Currently he serves as the Dean of School of Graduate Studies of the same university. His special area of interest is in the field of geotechnical and geological engineering, and slope engineering, and has authored and co-authored 18 books, edited 10 conference proceedings, and published more than 100 journal and conference proceedings papers in the field of soil mechanics and foundation engineering.

Professor David G. Toll is Professor of Geotechnical Engineering in the School of Engineering and Computer Sciences at Durham University, UK. After graduating from Cardiff University with a BSc in Civil Engineering in 1979 he worked for Soil Mechanics Ltd and Engineering & Resources Consultants, before joining Imperial College, London as a Research Assistant, where he gained his PhD. He joined Durham University in 1988. He has been Visiting Professor at the National University of Singapore and Tongji University, China and held Research Fellowships at Nanyang Technological University, Singapore, University of Western Australia and Newcastle University, Australia. His research areas are unsaturated & tropical soils and geoinformatics. He has co-authored 7 edited books and conference proceedings and published more than 150 journal and conference proceedings papers. He is Chair of Joint Technical Committee 2 on Data Representation for the three International Societies (ISSMGE, ISRM, IAEG), a member of TC 106 on Unsaturated Soils of ISSMGE and Chair of the Northern Geotechnical Group of the Institution of Civil Engineers, UK.

Associate Professor Arun Prasad is Associate Professor of Geotechnical Engineering in the Institute of Technology at Banaras Hindu University, India. He graduated with a BSc in Civil Engineering in 1986 from Utkal University, India. He obtained MSc and PhD from Sambalpur and Devi Ahilya University, India in 1989 and 2000 respectively. He has worked as Post-Doctoral Researcher at Universiti Putra Malaysia during 2009–10. His special areas of research are in the soil stabilization of soft and contaminated soils. He has co-authored one book on Geotechnical Engineering and has published more than 50 papers in journals and conference proceedings.

List of Contributors

Professor Faisal Haji Ali has been actively involved in research and consultancy in the field of geotechnical engineering particularly related to ground improvement, partially saturated soils, reinforced earth, piled foundations and slope instability. His current research activities include new and creative ideas involving development and application of a new pile instrumentation technique, development of a new keystone system for soil retention, use of waste material/soil mixtures as backfill materials, application of vegetation for slope stabilization, development of a new landfill liner and a new ground improvement technique. He has served on a number of editorial boards and many committees related to his area of expertise and been invited to present keynote papers at international and local conferences. He has published nine books and more than 250 publications in journals and conference proceedings. Currently he is the Director of Research and Innovation Management Centre and a professor of geotechnical engineering at the National Defence University of Malaysia.

Stephen Buttling has over 35 years' experience in geotechnical engineering on a wide range of infrastructure projects through the UK, South East Asia and Australia. Stephen has worked for contractors, consultants, owners and government agencies, both in offices and on site, and most of his work in the last 30 years has been on major projects in areas where tropical residual soils are prevalent. These have included elevated highways and cable stayed bridges in Bangkok, balanced cantilever bridges in Bangkok and Brisbane, ports and airports in Thailand and Australia, power projects, both thermal and hydro-electric, in Thailand, and energy related projects in Australia. His main interests are deep foundations, especially the interaction with the structure, deep excavations, and soft ground engineering.

Dr. Roberto Quental Coutinho was a Graduate in Civil Engineering (1969-1973) from the Federal University of Pernambuco, Department of Civil Engineering, after which he earned an MSc (1974-1976) and later a PhD (1981-1986) in Geotechnical Engineering from the Federal University of Rio de Janeiro, COPPE – Civil Engineering Program. He is currently Head of the Civil Engineering Graduate Program at the Federal University of Pernambuco. He has supervised or is supervising 14 PhD theses 33 MSc theses.

His expertise lies in laboratory and in situ testing, field observations/instrumentation, characterization and evaluation of soil parameters (saturated and unsaturated), studies of risk analysis and management, design of foundations, shallow and deep, slope stability and erosion studies and stabilization and embankment on soft soils.

He has coordinated a number of CNPq Program Universal Projects, was coordinator of the PRONEX Program Project (CNPq/FACEPE, period 2007-2011) and is currently sectorial coordinator (UFPE) of the REAGEO-INCT Project (MCT/CNPq/FAPERJ, 2009-2013: Federal Program for National Institute of Science and Technology) and coordinator of the CNPq Research Productivity Project, 2007 – 2013 (CNPq Researcher, since 1977).

He has (co-)authored papers in various refereed scientific journals (international and national), chapters in nationally published books, has presented papers at national and international conferences and courses (Brazil, Taiwan, USA, Norway, Italy, Great Britain, Mexico, Japan, Venezuela and Portugal) and has written numerous reports on research and technical activities, laboratory and field studies, design and consultancy.

Dr. Apiniti Jotisankasa is currently an Assistant Professor at the Department of Civil Engineering, Kasestsart University Bangkok, Thailand. After obtaining his BEng degree in Civil Engineering from Kasetsart University in 1999, he went on to pursue his MSc and PhD in Soil Mechanics at Imperial College London, UK, with the generous support of the Anandamahidol Scholarship from Thailand.

After being awarded the PhD degree in 2005, he started working for Kasetsart University as a lecturer in geotechnical engineering, focusing his main research activities on application of unsaturated soil mechanics on practical geotechnical engineering problems, such as rainfall-induced landslide, embankment stability and bio-slope engineering. In 2011, he received the Young Technologist Award from the Foundation for the Promotion of Science and Technology under the Patronage of His Majesty the King of Thailand. Dr. Apiniti has been secretary general of the Thai Geotechnical Society since 2009 and is currently a member of the TC106 (Unsaturated Soils) of the International Society of Soil Mechanics and Geotechnical Engineering.

Dr. E.C. Leong obtained his Bachelor in Engineering with first class honours and Masters in Engineering from the National University of Singapore. He obtained his PhD from the University of Western Australia. Currently he is an Associate Professor in the Division of Infrastructure Systems and Maritime Studies Division at the School of Civil and Environmental Engineering, Nanyang Technological University, Singapore. His area of specialization is in geotechnical engineering and his research interests include laboratory and field testing, unsaturated soil behaviour, soil dynamics, foundation engineering and numerical modeling. He has authored and co-authored more than 200 publications in international journals and conferences. He is actively involved in laboratory accreditation and standards programs in Singapore. He currently chairs the SPRING Technical Committee on Civil and Geotechnical Works and is the deputy chair of the Singapore Accreditation Council Committee for Inspection Bodies.

Professor Charles W.W. Ng is Chair Professor at the Department of Civil and Environmental Engineering, the Director of the Geotechnical Centrifuge Facility and a Senior Fellow for the Institute of Advanced Study at the Hong Kong University of Science and Technology. He obtained his PhD from the University of Bristol, UK in 1992 and subsequently joined the University of Cambridge as a Research Associate before returning to Hong Kong in 1995. He was elected as an Overseas Fellow at Churchill College, Cambridge, in 2005. Professor Ng is a Chartered Civil Engineer (CEng) and a Fellow of the Institution of Civil Engineers (FICE), the American Society of Civil Engineers (FASCE), the Hong Kong Institution of Engineers (FHKIE) and the Hong Kong Academy of Engineering Sciences (FHKEng). He is also a Changjiang Scholar (Chair Professorship) appointed by the Ministry of Education in China.

Professor Ng is interested in unsaturated and saturated soil behaviour and modelling, slope instability analyses and soil-structure interaction problems such as deep excavations, piles and tunnels. He has published over 250 international journal and conference papers. He is the leading author of two reference books: "*Advanced Unsaturated Soil Mechanics and Engineering*" published by Taylor & Francis and "*Soil-Structure Engineering of Deep Foundations, Excavations and Tunnels*" published by Thomas Telford. Currently Professor Ng is an Associate Editor of the Canadian Geotechnical Journal and a Board Member of the International Society of Soil Mechanics and Geotechnical Engineering.

Dr. Mohd Jamaludin Md Noor obtained his first degree, a B.Sc. in Civil Engineering (Hons) from the University College of Swansea, U.K. in 1984. He obtained a M.Sc. in Geotechnics from University MALAYA, Malaysia in 1996 and a Ph.D. from the University of Sheffield, U.K. in 2006. He is currently Associate Professor in the Faculty of Civil Engineering at the Universiti Teknologi MARA (UiTM), Malaysia.

His focal research areas include shear strength behaviour of soil, inundation settlement and landslide. He has introduced the non-linear envelope soil shear strength model relative to net stress and suction and incorporated that in his state-of-the-art infiltration type slope stability method which can back-analyse shallow landslide at ease. He has also introduced a state-of-the-art early warning slope monitoring system which is the first to monitor rainfall-induced instability base on the advancement of wetting front.

He has been actively involved in the development of new geotechnical products like the V-armed tip soil nail anchor. At international level he has won a Best Award from the Malaysian Technology Exhibition 2010, a Best Design Award from Japan Intellectual Property Association 2010, three gold medals and three silver medals. He also received the UiTM Deputy Vice Chancellor (Research) Innovation Award in 2010 and the UiTM Academic Award for book publication in 2011.

He served as a member of the International Advisory Committee for the Asia Pacific Unsaturated Soils Conference in Newcastle, Australia, 2009 and in Pattaya, Thailand, 2012. Besides, he is also a member of Technical Committee for the drafting of Malaysian Annex to Eurocode 7 for Geotechnical Works under SIRIM Berhad.

Beto Ortigao is Terratek's technical director, working on geotechnical engineering consultancy and design, as well as in situ testing and monitoring. He is involved in foundations, excavations, slopes, tunnels and dams.

He received his first degree in Civil Engineering in 1971 from the Federal University of Rio de Janeiro (UFRJ), followed by a PhD in 1980. From 1978 to 2003 he was an academic at the Federal University of Rio de Janeiro.

From 1982 to 1984 he was a researcher on offshore foundations at the Building Research Establishment, UK, followed by work on offshore foundation design at Fugro UK Ltd.

He has been Visiting Professor at the following universities: UBC University of British Columbia, Canada, the City University of Hong Kong and the University of Western Sydney, Australia.

He authored the textbook "*Soil mechanics in the light of critical state theories*" published by Balkema in 1995 and, in 2004, the *Handbook of slope stabilisation*, published by Springer Verlag.

Professor Harianto Rahardjo has conducted extensive research on unsaturated soil mechanics to solve geotechnical problems associated with tropical residual soils. His research focus has been on rainfall-induced landslides, one of the major natural disasters occurring in many parts of the world. He has applied unsaturated soil mechanics in understanding the mechanism of rainfall-induced slope failures and the mechanical behaviour of unsaturated residual soils. These research activities have led to the development of an advanced unsaturated soil mechanics laboratory at Nanyang Technological University and numerous comprehensive instrumented slopes. Professor Rahardjo's recent research activities involved the application of unsaturated soil mechanics in capillary barrier system for slope stabilization through laboratory and numerical analyses and instrumented capillary barrier systems in the field. He has authored over 200 technical publications and is the co-author together with Professor D.G. Fredlund of the first textbook on unsaturated soils "*Soil Mechanics for Unsaturated Soils*", published by John Wiley in 1993 and in Chinese translation in 1997 and in Vietnamese translation in 2000. He has also presented his research works in many keynote/invited lectures and short courses and served as a consultant in various countries.

Dr. Sudhakar M. Rao is a Professor in the Department of Civil Engineering and Chairman of the Center for Sustainable Technologies, Indian Institute of Science, Bangalore. His research areas include unsaturated soil behaviour, hazardous waste management, groundwater geochemistry and remediation of contaminated groundwater. He has published 70 journal papers in leading journals and supervised 13 PhD and 6 MSc (Engineering) dissertations in the broad area of Geoenvironmental engineering. He has been a visiting scientist at McGill University and Ryerson University in Canada and Cardiff University in the UK. He has several sponsored research projects in Geoenvironment engineering funded by the Department of Science and Technology, Bureau of Research in Nuclear Sciences and by private industry to examine diverse problems of remediating fluoride contaminated groundwater, developing a bentonite barrier to contain high level nuclear wastes in geological repositories, re-use of hazardous waste in civil engineering applications and the impact of anthropogenic contamination on groundwater geochemistry.

Dr. R.B. Rezaur is a Water Resources Engineer with Golder Associates Ltd., Canada. He has over 20 years of experience in the field of Water Resources and Geotechnical Engineering involving numerical modelling, statistical analyses, hydrology, hydraulics, sediment transport processes, river engineering, hydrologic monitoring, hydraulic structures, storm water management, unsaturated flow and transport processes in porous media. Dr. Rezaur has a mixed background in teaching, research and consultancy. Prior to joining Golder Associates in Canada he worked as an academic staff member at the University Teknology Petronas and University Sains Malaysia in Malaysia. Dr. Rezaur has also worked with Professor Harianto Rahardjo at Nanyang Technological University, Singapore. In his early career, Dr. Rezaur was a member of staff at the Khulna University of Engineering and Technology, Bangladesh. Rezaur has co-authored over 50 publications in the form of book chapters and journal and conference articles.

Fernando Schnaid Born on 18 April 1957 in Brazil, Professor Schnaid holds an MSc in Civil Engineering from the Catholic University of Rio de Janeiro, Brazil (1983) and a PhD from Oxford University, UK (1990). He is the author or co-author of over one hundred peer-reviewed articles and conference papers, as well as 5 books, including the title *In Situ Testing in Geomechanics* published in 2009. He received the Terzaghi Prize from the Brazilian Society of Soil Mechanics and the 5th Schofield Award for the best published paper in the International Journal of Physical Modelling in Geotechnics. He is currently Professor of Civil Engineering at the Federal University of Rio Grande do Sul, Brazil. His key research focuses on in-situ testing, geotechnical site characterization, foundation systems and prediction of soil and rock properties. He had delivered international lectures including the State-of-the-Art Report in Soil Properties at the Osaka Conference for the ISSMGE and has given continual educational lectures, courses and technical seminars both nationally and internationally, which have taken him to Argentina, Australia, Belgium, England, Germany, Italy, Poland, Paraguay, Portugal, Scotland and the USA.

Dr. Harwant Singh is a Lecturer in Environmental Geochemistry at the University Malaysia Sarawak, Malaysia and has been in academics since the 1990s. After a stint in the mineral exploration industry he joined the fledgling Faculty of Resource Science and Technology in the newly established university where the research opportunity in the early years was involvement, for the earth science component, in its regional studies, as well as, collaboration, for the engineering geology aspect, in geotechnical engineering in the Faculty of Engineering. He also pioneered the teaching of engineering geology in the Faculty of Engineering.

Currently, his research interests are understanding environmental contamination and environmental systems in general forming the natural systems on Earth. His research focus areas are the environmental geochemistry of contaminated environmental media and by corollary pertinent aspects of its management. He has taught various courses, engaged in research projects and has participated in consultancy work. He has also authored various publications including journal articles.

Dr. Venkatesh holds a B.E. (Civil) from Bangalore University (1979), and a M.Sc. (Geotechnical, 1983) and a Ph.D. (Geotechnical, 1989) from IISc., Bangalore. He has extensive experience of over 22 years in material testing and geotechnical engineering problems in the field. He has undertaken prestigious projects all over India such as pressure grouting for stabilizing loose flowing soils to facilitate construction of tunnels for the Konkan Railway in Byndoor, Karnataka, and Goa, vertical and raker piles for the Kanteerava Indoor Stadium, Bangalore for L&T (ECC), the design and execution of a protection system for facilitating deep excavations for the NHAI project in Salem, and consultancy for a KUDCEM project in Mangalore. He has published many papers in international and national journals and conference proceedings.

He presently heads Aptech Foundations, Bangalore, a leading geotechnical engineering firm undertaking geotechnical investigations into pile foundations, ground improvement, soil stabilisation, grouting, soil nailing, micropiles and more, all over India.

Antonio Viana da Fonseca is Associate Professor of Civil Engineering with Full Professor Habilitation since December 2008. He is Director of the Geotechnical Division and the Geotechnical Laboratory of Civil Engineering Department of FEUP. He is a Geotechnical Specialist and Fellow (Counselor Member) of the Portuguese Institution of Engineers (OE-P). He is also President of the College of Geotechnical Specialists of the Portuguese Association of Engineers and Vice-Chairman TC-16/102 of "Ground Property Characterization by In-Situ Tests" of ISSMGE.

He was responsible and/or consultant for 92 projects, including bridge foundations, embankments on soft soils, large excavations and tunnels (in Portugal, Algeria, Brazil, Morocco, Mozambique, Poland and Spain). He coordinated 280 experimental processes in LabGeo-FEUP, with formal reports.

He is coordinator of several research programs, in the areas of experimental characterization and modeling of geomechanical behaviour of non-textbook soils, seismic analysis and cyclic liquefaction, monotonic and flow liquefaction in tailings, behaviour of foundations and earth retaining structures in residual soils, and static and dynamic properties characterization and modeling of stabilized soils for subgrade and foundations.

He supervised 16 PhD theses and of 41 MSc theses and has published extensively, authoring or co-authoring 23 articles in international refereed journals and 30 others in national journals; he has published 112 communications in proceedings of international conferences (8 keynote/special lectures in national and international conferences) and 42 in national conferences; 44 technical documents for workshops and courses. He has authored 7 book chapters and edited 2 proceedings published by international publishers; furthermore, he is the author of 3 nationally published books and of 2 chapters in nationally published books.

Dr. Jie Xu is currently a post-doctoral fellow at the Department of Civil and Environmental Engineering at the Hong Kong University of Science and Technology (HKUST). She obtained her PhD in Civil Engineering in 2011 from HKUST. Her research interests include mechanical and hydraulic behaviour of unsaturated soil, small strain stiffness of soil and its modelling and application, biochemical and climate effects on soil behaviour and water-gas flow mechanism in landfill cover. She has done a lot of research work on unsaturated soil behaviour, especially experimental and theoretical study of small strain stiffness of unsaturated soil. She has published around 10 international journal and conference papers.

1

Introduction

Bujang B.K. Huat
Universiti Putra Malaysia, Serdang, Malaysia

David G. Toll
University of Durham, Durham, United Kingdom

1.1 AIM AND SCOPE

Residual soils are found in many parts of the world and are extensively used in the construction of foundations and as a construction material. In tropical areas, residual soil layers are often extensive and may measure a few hundred meters until unweathered rock is reached. As the major soil and foundation engineering conditions are determined by this type of soil, evidently all aspects should be known before the engineering work is started.

The Handbook of Tropical Residual Soil Engineering is intended as a complete reference source and manual for every engineer working on or interested in soil and foundation engineering in tropical areas. Almost every aspect of tropical residual soils is treated, including major applications, and a dedicated part is focused on region- and country-specific sections, including typical characteristics and soil conditions. Some tables and charts with typical data are included. This book includes among others the following topics: the genesis and classification of residual soils, the role of climate in soil formation (in particular tropical residual soil and why residual soils are different), the sampling and testing of tropical residual soils, the behaviour of weakly bonded and unsaturated soils, and the volume change and shear strength of tropical residual soils. There is a section on engineering applications of tropical residual soils such as in slopes and foundations. The book also features regional/country case studies where these soils are found, such as Hong Kong, India and Southeast Asia.

This unique handbook will constitute an invaluable reference and should be a standard work in the library of any engineer involved in geological, foundation and construction engineering work in tropical residual soil.

1.2 SOILS

Soils constitute the multi-phased interface stratum formed by the interaction of the lithosphere with the atmosphere, hydrosphere and biosphere. This dynamism may be expressed through the metaphor of the Earth as an engine where the atmosphere is the working fluid of the Earth's heat engine (Ingersoll, 1983). This drives the dynamics on the terrestrial surface as expressed through the operation of physical and chemical processes at the Earth's surface resulting in the formation of soils.

Soils form a mantle of unconsolidated superficial cover that is variable in thickness. This is due to the fact that the depths to which the terrestrial materials, namely rocks, have been altered to form soils vary as a function of factors such as climate, topography and the nature of the subsurface. It is also due to the varying thickness of transported materials that come to form the unconsolidated superficial cover.

Soils may, therefore, be grouped into two broad categories: residual and transported soils. Residual soils form or accumulate and remain at the place where they are formed. Transported soils are formed from materials originating elsewhere that have moved to the present site, where they constitute the unconsolidated superficial layer.

The physical processes involved in the operation of their agents of transportation, i.e. gravity, wind, water etc., have dislodged, eroded and transported soil particles to their present location.

1.3 RESIDUAL SOILS

According to McCarthy (1993), residual soils are those that form from rock or accumulation of organic material and remain at the place where they were formed. This entire unconsolidated superficial cover is referred to as soil (Press and Siever, 1994). However, according to Bland and Rolls (1998) this mantle is also termed the *regolith*, which is separated into an upper part, referred to as *soil*, and the portion below it and above the bedrock, called the *saprolite*. The regolith is chemically altered, especially in humid tropical regions. There are, however, a variety of definitions of residual soils, indicating a diversity of understanding of what are perceived as residual soils.

Brand and Philipson (1985) define residual soil as 'a soil formed by weathering in place, but with the original rock texture completely destroyed'. This term is commonly used in a wider sense to include highly or completely decomposed rock, which is an engineering material and behaves like a soil in places such as Hong Kong. Blight (1985) gave the definition of residual soil in South Africa as all material of a soil-like consistency that is located below the local ancient erosion surface, i.e. below the pebble marker. The exceptions are the extensive deposits of ancient windblown desert sands with some cohesion. Materials of this type are also considered to be residual soils. Soil reworked in-situ by termites is also strictly residual and not transported material. Sowers (1985) defined residual soil as the product of rock weathering that remains in place above the yet-to-be-weathered parent rock. The boundary between rock and soil is arbitrary and often misleading. There is a graduation of properties and no sharp boundaries within the weathering profile. The Public Works Institute of Malaysia (1996) defines residual soil as 'a soil which has been formed in-situ by the decomposition of parent material and which has not been transported any significant distance' and defines tropical residual soil as 'a soil formed in-situ under tropical weathering conditions'. Other workers echo the above definitions as soil formed in-situ by the weathering of rocks whereby the original rock texture is completely destroyed, or soil formed by weathering in places where the original rock is completely destroyed nearer the surface. The rocks have totally disintegrated and the mass behaves like a soil.

Compositionally, the solid phase is constituted not only from the residue of unaltered terrestrial material but also from the products of the interaction of terrestrial material with the agents of surface processes. In addition, there are aqueous (water) and gaseous phases. That portion, normally the upper regolith that supports plant life, also includes organic matter (both living and dead).

1.4 GEOGRAPHICAL OCCURRENCE OF RESIDUAL SOILS

Residual soils occur in all parts of the world on various types of rocks as shown in Figure 1.1(a)–(d). Conditions conducive to the formation of chemically weathered residual soils are not present in temperate zones but extensive remnant deposits of such soils are found from periods when hot, humid conditions existed (Brand and Philipson, 1985).

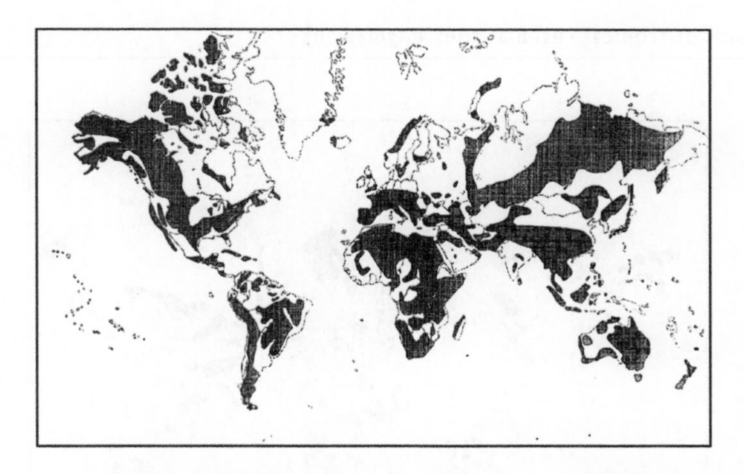

Figure 1.1a Residual soils developed from sedimentary rocks (*After* Rollings and Rollings, 1996)

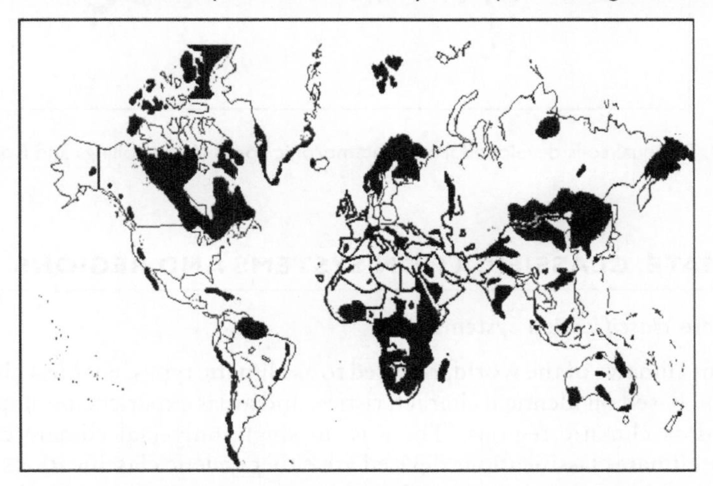

Figure 1.1b Residual soils developed from intrusive igneous rocks, primary granitic shields and mountains (*After* Rollings and Rollings, 1996)

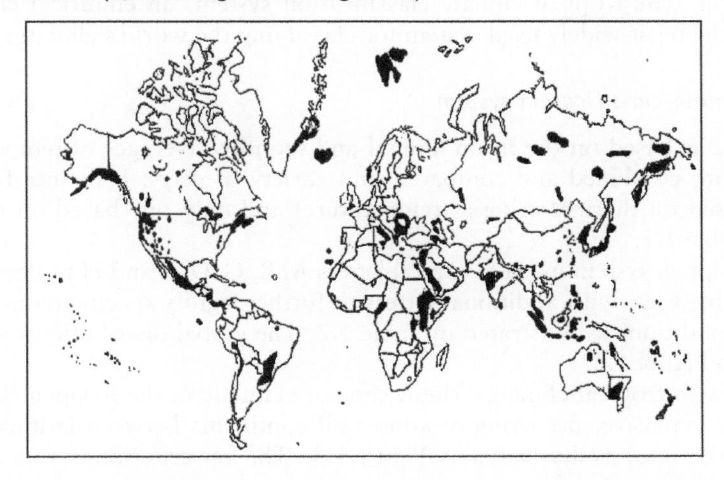

Figure 1.1c Residual soils developed from extrusive igneous rocks, primary basaltic plateaus and mountains (*After* Rollings and Rollings, 1996)

Figure 1.1d Residual soils developed from metamorphic rocks (*After* Rollings and Rollings, 1996)

1.5 CLIMATE, CLASSIFICATION SYSTEMS AND REGIONS

Global climate classification systems

The different climates of the world, referred to as climatic types, enable a global climate classification based on identical characteristics, and areas experiencing similar climates are grouped as climatic regions. There is no single universal climate classification system. The climate classifications devised are either genetic classifications based upon mechanisms such as net radiation, thermal regimes or air-mass dominance over a region, or empirical classifications based on recorded data such as temperature and precipitation. The Köppen climate classification system, an empirical classification system, is the most widely used system for classifying the world's climates.

Köppen climate classification system

This system is based on the mean annual and monthly averages of temperature and precipitation, combined and compared in a variety of ways. It defines five climatic regions based on thermal criteria (temperature) and only one based on moisture as listed in Table 1.1.

The Köppen system uses the capital letters A, B, C, D, E and H to designate these climatic categories, with additional letters to further signify specific temperature and moisture conditions as illustrated in Table 1.2. The global distribution of the system is shown in Figure 1.2.

Areas with tropical climates (the A category climate in the Köppen classification system) are extensive, occurring in almost all continents between latitudes 20°N to 20°S of the equator as illustrated in Figure 1.3. The key criterion for an A category

Table 1.1 The Köppen climatic regions

Category	*Climate region*
Based on temperature	
A	Tropical moist (equatorial regions)
C	Moist mid-latitude with mild winters (Mediterranean, humid subtropical)
D	Moist mid-latitude with cold winters (humid continental, subarctic)
E	Polar with extremely cold winters and summers (polar regions)
H	Highland (cool to cold, found in mountains and high plateaus)
Based on moisture	
B	Dry with deficient precipitation through most of the year (deserts and steppes)

Table 1.2 Modified Köppen climatic classification (*After* Bergman and McKnight, 1993)

Letters					
1st	*2nd*	*3rd*	*Description*	*Definition*	*Types*
A			Low-latitude humid climates	Average temperature of each month above 18°C	Tropical wet (Af) Tropical monsoonal (Am) Tropical savanna (Aw)
	f		No dry season	Average rainfall of each month at least 6 cm	
	m		Monsoonal; short dry season compensated by heavy rains in other months	1 to 3 months with average rainfall less than 6 cm	
	w		Dry season in 'winter' (low sun season)	3 to 6 months with average rainfall less than 6 cm	
B			Dry climates; evaporation exceeds precipitation		Subtropical desert (Bwh) Subtropical steppe (Bsh) Mid-latitude desert (Bwk) Mid-latitude steppe (Bsk)
	w		Arid climates; 'true deserts'	Average annual precipitation less than about 38 cm in low latitudes; 25 cm in mid-latitudes	
	s		Semiarid climates; steppe	Average annual precipitation between about 38 cm and 76 cm in low latitudes; between about 25 cm and 64 cm in mid-latitudes; without pronounced seasonal concentration	
		h	Low-latitude dry climate	Average annual temperature more than 18°C	
		k	Mid-latitude dry climate	Average annual temperature less than 18°C	

(*Continued*)

Table 1.2 Continued

Letters					
1st	*2nd*	*3rd*	*Description*	*Definition*	*Types*
C			Mild mid-latitude dry climates	Average temperature of coldest month between 18°C and −3°C; average temperature of warmest month above 10°C	Mediterranean (Csa, Csb) Humid subtropical (Cfa, Cwa)
	s		Dry summer	Driest summer month has less than 1/3 the average precipitation of wettest winter month	Marine west coast (Cfb,Cfc)
	w		Dry winter	Driest winter month has less than 1/10 the average precipitation of wettest summer month	
	f		No dry season	Does not fit either s or w above	
		a	Hot summers	Average temperature of warmest month more than 22°C	
		b	Warm summers	Average temperature of warmest month below 22°C; at least 4 months with average temperature above 10°C	
		c	Cool summers	Average temperature of warmest month below 22°C; less than 4 months with average temperature above 10°C	
D			Humid mid-latitude climates with severe winters	4 to 8 months with average temperature more than 10°C	
	2nd and 3rd letters same as in C climates				
		d	Very cold winters	Average temperature of coldest month less than −38°C	
E			Polar climates; no true summer	No month with average temperature more than 10°C	Humid continental (Dfa, Dfb, Dwa, Dwb) Subarctic (Dfc, Dfd, Dwc, Dwd)
	T		Tundra climates	At least one month with average temperature more than 0°C but less than 10°C	Tundra (ET)
	F		Ice cap climates	No month with average temperature more than 0°C	Ice cap (EF)
H			Highland climates	Significant climatic changes within short horizontal distances due to altitudinal variations	Highland (H)

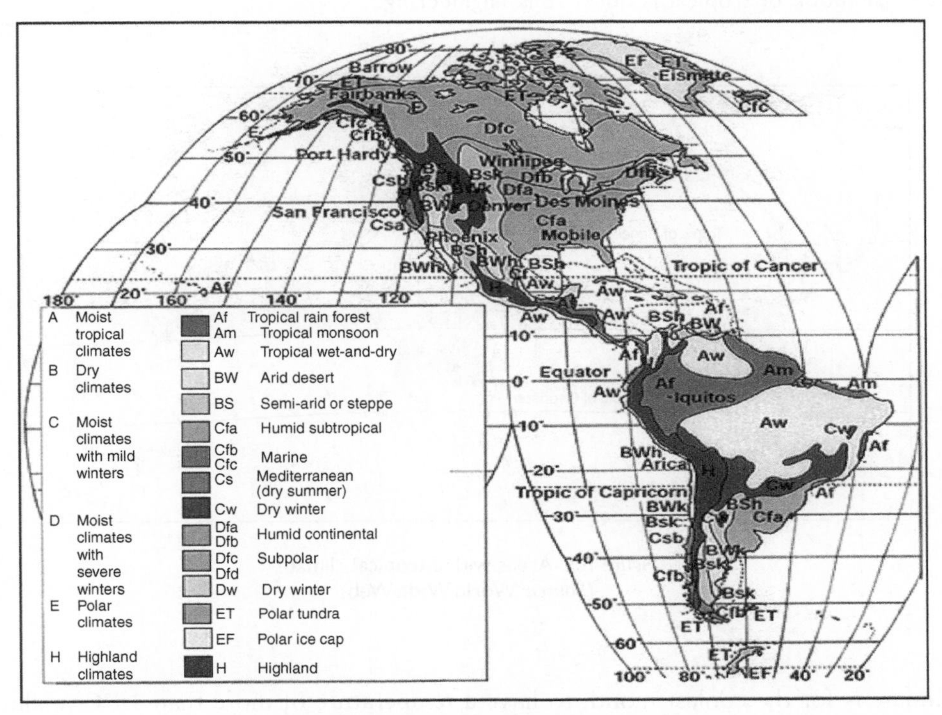

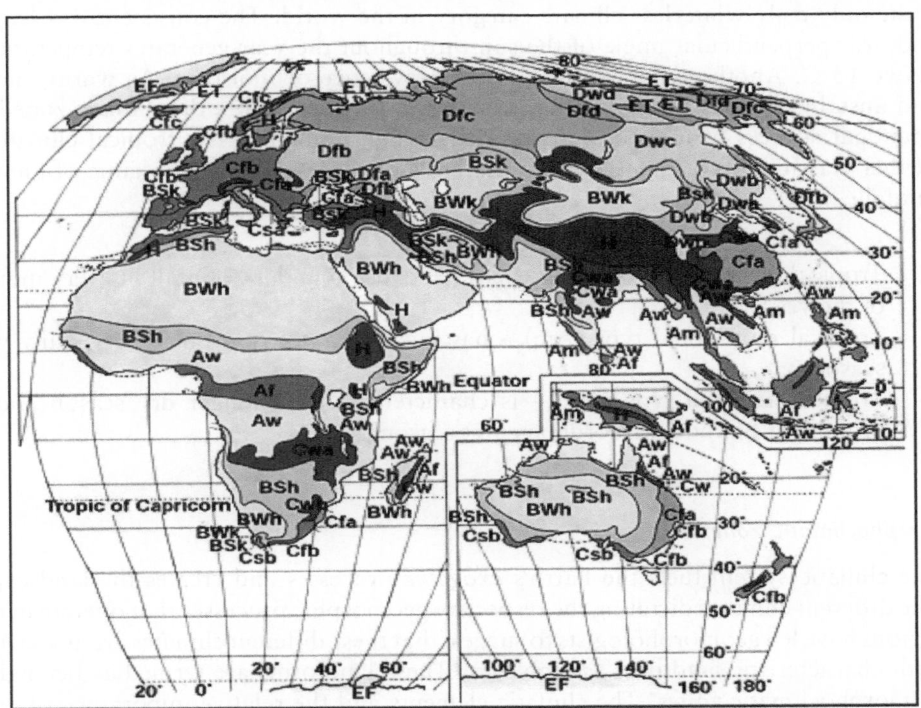

Figure 1.2 Global distribution of the Köppen classification system
(*Source*: http://calspace.ucsd.edu/virtualmuseum/climatechange1/07_1.shtml)

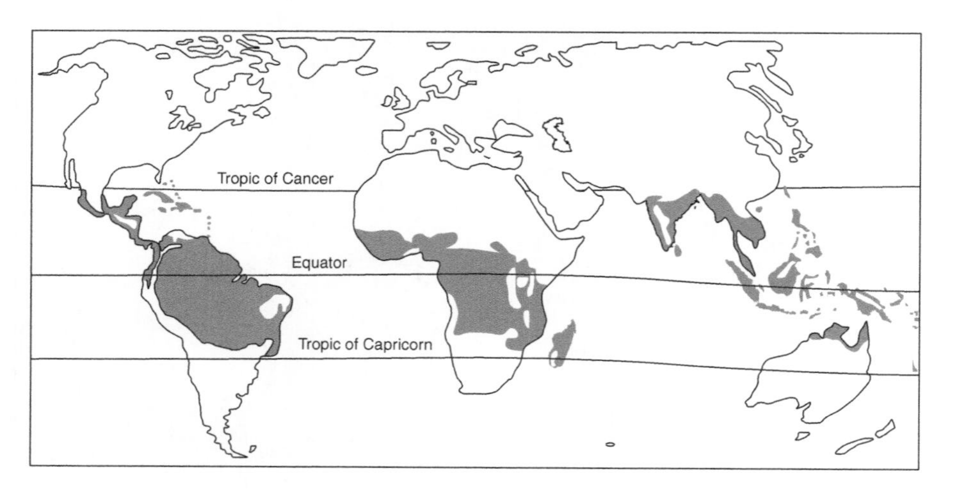

Figure 1.3 Areas with a tropical climate
(*Source*: World Wide Web)

climate is for the coolest month to have a temperature of more than 18°C, making it the only truly winterless climate category in the world. The consistent day length and near-perpendicular angle of the sun throughout the year generates temperatures above 18°C. Another characteristic is the prevalence of moisture. As warm, moist and unstable air masses dominate the oceans at these latitudes, this climate zone has abundant sources of moisture, giving rise to high humidity. The tropical climate is further classified into three types on the basis of the quantity and regime of annual rainfall:

a) Tropical wet type (Af) – experiences relatively abundant rainfall in every month of the year.
b) Tropical monsoonal type (Am) – has a short dry season but a very rainy wet season.
c) Tropical savanna type (Aw) – is characterised by a longer dry season and a prominent but not extraordinary wet season.

Morphoclimatic zones

The climatic system fuels the Earth's *exogenic* processes and creates the landscape. The different climates dictating the respective geomorphic processes that operate in the regions have led geomorphologists to suggest that these different climates are associated with characteristic landform assemblages. These landforms are termed as belonging to morphoclimatic zones. The climatic elements and the relative importance of geomorphologic processes for the various delineated morphological zones are described in Table 1.3. Figure 1.4 shows the geographical distribution of these zones.

Table 1.3 The Earth's major morphoclimatic zones (*After* Summerfield, 1996)

Morphoclimatic zone	*Mean annual temperature (°C)*	*Mean annual precipitation (mm)*	*Relative importance of geomorphic processes*
Tropical humid	20–30	>1500	High potential rates of chemical weathering; mechanical weathering limited; active, highly episodic mass movement; moderate to low rates of stream corrosion but locally high rates of dissolved and suspended load transport
Tropical wet-dry	20–30	600–1500	Chemical weathering active during wet season; rates of mechanical weathering low to moderate; mass movement fairly active; fluvial action high during wet season with overland and channel flow; wind action generally minimal but locally moderate in dry season
Tropical semi-arid	10–30	300–600	Chemical weathering rates moderate to low; mechanical weathering locally active, especially on drier and cooler margins; mass movement locally active but sporadic; fluvial action rates high but episodic; wind action moderate to high
Tropical arid	10–30	0–300	Mechanical weathering rates high (especially salt weathering); chemical weathering minimal; mass movement minimal; rates of fluvial activity generally very low but sporadically high; wind action at a maximum
Humid mid-latitude	0–20	400–1800	Chemical weathering rates moderate, increasing to high at lower latitudes; mechanical weathering activity moderate with frost action important in higher fluvial processes; wind action confined to coasts
Dry continental	0–10	100–400	Chemical weathering rates low to moderate; mechanical weathering, especially frost action, seasonally active; mass movement moderate and episodic; fluvial processes active in wet season; wind action locally moderate
Periglacial	<0	100–1000	Mechanical weathering very active with frost action at a maximum; chemical weathering rates low to moderate; mass movement very active; fluvial processes seasonally active; wind action rates locally high
Glacial	<0	0–1000	Mechanical weathering rates (especially frost action) high; chemical weathering rates low; mass movement rates low except locally; fluvial action confined to seasonal melt; glacial action at a maximum; wind action significant
Azonal mountain zone	Highly variable	Highly variable	Rates of all processes vary significantly with altitude; mechanical and glacial action significant at high elevations

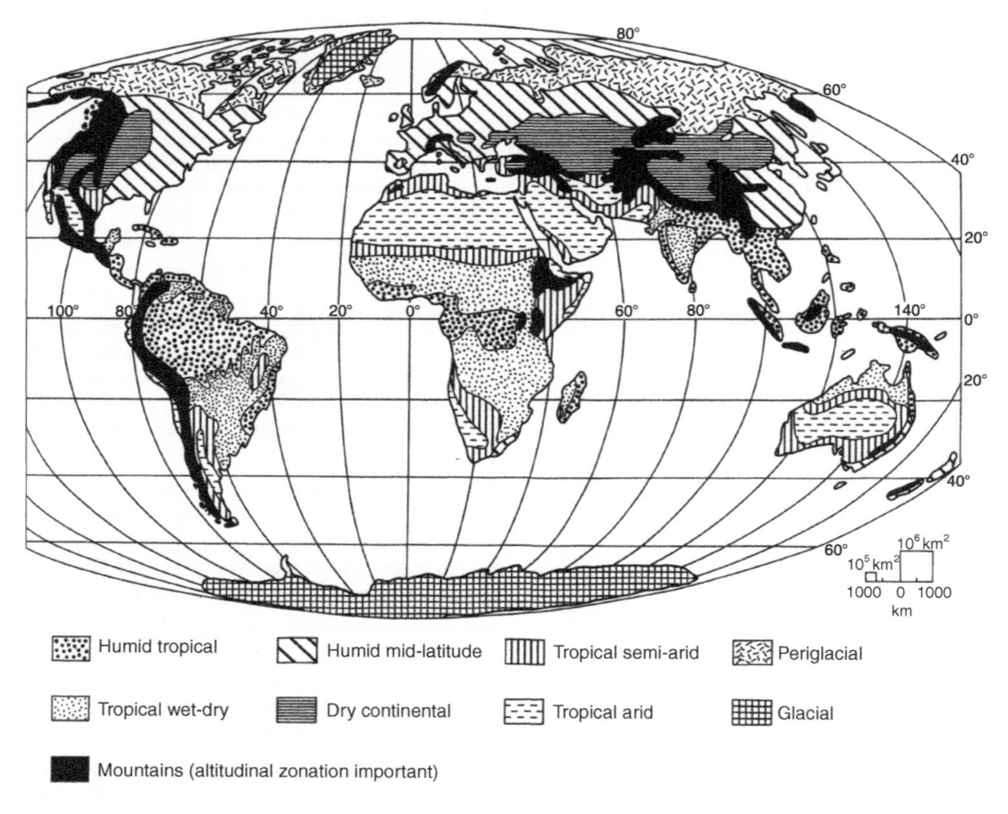

Figure 1.4 Global distribution of morphoclimatic zones (*After* Summerfield, 1996)

1.6 DISTRIBUTION OF TROPICAL RESIDUAL SOILS

The distribution of tropical residual soils is shown in Figure 1.5. Large expanses of residual soils to great depths are normally found in the tropical humid regions, e.g. Northern Brazil, Ghana, Malaysia, Nigeria, Southern India, Sri Lanka, Singapore and the Philippines, due to active weathering leading to residual soil formation.

1.7 ENGINEERING PECULIARITIES OF TROPICAL RESIDUAL SOILS

The term 'soil' in engineering is commonly used to refer to any kind of loose, unconsolidated natural material enveloping the surface that is relatively easy to separate by even gentle means (Terzaghi and Peck, 1967). According to Johnson and De Graff (1988), any mineral that lacks high strength is considered a soil. As a consequence, both the above two types of unconsolidated superficial cover are referred to as soil. The residual soils found in the tropics form the main subject matter of this book.

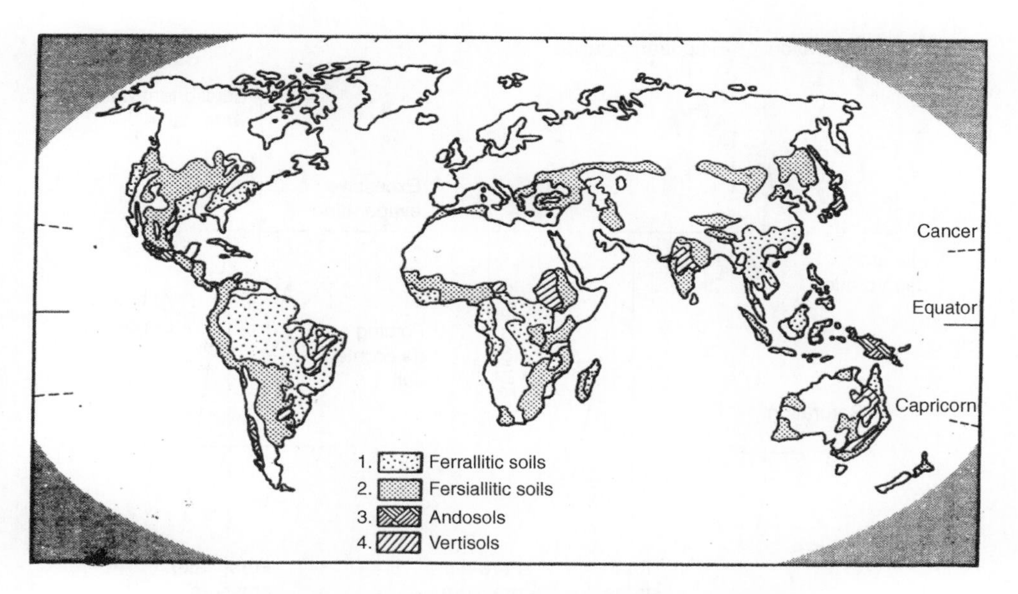

Figure 1.5 Distribution of tropical residual soils (*After* Bell, 2000)

Residual soils are generally located above the groundwater table. The soils are therefore generally unsaturated and possess negative pore water pressures. Climatic changes (i.e. evaporation and infiltration) and transpiration influence the water content and the negative pore-water pressure of the unsaturated soils, especially those located close to the ground surface. As a result, the hydraulic properties, shear strength and volume of the soil vary in response to the climatic changes. The traditional practices of soil mechanics have undergone significant changes during the past few decades. Some of these changes are related to increased attention being given to the unsaturated soil zone above the groundwater table. The computational capability available to the geotechnical engineer has strongly influenced the engineer's ability to address these complex problems. The unsaturated soil zone is subjected to a flux-type boundary condition for many of the problems faced by geotechnical engineers. Unsaturated soil mechanics has become a necessary tool for analyzing the behaviour of soils in this zone.

The behaviour of numerous materials encountered in engineering practice is not consistent with the principles and concepts of classical, saturated soil mechanics. Commonly, it is the presence of more than two phases that results in a material that is difficult to deal with in engineering practice. Soils that are unsaturated form the largest category of materials which do not adhere to the behaviour of classical, saturated soil mechanics. An unsaturated soil has more than two phases, and the pore-water pressure is negative relative to the pore-air pressure. Any soil present near the ground surface in a relatively dry environment will be subjected to negative pore-water pressures and possible desaturation. The process of excavating, remolding and recompacting a soil also results in unsaturated material. The resulting materials form a large category of

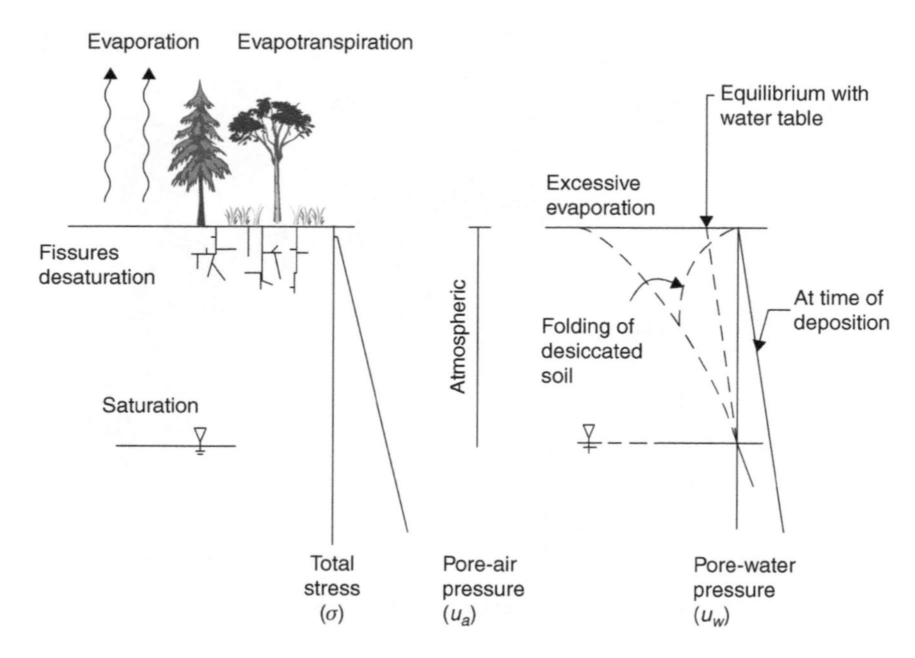

Figure 1.6 Stress distribution during the desiccation of a soil (*After* Fredlund and Rahardjo, 1993)

soils that have been difficult to consider within the framework of classical soil mechanics. Natural surficial deposits of soil have relatively low water content over a large area of the Earth. Residual soils have been of particular concern in recent years. Once again, the primary factor contributing to their unusual behaviour is their negative pore-water pressure. Attempts have been made to use design procedures on these soils based on saturated soil mechanics, with limited success.

Climate plays an important role in whether a soil is saturated or unsaturated. Water is removed from the soil either by evaporation from the ground surface or by evapotranspiration from a vegetative cover (Figure 1.6). These processes produce an upward flux of water out of the soil. On the other hand, rainfall and other forms of precipitation provide a downward flux into the soil. The difference between these two flux conditions on a local scale largely dictates the pore-water pressure conditions in the soil. A net upward flux produces a gradual drying, cracking and desiccation of the soil mass, whereas a net downward flux eventually saturates a soil mass. The depth of the water table is influenced, amongst other things, by the net surface flux. Grasses, trees and other plants growing on the ground surface dry the soil by applying a tension to the pore-water through evapotranspiration. Most plants are capable of applying 1–2 MPa of tension to the pore-water prior to reaching their wilting point. Evapotranspiration also results in the consolidation and desaturation of the soil mass.

Year after year, the deposit is subjected to varying and changing environmental conditions. These produce changes in pore-water pressure distribution, which in turn results in shrinkage and swelling of the soil deposit. The pore-water pressure distribution with depth can take on a wide variety of shapes as a result of environmental changes (Figure 1.6).

Arid and semi-arid areas usually have a deep groundwater table. Soils located above the water table have negative pore-water pressures. The soils are desaturated due to the excessive evaporation and evapotranspiration. Climatic changes greatly influence the water content of the soil close to the ground surface. Upon wetting, the pore-water pressure increases, tending toward positive values. As a result, changes occur in the volume and shear strength of the soil, with many soils exhibiting extreme swelling or expansion when wetted. Other soils are known for their significant loss of shear strength upon wetting. Changes in the negative pore-water pressures associated with heavy rainfall are the cause of numerous slope failures. Reductions in the bearing capacity and resilient modulus of soils are also associated with increases in pore-water pressures. These phenomena indicate the important role that negative pore-water pressure plays in controlling the mechanical behaviour of unsaturated soils.

The types of problems of interest in unsaturated soil mechanics are similar to those of interest in saturated soil mechanics. Common to all unsaturated soil situations are the negative pressures in the pore-water. Some of these problems are given below.

Natural slopes subjected to environmental changes

Natural slopes are subjected to a continuously changing environment (Figure 1.7). An engineer may be asked to investigate the present stability of a slope, and predict what would happen if the geometry of the slope were changed or if the environmental conditions should happen to change. Most or all of the potential slip surfaces may lie above the groundwater table. In other words, the potential slip surface may pass through unsaturated soils with negative pore-water pressures. Typical questions that might need to be addressed are (Fredlund and Rahardjo, 1993):

- What effect could changes in the geometry have on the pore pressure conditions?
- What changes in pore pressures would result from a prolonged period of precipitation?
- How could reasonable pore pressures be predicted?
- Could the location of a potential slip surface change as a result of precipitation?
- How significantly would slope stability analysis be affected if negative pore-water pressures were ignored?
- What would be the limit equilibrium factor of safety of the slope as a function of time?
- What lateral deformations might be anticipated as a result of changes in pore pressures?

Similar questions might be of concern with respect to relatively flat slopes. Surface sloughing commonly occurs on slopes following prolonged periods of precipitation. These failures have received little attention from an analytical standpoint. One of the main difficulties appears to have been associated with the assessment of pore-water pressures in the zone above the groundwater table. The slow, gradual, downslope creep of soil is another aspect which has not received much attention in the literature. It has been observed, however, that the movements occur in response to seasonal environment

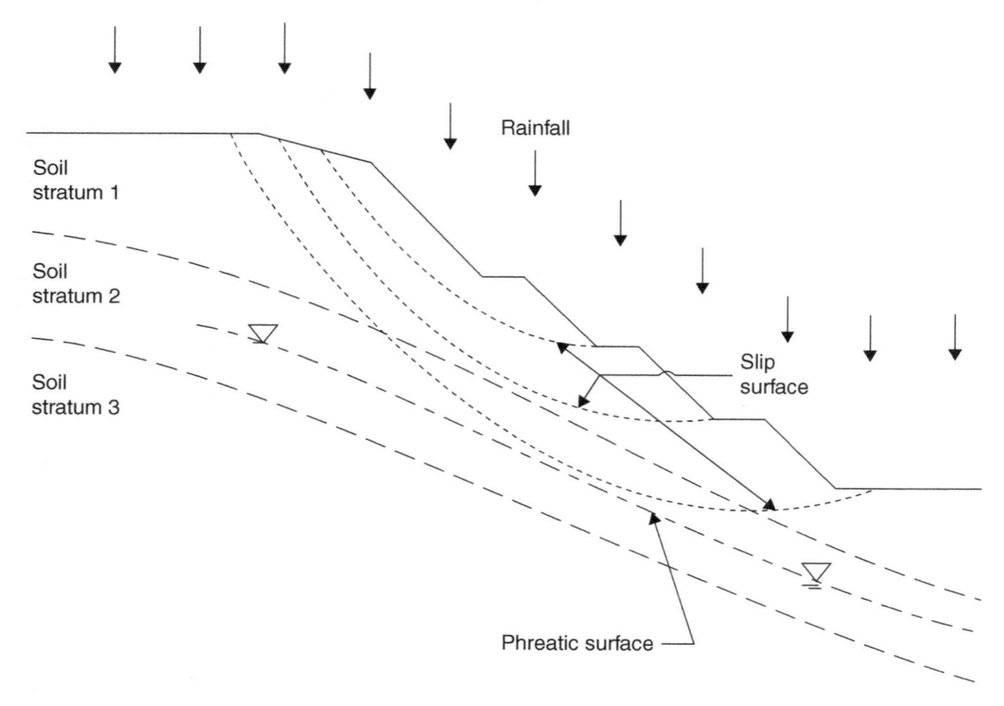

Figure 1.7 An example of the effect of excavations on a natural slope subjected to environmental change (*After* Fredlund and Rahardjo, 1993)

changes. Wetting and drying are known to be important factors. It would appear that an understanding of unsaturated soil behaviour is imperative in formulating an analytical solution to these problems.

Stability of vertical or near-vertical excavations

Vertical or near-vertical excavations are often used for the installation of a foundation or a pipeline (Figure 1.8). It is well known that the backslope in a moist silty or clayey soil will remain as a near-vertical slope for some time before failing. Failure of the backslope is a function of the soil type, the depth of the excavation, the depth of tension cracks and the amount of precipitation, as well other factors. In the event that the contractor should leave the excavation open longer than planned, or should a high precipitation period be encountered, the backslope may fail, causing damage and possible loss of life. The excavations being referred to are in soils above the groundwater table where the pore-water pressures are negative. The excavation of soil also produces a further decrease in the pore-water pressures. This results in an increase in the shear strength of the soil. With time, there will generally be a gradual increase in the pore-water pressures in the backslope, and a corresponding loss in strength. The increase in the pore-water pressure is the primary factor contributing to the instability of the excavation. Engineers often place responsibility on the contractor for ensuring backslope stability. Predictions associated with this problem require an

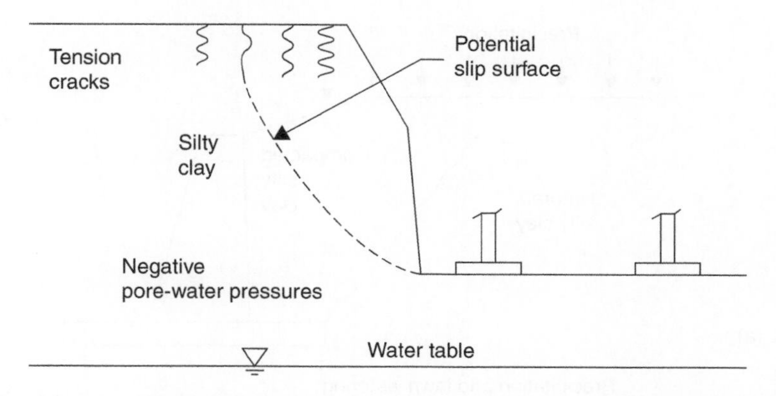

Figure 1.8 An example of potential instability in a near-vertical excavation during the construction of a foundation (*After* Fredlund and Rahardjo, 1993)

understanding of unsaturated soil behaviour. Some relevant questions that might be asked are (Fredlund and Rahardjo, 1993):

- How long will the excavation backslope stand prior to failing?
- How could the excavation backslope be modeled analytically, and what would be the boundary conditions?
- What soil parameters are required for the above modeling?
- What in-situ measurements could be taken to indicate incipient instability?
- Also, could soil suction measurements be of value?
- What effect would a ground surface covering (e.g. plastic sheeting) have on the stability of the backslope?
- What would be the effect of temporary bracing, and how much bracing would be required to ensure stability?

Lateral earth pressures

Figure 1.9 shows two situations where an understanding of lateral earth pressures is necessary. Some situations might involve lateral pressure against a grade beam placed on piles. Let us assume that in each situation a relatively dry clayey soil has been placed there and compacted. With time, water may seep into the soil, causing it to expand both vertically and horizontally. Although these situations may illustrate the development of high lateral earth pressures, they are not necessarily good design procedures. Some questions that might be asked are (Fredlund and Rahardjo, 1993):

- How high might the lateral pressures be against a vertical wall with wetting of the backfill?
- What are the magnitudes of the active and passive earth pressures for an unsaturated soil?
- Are the lateral pressures related to the 'swelling pressure' of the soil?

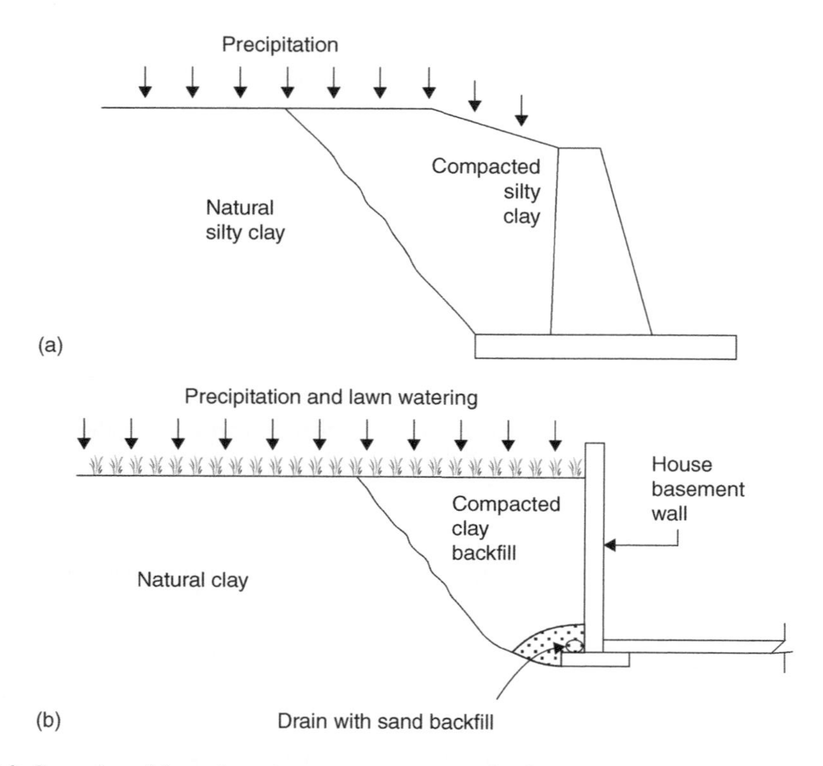

Figure 1.9 Examples of lateral earth pressures generated subsequent to backfilling with dry soils: (a) lateral earth pressures against a retaining wall as water infiltrates the compacted backfill; (b) lateral earth pressure against a house basement wall (*After* Fredlund and Rahardjo, 1993)

- Is there a relationship between the 'swelling pressure' of a soil and the passive earth pressure?
- How much lateral movement might be anticipated as a result of the backfill becoming saturated?

Foundations

The foundations for light structures are generally shallow spread footings (Figure 1.10). The bearing capacity of the underlying (clayey) soils is computed on the basis of the unconfined compressive strength of the soil. Shallow footings can easily be constructed when the water table is below the elevation of the footings. In most cases, the water table is at a considerable depth, and the soil below the footing has a negative pore-water pressure. Undisturbed samples, held intact by negative pore-water pressures, are routinely tested in the laboratory. The assumption is that the pore-water pressure conditions in the field will remain relatively constant with time, and therefore the unconfined compressive strength will also remain essentially unchanged. Based on this assumption, and on a relatively high design factor of safety, the bearing capacity of the soil is computed.

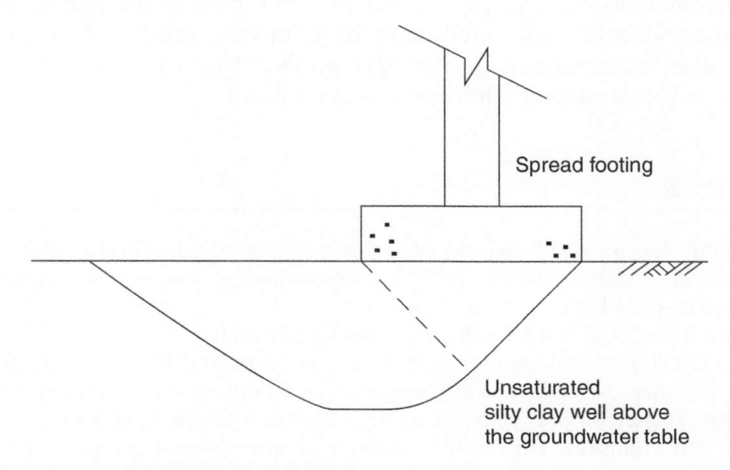

Figure 1.10 Illustration of bearing capacity conditions for a light structure placed on soils with negative pore-water pressure (*After* Fredlund and Rahardjo, 1993)

The above design procedure has involved soils with negative pore-water pressures. It appears that the engineer has been almost oblivious to problems related to the long-term retention of negative pore-water pressure when dealing with bearing capacity problems. Almost the opposite attitude has been taken towards negative pore-water pressures when dealing with slope stability problems. That is, the attitude of the engineer has generally been that negative pore-water pressures cannot be relied upon to contribute to the shear strength of the soil on a long-term basis when dealing with slope stability problems. These two, seemingly opposite, attitudes or perceptions give rise to the question, 'How constant are the negative pore-water pressures with respect to time?' Other questions related to the design of shallow footings that might be asked are (Fredlund and Rahardjo, 1993):

- What changes in pore-water pressures might occur as a result of sampling soils from above the water table?
- What effect does the in-situ negative pore-water pressure and a reduced degree of saturation have on the measured, unconfined compressive strength?
- How should the laboratory results be interpreted?
- Would confined compression tests more accurately simulate the strength of an unsaturated soil for bearing capacity design?
- How much loss in strength could occur as a result of watering the lawn surrounding the building?

The above examples show that there are many practical situations involving unsaturated soils that require an understanding of the seepage, volume change and shear strength characteristics. In fact, there is often an interaction between, and a simultaneous interest in, all three aspects of unsaturated soil mechanics. Typically, a flux boundary condition produces an unsteady-state saturated/unsaturated flow situation which results in a volume change and a change in the shear strength of the soil. The change in shear strength is generally translated into a change in the factor of safety. There may also be an interest in quantifying the change of other volume-mass soil properties (i.e. water content and degree of saturation).

REFERENCES

Bell, F.G. (2000). *Engineering Properties of Soils and Rocks*. 4th edn, Oxford: Blackwell Science.

Bergman, E.F. and McKnight, T.L. (1993). *Introduction to Geography*. Englewood Cliffs, New Jersey: Prentice Hall.

Bland, W. and Rolls, D. (1998). *Weathering*. London: Arnold.

Blight, G.E. (1985). Residual soils in South Africa, in Brand and Philipson (ed.) *Sampling and Testing of Residual Soils: Technical Committee on Sampling and Testing of Residual soils*. International Society for Soil Mechanics and Foundation Engineering, 159–168.

Brand, E.W. and Philipson, H.B. (1985). Review of international practice for the sampling and testing of residual soils, in Brand, E.W. and Philipson, H.B. (eds) *Sampling and Testing of Residual soils: A Review of International Practices*. Technical Committee on Sampling and Testing of Residual Soils, International Society for Soil Mechanics and Foundation Engineering, 7–22.

Fredlund, D.G. and Rahardjo, H. (1993). *Soil Mechanics for Unsaturated Soils*. New York: John Wiley & Sons.

Ingersoll, A.P. (1983). The atmosphere. *Scientific American*, Sept., 114–130.

Johnson, R.B. and De Graff, J.V. (1988). *Principles of Engineering Geology*. New York: John Wiley.

McCarthy, D.F. (1993). *Essentials of Soil Mechanics: Basic Geotechnics*. Englewood Cliffs, New Jersey: Regents/Prentice Hall.

Press, F. and Siever, R. (1994). *Understanding Earth*. New York: W. H. Freeman.

Public Works Institute Malaysia (IKRAM) (1996). *Geoguides* 1–5, Tropical weathered in-situ materials.

Rollings, M.P. and Rollings, R.S. (1996). *Geotechnical Materials in Construction*. McGraw-Hill.

Sowers, G.F. (1985). Residual soils in the United States, in Brand and Philipson (eds) *Sampling and Testing of Residual Soils: A Review of International Practices*. International Society for Soil Mechanics and Foundation Engineering.

Summerfield, M.A. (1996). *Global Geomorphology: An Introduction to the Study of Landforms*. Singapore: Longman.

Terzaghi, K. and Peck, R.B. (1967). *Soil Mechanics in Engineering Practice*. New York: John Wiley.

2

Formation and classification of tropical residual soils

Harwant Singh
Universiti Malaysia Sarawak, Kota Samarahan, Malaysia

Bujang B.K. Huat
Universiti Putra Malaysia, Serdang, Malaysia

2.1 INTRODUCTION

Soils in engineering commonly refer to any kind of loose, unconsolidated natural material enveloping the surface that is relatively easy to separate by even gentle means (Terzaghi and Peck, 1967; Johnson and De Graff, 1988). Soils, three-phased materials consisting of solids, water and air, may be grouped into two broad categories of which one is residual soil, the other being transported soils of materials dislodged, eroded and moved via agents of transportation i.e. gravity, wind, water etc. and deposited at a different location. This chapter describes the residual soils found in the tropics, firstly, by looking at what they are and then the engineering approach in working with them, with the latter also requiring an understanding of their nature.

2.2 RESIDUAL SOILS

Residual soils form or accumulate and remain where they are formed. As the customised engineering definition of residual soils is derived from the conventional characterisation of residual soil, and major engineering parameters depend on the behavior of its constituents, a comprehensive understanding of its nature is first necessary.

This entire *in situ* unconsolidated superficial cover they constitute is also termed the *regolith*, which is a geological term. Pedologically, however, only an upper part of a thickness of 0.3–2.0 m or more is referred to as *soil* and the remaining lower portion progressively grading into the bedrock is referred to as the *saprolite*, that is, in-place bedrock that is chemically altered and coherent retaining its original texture (Bland and Rolls, 1998).

2.2.1 Origin and general features of residual soils

Soils may be thought of as the corollary of the Earth's *exogenic* processes. They form as a result of the dynamics of geomorphic systems through the interaction of climatic elements with materials forming the subsurface. The origin of residual soils as an *in situ* mantle is the part of the geological rock cycle, shown in Figure 2.1, where an exposure of the three rock types to *exogenic* processes at the interface with the atmosphere resulted in their weathering.

Residual soils have distinct characteristics that are quite ubiquitous. These prominent features may be summarised as follows:

a) *Thickness*: residual soils form a layer of significant thickness above the bedrock which varies from place to place (Bergman and McKnight, 2000) depending upon the efficiency or persistence of soil forming processes in the absence of removal by erosion. The bedrock at some locations is found at depths exceeding 30 m (Tan, 2004).

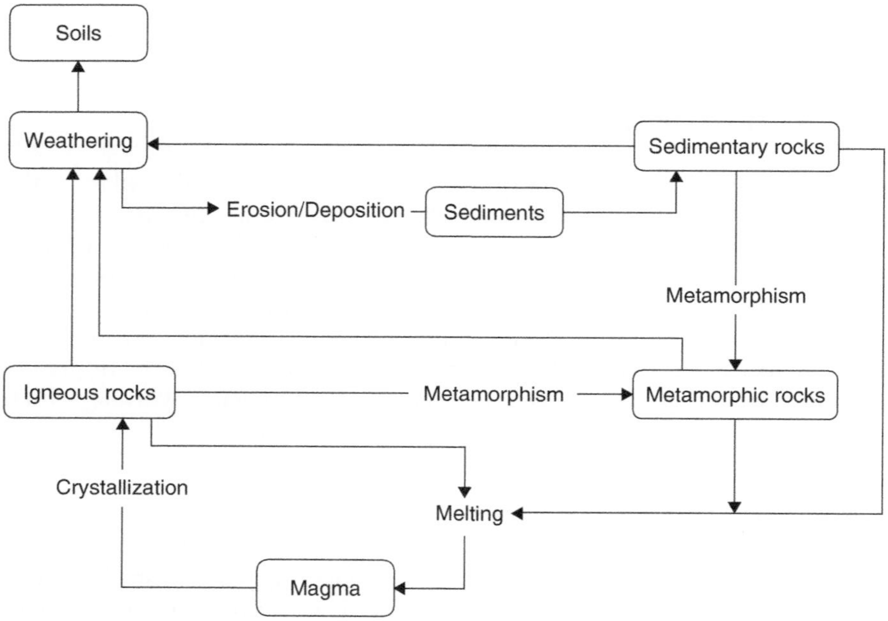

Figure 2.1 The geologic rock cycle and formation of residual soil

b) *Horizons*: residual soils display identifiable *in situ* differentiated horizontal horizons. This is due to the process of leaching by water, for instance, as rainwater percolating through it dissolves minerals, transports the solutes and precipitates these further down the soil profile resulting in the leached and precipitation levels forming visible horizons.

c) *Composition*: the composition of residual soils is initially defined by the type of parent rocks over which they arise. Residual soils are derived from various rock types such as granite, basalt, limestone, sandstone, schist and others. Over time in the case of well developed old residual soils, the advanced chemical decay, alteration and leaching may render the recognition of the parent rock materials very difficult or they may be obliterated.

2.2.2 Formation of residual soils

Residual soils are the residua that result from the *in situ* weathering of rocks over a period of time as the weathering caused by the interaction of state factors transforms the parent rock mass. They can be products of weathering of different parent rocks, igneous, sedimentary or metamorphic. The defining factor for residual soils is that there is very little or no transport during or after formation (Singh and Kataria, 1980).

(a) State factors

The development of residual soils depends on the interaction of three natural features, that is, chemical composition of the rock, environmental conditions and time, that give

Table 2.1 Major factors affecting soil formation (*After* Bergman and McKnight, 2000)

Factor	*Description*
Climatic	Refers to the effects on the surface of temperature and precipitation.
Geologic	Refers to the parent material (bedrock or loose rock fragments) that provide the bulk of most soils.
Geomorphic/Topographic	Refers to the configuration of the surface and is manifested primarily by aspects of slope and drainage.
Biotic	Consists of living plants and animals, as well as dead organic material incorporated into the soil.
Chronological	Refers to the length of time over which the other four factors interact in the formation of a particular soil.

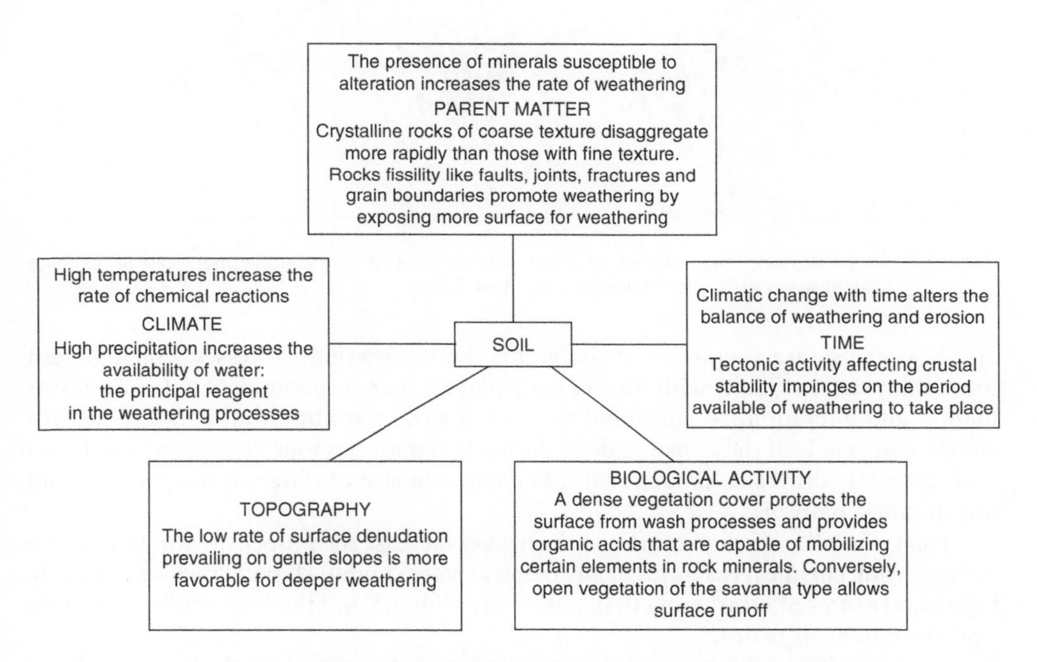

Figure 2.2 Factors responsible for soil formation (*After* Thomas, 1974)

rise to state factors responsible for the formation of soils. Table 2.1 lists the various factors which are briefly reviewed before being considered in detail.

The five soil forming factors and some of their aspects are also given in Figure 2.2.

(i) Climate

Climate is perhaps the most important factor. The climate is the average weather conditions (atmospheric conditions at any time in a given place) over the long-term (>30 years) taking into account the extremes, means and frequencies of departures from these means (Whittow, 1988). It essentially determines the amount of precipitation

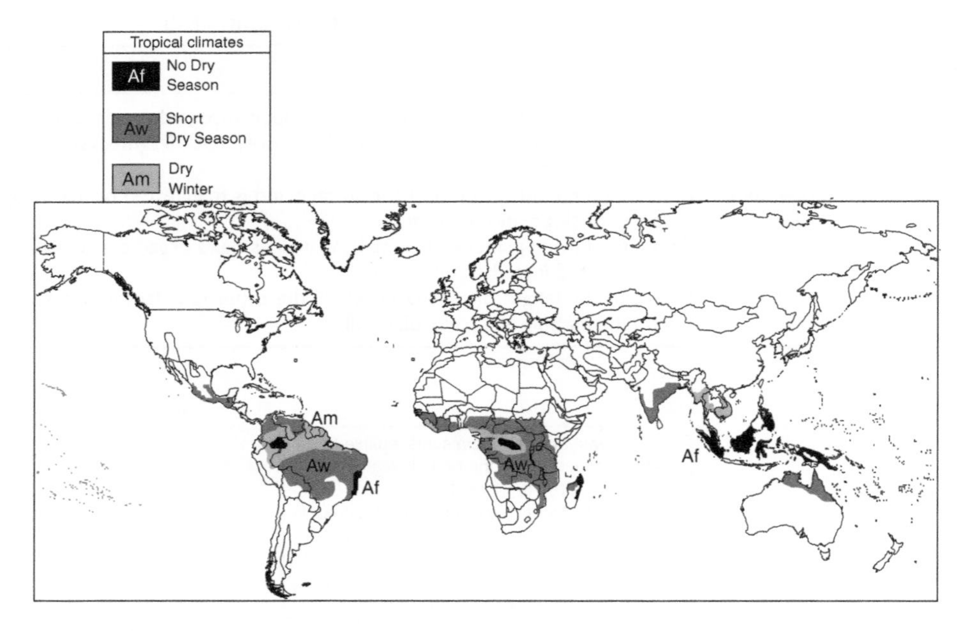

Figure 2.3 Areas showing the various tropical sub-climates (from the Köppen World Climate Classification Map; *after* McKnight and Hess, 2000)

and heat energy (temperature) available for the weathering processes which govern the soil formation. Thus, with time climate plays a more predominant role in soil formation and alteration, so soils tend to show a strong geographical correlation with climate, especially at the global scale as different climatic regions develop residual soils with distinctly different characteristics as a consequence of different temperature and precipitation patterns.

High rainfall and temperatures in the tropics increase the propensity for weathering by increasing chemical reactions. Thus tropical warm humid climatic regions generally have higher rates of weathering than colder dry climates and thicker weathered mantle over the parent or bedrocks.

The key criterion for tropical climate (the A category climate in the Köppen Classification System) extensively occurring in almost all continents between latitudes 20°N to 20°S of the equator, as illustrated in Figure 2.3, is for the coolest month to have a temperature of more than 18°C and the prevalence of moisture from warm, moist and unstable air masses frequenting the oceans at these latitudes giving rise to high humidity. The tropical climate is further classified into three types on the basis of the quantity and regime of annual rainfall as follows:

a) Tropical wet type (Af): experiences relatively abundant rainfall in every month of the year.
b) Tropical monsoonal type (Am): has a short dry season but a very rainy wet season.
c) Tropical savanna type (Aw): characterised by a longer dry season and a prominent but not extraordinary wet season.

These different sub-tropical climates also cause variations in tropical soils as residual soil; they cause the development of residual soils with distinctly different characteristics as a consequence of different temperature and precipitation patterns.

(ii) Parent materials

The formation of soils commences by chemical alteration and physical disintegration at the exposed surface of rocks which form the parent material for the soils. Gerrard (2000) states that parent material influences soils through the process of weathering and subsequently through the weathered material but its greatest influence is in the early stages on the incipient soils and in the drier regions (Birkeland, 1984) but lessens over time as other soil forming factors become more active. There are a variety of parent materials as there are a variety of rock types with different mineral phases and chemical compositions. Rocks have been divided into three main types based upon their method of formation: igneous rocks, sedimentary rocks and metamorphic rocks. Each group contains a wide variety of rock types that differ from one another by either their chemical composition, texture or both.

Igneous rocks: igneous rocks, as shown in Figure 2.1, form from molten magma from the Earth's interior which on cooling crystallises into minerals that form rocks. The two main types of igneous rocks distinguished by their texture (appearance of a rock based on the size, shape and arrangement of their interlocking mineral crystals) are:

a) Extrusive rocks (volcanic rocks) – These form when magmas reach and intrude onto the surface of the Earth through fissures or vents. The rapid cooling of the magma produces crystals invisible to the naked eye. The rocks normally have a fine-grained or aphanitic texture.
b) Intrusive rocks (plutonic rocks) – These form below the Earth's surface from magma that cools slowly resulting in a coarse-grained or phaneritic texture and are exposed to the surface by erosion or when uplifted.

Figure 2.4 shows the classification of igneous rocks (based on chemical composition and mineral composition) while Table 2.2 gives the field classification of igneous rocks (according to texture, mineral composition and color).

Sedimentary rocks these are formed from materials derived from pre-existing rocks by erosive and sedimentary process, and upon deposition undergo consolidation, lithification (cementing together) and diagenesis (chemical, physical or biological changes) to form sedimentary rocks. Sedimentary rocks may also form by chemical precipitation. Figure 2.5 illustrates the sedimentary process leading to the formation of different types of sedimentary rocks.

Sedimentary rocks, shown in Figure 2.5, can be classified according to their composition or mode of origin into three different types: clastic, chemical and organic rocks, as described below.

a) Clastic sedimentary rocks – These form from sediments i.e. fragments or clasts of pre-existing rocks broken and fragmented through erosion and transportation. These clasts are differentiated by their grain size and the clastic sedimentary rocks

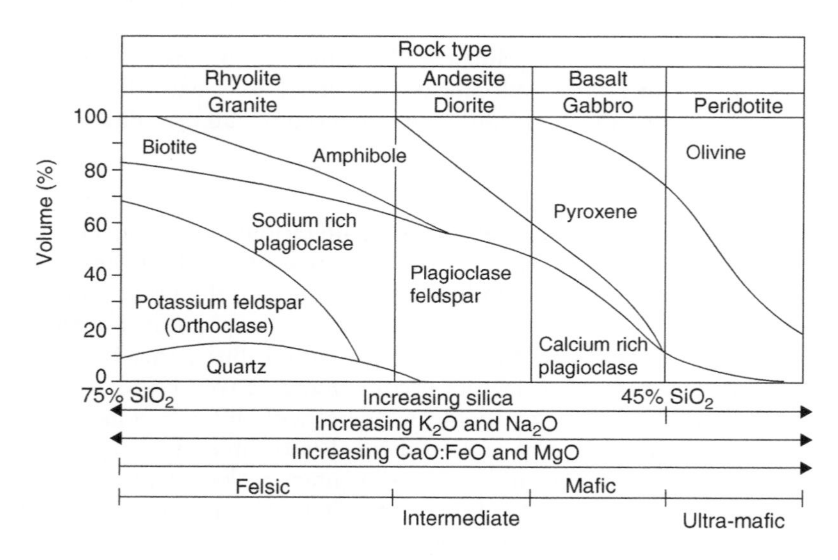

Figure 2.4 Classification of igneous rocks

Table 2.2 Identification of igneous rocks

		Rock type			
Texture	Coarse-Grained Fine-Grained	Granite Rhyolite	Diorite Andesite	Gabbro Basalt	Ultramafic rocks –
Minerals		Quartz, feldspars, minor ferromagnesian minerals	Feldspars, ferromagnesian minerals (30–50%), no quartz	Predominantly ferromagnesian minerals. The other mineral is feldspar	Entirely ferromagnesian minerals (normally olivine and pyroxene)
Color		Light	Intermediate	Dark	Dark

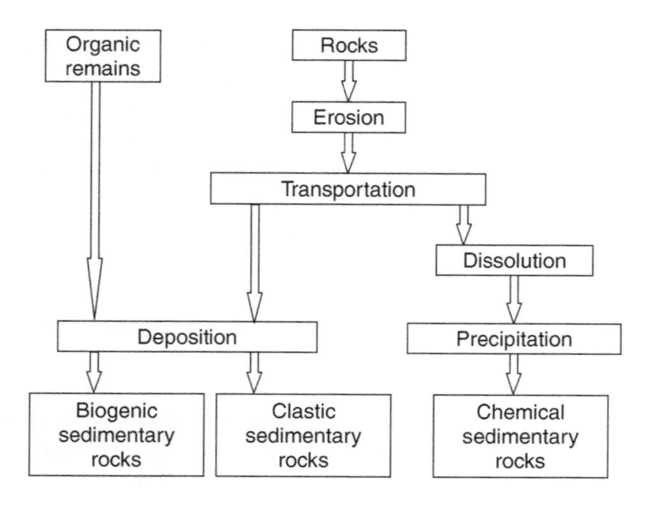

Figure 2.5 The formation of sedimentary rocks

Table 2.3 Clastic sedimentary rocks

Size range (mm)	*Particle name*	*Sedimentary rocks*
	Boulder	Conglomerates (rounded particles)
256		or
	Cobble	Breccia (angular particles)
64		
	Pebble	
2		
	Sand	Sandstone
1/16	Silt	Siltstone
1/256	Clay	Shale or Mudstone

Table 2.4 Chemical and biogenic sedimentary rocks

Chemical sedimentary rocks	
Cherts	Formed from the chemical precipitation of SiO_2
Evaporites	Formed by evaporation of sea water which produces chemically precipitated halite (salt) and gypsum ($CaSO_4.2H_2O$) deposits
Limestone	Formed by precipitation of calcite ($CaCO_3$)
Biogenic sedimentary rocks	
Limestone	Formed from the cementation of skeletal remains of calcite ($CaCO_3$) precipitated by organisms to form shells
Diatomite	Formed from the siliceous remains of radiolarians or diatoms
Coal	Formed from the deposition of dead plant matter in a reducing environment (lack of oxygen)

are classified according to the sediment or grain size they are constituted from as given in Table 2.3.

b) Chemical sedimentary rocks – These originate from the dissolved materials precipitated inorganically. The chemical sedimentary rocks are given in Table 2.4.

c) Organic sedimentary rocks – Also called biogenic rocks, form from the accumulation of organic remains e.g. coal is an organic rock that forms from plant remains while limestone may also form from shells and corals. Organic sedimentary rocks are also listed in Table 2.4.

The clastic sedimentary rocks usually contain internal layering, called bedding or stratification, separated by bedding planes which may also mark a change in sedimentary rock type.

Metamorphic rocks: metamorphic rocks are formed when pre-existing igneous and sedimentary rocks (or even metamorphic rocks) are forced to undergo change in the solid state by changes in temperature and pressure. These changes occur beyond diagenetic (low temperature and pressure) conditions (see sedimentary rocks above) and stop short of melting or anataxis. Material may be added or lost during the metamorphism, causing new minerals to form and others to re-crystallise producing completely new types of rocks. Common types of metamorphic rocks are given in Table 2.5. Metamorphic rocks are classified into two textural categories i.e. *foliated* (with preferred orientation of minerals within a rock) and *non-foliated*. The most intense form of

Table 2.5 Common types of metamorphic rocks

Texture			*Rock Name*	*Precursor rock*	*Metamorphism*		*Mineral composition**						
					Type	*Grade*	*a*	*b*	*c*	*d*	*e*	*f*	*g*
Texture	Foliated	Fine grained	Slate	Shale	Regional	Low	↕	↕	↕	↕			
			Phyllite			Medium		↕	↕	↕			
		Coarse grained	Schist					↕	↕	↕			
			Gneiss			High		↕	↕	↕	↕	↕	
	Non foliated	Fine grained	Hornfels	Shale	Contact				↕	↕	↕	↕	
		Coarse grained	Quartzite	Quartz sandstone	Contact or regional					↕	↕	↕	
			Marble	Limestone or dolomite	Contact or regional								↕

*a – clay; b – chlorite; c – mica; d – quartz; e – amphibole; f – feldspar; g – calcite

foliation results in mineral banding when minerals segregate into separate bands. Metamorphism can occur by various processes i.e. regional and contact which is referred to as the type of metamorphism.

The structural features of the various rock types are observed as morphological features where decomposition and disintegration is incomplete, or as relict features in the weathered profile. This includes sedimentary bedding and foliation in sedimentary and metamorphic rocks respectively. The altered mineralogical composition is also dependent upon the parent rocks.

(iii) Time

The extent of soil formation is dependent on time. Time is a critical factor as it determines the degree to which other factors are able to either undergo change or operate, therefore the thickness and alteration seen in the soil depends on the time the soil forming processes have been occurring. In the tropics the appearance of soils shows the effects of time with older soils being generally thicker and more highly weathered i.e. altered and leached.

Table 2.6 Essential topographic attributes

Attribute	*Definition*	*Effect*
Altitude	Elevation	Elevation affects the soil wetness
Slope	Angle of the land surface or gradient	Steepness of the slopes affects the rates of surface-water runoff and erosion, as on steep slopes weathering products may be quickly washed away by rains but accumulate on gentle slopes
Aspect	Direction the slope faces or azimuth	This deals with the direction the slope faces, south-facing will generally be drier, warmer, and less moist than north-facing slopes because they get more direct sunlight. This slope idea can be related to mountainsides and hillsides

(iv) Topography

Topography is important as it exercises a significant control on surface processes like erosion and drainage, and influences the soils through the orientation and aspect of relief, which affects the microclimate and surface processes on slopes. Different topographic terrains with different landscapes uniquely affect soil development. The essential components of topography are given in Table 2.6.

(v) Biological activity

Plants and animals play an important role in soil formation. Their primary contribution is organic matter to the upper part of the regolith. Organic matter or humus in soils originates from the decay of dead vegetation by soil microorganisms that also produce soil acids which react chemically to weather rocks. The mixing of rock fragments and soil particles by burrowing organisms also creates pore space for aeration of the soil. Soils are reported to be thicker and form more rapidly in areas where there is an abundance of microorganisms. Plants also extend pore spaces into deeper soil layers with their penetrating roots. This enables moisture to reach deeper material and fresh rock surfaces thus extending chemical weathering.

(b) Formation process: Weathering

Weathering can involve three processes: chemical weathering (involving chemical alteration), physical weathering (involving physical disintegration) and biological weathering (consisting of chemical or physical changes through the agency of biological organisms). Chemical and physical weathering are quite distinct but often operate simultaneously, abetting each other as in the case of the physical disintegration of a larger rock into smaller parts thereby increasing the total surface area and accelerating chemical decomposition. In certain situations, one type of weathering may dominate the other. Thermodynamically, the rocks formed at greatly different temperatures and pressures within the crust, and are inherently unstable when exposed at the surface environment, and so have a propensity to respond to weathering. This makes them susceptible to weathering processes whereby they break down and alter to form new phases that are closer to equilibrium at the surface environment.

(i) Types of weathering

Physical weathering: physical or mechanical weathering is the physical disintegration of a rock into smaller fragments. The different types of physical weathering include the following phenomena:

- *Frost wedging* – this occurs when water freezes and expands in cracks in rocks. The expansion pushes the cracks apart causing further breakup of the rock.
- *Wetting and drying* – this can cause slaking of the rock. Slaking occurs when layers of water molecules accumulate within the mineral grains of a rock and push the rock grains apart due to their increasing thickness which can induce great tensional stress.
- *Thermal expansion and contraction* – as heating causes a rock to expand and cooling results in contraction, the repeated expansion and contraction of different minerals at different rates results in stresses along mineral boundaries leading to breakdown of the rock.
- *Mechanical exfoliation* – also called unloading and this happens when the rock breaks into flat sheets along joints parallel to the ground surface as the rock is exposed. This phenomenon is caused by expansion of rock due to the pressure released by the removal of overlying rock by erosion. Exfoliation commonly occurs when plutonic igneous rocks are exposed as these, which are cooled at depth under great pressure, undergo pressure release once the overburden is removed resulting in sheets of rock peeling off.
- *Abrasion* – this is the physical grinding down of rock fragments. It occurs when two rock surfaces in contact undergo mechanical wearing or grinding at their surfaces.

Chemical weathering: chemical weathering is the chemical dissolution or alteration of the mineral constituents of rocks into new compounds. The rock forming minerals are vulnerable to attack by water, oxygen, chemical agents (e.g. acid in the air, in rain, etc.) and carbon dioxide. In the surface environment, chemical phases undergo chemical changes to form new stable minerals. Gidigasu (1976) gives the main sequence of the chemical weathering process as follows:

1. Firstly, the rock structure is made weaker by a selective attack on the constituent minerals in igneous and metamorphic rocks, a more general attack on calcareous rocks, and an attack on the cements of some sedimentary rocks.
2. Secondly, stresses are created within a rock by varying expansion of certain minerals.
3. Thirdly, compounds are produced which can be removed in solution, leaving behind residual deposits. These include residual decomposition products e.g. clay minerals.

Chemical weathering is dependent on the surface area, temperature and availability of chemically active fluids. Smaller sized particles weather more rapidly due to their larger surface area. Figure 2.6 illustrates the increased total surface area available for chemical weathering with smaller particles sizes.

Water and dissolved materials like ammonia, oxygen, carbonic acid, chlorides, sulphates and others cause chemical weathering. An increase in temperature increases the

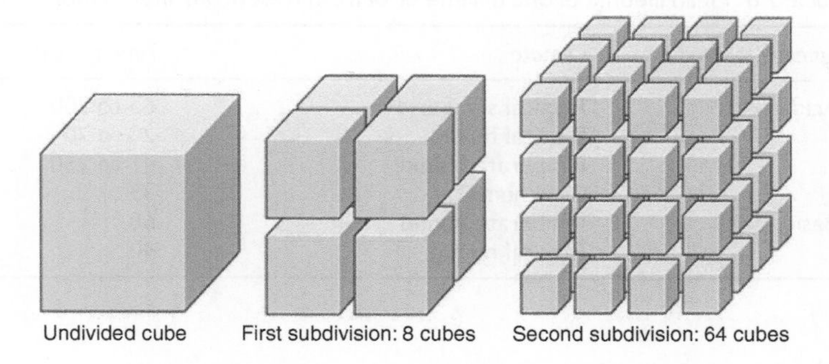

Figure 2.6 Surface area available for weathering

Table 2.7 Common chemical processes

Process	*Description*
Solution	Decomposition of minerals involving water as a solvent and in rainwater aided by carbonic acid (H_2CO_3) formed due the presence of CO_2.
Oxidation	The combination with oxygen, the most common oxidizing agent, to form oxides and hydroxides or any other reaction which causes the loss of an electron that results in an increase in the positive charge.
Hydration	Hydration involves the absorption of water into the mineral causing the expansion of the mineral leading to eventual weakening.
Hydrolysis	Reaction of water directly with minerals where metal cations, replaced by hydrogen (H^+) ions, combine with hydroxyl (OH^-) ions.
Carbonation	The carbon dioxide (CO_2) dissolves in water to form carbonic acid (H_2CO_3) that reacts with minerals e.g. to dissolve calcite into bicarbonate ions and calcium.
Leaching	This involves the migration of ions by dissolution into water. The mobility of ions depends upon their ionic potential (charge/radius = charge density). For example, Ca^{2+}, Mg^{2+}, Na^+, K^+ are easily leached by moving water; Fe^{3+} is more resistant, Si is difficult to leach while Al^{3+} is almost immobile.

effectiveness of carbonic acid and oxygen as seen in humid climates where the intensity of chemical weathering increases because of higher temperatures; consequently, the maximum intensity of chemical weathering occurs in tropical regions (Jumikis, 1965). Chemical weathering takes place through a number of different processes. The most common ones are hydrolysis, oxidation, reduction, hydration, carbonation and solution described in Table 2.7.

Biological weathering: Biological weathering occurs with the disintegration of rock by biological organisms acting both as chemical and/or physical agents of weathering. A wide range of organisms ranging from microorganisms to plants and animals can cause weathering. These organisms release acids that react with rocks or mechanically break them up. Some of the more important processes involved are:

1 Rock fracture due to animal burrowing or the pressure from growing roots.
2 Movement and mixing of materials resulting from movement of soil organisms.
3 Carbon dioxide produced by organism respiration mixes with water to form carbonic acid that augments chemical processes.

Table 2.8 Mean lifetime of one millimeter of fresh rock (from after Nahon, 1991)

Igneous rock type	*Climate*	*Time (years)*
Acid rocks	Tropical semi-arid	65 to 200
	Tropical humid	20 to 70
	Temperate humid	41 to 250
	Cold humid	35
Basic rocks	Temperate humid	68
	Tropical humid	40

4 Organic substances from organisms called *chelates* decompose minerals and rocks by the removal of metallic *cations*.
5 The moisture regime in soils influenced by organisms induces weathering as water is a necessary component in the weathering processes.

(ii) Rate of weathering

The rate of weathering is influenced by the above mentioned factors such as climatic conditions; respective resistances of the different rocks and minerals to weathering; topography; size and relative exposed surface area of rocks and the permeability of rock mass to mention a few. The temperature and mean annual precipitation rates resulting in different soil moisture (soil water) content affect the rate of weathering, as can be gauged from the mean lifetime of one millimeter of each of the different igneous rocks weathered into a kaolinitic saprolite as they are exposed to the different climatic conditions shown in Table 2.8. These rates reveal that in cold, temperate or tropical humid zones, the average rainfall (the flushing of water) probably controls the rate of weathering (Tardy, 1969) climate (temperature and precipitation) controls the rate of weathering.

Generally, the following conclusions for the influence of climate are valid:

1 Hot and wet climate increases the rate of weathering dramatically.
2 Cold and dry climates decrease the rate of weathering.

(iii) Rock forming minerals and weathering

Minerals, naturally formed crystals composed of one or more chemical elements, are distinguished from other natural materials by their crystalline structure. Different rocks and minerals display varying resistance to weathering as a function of their structure and mineral composition. Some minerals weather more quickly than others depending on the mineral stability in the weathering environment. A few minerals readily dissolve in slightly acidic water whereas others are very resistant and persist for a long time without alteration. Minerals formed at high temperatures and pressures are least stable, weathering most quickly because they are furthest away from equilibrium or the conditions under which they formed. Conversely, those formed at lower temperatures and pressures are most stable under surface conditions. The Goldich's mineral stability series, as shown in Figure 2.7, gives the order of weathering for silicate minerals.

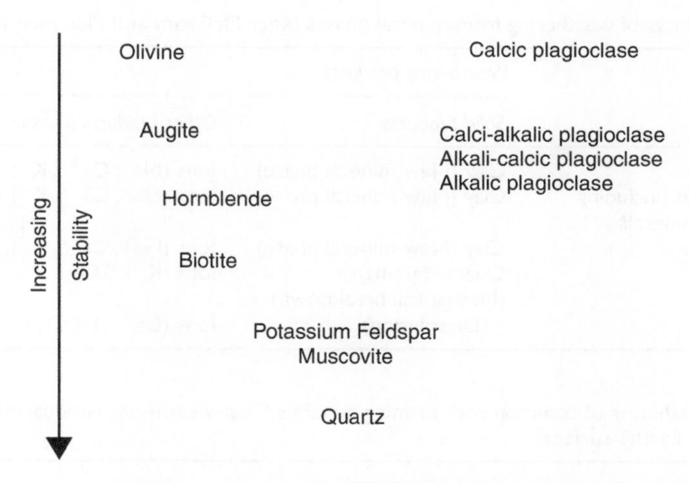

Figure 2.7 Goldich's mineral stability series (*After* Goldich, 1938)

The behavior of main rock forming minerals through some selected examples is cited below:

1 The highly resistant mineral quartz is one of the most stable minerals and remains such, as it was formed, for a considerable time. It is, however, eventually broken into sedimentary particles of various sizes over a long period of time.
2 Feldspars transform quite rapidly into clay minerals with some constituent elements released into solution.
3 Mica minerals may remain relatively unaffected by weathering.
4 Ferromagnesian minerals weather rapidly to form clays and iron oxides. The latter are often the cause for the rich red color stains observed in weathering rocks.
5 Marbles and limestone are much more susceptible to chemical attack by acids in rainwater.

Silicate minerals are the most common minerals in rocks and also in weathered products like clays etc. The basic structural unit in silicates is the SiO_4 tetrahedron with four oxygen atoms surrounding a silicon atom in a tetrahedral arrangement. The resistance of silicate minerals to weathering is attributed to an increase of the degree of sharing of oxygen atoms between adjacent SiO_4 tetrahedra. The order of increasing stability of silicates is shown in Figure 2.7. Olivine undergoes rapid weathering as the silicon tetrahedra are joined together only by oxygen-metal cations. Quartz, on the other hand, is very resistant as it is formed of silicon tetrahedra linked together. In amphiboles and pyroxenes (the chain silicates) and micas (phyllosilicates or sheet structures) the oxygen-metal cations are the weakest points in their structures. The calcium feldspars have a lower stability in comparison to sodium and potassium feldspars as the substitution of Al^{3+} for Si^{4+} also contributes to instability as the Si-O bonds decrease and more oxygen-metal cation bonds are necessary.

Table 2.9 Products of weathering from mineral phases (After McGeary and Plummer, 1998)

	Weathering products	
Mineral	*Solid products*	*Other products (mostly soluble)*
Feldspar	Clay (New mineral phase)	Ions (Na^{+}, Ca^{++}, K^{+}), SiO_2
Ferromagnesian (including biotite mica) minerals	Clay (New mineral phase)	Ions (Na^{+}, Ca^{++}, K^{+}), SiO_2, Fe oxides
Muscovite mica	Clay (New mineral phase)	Ions (Na^{+}, Ca^{++}, K^{+}), SiO_2
Quartz	Quartz (grains) (Mechanical breakdown)	Ions (K^{+}), SiO_2
Calcite	– (Dissolution)	Ions (Ca^{++}, HCO_3^{-})

Table 2.10 Weathering of common rock forming minerals (Post-weathering residual minerals stable at the Earth's surface)

Primary minerals	*Residual minerals**	*Leached ions*
Feldspars	Clay minerals	Na^{+}, K^{+}
Micas	Clay minerals	K^{+}
Quartz	Quartz	–
Fe-Mg minerals	Clay minerals + Hematite + Goethite	Mg^{2+}
Feldspars	Clay minerals	Na^{+}, Ca^{2+}
Fe-Mg minerals	Clay minerals	Mg^{2+}
Magnetite	Hematite, Goethite	–
Calcite	None	Ca^{2+}, CO_3^{2-}

*Residual Minerals = Minerals stable at the Earth's surface and left in the rock after weathering.

(iv) Products of weathering

The products of rock and mineral weathering constitute the residual soils. The weathering process can result in the following outcomes as shown for selected minerals in Table 2.9:

a) The complete loss of elements or compounds from rocks and minerals.
b) The addition of new elements or compounds to form new phases.
c) Only a mechanical breakdown of one mass into two or more parts without any chemical change of minerals or rocks.

Weathering products from rock-forming minerals: Quartz and clay minerals are commonly left after complete chemical weathering of rock-forming minerals. Others, such as iron oxides, are also precipitated. Table 2.10 summarises the residual minerals after weathering of common minerals.

Weathering products from parent rocks: Rock weathering involves the weathering of their respective constituent minerals. The mineral suite forming the respective major rock type (igneous, sedimentary and metamorphic) determines the residual minerals left or formed from weathering. The character of a residual soil, therefore, depends on the parent rock it develops from. For example, residual soil on weathered granite will initially be sandy, as sand-sized particles of quartz and partially weathered feldspar are released. The partially weathered feldspar grains will gradually completely weather

Table 2.11 Weathering products of some common rock types

Rock type	*Product*
Igneous	
Granite	Quartz sand and clay minerals
Basalt	Clay minerals
Sedimentary	
Shale	Clay minerals
Sandstone	Quartz sand
Limestone	Dissolved ions and residual clay sized particles
Metamorphic	
Metasediments (schist/phyllite/amphibolite/slate)	Clay minerals (from biotite and muscovite), micaceous silt and clay size particles
Gneiss, granulites and other quartz rich rocks	Quartz sand

into fined-grained clay minerals but the quartz will be resistant resulting in soil with sand-sized quartz and clay. The presence or absence of coarse grained quartz from the parent rock becomes the only vestige that survives. The weathering products of some common rock types are listed in Table 2.11.

The chemical alteration and the ensuing mineralogy and other features are discussed further in Section 2.4.

2.3 FORMATION OF TROPICAL RESIDUAL SOILS

On a global scale, chemical weathering is more prevalent and effective in breaking down rocks than mechanical disintegration and the dominance of this activity is very pronounced in some regions. As mentioned, chemical weathering is especially effective in the presence of water (due to its reactivity) and high temperature (which accelerates the rates of chemical reaction). The predominance of chemical or mechanical weathering as a function of climatic elements, primarily rainfall and temperature, is as shown in Figure 2.8.

The residual soils formed from weathered material are normally very thick in the tropics as this is an area with conditions favorable for intense weathering. Conversely, they may be very thin or absent in areas of unfavorable conditions like arid regions or steep mountain slopes subjected to erosion by mass movements. Soils formed by weathering in the tropics are generally referred to as tropical residual soils. Figure 2.9 shows the geographical distribution of the types of weathering with their expanse and depth.

2.4 CHARACTERISTICS OF TROPICAL RESIDUAL SOILS

2.4.1 Development of a weathered profile

The weathering process causes the incorporation of humus and leaching of insoluble chemical species and their subsequent accumulation, leaving the remaining insoluble

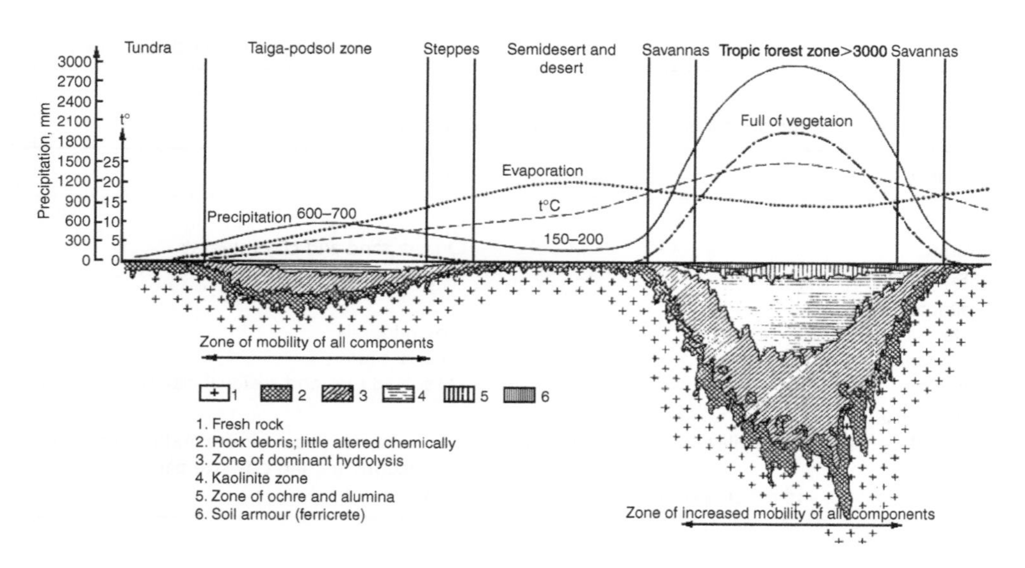

Figure 2.8 The zones of weathering in relation to latitude (*After* Strakhov, 1967)

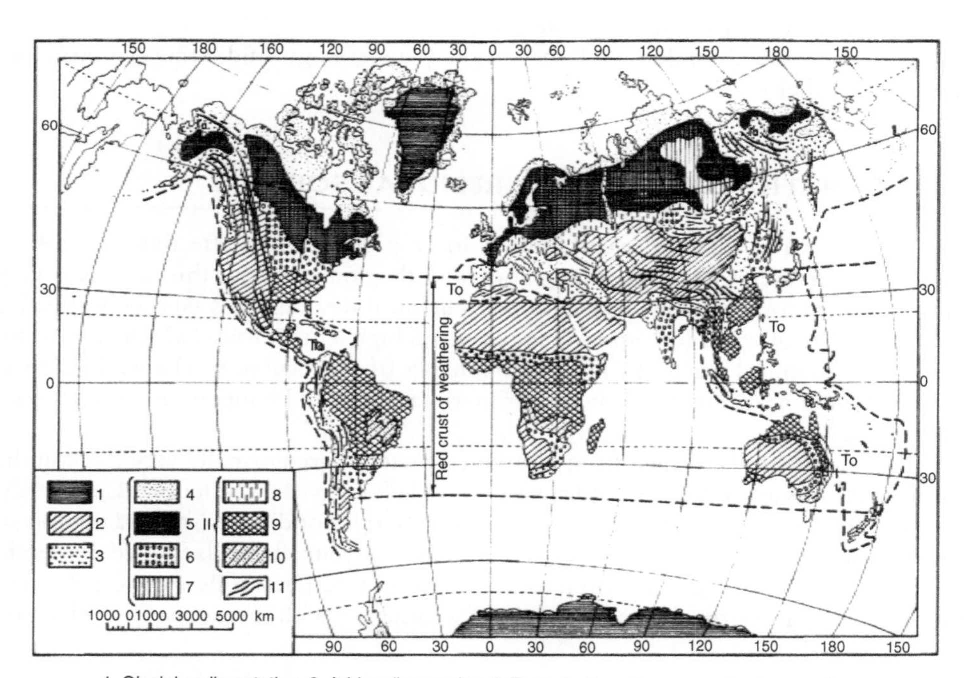

1. Glacial sedimentation. 2. Arid sedimentation. 3. Tectonically active-no weathering mantle
I. Region of temperate moist climate: 4. Chemical weathering reduced by low temperatures.
5. Normally developed weathering. 6. Chemical weathering reduced by low precipitation.
7. Chemical weathering reduced by high relief.
II. Region of tropical moist climate: 8. Chemical weathering slight because of low precipitation.
9. Chemical weathering intense. 10. Periphery of zone. 11. Mountain ranges.
Ta – Tectonically active areas

Figure 2.9 The geographical distribution of weathering types (*After* Strakhov, 1967)

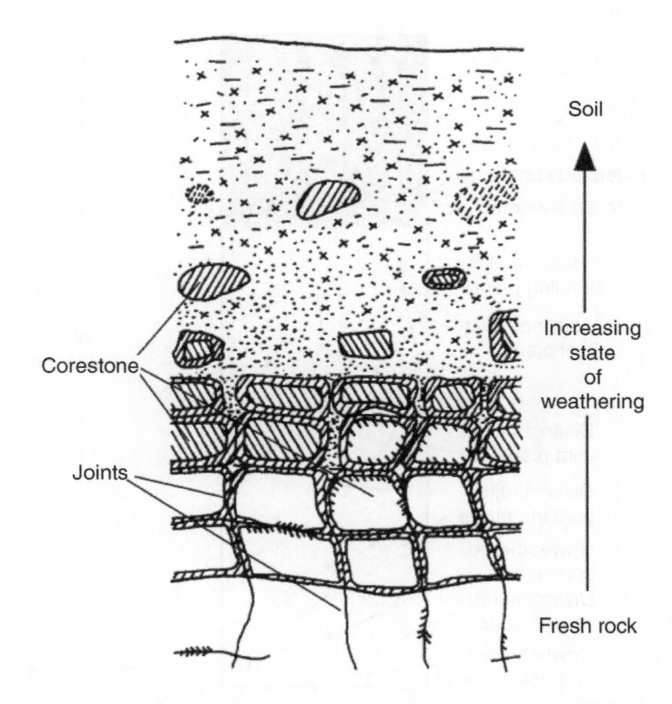

Figure 2.10 Typical weathering in granitic rock soil profile (modified after Little, 1969)

residue coupled with the physical disintegration and downward washing of finer particles by percolating waters (lessivage).

The above results in the compositional changes that are reflected in the vertical soil section called the soil profile. Residual soils in the tropics develop a soil profile consisting of a sequence of distinct layers called soil horizons, more or less paralleling the ground surface. The study of the soil horizons, however, is the domain of soil science. Simultaneously, the soil profile also develops a physical/morphological vertical weathered profile which is its pertinent aspect from the engineering perspective.

The weathering profile, generally, indicates a gradual change from fresh rock to a completely weathered soil as illustrated in Figures 2.10 and 2.11 for two typical rock types in a tropical terrain.

The weathered profiles, above these two types of rock, display a different morphological configuration as a consequence of their different nature and structure. The weathering profile in igneous rocks, which are massive, shows rounded, irregular masses called corestones; which are non-weathered masses left behind as the weathering front progresses downward. This is probably on account of the presence of joints (fractures in rock) in the fresh rock where weathering can proceed along the joints. Foliations in the rocks enhance the variability that can be found in the weathering profiles in metamorphic rocks. The weathering profile over the metamorphic rock shows relict banding planes, joints and also corestones.

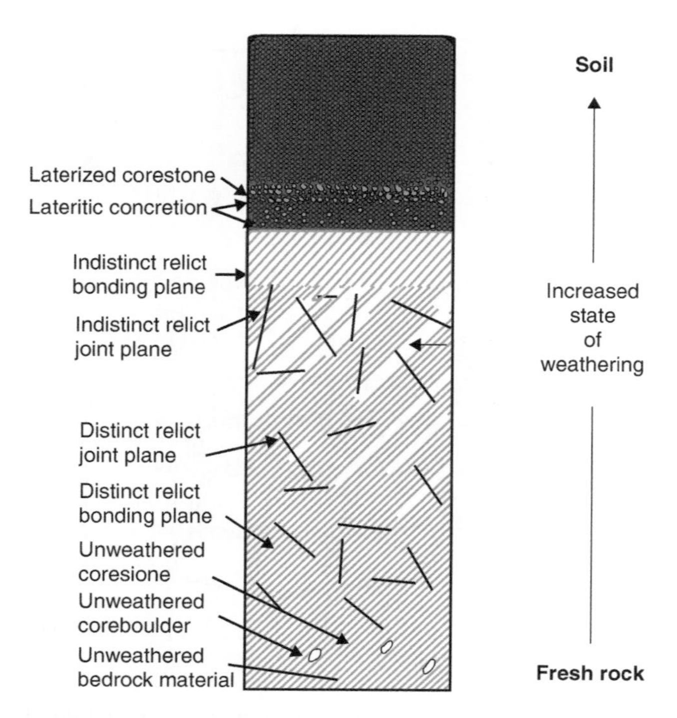

Figure 2.11 Weathering in a metamorphic rock (amphibole schist) soil profile (modified after Raj, 1994)

2.4.2 Chemical alteration and composition of the weathered profile

(a) Chemical alteration

The hot, humid environment speeds up the chemical dissolution of rock-forming silicate minerals, the formation of new and residual mineral species and the decay of organic matter. The soluble ions are translocated by leaching while small particles are also eluviated by physical downwashing. In addition, the leaching due to chelating agents like humic acids, however, is rare as the buildup of acidic compounds due to the slow decomposition of vegetation is restricted. As a consequence, generally, the bases (K, Na, Ca, Mg) and silica are leached downwards in solution to be redeposited lower down in the soil profile, while Fe and Al remain insoluble as the decomposition takes place in near-neutral conditions below the depth of acidic chelating agents confined nearer to the small surface humus cover.

This results in breakdown products like clay minerals (kaolinite, smectite, illite, montmorillonite), other phyllosilicates (mica, chlorite and muscovite) and the accumulation of residual materials like oxides, hydroxides and free silica. This process is also referred to as ferralitisation, a term indicating the importance of domination by the insoluble iron often responsible for the red color of the soil. Quartz is resistant to chemical attack and so is resistant to weathering processes.

Table 2.12 Types of clay minerals and leaching intensity (*After* Summerfield, 1991)

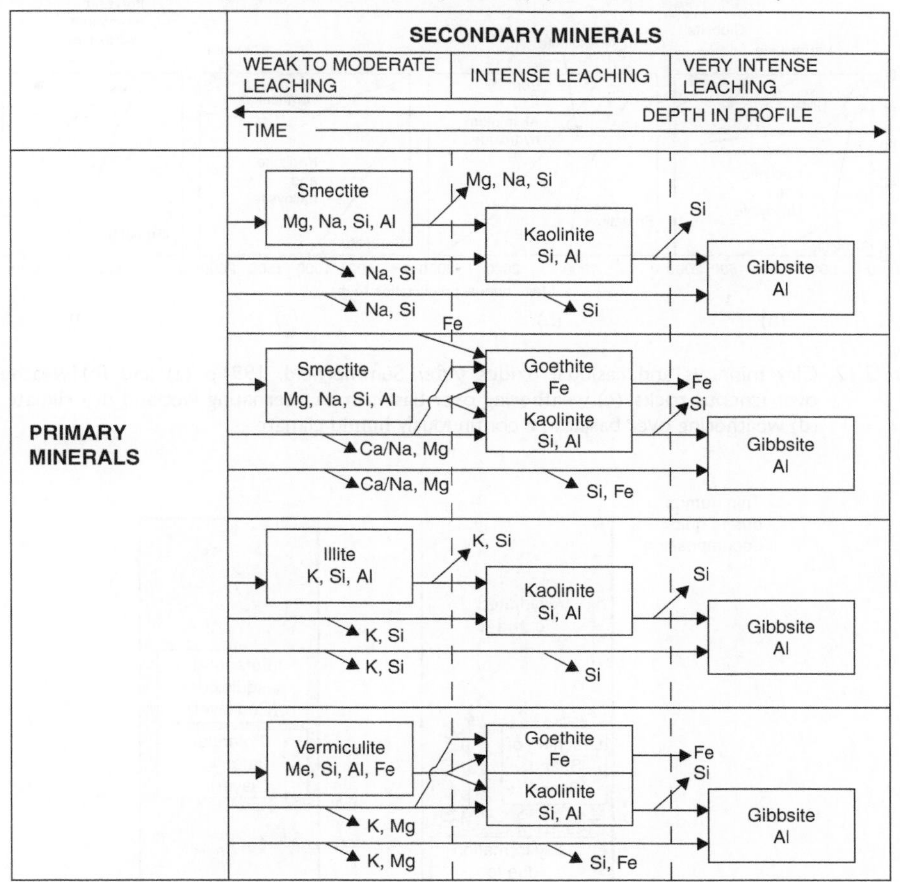

The secondary minerals that result from primary minerals upon weathering depend on the climatic conditions. Table 2.12 shows the sequence of alteration with respect to rainfall and the subsequent leaching showing the progressing development of different clays and oxides.

Figure 2.12 shows again the variations in clay mineralogy and residual oxide compositions as a function of mean annual precipitation in different types of tropical climatic zones.

The areas in the tropics affected by alternating wet and dry seasons cause the soils to experience a different translocation of elements. During the wet season, weathering of the primary silicate minerals in the upper soil horizon causes leaching of the silica and bases away leaving behind the iron and aluminium oxides. During the dry season, however, due to water rising up from below due to capillary rise, the leaching of silicates progresses below the hydrated Fe and Al oxides creating a layer of insoluble oxidised iron and other minerals. This gives rise to two different soil profiles as shown in Figure 2.13.

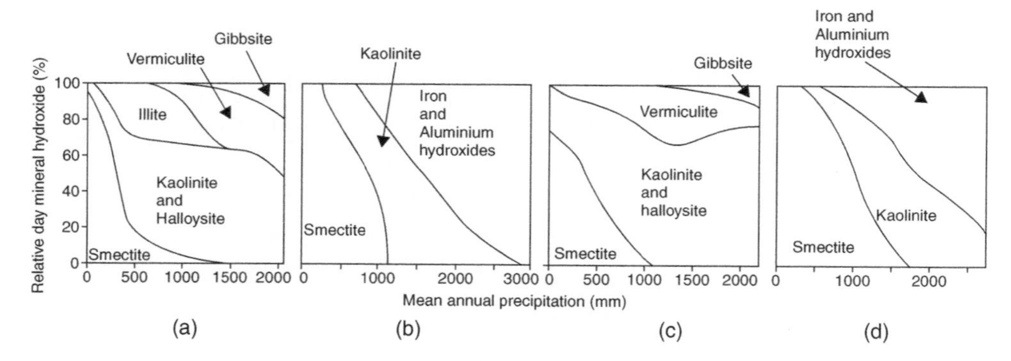

Figure 2.12 Clay minerals and residual oxides (*After* Summerfield, 1991). (a) and (b) weathering over igneous rocks, (c) weathering over basalt in an alternating wet and dry climate and (d) weathering over basalt in a continuously humid climate

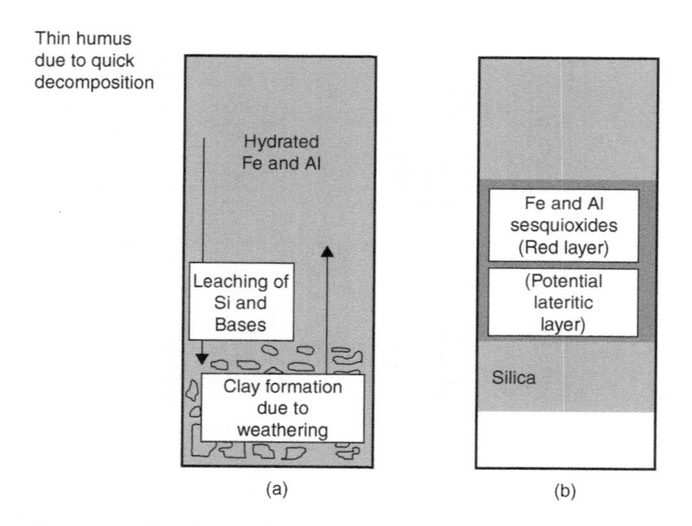

Figure 2.13 Tropical soil profiles (*After* Nagle, 2000). (a) in continuously humid climate and (b) in alternating wet and dry climate

(b) Mineralogy

(i) Clay minerals

Clays belong to the larger class of sheet silicates known as phyllosilicates with a structure composed of two basic units i.e. (1) The tetrahedral sheet composed of silicon-oxygen tetrahedra linked to neighboring tetrahedra by sharing three corners to form a hexagonal network and (2) the octahedral sheet is usually composed of Al or Mg in six-fold coordination with oxygen forming the tetrahedral sheet and with hydroxyl. These tetrahedral and octahedral sheets taken together form a layer, and the individual layers may be joined to each other in a clay crystallite by interlayer cations,

Table 2.13 Clay mineral groups (*After* Berry *et al.*, 2004)

Group	*Layer Type*	*Type of Structural Formula*
Kaolinite	1:1	$Al_4Si_4O_{10}(OH)_8$
Illite	2:1	$(K,H_3O)(Al,Mg,Fe)_2(Si,Al)_4O_{10}[(OH)_2,H_2O)$
Smectite	2:1	$(Na,Ca)_{0.33}(Al,Mg)_2Si_4O_{10}(OH)_2{\cdot}nH_2O$
Vermiculite	2:1	$(Mg,Fe,Al)_3(Al,Si)_4O_{10}(OH)_2{\cdot}4H_2O$
Chlorite	2:1:1	$(Mg,Fe,Al)_6(Al,Si)_4O_{10}(OH)_8$

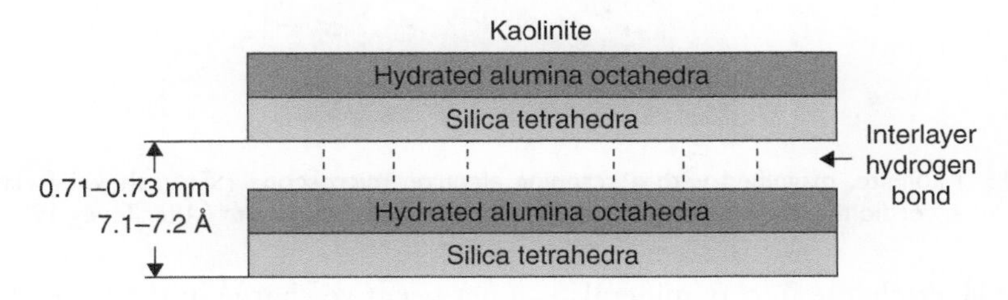

Figure 2.14 Structure of a 1:1 clay

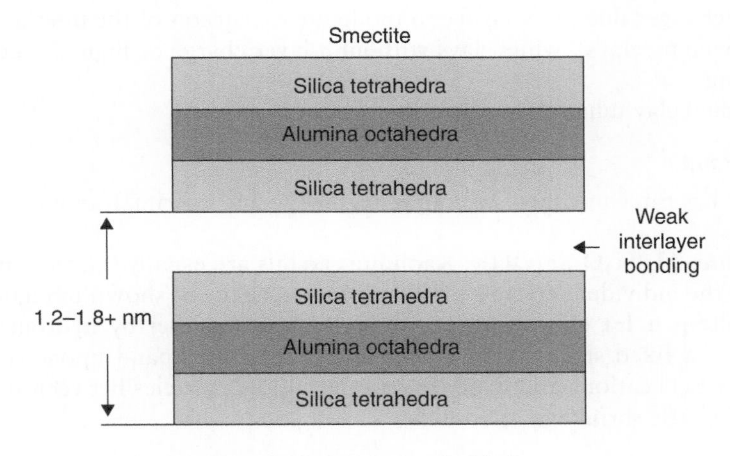

Figure 2.15 Structure of a 2:1 clay

by van der Waals and electrostatic forces, or by hydrogen bonding. The various clay groups are classified by the arrangement of the tetrahedral and octahedral sheets into layers as 1.1 layered clays and 1.2 layered clays. The first group of 1:1 clay minerals, as shown in Figure 2.14, consists of two layers each with one tetrahedral and one octahedral sheet. The layers are bound together tightly by a hydrogen bond.

The second group of 2:1 clay minerals, as shown in Figure 2.15, consists of two layers each with two tetrahedral sheets with an octahedral sheet between them. These two clays are not bound together tightly and swell when wet as hygroscopic surfaces between layers adsorb water forcing the layers apart. There are also 2:1:1 clay minerals containing an octahedral sheet adjacent to the 2:1 layer.

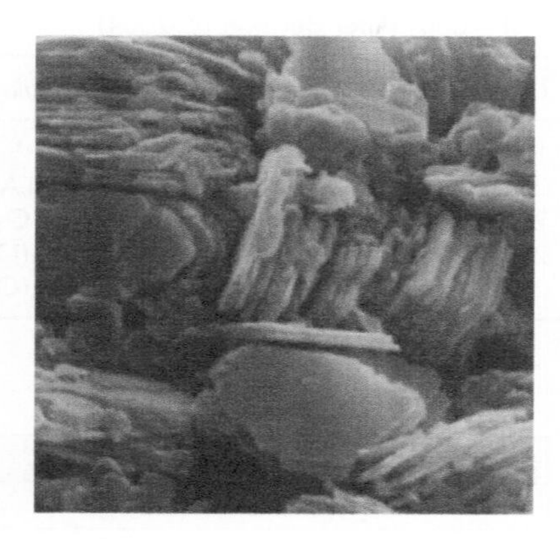

Figure 2.16 Kaolinite, magnified with a scanning electron microscope (SEM), showing classic platy, accordion- or book-like mineral form; picture width is 17 μm (*After* Tovey, 1971)

The layer charge in clay minerals is a net negative charge in the layer. The layer charge variation is important for the second group of 2:1 clays as it affects their behavior. The clays of low charge (smectite) and moderate charge (vermiculite) are prone to volume changes due to extensive to moderate expansion of the interlayer distance forming 'swelling clays', while clays without a layer charge or high charge (illite) are non swelling.

Individual clay minerals are described below.

Kaolinite group

This group has three members i.e. kaolinite, dickite and nacrite. Kaolinite is described below.

Kaolinite; $Al_2(Si_4O_{10})\ (OH)_8$: Kaolinite crystals are usually 0.2 to 2 micrometers in size and the individual crystals are hexagonal in shape as shown in Figure 2.16.

Kaolinite is a 1:1 clay mineral with layers held together by hydrogen bonding, allowing for a fixed structure that does not ordinarily expand upon wetting. This bonding restricts cations and water from entering the particles between the layers. It also exhibits little shrinkage or swelling.

Illite group

This group consists of a number of clay minerals that have a layer charge of approximately 1 per 1/2 unit cell. One of the minerals constituting this group i.e. illite is described below.

Illite; $K(Al_{2-x-y}Fe_xMg_y)O_{10}(Si_{4-z}Al_z)(OH)_2$: Illite or Hydrous Mica, a potassium aluminum silicate hydroxide hydrate mineral, is usually grayish-white to silvery-gray, sometimes greenish-gray in color (Miller and Gardiner 2002).

Smectite group

Smectite is a term used for a number of 2:1 clay minerals that have a layer charge between 0.2–0.6 per 1/2 unit cell. One of the most common smectites is montmorillonite which is described below.

Montmorillonite; $(Al_{2-x}Mg_x)(Si_{4-y}Al_y)(O_{10})(OH)_8$: Montmorillonite (or Hydrated Sodium Calcium Aluminum Magnesium Silicate Hydroxide) is usually white, grey or pink with tints of yellow or green in color. It typically forms microscopic or at least very small platy micaceous crystals. In Montmorillonite, water easily penetrates between planes of adjacent oxygen ions, causing the individual layers of clay particles to separate and swell. The water content is variable, and in fact when water is absorbed by the crystals, they tend to swell to several times their original volume. The swelling property of montmorillonite creates geotechnical and structural problems.

Vermiculite group

This group consists of 2:1 clays that have a layer charge between approximately 0.6 and 0.9 per 1/2 unit cell. They are formed from weathered micas when the K^+ ions between the molecular sheets are replaced by Mg^{2+} and Fe^{2+} cations.

Chlorite group

These are a group of phyllosilicate minerals described by four end members based on their substitution of the following four elements Mg, Fe, Ni, and Mn in the silicate lattice as follows: Clinochlore: $(Mg_5Al)(AlSi_3)O_{10}(OH)_8$, Chamosite: $(Fe_5Al)(AlSi_3)O_{10}(OH)_8$, Nimite: $(Ni_5Al)(AlSi_3)O_{10}(OH)_8$ and Pennantite: $(Mn,Al)_6(Si,Al)_4O_{10}(OH)_8$. This group is not always considered a part of the clays and is sometimes left alone as a separate group within the phyllosilicates.

(ii) Other phyllosilicates:

Mica: Mica consists of a group of minerals characterised by highly perfect cleavage as they readily separate into very thin leaves. They differ widely in composition, and vary in color from pale brown or yellow to green or black. The important species of the mica group are: *muscovite*, common or potash mica, pale brown or green, often silvery; *biotite*, iron-magnesia mica, dark brown, green, or black; *lepidomelane*, iron mica, black; *phlogopite*, magnesia mica, colorless, yellow, brown; *lepidolite*, lithia mica, rose-red, lilac. Mica (usually muscovite, also biotite) is an essential constituent of granite, gneiss, and mica slate; biotite is common in many eruptive rocks.

Muscovite; $KAl_2(AlSi_3O_{10})(OH_2)$: Muscovite is commonly occurring mica. It forms fine to coarse platy books and scales ranging in color from clear to silvery white to gray to green to, less frequently, yellow or pale pink. Muscovite is common in granitic rocks. In granitic pegmatites, sheets may grow up to several meters in diameter. It is also common in metamorphic rocks, especially phyllites and schist derived from potassic, aluminous protoliths such as shales. A green chromium-rich variety called 'fuchsite' occurs in certain quartzites. It also survives weathering where it occurs as detrital flakes in sandstones such as arkose. Muscovite also forms from the breakdown of feldspars where it occurs with other minerals such as paragonite as fine aggregates referred to as sericite.

(iii) Oxides and hydrous metal oxides

The oxides and hydrous metal oxides are common components of soils and are formed by the weathering of primary silicate minerals. Al, Fe, and Mn hydroxides are some of the common minerals. Gibbsite is the most common of the Al (hydr)oxides and the only one stated to commonly occur in soils. Several species of iron (hydr)oxides occurring in soils include goethite, hematite, magnetite and lepidocrocite.

Gibbsite; $Al(OH)_3$: Gibbsite is one of the mineral forms of aluminium hydroxide. The basic structure is formed from stacked sheets of linked octahedrons of aluminum hydroxide. The sheets are only held together by weak residual bonds and these result in a very soft easily cleaved mineral. It is one of the minerals that make up bauxite, a rock formed from intense weathering in richly forested, humid, tropical climates.

Goethite; $FeO(OH)$: Goethite, a common iron mineral, is a hydrous oxide of iron. The colour varies from yellowish to blackish brown. As it often forms by weathering of other iron-rich minerals, it is a common component of soils. It may form excellent pseudomorphs after the original minerals particularly pyrite or marcasite.

Hematite; (Fe_2O_3): An iron oxide, is the mineral form of iron (III) oxide. It is colored black to steel or silver-grey, brown to reddish brown and is more commonly observed in highly weathered soils from the tropics and subtropics imparting a bright red colour to soils.

(iv) Quartz, SiO_2

Quartz or silicon dioxide is the most common mineral on the face of the Earth. It is found in nearly every geological environment and is a component of almost every rock type. It is a tectosilicate consisting of interconnected silica (SiO_2) tetrahedra forming an intricate framework.

Table 2.14 shows the mineralogical composition of some Malaysian tropical residual soils.

Kaolinite and quartz are two major minerals while muscovite, illite, montmorillonite and gibbsite are minor minerals (Aminaton *et al.*, 2002).

2.5 PEDOGENETIC CLASSIFICATION OF TROPICAL RESIDUAL SOILS

Various pedogenetic (soil science) classifications have been devised for tropical residual soils but these, however, are not useful for engineering purposes (Netterberg, 1978). Other purpose specific classifications that have been attempted are listed in Fookes (1997). However, a classification by Duchaufour (1982), expounded in Fookes (1997) and White (2006), distinguishes three residual soil phases in relation to climatic factors. This has relevance as it relates to the breakdown of parent materials and composition of residual soil and is related to a particular tropical climatic zone. Duchaufour (1982) describes the three phases of weathering with increasing temperature and rainfall as: fersiallic soils (Phase 1), ferruginous soils (Phase 2) and ferrallitic soils (Phase 3) respectively. In subtropical climates, with a marked dry season, phase (1) is rarely exceeded. In a dry tropical climate, development stops at phase (2). Only in humid equatorial climates is phase (3) reached. These are summarised in Table 2.15.

Fersiallitisation: This occurs in a subtropical climate with a marked dry season. The dominant clay minerals formed are the 2:1 clays, like smectite, as much of the silica and bases released by the weathering are retained in the profile. The other clay formed

Table 2.14 Mineralogical composition of some Malaysian tropical residual soils (*After* Tessens and Jusup, 1983)

Parent material	*Dominant mineral*	*Present*	*Marginal*
Igneous rocks			
Basalt	Gibbsite	K, Go	
Granite	Kaolinite	M	
Granite	Kaolinite	C	M
Granite	Kaolinite	Gi, Go	C, Feldspar
Granite	Kaolinite	Gi	Go, Q
Granite	Kaolinite	Gi	
Granodiorite	Kaolinite	Go	Gi
Ryolites	Kaolinite	Q, M	
Tuffs	Mica	K, Q	
Metamorphic rocks			
Schist	Kaolinite	Gi	
Schists	Kaolinite	Gi	Go, M, M-C
Sedimentary rocks			
Sandstone	Kaolinite		Go, C
Sandstone	Kaolinite	M	
Shale	Kaolinite	Go, Gi	

K = Kaolinite; M = Mica; Gi = Gibbsite; Go = Goethite; C = Chlorite; ML = Mixed Layers; Q = Quartz.

Table 2.15 Phases of Residual Tropical Soils (*After* Duchaufour, 1982)

Phase	*Climatic Zone*	*Mean annual temp (C)*	*Annual rainfall (m)*	*Dry season*	*Soil type*
3	Tropical	>25	>1.5	no	Ferrallitic
2	Subtropical	20–25	1.0–1.5	sometimes	Ferruginuous Ferrisols (transitional)
1	Subtropical	13–20	0.5–1.0	yes	Fersiallitic

is kaolinite. The liberated Fe^{2+} is rapidly oxidised and precipitated as ferric hydroxide and the clays undergo rubification (reddening) as the hematite formed forms a red colored coating on clays (White, 2006). The clays, upon being lessivaged (removed by the action of a percolating liquid), are carried downwards to form clay enriched layers appearing red or red-mottled with iron oxide. When freed, bases are forced upwards by capillary rise during the dry season and re-precipitated e.g. when carbonates are present, these first undergo decalcification after which the dissolved calcium carbonate is precipitated to form a calcrete layer (duricrust) thickening to develop a calcareous if favorable conditions (like prolonged dry conditions) persist.

Ferrugination: This occurs in a tropical climate that is more humid than fersiallitic conditions and with a less pronounced dry season. The dominant clay minerals are the 1:1 clays like kaolinite. Smectite may also be present. The Fe oxides are also formed but may or may not be rubified on the translocated clays (White, 2006).

Ferrallitisation: Almost all primary minerals are weathered leaving quartz, kaolinite (neo-formed) and Fe and Al oxides e.g. goethite, gibbsite and hematite. The dominant clay mineral kaolinite is formed from the remaining silica and alumina. The latter, occurring in excess, also forms gibbsite. They are, depending on the amount of iron and aluminium oxides present, divided into iron oxide dominant ferrites and aluminium oxides (usually gibbsite) predominant allites.

2.6 DEFINITION AND CLASSIFICATION OF TROPICAL RESIDUAL SOILS IN CIVIL ENGINEERING PRACTICE

Residual soils are extensively encountered in engineering practice in tropical areas. These, for geotechnical engineering purposes, are the loose, unconsolidated natural material enveloping the surface that is relatively easy to separate by even gentle means. This, as mentioned earlier, is the regolith including the saprolite i.e. in-place bedrock that is chemically altered and coherent, retaining its original texture. The aspects considered in examination of the regolith for engineering purposes and the delineation of the section 'definition and classification of tropical residual soil in civil engineering practice' are explained further below.

2.6.1 Definitions of residual soils

Their definitions re-iterated below, showing the diverse understandings of what are perceived as residual soils.

1. Brand and Phillipson (1985) definition is 'a soil formed by weathering in place, but with the original rock texture completely destroyed' and is comprehensive. It is commonly used in a wider sense to include highly and completely decomposed rock which as an engineering material behaves like a soil in places like Hong Kong.
2. Blight (1985) definition is 'as all material of a soil consistency that is located below the local ancient erosion surface i.e. below the pebble marker'.
3. Sowers (1985) definition is 'as the product of rock weathering that remains in place above the yet-to-be weathered parent rock' which is also comprehensive as it includes all the weathered material.
4. The Public Works Institute of Malaysia (1996) definition of a residual soil is 'a soil which has been formed *in situ* by decomposition of parent material and which has not been transported any significant distance' and defines tropical residual soil as 'a soil formed *in situ* under tropical weathering conditions'.

In reviewing the international practice for the sampling and testing of residual soils, Brand and Phillipson (1985) also found that authors from different regions varied in the interpretation of what might be defined as 'residual soil' as shown in Table 2.16.

Generally, the majority of the researchers primarily defined residual soils as a 'highly weathered *in situ* material with >50% soil' (this corresponds to weathering grades IV, V and VI of the classification scheme as explained further below).

Table 2.16 Materials categorised as residual soil by workers from different regions (*After* Brand and Phillipson, 1985)

		Degree of weathering			*Transported soils*	
Country	*Parent rock All types*	*Completely: Original structure destroyed?*	*Completely: Original structure intact?*	*Highly: Structure intact, Soil >50%?*	*Colluvium included?*	*Others included?*
Australia	Yes	?	Yes	Yes	Yes	Yes
Brazil	No	Yes?	No	No	No	No
China	Yes	?	Yes	Yes	?	No?
Germany	Yes	?	Yes	?	No	No
Ghana	Yes	?	Yes	Yes	?	?
Hong Kong	Yes	Yes	Yes	Yes	Yes	No
India	Yes	?	Yes	Yes	No	No
Japan	Yes	?	No	Yes?	Yes	?
Malaysia	Yes	Yes	Yes?	Yes?	No?	No
New Zealand	Yes	Yes	Yes	Yes	Yes?	?
Nigeria	Yes	Yes	Yes	Yes	Yes?	Yes
Pakistan	Yes	No	Yes	Yes?	No	No
Philippines	Yes	No	Yes	Yes	Yes	No
Singapore	Yes	Yes	Yes?	Yes	?	No?
South Africa	Yes	Yes?	Yes	Yes	?	Yes
Sri Lanka	Yes	Yes	Yes	Yes	Yes	No
U.K.	Yes	Yes	Yes	Yes	No	No
U.S.A.	Yes	Yes?	Yes	Yes	No	No

? Indicates that the authors have not clearly defined these terms.

2.6.2 Pertinent aspects of classifications of tropical residual soils for engineering practice

The alteration of the parent rocks observed in the residual weathered profile is described by taking into account various alterations and changes, namely, the disintegration, mineralogical changes and cementation including the development of duricrust. Each of these is used for characterizing the residual component of the parent rock, thus, the limit of what is considered as residual soil.

(a) Morphology and classification of the weathered profile

The weathering profile reflects the state of weathering along the soil profile from the bedrock (unaltered parent rock) to the ground surface. It shows progressive stages of transformation or 'grading' from fresh rock to completely weathered material upwards. The degree of weathering describes the disintegration of the parent rock with depth. Continuous attempts have been made, since Vargas (1953), to devise methods for the description or classification of this weathered profile. Parent rocks are increasingly modified with the increase in the state of their weathering till the completely weathered portion may cease to resemble the parent rock. Amongst the methods devised for the description or classification are those based on this state of weathering observed in the soil profile. Variations from place to place, due to the local variation in rock type

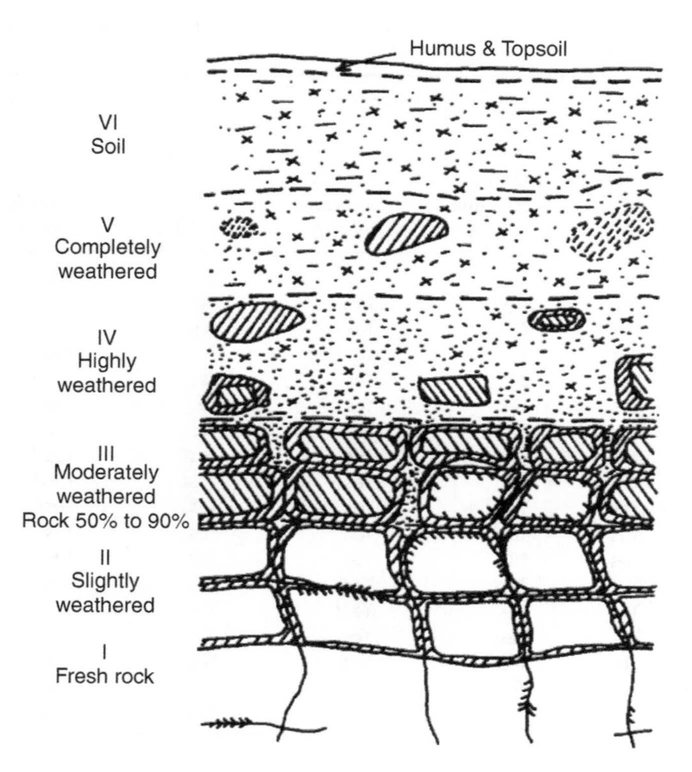

Figure 2.17 Engineering weathering grades in a residual soil profile in granitic rock (*After* Little, 1969)

and structure, topography and rates of erosion because of regional climatic variation, particularly rainfall, are amongst the difficulties in attaining a broad perspective from which to view the weathered profile. These weathered profiles have been investigated by several researchers like Little (1969), Fookes *et al.* (1971), Fookes (1997), Dearman *et al.* (1978), Irfan and Dearman (1978), Anon (1981a) and Anon (1981b) for a descriptive scheme for grading the degree of weathering. The classifications devised are generally suitable for soil profiles over most igneous rocks.

The disintegration scheme mostly used is the six weathering grades classification of Table 10 of BS 5930 (British Standards Institute, 1981) developed from the working parties, namely, Core Logging (Anon, 1970), Engineering Geology Mapping (Anon, 1972) and Rock Masses (Anon, 1977) as shown in Figure 2.17. The classification is described in Table 2.17.

Fookes (1997) defines the 'weathered bedrock' as consisting of the fresh 'bedrock' (Grade I) and the slightly weathered and moderately weathered zones (Grades II and III) as these tend to behave as rock in engineering terms, and soil as the material consisting of the highly weathered zone and above (Grade IV and above) as these tend to behave as soil in engineering terms.

The use of classification by an adaptation to suit the weathered profiles of the above specific rock types, is shown by a modified version in Table 2.18 for a particular type of metamorphic rock using the zones of weathering.

Table 2.17 Classification of the engineering weathering profile (*After* Anon, 1970; McLean and Gribble, 1979)

Weathering classification		
Term	*Grade*	*Description*
Residual soil	VI	All rock material is converted to soil; the mass structure and material fabric are destroyed; there is a large change in volume but the soil has not been significantly transported
Completely Weathered	V	All rock material is decomposed and/or disintegrated to soil; the original mass structure is still largely intact
Highly weathered	IV	More than half of the rock material is decomposed and/or disintegrated to soil; fresh or discolored rock is present either as a discontinuous framework or as corestones
Moderately weathered	III	Less than half of the rock material is decomposed and/or disintegrated to soil; fresh or discolored rock is present either as a discontinuous framework or as corestones
Slightly weathered	II	Discoloration indicates weathering of rock material and discontinuity surfaces; all the rock material may be discolored by weathering
Fresh rock	I	No visible sign of rock material weathering; perhaps slight discoloration on major discontinuity surfaces

Table 2.18 Classification of weathering profile over metamorphic rock (Clastic Metasediment) in Peninsular Malaysia (*After* Komoo and Mogana, 1988)

Weathering classification		
Term	*Zone*	*Description*
Residual soil	VI	All rock material is converted to soil. The mass structure and material fabric (texture) are completely destroyed. The material is generally silty or clayey and shows homogenous color
Completely weathered	V	All rock material is decomposed to soil. Material partially preserved. The material is sandy and is friable if soaked in water or squeezed by hand
Highly weathered	IV	The rock material is in the transitional stage to form soil. Material condition is either soil or rock. Material is completely discolored but the fabric is completely preserved. Mass structure partially present.
Moderately weathered	III	The rock material shows partial discoloration. The mass structure and material texture are completely preserved. Discontinuity is commonly filled by iron-rich material. Material fragment or block corner can be chipped by hand
Slightly weathered	II	Discoloration along discontinuity and may be part of rock material. The mass structure and material texture are completely preserved. The material is generally weaker but fragment corners cannot be chipped by hand
Fresh rock	I	No visible sign of rock material weathering. Some discolouration on major discontinuity surfaces

	Weathering classification	
	Term	*Grade*
Residual soil	Residual soil	VI
	Completely weathered	V
	Highly weathered	IV
Semi-residual soil	Moderately weathered	III
	Slightly weathered	II
	Fresh rock	I

Figure 2.18 Weathering classification

The weathered residual profile is also categorised into two distinct parts, as shown in Figure 2.18. The first incorporating the top three weathering grades: VI (residual soil), V (completely weathered) and IV (highly weathered) considered as 'Residual Soil' where the grades V and VI together are termed as 'Saprolite' and the second incorporating the remaining three grades are considered as 'Semi-Residual Soil'. The term 'Saprolite' comes from the word *Sapros* or 'Rotten' in Greek and is often used in the general sense to mean weathered or partially weathered rock *in situ*. The common use of 'Saprolite' in engineering contexts is also often restricted to the middle part of the weathering mantle.

There is, however, no unanimity on the demarcation of tropical residual soils as there are differing views regarding which weathering grades constitute residual soil ranging from exclusively Grade VI, to both Grades VI and V, to Grades VI, V and IV as by Fookes (1997).

The description of residual soils is also approached from the perspective of alternate weathering classifications. This is illustrated by another weathering classification as given in Table 2.19. This, as can be seen, has different weathered classes and the uppermost is considered as residual soil.

A description of a weathering profile for igneous and metamorphic rocks is given in Table 2.20 and shown in Figures 2.19(a) and (b).

Another weathering classification of the weathered profile, over a metamorphic rock, found in the literature is shown in Figure 2.20.

(b) Engineering weathering grades versus pedological classification

Figure 2.21 shows the relationship between engineering weathering grades and pedological classification of tropical residual soils. The residual soils, delimited from the weathering grades by Fookes (1997), are juxtaposed with the phases of weathering and the consequential alterations in the soil profile. Also listed are other terms used to categorise the weathering grades.

Table 2.19 Rock Material Weathering Classification (Australian Standard: AS1726-1993) (Standards Australia, 1998)

Term	*Symbol*	*Definition*
Residual soil	RS	Soil developed on extremely weathered rock, the mass structure and substance fabric are no longer evident; there is a large change in volume but the soil has not been significantly transported
Extremely weathered rock	XW	Rock is weathered to such an extent that it has 'soil' properties, i.e. it either disintegrates or can be remolded, in water
Distinctly weathered rock	DW	Rock strength usually changed by weathering. The rock may be highly discolored, usually by iron staining. Porosity may be increased by leaching, or may be decreased due to deposition of weathering products in pores
Slightly weathered rock	SW	Rock is slightly discolored but shows little or no change of strength from fresh rock
Fresh rock	FR	Rock shows no sign of decomposition or staining

Table 2.20 Weathering profile for igneous and metamorphic rocks (*After* Deere and Patton, 1971)

Zone		*Description*
I. Residual soil	IA – A horizon	• top soil, roots, organic material • zone of leaching and eluviation • may be porous
	IB – B horizon	• characteristically clay-enriched • accumulations of Al, Fe, and Si • may be cemented • no relict structures present
	IC – C horizon (saprolite)	• relict rock structures retained • silty to sandy material • <10% corestones • often micaceous
II. Weathered rock	IIA – transition from residual soil or saprolite to partly weathered rock	• highly variable, soil-like to rock-like • fines commonly fine to coarse sand (gruss) • 10 to 95% corestones • spheroidal weathering common
	IIB – Partly weathered rock	• rock-like, soft to hard rock • joints stained to altered • some alteration of feldspars and micas
III. Unweathered rock		• traces or no iron stains or joints • no weathering of feldspars and micas

(c) Cementation and development of duricrust

Soil development in the tropics may have begun over 2–3 million years ago in the Tertiary era (Courtney and Trudgill, 1999). This prolonged process of soil development over geological time has also been coupled with climatic changes as shown in Figure 2.22. These climatic swings resulting in areas under subtropical and dry tropical climates at present have caused the development of distinctive soil profiles.

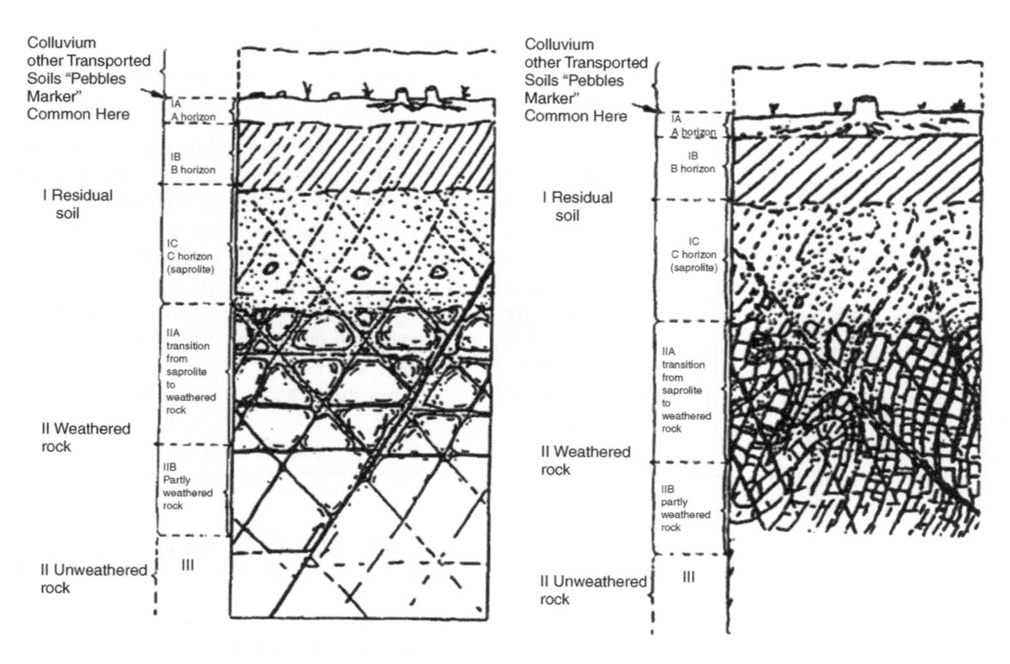

Figure 2.19 (a) Weathering profile for intrusive igneous rocks (left) and metamorphic rocks (right) (*After* Deere and Patton, 1971)

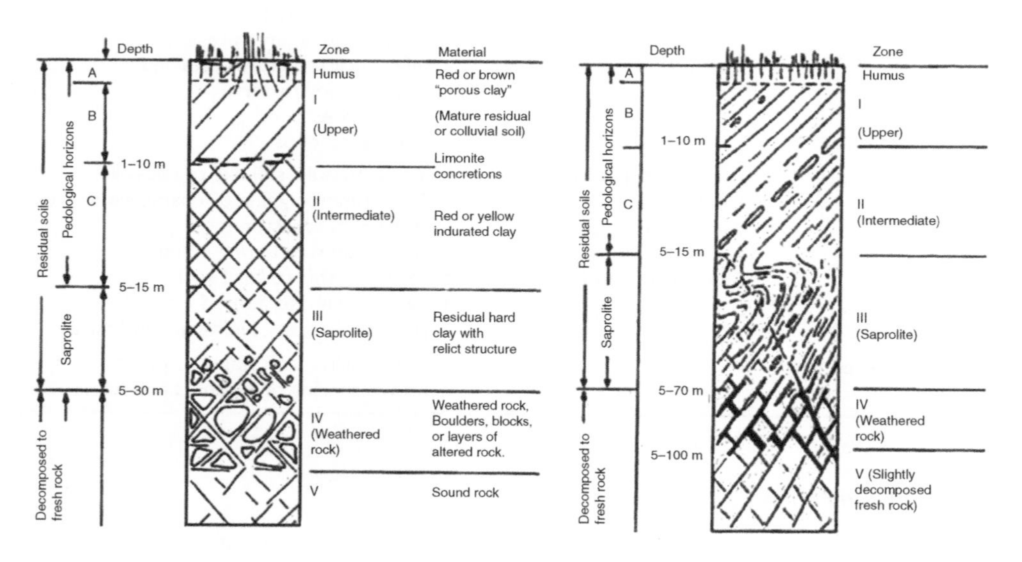

Figure 2.19 (b) Weathering profile for basaltic igneous rock (left) and metamorphic rock (right) (*After* Vargas, 1974)

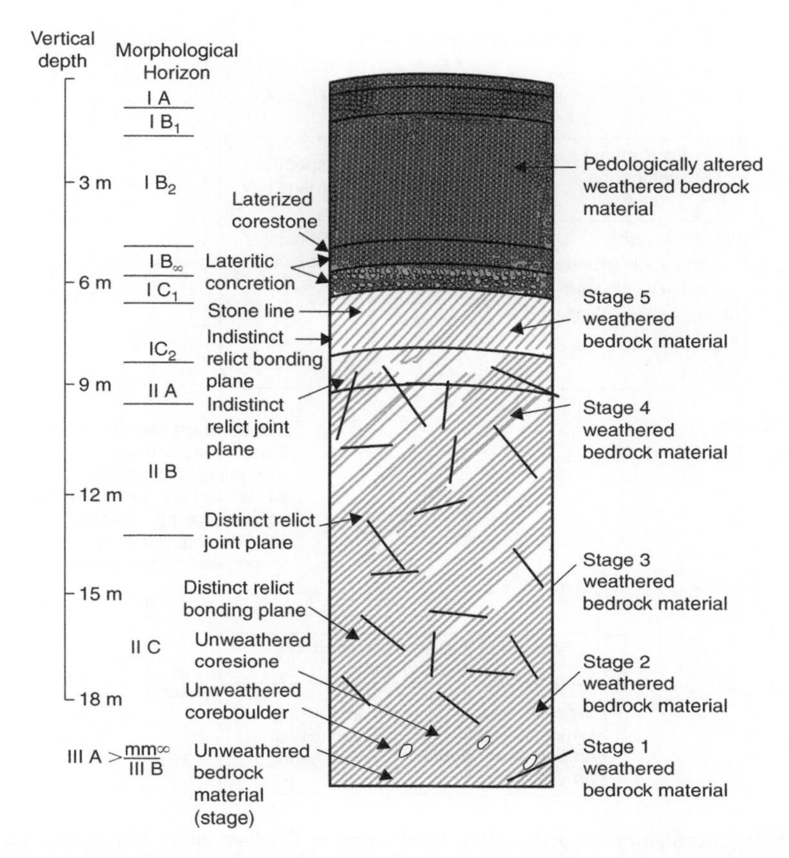

Figure 2.20 Weathering profile in a metamorphic rock (amphibole schist) soil profile (*After* Raj, 1994)

The prolonged leaching has resulted in profound transformation of the parent bedrock and the formation of duricrusts. Duricrust is a term applied to all types of surface or near-surface, hard, roughly horizontal soil horizons due to the occurrence of residual accumulation of iron and alumina or precipitation of calcite, dolomite or gypsum. Examples of duricrusts include laterites or ferricrete, silcretes, calcretes, gypcrete and alucrete.

The term laterite, as originally used by Buchanan (1807), defines residual material arising due to the leaching of silica and enrichment with aluminum and iron oxides, especially in humid climates, which has been hardened upon exposure to alternate wetting and drying. However, over time, laterite has also been used in a broad sense to indicate all kinds of tropical iron dominated red soils. It has also, according to Tardy (1992), been used as an overall term for all the different types of duricrusts.

The development of different types of duricrusts with respect to pH and rainfall is given in Figure 2.23.

Although Goudie (1973) states that the nomenclature for hardened (indurated) crusts remains confused, Fookes (1997) has attempted a classification given in

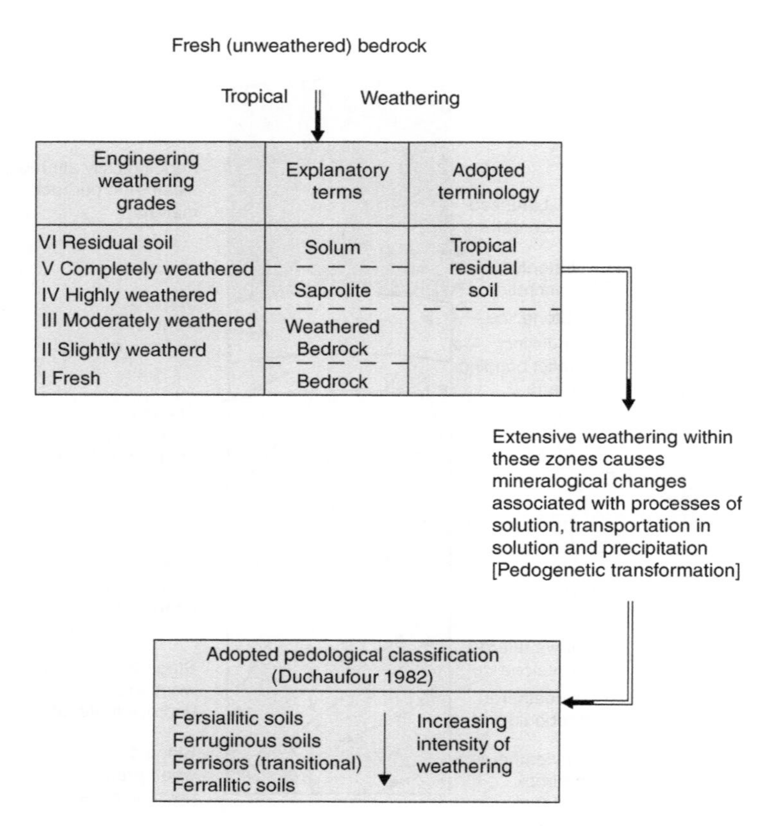

Figure 2.21 Relationship of the pedological classification by Duchaufour (1982) to weathering grades (*After* Fookes, 1997)

Table 2.21 in order of increasing aridity. However, as the classification was broadened to include powder/nodular forms, materials considered as Duricrust are not limited to hardpan/cuirasse.

The classification for Calcrete, included in the classification by Fookes (1997), is based on Netterberg and Caiger (1983). Another classification for Laterite by Charman (1988), that is referred to is given in Table 2.22, while possible conditions for the development of concretionary laterite put forward by Ackroyd (1967) are given in Table 2.23.

2.7 EXAMPLES OF RESIDUAL SOILS OVER DIFFERENT ROCK TYPES

The weathering profiles are formed over different rock types. The different types of weathering profiles with different thicknesses are formed over different rock types. These develop from the chemical weathering of the respective mineral components

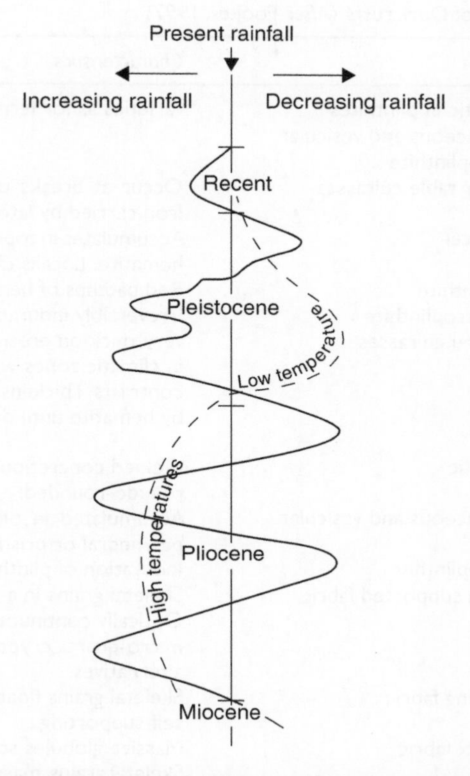

Figure 2.22 Climatic changes since the Miocene (*After* Courtney and Trudgill, 1999)

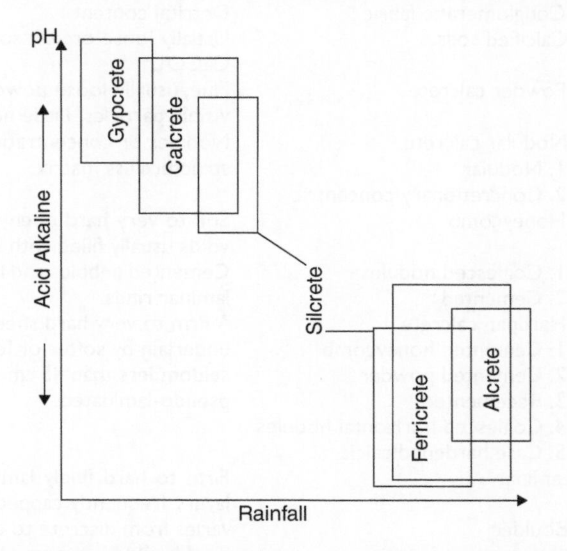

Figure 2.23 The formation of different types of duricrusts (*After* Summerfield, 1991)

Table 2.21 Classification of Duricrusts (*After* Fookes, 1997)

Types	*Groups*	*Characteristics*
Alcrete (Alucrete) (Al_2O_3)	a. Pisolitic in plinthites b. Scoraceous and vesicular c. Petroplinthite	Varieties as for ferricretes.
Ferricrete (Fe_2O_3)	a. Water table cuirasses	Occur at breaks of slope and basin margins. Iron carried by lateral circulation.
	1. Local	Accumulates in topographic lows as goethite or hematite. Localised segregation.
	2. Plinthite	Red patches of hematite.
	3. Petroplinthite	Irreversibly indurated by water table fall.
	b. Plateau cuirasses	Very thick on erosion surfaces. Concentrated in climatic zones with marked seasonal contrasts. Thickens laterally. Goethite replaced by hematite until dominant.
	or	
	a. Pisolitic	Welded concretions unbanded to pseudo-rounded.
	b. Scoraceous and vesicular	Accumulated in old network of fissures of a polyhedral or prismatic horizon.
	c. Petroplinthite	Induration of plinthites often vesicular.
Silcrete (SiO_2)	a. Grain supported fabric.	Skeletal grains in a self-supporting matrix. Optically continuous overgrowths chalcedony, micro-quartz cryptocrystalline and opaline alternatives.
	b. Floating fabric	Skeletal grains floating in matrix, not self-supporting.
	c. Matrix fabric	Massive globules sometimes present. Skeletal grains, massive globules absent in some forms, common in others.
	d. Conglomeratic fabric	Detrital content.
Calcrete ($CaCO_3$)	a. Calcified soils	Usually loose or soft soil weakly cemented by $CaCO_3$.
	b. Powder calcrete	Fine, usually loose powder ($CaCO_3$) with few visible particles. Little nodular development.
	c. Nodular calcrete 1. Nodular 2. Concretionary/concentric	Nodules or concentrations in a loose structureless matrix.
	d. Honeycomb	Stiff to very hard open textured calcrete with voids usually filled with soil.
	1. Coalesced nodules 2. Cemented	Cemented pebbles and fragments coalesced by laminar rinds.
	e. Hardpan calcrete 1. Cemented honeycomb 2. Cemented powder 3. Recemented 4. Coalesced horizontal nodules 5. Case hardened calcic	A firm to very hard sheet-like layer always underlain by softer or looser material and seldom less than 45 cm thick. May be pseudo-laminated.
	f. Laminar	Firm to hard finely laminated undulose sheet layers, frequently capped by hardpan.
	g. Boulder	Varies from discrete to coalesced hard to very hard boulders in a usually red sand matrix.

(*Continued*)

Table 2.21 Continued

Types	*Groups*	*Characteristics*
Gypcrete ($CaSO_4 \cdot 2H_2O$)	a. Desert rose	Individual, twinned and ingrown crystal clusters in loose sand matrix at capillary level. Subsurface crusts, euhedral or lenticular.
	b. Mesocrystalline	Surface crusts-microcrystalline.
	c. Indurated	
	1. Powder	
	2. Indurated alabastine	
	3. Alabastine cobbles	Horizontal laminae or occasionally layer bedding forms.
	d. Evaporitic	
	1. Laminated	
	2. Bedded	Loose or lightly cemented, often in dune form.
	e. Gypsum rich dune sand	

Table 2.22 Classification of Laterite (*After* Charman, 1988)

Age	*Recommended name*	*Characteristic*	*Equivalent terms in the literature*
Immature (young)	Plinthite	Soil fabric containing significant amount of lateritic material. Hydrated oxides present at expense of some soil material. Unhardened, no nodules present but may be slight evidence of concretionary development	Plinthite Laterite Lateritic clay
	Nodular Laterite	Distinct hard concretionary nodules present as separate particles	Lateritic gravel Ironstone Pisolitic gravel Concretionary gravel
	Honeycomb Laterite	Concretions have coalesced to form a porous structure which may be filled with soil material	Vesicular laterite Pisolitic ironstone Vermicular ironstone Cellular ironstone Spaced pisolitic laterite
Mature (old)	Hardpan Lat-erite	Indurated laterite layer, massive and tough	Ferricrete Ironstone Laterite crust Vermiform laterite Packed pisolitic laterite
	Secondary Laterite	May be nodular, honeycomb or hardpan, but is the result of erosion of pre-existing layer and may display brecciated appearance	

Table 2.23 Possible conditions for development of concretionary laterite

Annual rainfall (mm)	750–1000	1000–1500	1500–2000
Thornthwaite moisture index	−40 to −20	−20 to 0	0 to +30
Length of dry season (months)	7	6	5
Type of product	Rock laterite or cuirasse	Hard concretionary gravels	Minimum requirements for concretions to develop

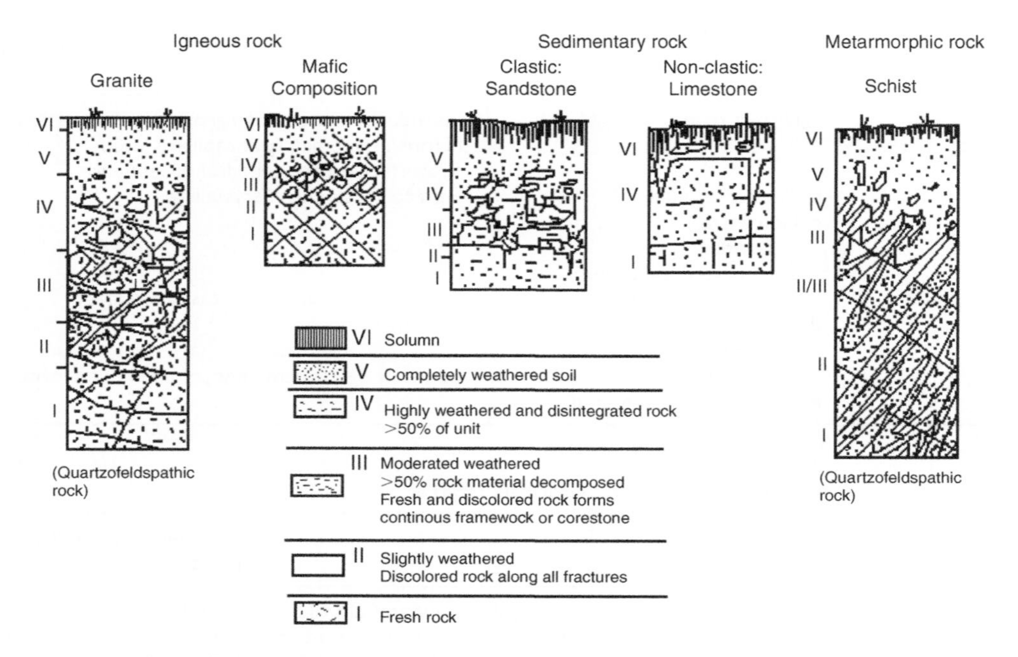

Figure 2.24 Weathering profiles over different rock types (*After* Selby, 1993)

present in the different rock types. As mentioned, secondary minerals are created from primary minerals that constitute these rocks. The weathering profiles are formed over different rock types as shown in Figure 2.24.

2.7.1 Profiles over igneous rocks

Granitic rocks: The most abundant minerals are quartz and feldspars. The quartz (and muscovite, if present) will remain as residual minerals because they are very resistant to weathering while feldspars alter to clays. The weathering is isovolumetric as the relatively inert skeletal framework of quartz and potassium feldspars is retained even after other minerals have been altered to clays, or chemical species removed by dissolution leaving the same volume of material as the parent rock.

Mafic rocks: The most abundant minerals are quartz and feldspars and ferromagnesian minerals. The basalts and gabbros are similar as they have a similar mineralogical composition but vary only in grain size, one being the fine-grained volcanic equivalent (basalt) of the other (gabbro). The weathering profiles are found to be thin as the structurally weak minerals weather easily.

2.7.2 Profiles over sedimentary rocks

Clastic Sedimentary Rocks: Sandstones, having a percentage of quartz higher than 75%, consist primarily of quartz grains. As these are composed of the more resistant mineral quartz, they weather to produce sandy soils and the weathering profiles are found to be thin. Shales (mudrocks) are primarily composed of clay minerals with

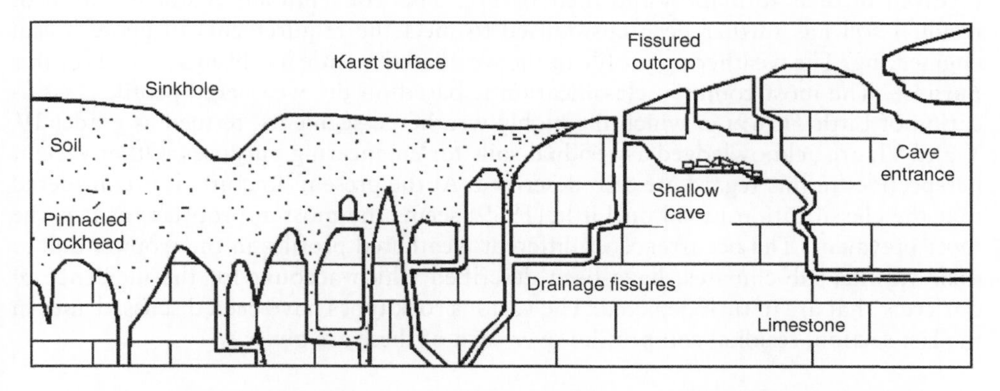

Figure 2.25 Fresh rock at the lower part of the weathered profile (*After* Waltham, 1994)

lesser amounts of quartz and feldspars. They may be altered further to more stable clay minerals.

Non-clastic Sedimentary Rocks: Limestones are high in calcium while dolomites are high in calcium and magnesium. Soils developed from limestone are usually fine-textured soils, and in humid regions the carbonates are readily dissolved and carried away, leaving a thin weathering profile, and at places bare rock outcrops. Impurities, like quartz sand grains and clay particles, even amounting to less than 10% on an average accumulate quickly (Jenny, 1994). The soil water seeping into the limestone causes rock dissolution, creating wide fissures and caves forming an uneven ground with pinnacled fresh rockhead and large voids as shown in Figure 2.25.

2.7.3 Profiles over metamorphic rocks

Examples of metamorphic rocks are marble (metamorphic equivalent of limestone), schist (metamorphic equivalent of shale) and gneiss (metamorphic equivalent of shale or igneous rocks) and quartzite (metamorphic equivalent of sandstone). Gneiss consists of layers of coarse grains of quartz or feldspars alternating with layers of fine-grained minerals. Schist consists of micaceous layers. The changes in mineralogy and rock structure generally render the metamorphosed rock more resistant to weathering. Since there is such a wide range of metamorphic rocks, there is also a wide range of weathering products.

The gneiss, granulites and other quartz rich rocks like quartzites disintegrate into sandy residual soils similar to those over sandstones. The slate, phyllite and schist alter to the more stable clay minerals. The weathering is also isovolumetric as a relatively inert skeletal framework of quartz etc. is retained even after mineral alteration to clays or dissolution of chemical species, leaving the same volume of material as the parent rock.

2.8 CONCLUSIONS

Residual soils are invariably encountered in engineering practice in tropical areas. The conventional perspective of residual soils first given described the changes that are involved in their formation and their nature. This conventional characterisation of residual soil has further been customised to meet the requirements of geotechnical engineering. The weathering profile of the weathered mantle has been adopted for this purpose. The most common classification is based on the weathered profile classification of Little (1969) in which the highly weathered segments, termed as grades IV, V and VI, are acknowledged as residual soils for engineering purposes. Other variant perspectives in this regard are also described. At the present juncture, it is considered that the classification based on Little (1969) seems the most appropriate as it is the most prevalent. The occurrence of different weathered profiles in the tropics, due to their various sub-climates, have been described which account for the incidence of duricrust that are in turn depicted. The various rock types have been discussed and, in the last section, residual soil profiles over each of these are given.

REFERENCES

Ackroyd, L.W. (1967). Formation and properties of concretionary and non-concretionary soils in western Nigeria, *Proc. 4th African Conf. on Soil Mechanics and Foundation Engineering*, pp. 47–52, Balkema, Rotterdam.

Aminaton, M., Fauziah, K. and Khairul Nizar, M.Y. (2002). Mineralogy, microstructure and chemical properties of granitic soils at the central regions of Peninsular Malaysia – 1, Unpublished: 15.

Anon. (1970). The logging of rock cores for engineering purposes: Engineering Group Working Party Report, *Quarterly Journal of Engineering Geology*, 3, 1–24.

Anon. (1972). The preparation of maps and plans in terms of engineering geology: Working Party Report, *Quarterly Journal of Engineering Geology*, 5, 193–381.

Anon. (1977). The description of rock masses for engineering purposes: Working Party Report, *Quarterly Journal of Engineering Geology*, 10, 355–388.

Anon. (1981a). *Code of Practice for Site Investigation (BS5930)*. British Standards Institute, London.

Anon. (1981b). Rock and soil description and classification for engineering geological mapping. Report by the IAEG commission on engineering geological mapping. *Bulletin of the International Association of Engineering Geology* 24: 235–274.

Berry, L.G., Mason, B. and Dietrich, R.V. (2004). *Mineralogy*, New Delhi, India, CBS Publishers & Distributors.

Blight, G.E. (1985). Residual soils in South Africa, In Brand & Philipson (ed.) *Sampling and Testing of Residual Soils*. Technical Committee on Sampling and Testing of Residual soils. International Society for Soil Mechanics and Foundation Engineering, 159–168.

Bland, W. and Rolls, D. (1998). *Weathering*. London: Arnold.

Buchanan, (1807). A journey from Madras through the countries of Mysore, Kanara and Malabar, *East India Company, London*, 2, pp. 440–441.

Charman J.G. (1988) *Laterite in Road Pavements,* CIRIA Special Publication 47, Construction Industry Research and Information Assoc., London.

Courtney, F.M. and Trudgill, S.T. (1999). *The Soil*. Hodder and Stoughton.

Dearman, W.R. Baynes, F.J. and Irfan, T.Y. (1978). Engineering grading of weathered granite. *Engineering Geology*. 345–374.

Deere, D.U. and Patton, F.D. (1971). Slope stability in residual soils: Panamerican Conf. Soil Mechanics and Foundation Engineering, *4th, Caracas 1971, Proc.*, pp. 87–170.

Duchaufour, P. (1982). *Pedology:Pedogenesis and Classification,* George Allen and Unwin, London.

Fookes, P.G. (1997). *Tropical Residual Soils.* A Geological Society engineering group working party revised report. London: The Geological Society.

Fookes, P.G., Dearman, W.R. and Franklin, J.A. (1971). Some engineering aspects of rock weathering, *Quarterly Journal of Engineering Geology* 4: 139–185.

Gerrard, J. (2000). *Fundamentals of Soils.* London: Routledge/Taylor & Francis.

Gidigasu, M.D. (1976). *Laterite Soil Engineering. Pedogenesis and Engineering Principles.* Elsevier Scientific Publishing Company.

Goldich, S.S. (1938). A study of rock weathering. *Journal of Geology* 46: 17–58.

Goudie, A.S., (1973). *Duricrusts in Tropical and Sub-tropical Landscapes.* Clarendon, Oxford.

Ingersoll, A.P. (1983). The atmosphere. *Scientific American*, Sept: 114–130.

Irfan, T.Y. and Dearman, W.R. (1978). Engineering classification and index properties of a weathered granite, *Bulletin International Association of Engineering Geology* 17: 79–80.

Jenny, H. (1994), *Factors of Soil Formation*, Dover Publications.

Johnson, R.B. and De Graff, J.V. (1988). *Principles of Engineering Geology.* John Wiley.

Jumikis, A.R. (1965). *Soil Mechanics.* New York: Van Nostrand-Reinhold.

McKnight, T.L. and Hess, D. (2000), *Physical Geography: A Landscape Appreciation*, Prentice Hall.

Komoo, I. Mogana, S.N. (1988). Physical characterization of weathering profiles of clastic metasediments in Peninsular Malaysia. *Proc. 2nd Conf. on Geomech. in Tropical Soils, Singapore* 1: 37–42.

Little, A.L. (1969). The engineering classification of residual tropical soils. *Proc. Seventh International Conference on Soil Mechanics and Foundation Engineering, Mexico City*, 1: 1–10.

Marto, A. and Kasim, F. (2003). Characterisation of Malaysian Residual Soils for Geotechnical and Construction Engineering. *Project Report. Universiti Teknologi Malaysia.* (Unpublished).

McLean, A.C. and Gribble, C.D. (1979). *Geology for Civil Engineers.* London: E & FN Spon.

McGeary, D. and Plummer, C.C. (1998). *Physical Geology.* WCB McGraw-Hill.

McCarthy, D.F. (1993). *Essentials of Soil Mechanics: Basic Geotechnics.* Englewood Cliffs, New Jersey: Regents/Prentice Hall.

Miller, R.W. and Gardiner, D.T. (2002). *Soils in our Environment.* Prentice Hall.

Nagle, G. (2000). *Advanced Geography.* Oxford University Press.

Nahon, D.B. (1991). *Introduction to the Petrology of Soils and Chemical Weathering.* New York: John Wiley.

Netterberg, F. (1978). Dating and correlation of calcretes and other pedocretes. *Geol Soc. South Africa Transactions*, 81, 379–391.

Netterberg, F. and Caiger, J.H. (1983), A Geotechnical classification of calcretes and other pedocretes, *Geological Society, London, Special Publications.* 11: 235–243.

Press, F. and Siever, R. (1994). *Understanding Earth.* New York: W. H. Freeman and Co.

Public Works Institute Malaysia (IKRAM). (1996). *GEOGUIDES* 1–5, Tropical weathered in-situ materials.

Raj, J.K. (1994). Characterization of the weathered profile developed over an amphibole schist bedrock in Peninsular Malaysia. *Bulletin of the Geological Society Malaysia* 35: 135–144.

Rollings, M.P. and Rollings, R.S. (1996). *Geotechnical Materials in Construction.* McGraw-Hill.

Selby, M.J. (1993). *Hillside Materials and Processes.* London: Oxford University Press.

Singh, P. and Kataria, D.S.K. (1980). *Engineering and General Geology.* Ludhiana, India: Katson Publishing House.

Sowers, G.F. (1985). Residual soils in the United States. In Brand & Philipson (ed.) *Sampling and Testing of Residual Soils: A Review of International Practices.* International Society of Soil Mechanics and Foundation Engineering.

Standards Australia, (1998). Australian Standard ASI 1726–1993, *Rock Material Weathering Classification.*

Strakhov, M.N. (1967). *Principle of Lithogenesis*. Edinburgh: Oliver and Boyd.
Summerfield, M.A. (1996). *Global Geomorphology. An Introduction to the Study of Landforms*. Singapore: Longman.
Summerfield, M.A. (1991). *Global Geomorphology*. Harlow: Longman.
Tan, B.K. (2004). Country case study: engineering geology of tropical residual soils in Malaysia. In: *Tropical Residual Soils Engineering*, pp. 237–244. Balkema: Rotterdam, The Netherlands.
Tardy, Y. (1992). Diversity and terminology of lateritic profiles in Martini, I.P. and Chesworth, W., Eds., *Weathering, Soils and Paleosols, Developments in Earth Surface Processes* 2, Elsevier, Amsterdam.
Tardy,Y. (1969). Geochimie des alterations. Etude des arenes et des eaux de quelques massifs cristallins d'Europe et d'Afrique. *Mem. Serv. Carte Geol. Alsace Lorraine*, Strasbourg, 31, p. 199
Tessens, E. and Jusop, S. (1983). *Quantitative Relationships Between Mineralogy and Properties of Tropical Soils*, University Pertanian Malaysia (University Putra Malaysia).
Tovey, N.K. (1971). A selection of scanning electron micrographs of clays, *CUED/CSOILS/Tr5a*. University of Cambridge, Departure of Engineering.
Thomas, M.F. (1974). *Tropical Geomorphology. A Study of Weathering and Landform Development in Warm Climates*. Macmillan.
Terzaghi, K. and Peck, R.B. (1967). *Soil Mechanics in Engineering Practice*. New York: John Wiley.
Vargas, M. (1953). Some engineering properties of residual clay soils occurring in southern Brazil. *Proc. 3rd International Conference on Soil Mechanics and Foundation Engineering, Zurich* 1: 259–268.
Vargas, M. (1974). Engineering properties of residual soils from South Central region of Brazil, *Proc. 2nd International Congress of the International Association of Engineering Geology*, São Paulo, Vol. I.
Waltham, A.C. (1994). *Foundations of Engineering Geology*. Blackie Academic & Professional.
White, R.E. (2006). *Principles and Practice of Soil Science*. Blackwell Pub.
Whittow, J. (1988). *The Penguin Dictionary of Physical Geography*. Penguin Books.

3

Sampling and testing of tropical residual soils

Fernando Schnaid
Federal University of Rio Grande do Sul Porto Alegre, Brazil

Bujang B.K. Huat
Universiti Putra Malaysia Serdang, Malaysia

3.1 INTRODUCTION

The characterisation and assessment of geotechnical properties of residual soils is a complex subject given the fact that these soil formations are a product of the physical, chemical and biological weathering processes of the rock. This in-situ decomposition of the parent rock and rock minerals produces characteristic features of mechanical behaviour that cannot necessarily be approached by conventional geotechnical design methods due to one or more of the following reasons (Schnaid *et al.*, 2004a, 2004b):

1. The soil state is variable due to complex geological conditions.
2. Classical constitutive models do not offer a close approximation of its true nature.
3. These soil formations are difficult to sample and the soil structure cannot be reproduced in the laboratory. As a consequence, the mechanical behaviour and geotechnical properties are assessed directly from in-situ testing data in most geotechnical design problems.
4. Little systematic experience has been gathered and reported, and values of parameters are outside the range that would be expected for more commonly encountered soils such as sand and clay formations of sedimentary soils.
5. The deposits are often unsaturated and the role of matrix suction and its effect on soil permeability and shear strength has to be acknowledged and accounted for.

Ground investigation of residual soils often reveals weathered profiles exhibiting high heterogeneity in both vertical and horizontal directions, complex structural arrangements, expectancy of pronounced metastability due to decomposition and lixiviation processes, presence of rock blocks and boulders, among others (e.g. Novais Ferreira, 1985; Vargas, 1974). The process of *in situ* weathering of parent rocks (which creates residual soils) gives rise to a profile containing material ranging from intact rocks to completely weathered soils. Rock degradation generally progresses from the surface and therefore there is normally a gradation of properties with no sharp boundaries within the profile. Lateritic and saprolitic horizons have to be distinguished because of their different geological history. Lateritic soils are formed under hot and humid conditions involving a high permeability profile which often results in a bond structure with a high content of oxides and hydroxides of iron and aluminium. In saprolitic profiles, the original disposition of the decomposed crystals of the parent rock is retained. This gives to the soil a peculiar relic structure where the soil grains are well arranged and orientated (e.g. Novais Ferreira, 1985; Vaughan, 1985).

In this highly heterogeneous environment, site investigation campaigns are generally implemented from a mesh of boreholes associated to either standard penetration test (SPT) or cone penetration test (CPT), to depths defined by the capacity of the penetration tool, and followed by continuous rotational coring below the soil-rock interface for a global geological characterisation of weathering patterns. To enhance consistency, recommendations are made to encompass geophysical surveys. In this highly variable environment, both laboratory and in-situ tests still assist in characterizing stress-strain and strength properties. However, the geotechnical practice in residual deposits has tended to diverge from the practice established for sedimentary deposits, both in terms of the applicability of commercial investigation tools and the interpretation of testing data in order to assess geotechnical design parameters. One of the aims of this Chapter is to highlight the differences in practice, procedures and theoretical background.

3.2 SAMPLING

Sampling and testing are prerequisites in determining the index and engineering properties of soils. The first collection of state-of-the-art papers in sampling of residual soils was presented at a symposium in Singapore organised by International Society for Soil Mechanics and Foundation Engineering (ISSMFE) in 1979 (Nicholls, 1990). International practice on sampling and testing of residual soils was reviewed by ISSMFE in 1985. ISSMFE emphasises the need to take care in sampling residual soil, which usually behave differently from other conventional soils particularly in relation to properties and structure. Further description on how sampling will be done is given below.

Criteria of good quality samples

Soil samples provide some of the most important evidence from a subsurface investigation. The preparation of soil samples for testing demands special care, and accepted practices may not be suitable and should not be applied both for the preparation of disturbed samples for classification testing or for the handling of disturbed samples for shear strength, compressibility and permeability tests. Conventional index tests should be used with caution when correlating with engineering behaviour. The void ratio of these soils varies considerably and strongly influences their engineering properties.

As stated in Geoguide 3 (1996), assessment for in-situ material is essential. The assessment should be done with great care and should not be influenced either by optimistic estimations of a lack of disturbance or by the requirements of any analytical programs to be used for design. A project requirement for a particular undisturbed parameter value does not necessarily mean that it can or should be supplied by direct sampling and testing methods. Disturbance needs to be assessed both with respect to sampling in the field, transport of the sample and transfer of the recovered sample to the test apparatus in the laboratory. Geoguide 3 (1996) also states that the test procedures require to be selected on the basis of project requirements, general material types, sample quality and the capability of the test laboratory.

The 'Good Quality Sample' or 'Perfect Sample' is defined as a 'sample which has not been disturbed by boring, sampling and trimming but has experienced stress released'. In actual conditions, there is no truly undisturbed sample, as all drilling techniques will eventually initiate some mechanical disturbance and stress relief. Samples are classified as undisturbed or disturbed depending on how much alteration there is to the soil structure after it is removed from its in-situ state. The validity of investigations carried out in laboratory tests rests solely on the quality of the samples and on how far they are representative of the stratum from which they are taken. The samples taken are said to be 'Good Quality Samples' depending on various factors such as purpose of sampling (what type of testing will be performed?), the location of samples taken (must be representative of an area), the method of sampling applied, and how the sample taken was handled before (including preparation for testing) and during testing.

Purpose of sampling

The samples must be taken according to the type of test to be performed, and whether an undisturbed sample is needed. If an undisturbed sample is needed, it must be taken and handled accordingly before and during testing to preserve the sample. Undisturbed samples should represent the actual condition on site where the structure and water content is preserved as far as possible and can be obtained by using a suitable coring method. As far as residual soils are concerned, the disturbance of mechanical force can alter the structure and soil fabric. Therefore, to handle this type of soil, special mechanical tools are used. It is important to minimise disturbance of the samples as far as possible especially for tests of shear strength, compressibility and permeability. The number and spacing of disturbed samples usually depend on the anticipated testing programs and design problems. Some of the methods used to preserve the actual condition of the soil will affect the location of samples taken, techniques for sampling, and sample storage.

For the purpose of sampling which involves remolding soil or changing of moisture condition from the field condition, there is no need to preserve the samples in an undisturbed state. However, the need for undisturbed samples is for soil identification and classification and quality tests, in which case the samples are usually collected and sealed in glass or plastic containers, tins or plastic bags. The representative disturbed samples should be taken vertically less than 1.5 m and at every change in strata.

The location of samples taken

The samples must be representative of an area for the purpose of generalisation of the properties for identification and use in engineering design. Generally, obtaining representative soil samples is still a challenge since, by nature, soil types and their properties vary greatly both vertically and horizontally. It is more difficult when dealing with tropical residual soils due to their non-homogeneous and anisotropic condition. It is important to establish a well-planned sampling program that ensures representativeness of the area.

The correct method of sampling applied is important for good quality samples. The sample for tests which require undisturbed samples should be obtained in the way prescribed, and likewise for disturbed samples. Personnel who are given the task of sampling should be familiar with various sampling techniques, and should be able to decide the right technique for undisturbed sample purposes. The soil fabric, bonding between particles, void ratio and moisture content of the soil must be preserved because these factors have a significant influence on their shear strength. The only way to get good quality samples of tropical residual soils is by fully utilizing the experience, competence and capabilities of personnel involved in the sampling and testing processes. Besides personnel capabilities, the laboratory and testing equipment must also be capable of producing good quality results. It is important to note that some of the conventional testing methods might not be reliable for tropical residual soils which may need some calibration and special tools.

Drilling and sampling techniques

The most popular sampling technique is by drilling. There are various methods of drilling widely used in obtaining soil samples, namely hollow-stem auger, solid flight and bucket augers, direct air rotary, cable and rotary diamond drilling. Different drilling techniques are necessary depending on whether the goal of the investigation is to collect undisturbed samples or if disturbed samples will suffice. A principal drilling requirement is that the driller must be prepared to encounter different consistencies of material from very soft to extremely hard rock, and for sudden and repeated changes from one to another. Other alternative sampling techniques are by open hand-dug pit or machine-dug pit. Brand and Phillipson (1985) stated the main reason for using this method is that the detailed inspection of a relatively large exposure of residual soil is highly desirable to enable an examination to be made of the occurrence of relict joints and other structural features, which can often dominate the engineering behaviour of the material. This method is most suitable at shallow depths in the completely weathered material. The pit must have enough space for person access. Through this technique, a visual inspection of the soil profile can be made. Hand-cut samples can be taken and other in-situ testing can be performed as required. As mentioned above, several sampling techniques can be applied to tropical residual soil, but with careful drilling techniques. Some practices use core barrels for high quality drilling and sampling. Types of core barrels used include a 63 mm diameter standard triple tube core barrel with retraction shoe and split steel liners (HMLC), an 83 mm diameter non-rectractable triple tube core barrel incorporating a wireline mechanism for withdrawing the liner barrel (PQ-Wireline) and a 74 mm Mazier automatic core barrel. Use of water in drilling, especially in dry tropical residual soils above the water table, must be avoided because it can disturb the samples. Sometimes, water from drilling activities in the hole is misunderstood as groundwater table. The water flush during coring causes erosion and loss of core. Water also increases moisture content on the surface of the sample. It is suggested that foam, mud, air or special drilling fluid is used. In the case of water being used in the drilling process, it is suggested that a Mazier automatic core barrel be used. A Mazier automatic core barrel using triple wall core barrel permits removal of the sample as it is taken from the ground, guaranteeing the 'in-situ condition' of the core. Figure 3.1 shows the diagram of a typical Mazier automatic core barrel.

A driven sample is also another alternative in obtaining undisturbed samples. The quality varies depending on the drilling tube used and its material, the condition of the cutting shoe, the area ratio and the method of driving. Samples of 35 mm diameter are frequently obtained in conjunction with the Standard Penetration Test. A strong open-driver sampler which is sometimes used, and which produces samples of much higher quality than the SPT, is the British U100. This gives 100 mm diameter samples up to 450 mm long. In order to obtain least disturbed samples, thin-walled stainless steel tubes of low area ratio are used on clayey saturated materials.

Sample storage

As suggested by Head (1992), samples should be kept in a cool room to protect them from extremes of cold and heat. Disturbed samples in glass jars can be conveniently stored in milk bottle crates. Large disturbed samples in polythene bags should not be

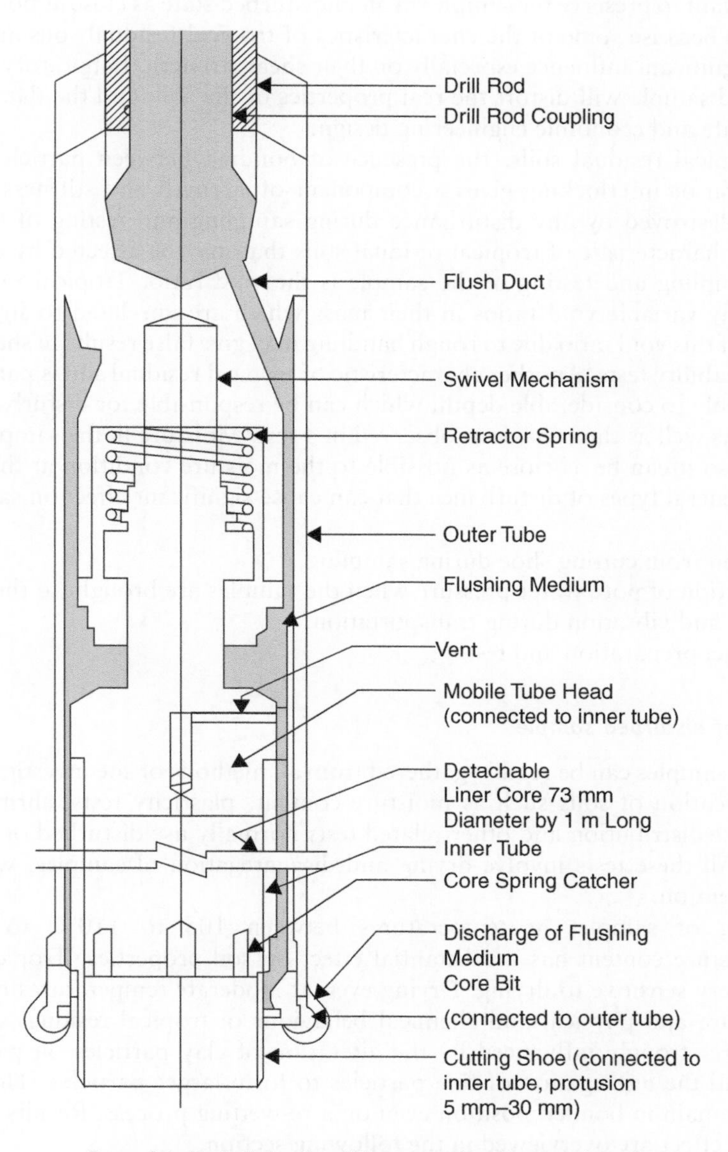

Figure 3.1 Typical details of Mazier triple-tube core barrel

piled one on top of another, but placed individually on shelves or racks. Undisturbed samples must be laid on racks designed for storage; however tubes containing wet sandy or silty soil should be stored upright to prevent segregation of water.

Sampling of undisturbed samples

Undisturbed samples are needed to obtain properties of tropical residual soils from laboratory testing. A complete disturbed sample is better than a bad undisturbed sample.

It is important to preserve the samples in an undisturbed state as close as possible to site conditions because some of the characteristics of tropical residual soils are sensitive, or have significant influence especially on their shear strength. Laboratory test results from a bad sample will distort the real properties of the soil, and the data cannot be used for safe and economic engineering design.

In tropical residual soils, the presence of bonding between particles either by cementation or interlocking gives a component of strength and stiffness which can be easily destroyed by any disturbance during sampling and testing of the sample. A second characteristic of tropical residual soils that may be affected by disturbance during sampling and testing of the sample is the void ratio. Tropical residual soils have widely variable void ratios in their mass which are unrelated to stress history. Changes of this void ratio due to rough handling may give false results in shear strength and permeability tests. Another characteristic of tropical residual soil is partial saturation, possibly to considerable depth, which can be responsible for disturbance during sampling as well as the behaviour observed in a test. Moisture in the samples must be preserved so it can be as close as possible to the moisture condition at the site upon testing. General types of disturbance that can cause significant effect on samples are:

1 Friction from cutting shoe during sampling,
2 Reduction of pore water pressure when the samples are brought to the surface,
3 Shock and vibration during transportation,
4 Storage, preparation and testing.

Sampling of disturbed sample

Disturbed samples can be readily gathered from all methods of site investigation. Tests for classification of soils such as moisture content, plasticity tests, shrinkage tests, particle size distribution and other related tests normally use disturbed or remoulded samples. All these tests involve drying and disaggregation of samples, which needs special attention.

Drying of samples at temperatures between 105 to 110°C to determine water/moisture content has a substantial effect on soil properties. Tropical residual soils are very sensitive to drying. Drying even at moderate temperatures may change the structure and physical and chemical behaviour of tropical residual soils. These changes are strongly influenced by the alteration of clay particles on partial dehydration and the aggregation of fine particles to form larger particles. The reformed particles remain in bonded position even on a re-wetting process. Results that reflect the drying effect are overviewed in the following section.

Crushing or splitting technique must be avoided in disaggregating tropical residual soils for purposes of the classification test (Blight, 1997; Fookes, 1997). The disaggregation process must be done with care and with regard for what is meant by 'individual particles'. The samples should be soaked in the water overnight. For particles which have a cemented characteristic, the disaggregation should be done by applying finger pressure only. The need for dispersion of fine particles is discussed in the particle size distribution test section below.

The degree of disturbance that occurs in a fabric-influenced material may vary considerably depending on the project or sampling conditions. A fundamental aspect

Table 3.1 General divisions of geotechnical behaviour (*After* Nik Ramlan *et al.*, 1994 in Geoguide 3, 1996)

Behaviour pattern	*Description*	*Laboratory modelling*	*Project activity*
Instrinsic remoulded, de-structured material	Behaviour a function of particle type (mineralogy), shape and size (texture). Dependent on moisture condition	Completely remoulded index tests	Well compacted fill, road haul performance, erosion
Meso-structured undisturbed material	Behaviour is a function of intrinsic properties and the material fabric and meso-structure	Standard 'undisturbed' testing, triaxial, shear box, oedometer etc.	Possibly lightly compacted fill, erosion, aggregates
In-situ mass macro-structured mass	Behaviour a function of intrinsic, meso- and macro-structural properties of the mass and component materials, allied to the influence of relict mega-discontinuities and material boundaries	Only possible directly by combining relevant material tests with macro-structural data to give a mass character. Indirectly by semi-empirical, terrain correlation or back analysis procedure	Cut slopes, foundations

Notes:

Texture: The morphology, type and size of component particles
Fabric: The spatial arrangement of component particles
Discontinuities: The nature and distribution of surfaces separating elements of fabric, material or soil-rock mass
Structure: The fabric, texture and discontinuity patterns making up the soil-rock material, mass or unit.

The above may be described at number of scale levels:

Micro: <0.5 mm Generally only described with the aid of SEM or petrographic microscope
Meso: 0.5–5 mm Generally seen with the aid of field microscope or a good hand lens
Macro: 5–50 mm Patterns visible to the naked eye in the field
Mega: >50 mm Patterns that become apparent by means of maps or remote sensing, although individual elements may be visible at field level.

of the characterisation of materials for civil engineering construction is in recognizing and establishing various levels of behaviour based not only on scale but also upon the influencing elements. The recognition of these levels of behaviour allows an important distinction to be made between inherent in-situ characteristics and those that become apparent during various aspects of construction. This facilitates a more relevant correlation between in-situ character, laboratory testing and likely project performance as shown in Table 3.1.

The following aspects should be considered in preparing remoulded residual soil; drying, disaggregation and sub-dividing. As soil is formed by the decomposition of rock in-situ by chemical decay, and may retain signs of its original structure, residual soils are likely to be highly variable and testing programs need to consider the use of both soil mechanics and rock mechanics testing procedures. The material might be considered in terms of being soil, rock or a soil-rock mixture. General classification should give an early indication of the general range of tests methods that it will be appropriate to conduct (Geoguide 3, 1996).

Drying

Partial drying at moderate temperatures may change the structure and physical behaviour of tropical residual soils. Some of the structures are changed by chemical means and not reversed when re-mixed with water. Physical changes can be seen according to these aspects:

1 Alteration of the clay minerals
2 Aggregation of fine particles to become larger particles that remain bonded even on re-wetting.

Fookes (1997) reported that clay soils often become more silt- or sand-like with a lower plasticity; although in some instances the opposite can occur. Oven drying from 105 to 110°C frequently has a substantial effect on soil properties, but drying at a lower temperature (e.g. 50°C) and even partial air-drying at ambient laboratory temperature can also produce significant changes. Blight (1997) and Fookes (1997) both agreed that generally all tropical residual soils will be affected in some way by drying. In preparation for a classification test, natural soil with as little drying as possible should be applied, at least until it can be established from comparative tests that drying has no significant effect on the test result. The method of preparation should be reported from time to time.

Disaggregation

Fookes (1997) stated that disaggregation should be handled with care. The objective of this process is to separate the discrete particles without the act of crushing and splitting. It is suggested that some soil should be soaked overnight without any interference from mechanical force to obtain better results. Particles with cemented bonds should be split only by using finger pressure.

Sub-dividing

Samples are sub-divided by a riffling box and poured evenly by using a scoop or shovel. The sub-dividing procedures must be acceptable for residual soil use so that representative samples are obtained and ready for testing. The soil should be evenly distributed along all or most of the slots to ensure that each container receives an identical sample. Other accepted quartering procedures can also be used to obtain required samples.

3.3 LABORATORY TESTING

Only routine laboratory tests will be described in this chapter. The specialised laboratory tests will be described in Chapters 5 and 6 of the Handbook.

Index and engineering properties test

It is accepted that residual soils behave differently from conventional soils (e.g. transported soil) as the soils are formed in-situ and have only slight changes in stress

Table 3.2 Categories of water (*After* Geoguide 3, 1996)

Category	*Description*	*Availability*
Structural water	Water held within the structure of component minerals	Generally not removable below 110°C except for clays such as halloysite, allophane and gypsum
Strongly adsorbed water	Held on particle surface by strong electrical attraction	Not removed by drying at 110°C
Weakly adsorbed water	Held on particle surface by weak electrical attraction	Can be removed by drying at 110°C but not by air drying
Capillary water (Free water)	Held by surface tension	Removed by air drying
Gravitational water (Free water)	Moveable water held in the materials	Removed by drainage

history. Any changes strongly depend on mineral bonding and soil suction. Geoguide 3 (1996) distinguished major differences in undertaking and interpreting geotechnical laboratory test results in tropically weathered soil as opposed to sedimentary soils as follows:

1 The materials are chemically altered and sometimes bonded rather than produced by a physical sedimentation process.
2 The materials are in many cases non-saturated, exhibiting negative pore water pressures (soil suction).
3 There is difficulty in obtaining high quality undisturbed samples in these materials which may have a sensitive fabric.
4 There is difficulty in obtaining truly representative geotechnical parameters from these heterogeneous materials and masses.
5 Therefore, modelling residual soil testing might have to take into consideration these aspects which would require calibration of test equipment.

Moisture content

The conventional definition of moisture content is based on the amount of water within the pore space between the soil grains when a soil or rock material is dried to a constant mass at a temperature between 105 and 110°C, expressed as a percentage of the mass of dry soil. This loss in weight due to drying is associated with the loss of the 'free water' as listed in Table 3.2.

For some tropical residual soils, in addition to 'free water' that is available to influence engineering behaviour there may exist crystallised water within the structure of minerals that is released at these drying temperatures. As suggested by Fookes (1997), in order to identify this type of soil, comparative tests should be carried out on duplicate samples taking measurements of moisture content by drying to a constant mass at between 105 and 110°C and at a temperature not exceeding 50°C until successive weighing shows no further loss of mass. A significant difference should indicate that intraparticle water is present. This water exists as a part of solid particles and should be excluded from the calculation of moisture content. The releasibility of this additional

Table 3.3 Some examples of the effects on index properties of tropical residual soils (*After* Fookes, 1997)

		Liquid limit			*Plasticity index*		
Soil Location	*Soil type*	*AR*	*AD*	*OD*	*AR*	*AD*	*OD*
Costa Rica	Laterite	81	–	56	29	–	19
	Andosol	92	–	67	66	–	47
Dominica	Allophane	101	56	–	69	43	–
	Latosolic	93	71	–	56	43	–
	Smectoid	68	47	–	25	21	–
Hawaii	Humic Latosol	164	93	–	162	89	–
	Hydrol Latosol	206	61	–	192	NP	–
Java	Andosol	184	–	80	146	–	74
Kenya	Red Clay, Sasumua	101	77	65	70	61	47
Malaysia	Weathered Shale	56	48	47	24	24	23
	Weathered Granite	77	71	68	42	42	37
	Weathered Basalt	115	91	69	50	49	49
New Guinea	Andosol	145	–	NP	75	–	NP
Vanuatu	Volcanic Ash, Pentecost	261	192	NP	184	121	NP

AR: As Received, AD: Air Dried, OD: Oven Dried, NP: Indicates Non-Plastic.

water varies with mineral types and in some cases results in highly significant differences in moisture content between conventional testing temperature and engineering working temperatures.

Plasticity

Plasticity or Atterberg limits, namely the Liquid Limit (LL) and Plastic Limit (PL) are often employed in the classification of fine-grained soils. Although water in a soil sample can be removed by oven-drying at a temperature of $110 \pm 5°C$ as in normal practice, it can also be removed by air-drying or it can be tested at its natural moisture condition as suggested in BS 1377:1990 (BSI, 1990). There are significant effects in determining plasticity of the soil with regards to pre-test drying, duration and mixing methods. Many researchers and writers state that tropical residual soils are very sensitive to drying because drying may change the physical and chemical properties of the affected soils. Table 3.3 shows some examples of the effects on index properties of tropical residual soils (Fookes, 1997).

Shrinkage

Some tropical residual soils exhibit considerable volume change in response to wetting or drying and the shrinkage limit test may provide an indication of an intrinsic capacity for shrinking or swelling. The shrinkage limit test, as in BS1377 (1990), was initially intended for undisturbed samples although remoulded material can be used. Linear shrinkage to BS1377 (1990) is a simpler test on remoulded materials, which gives a

Table 3.4 Effect of sample preparation on particle size distribution test (*After* Fookes, 1997; Blight, 1997)

Aspect	*Description*
Drying	Drying can cause a reduction in the percentage of clay fraction. It is accordingly recommended that drying of the soil prior to testing be avoided. The initial soil sample should be weighed and a duplicate sample taken for moisture content determination so that the initial dry mass can be calculated.
Chemical pre-treatment	This treatment should be avoided wherever possible. Pretreatment with hydrogen peroxide is only necessary when organic matter is present. To eliminate carbonates or sesquioxides, pretreatment with hydrochloric acid is used.
Sedimentation	The dispersion of fine particles should be carried out by using alkaline sodium hexametaphosphate. In some instances a concentration of twice the standard value may be required. The solution must be made before conducting any sedimentation process.
Sieving	If accepted standard procedure is to be used, experience is needed to judge carefully. Care is necessary at every stage, especially to avoid breakdown of individual particles.

linear rather than volumetric shrinkage. The established relationship between linear shrinkage and the Plasticity Index for sedimentary soils may not hold true for tropical residual soils. It is important to differentiate between materials that shrink irreversibly and those that expand again on re-wetting.

Studies by Mutaya and Huat (1993) on the effect of the degree of drying on linear shrinkage of Malaysian tropical residual soils show that when the degree of drying is increased, the linear shrinkage value reduces. This is due to the presence of moisture and clay content. A soil having higher moisture and clay content tends to shrink more compared to that of lower moisture and clay content.

Particle size distribution

A particle size distribution test is done to determine the range of soil particles within a mass of coarse grained soil sample by the act of sieving. The complete test procedure can be reviewed in BS1377 (1990). Residual soil can be visualised as complex functions of particle size, fabric and the nature of the particles. The standard particle size distribution test should be applied with extra care concerning the aspects discussed and described in Table 3.4.

In one soil sample, the variation of particles sizes encountered may vary widely. Although natural soils are mixtures of various-sized particles, it is common to find a high proportion of types of soil occurring within a relatively narrow band of sizes. In some test procedures, some of the soils might undergo a sedimentation process by applying dispersants into the solution. The need for sedimentation of fine particles is to ensure that discrete particles are separated. Geoguide 3 (1996) explains that soils which undergo a sedimentation process should have proper dispersion of the fine particles. Alkaline sodium hexametaphosphate has been found to be suitable for a wide range of soils. Alternative dispersants such as trisodium phosphate may be more effective.

Table 3.5 Methods of density measurement (*After* Geoguide 3, 1996)

Method	*Reference*	*Comment*
Measured dimensions. Hand trimmed from block or tube	Part 2: 7.2 BS 1377 (1990)	Material has to be suitable for trimming; e.g. robust soil or weak rock.
Measured dimensions. Sample within tube	Part 1: 8.4 BS 1377 (1990)	Used where extrusion may disturb sample; e.g. loose or weakly bonded soil material.
Water displacement (waxed sample)	Part 2: 7.4 BS 1377 (1990)	Simple test used for irregular shaped water sensitive samples.
Weighed in water (waxed sample)	Part 2: 7.3 BS 1377 (1990)	As above, generally more accurate.
Weighed in water (non waxed sample)	ISRM (1981), Part 1	Used for irregular lumps of rock-like material not susceptible to swelling or slaking.

Density

Measurement of the quantity of material related to the amount of space it occupies is referred to by the term density. It is also normally understood as mass per unit volume. Density is widely used to obtain the relation between density and moisture content in the determination of compaction characteristics. Another test, the in-situ density, is another requirement for the assessment of structural stability and determination of the void ratio. This value can prove to be a useful index test, particularly as it may be used to correlate between soil and rock materials. Bulk density may be recommended using a variety of test procedures as summarised in Table 3.5.

Specific gravity

Specific gravity refers to the ratio of the mass of a given volume of a material to the mass of the same volume of water. This term however has been replaced by particle density as described below.

Particle density

The term 'particle density' (Ps) is replacing the previously used term 'specific gravity' (Gs) in current British practice, BS 1377:1990 (BSI, 1990) to comply with international usage. This term refers to the average mass per unit volume of the solid particles in a sample of soil, where the volume includes any preserved voids contained within solid particles. In other words, particle density is a measure of the average density of the solid particles which make up a soil mass. Particle density has the same numerical value as specific gravity although it has the units Mg/m^3 rather than being dimensionless. Particle density value is needed as the value can be used to determine other soil properties such as void ratio, clay fraction and porosity, which can be related to fabric structure.

Tropical residual soils may have highly variable particle densities. Some soils indicate unusually high, and some unusually low, densities. For this reason the value should be measured whenever it is needed in the calculation. An assumed valued is

Table 3.6 Typical particle densities of minerals in tropical residual soils (*After* Geoguide 3, 1996)

Mineral	*Particle density (Mg/m³)*
Calcite	2.71
Feldspar-orthoclase	2.50–2.60
Feldspar-plagioclase	2.61–2.75
Gibbsite	2.40
Haematite	4.90–5.30
Halloysite	2.20–2.55
Kaolinite	2.63
Magnetite	5.20
Quartz	2.65

not encouraged. The test should be conducted at its natural moisture content and due regard should be taken of moisture availability problems discussed above. Natural moisture content is to be used in obtaining particle density. Any pre-treatment is not advisable as the value could be distorted and it tends to reduce the measured value as compared with the natural moisture content of samples. The dry mass of solid particles should be taken after the particle density test has been conducted. The dry mass is taken by oven drying at 105 to 110°C. Note that whenever coarse grain particles are present (gravel size), the gas-jar method is to be used so that the whole sample will be presented (Head, 1992). Some particle density values gathered from regional studies by Geoguide 3 (1996) are shown in Table 3.6.

Samples for compaction test

Fookes (1997) reported that tropical residual soils with coarser particles are susceptible to crushing and therefore a separate sub-sample of soil is needed for compaction at each value of moisture content to avoid successive degradation. This also applies to the breakdown of the <425 μm fraction.

Care should be taken with soil samples that are sensitive to drying methods. Problems can arise when these are applied for compaction field control. The soil should not be dried before testing in the laboratory. If it is necessary to compact the soil lower than the natural moisture content, partial drying at room temperature is essential until the desired moisture content is achieved. Excessive drying, requiring re-wetting, should be avoided. For a given degree of compaction, drying has generally been found to increase the maximum dry density and to reduce the optimum moisture content. Data obtained from tests on dried soil are not applicable to the field behaviour and could result in inappropriate criteria being applied in field conditions (Fookes, 1997).

Shear strength tests

Shear strength problems are usually encountered in calculation and analysis for foundation and earthwork stability. The shear strength parameters can be found by using laboratory and field tests, and by approximate correlations with size, water content, density, and penetration resistance. Some of the laboratory shear strength tests

Table 3.7 Types of shear box tests

Soil sample	*Type of shear box test*
Coarse-grained soil	CD test – strength parameter c' and ϕ'
Fine-grained soil, cohesive soils (clays and clayey silts)	UU, CU and CD test For UU and CU tests, the strength parameters in terms of total stress and the shearing rate have to be as rapid as possible

applicable to residual soil are discussed below. The relevant in-situ tests are described in the following section.

Shear box test

The most common test used for determining shear strength of a soil in the laboratory is the shear box test. The shear box test is preferred because of its simplicity, ease of conducting compared with other tests, and less potential to disturb the sample preparation procedure than in the triaxial test. Another advantage is that the test can be used in fabric-sensitive materials by reverting to the use of a circular shear box which eliminates problems of sample disturbance at box corners. Despite its advantages, it does have disadvantages in some respects: drainage conditions cannot be controlled, determination of pore water pressure is not available, and thus only total normal stress can be determined. Total stress is equal to effective stress if full drainage is allowed and this requires the adoption of a suitable strain rate. This test however, enables relatively large strains to be applied and thus the need to determine both peak and residual soil strength. The principle is the action of a sliding movement to the soil sample while applying a constant load to the plane of relative movement. The soil sample can be directly sheared unconsolidated and undrained, but can also be consolidated and test drained or undrained. Various shear tests can be conducted depending on the designated model to present the actual condition. The tests are mainly unconsolidated undrained (UU), consolidated undrained (CU) and consolidated drained (CD) direct shear tests. Table 3.7 shows variations of test that can be conducted in the laboratory depending on the soil sample.

In slow 'drained' shear tests, the specimen is consolidated prior to shearing and a slow rate of displacement is applied during shearing. This method enables the consolidated drained or effective stress shear strength parameters to be determined.

Sampling of residual soils to verify shear strength of weathered zones should be carried as an additional concern. In almost all cuttings in tropical residual soils, there exist weathering zones which exhibit various degrees of alternating weathering grades, which is due to the nature of the parent rock. It is therefore inevitable that failures occur at the interfaces or relict joints. In anticipation of this situation specimens are obtained, as far as possible, at the positions of these discontinuities where the joint coincides with the sliding plane of the shear box. Undisturbed samples were obtained from the site using different types of sampling, from which test specimens were prepared. Briefly, the following test procedures are suggested (Blight, 1997):

- Test specimens are prepared and placed in the shear box with minimum disturbance. The specimens are then inundated with water.

- The normal stress is applied. In certain cases, the normal stress is equivalent to the overburden pressure of the specimen. When relatively soft specimens are tested, the load is applied in increments.
- Prior to shearing, in certain cases, the normal stress is reduced to the design overburden. Readings are taken until a stabilised condition is achieved (swelling has ceased).

Typical sizes of the square box sample for shear box test are 60 mm, 100 mm and, rarely, 300 mm or more. For circular shear boxes, common sizes are 50 mm and 75 mm diameter. The maximum particle size of the soil dictates the minimum thickness of the test sample. A study by Brenner *et al.* (1997) found that the drained strength of fissured dense soil (residual basalt) from 500 × 500 mm and 290 mm high shear box samples was 1.5 to 3 times less than the strength from 36 mm diameter triaxial samples in the normal stress range of 50 to 350 kPa. With relatively uniform samples, the size of the shear box was found to be less significant.

Triaxial tests

Triaxial test procedures play a large role in geotechnical testing programmes but have largely been derived for use on traditional sedimentary soils in temperate climates. The triaxial test is beneficial for obtaining a variety of test results such as triaxial strength, stiffness and characteristics of the stress ratios of soil specimens. The samples are either remoulded or from undisturbed samples trimmed and cut into cylindrical shape. Commonly used samples have a height/diameter ratio of 2:1. Most common sizes used are 76 mm × 38 mm and 100 mm × 50 mm. Samples will be sealed in a thin rubber membrane and subjected to fluid pressure. Axial load is then applied through a piston acting on the top cap and controlling the deviator stress.

Types of triaxial tests conducted in the laboratory are unconsolidated undrained (UU) test with or without pore pressure measurement, isotropically consolidated undrained compression (CIU) test with or without pore pressure measurement, and isotropically consolidated drained compression (CID) test. Unconfined compression (UCS) test is also an accepted method to test the strength of the more robust tropically weathered materials, from hard soils to strong rocks. Good samples recovered from high quality drilling techniques are particularly adaptable to this method provided steps are taken to preserve the in-situ moisture condition.

Since a variety of conditions of residual soil exist in reality, especially the partially saturated condition, an erratic result will be given if common methods of triaxial test are carried out. Some of the routine triaxial tests require the sample to be fully saturated to present a saturated condition but this is rather controversial for residual soils. Application to partially saturated fabric-influenced materials in climatic environments that impose rapid changes in moisture condition can cause difficulties both in establishing relevant test procedures and in the modelling of site conditions. The standard procedure of imposing saturation on under-saturated materials appears difficult to justify on the grounds of modelling site conditions. As reviewed by Brand and Phillipson (1985), pre-saturation which normally needs high back pressure is actually severe compared to the actual conditions. This technique can only be applied to achieve consistent effective strength parameters. However, Blight (1985) stated that the saturation condition can be applied considering that in unusually wet weather years the water table can rise for several meters. Saturation therefore represents the least favourable condition of the residual soil.

A soil that is partly saturated consists of a three phase system: gas (including air and water vapour), water and solid particles. Analysis of partial saturation is complex. The determination of effective stresses in partly saturated soils requires measurement of air pressures as well as pore water pressure. The following extended Mohr-Coulomb equation has been proposed for the solution of partial saturation problems (Fredlund, 1987):

$$\tau = c' + (\sigma_n - u_a)\tan\sigma_n + (u_a - u_w)\tan\phi^b \tag{3.1}$$

where τ = shear strength; c' = effective cohesion, σ_n = normal stress; $\tan\sigma_n$ = effective friction angle; ϕ^b = angle of shear strength change with a change in matric suction; u_a = pore air pressure; u_w = pore water pressure.

Undisturbed triaxial testing of suitable samples can be of practical use in a tropical residual soil environment although the resulting parameter must be interpreted in the light of field data and may in some cases serve only as a back up to empirical established figures (Geoguide 3, 1996). The following comments apply to the general use of triaxial testing of tropical residual soils:

a. Multistage triaxial testing is not recommended for tropical residual soils especially in those with an unstable fabric liable to collapse, brittle soils and those that show strain-softening characteristics. Multi-stage tests might give misleading strength values for design.
b. Quick undrained tests are not suitable for unsaturated materials. However, the test is appropriate for partially saturated soils.
c. Special procedures are likely to be required for high void ratio or bonded materials, e.g. low confining pressure, slow loading rate.
d. Significant numbers of slope failures in tropical environments are shallow in nature and analysis of these would require parameters derived at appropriate (low) confining pressures.
e. For undrained tests with pore pressure measurements, the rate of deformation must be slow enough to allow the non-uniform pore pressure to equalise, and in drained tests complete drainage condition must be achieved.

Sample size for triaxial testing in tropical residual soils should be about 75 mm in diameter. The common test specimen 38 mm in diameter is not applicable due to the disturbance caused in extruding or trimming small diameter specimens from borehole samples. Brand and Phillipson (1985) stated that a specimen of 100 mm is commonly used in Australia, Brazil, Germany, Hong Kong and UK. Samples with smaller diameters are not considered representative, because of the scale effect relating to fissures and joints in the soil (Brenner *et al.*, 1997). In addition, the sample diameter should not be less than eight times the maximum particle size. The ratio of sample length to diameter must be at least 2 to 1.

Permeability tests

Permeability of undisturbed samples can be derived from data obtained from consolidation tests; either triaxial, standard oedometer or Rowe cell. It may also be obtained from specific procedures using the permeameter equipment. Applications of the value of permeability are for drainage, analysing influence of seepage on slope stability, consolidation analysis and design of foundations for dams and excavations. Extrapolation to mass in-situ permeabilities from laboratory-derived material permeameters for tropical residual soils should be viewed with extreme caution. Despite conducting laboratory permeability tests, in-situ tests are more likely to represent the actual condition of residual soil (Brand and Phillipson, 1985). Field permeability will consider the relict joint and other preferential drainage paths that are most likely not to be identified in laboratory testing. Field permeability will be described in further detail in the following section of this chapter.

The laboratory permeability test for residual soils is observed to be best performed under back pressure in a triaxial cell, but should have a diameter of more than 75 mm and be 75 mm high as reviewed by Brand and Phillipson (1985).

A laboratory permeability test for residual soil can be conducted on compacted soil and more uniformly structured mature residual soils, particularly when the coefficient of permeability is determined in both directions, for vertical and horizontal trimmed samples. The test will then result in an estimate of the mass permeability of uniformly textured soils. Other advantages of using laboratory tests are for indications of the variation in the coefficient of permeability with changes in effective stress. Results obtained from the test can be used in designing earthworks.

Pore water pressure and suction test

Many soil-rock profiles of tropical residual soils, particularly on slopes, are known to be in an unsaturated condition. The stability of slope is a major concern in most tropical residual soil countries due to the frequent periods of heavy rainfall. Gasmo *et al.* (1999) stated that rainfall has a detrimental effect on the stability of residual soil slopes.

Negative pore pressures or matric suction is found to play an important role in the stability of slopes. These suctions have an important bearing on water entry, structural stability, stiffness, shear strength and volume change. The additional shear strength that exists in unsaturated soils due to negative pore-water pressures is lost as a result of rainwater infiltration into the soil. As a result, their in-situ geotechnical performance is likely to be influenced by variations in soil suction, in response to rainfall infiltration. Another study by Richards (1985) found that fine grained residual soils have high solute contents which also affect the physical properties of the soil, and the total soil suction, due to the solute component. The total soil suction, the water content and the solute content and how they vary with time are often the most important variables in soil engineering design.

In-situ soil suction can be measured with suitable sophisticated methods and equipment such as various types of tensiometers. The installation of a tensiometer is also incorporated with the use of piezometers to measure groundwater level, and a rain gauge to estimate rainfall intensities on the slope. A tensiometer generally comprises a water-filled plastic tube with a high air entry ceramic cup sealed at one end and a vacuum pressure gauge and a jet-fill cup sealed at the other end. When installed in soil, the pore-water pressure in the soil equilibrates with the water pressure in the tube and the pressure is measured by the vacuum pressure gauge or by a pressure transducer. The jet-fill cup is used as a reservoir to allow for easy refilling and de-airing of the tensiometer. Details of field suction measurements are included in Chapter 4 of this Handbook.

Another alternative is to measure suction indirectly in the laboratory by means of the filter paper method. This method involves placing Whatman's No. 42 filter paper in contact with the soil for a period of 7 days and measuring the amount of moisture taken up by the paper (*Wfp*). Matrix suctions may be arrived at by the following empirical relationships: (Chandler *et al.*, 1992).

$$\text{Suction (kPa)} = 10^{(4.84-0.0622Wfp)}; \quad \text{for } Wfp < 47\% \tag{3.2}$$

$$\text{Suction (kPa)} = 10^{(6.05-2.48 \log Wfp)}; \quad \text{for } Wfp > 47\% \tag{3.3}$$

Further descriptions on tests for soil in unsaturated conditions are included in the following section of the Handbook.

Compressibility and consolidation

There are a number of methods, both in-situ and laboratory, that can be used to determine the compressibility of tropical residual soils. The oedometer and triaxial compression tests are the main laboratory methods of testing, while for in-situ testing the standard penetration test, pressure meter test and plate loading test have been used. These field tests are described later in this chapter.

The oedometer is a one-dimensional consolidation test where complete lateral confinement is used to determine total compression of fine-grained soil under an applied load. The test is also used to determine the time rate of compression caused by a gradual volume decrease that accompanies the squeezing of pore water from the soil. An undisturbed sample is usually used for the consolidation test to obtain high quality results. The test first requires samples representative of principal compressible strata. Some two to eight tests should be conducted depending on the complexity of conditions (Brand and Phillipson, 1985).

The oedometer test is accepted for direct application for full analysis of amounts and rate of settlement only for intact clays. Oedometer tests are not suitable for measuring the compressibility of coarse grained soil and are thus not advisable for testing predominantly coarse grained residual soils. Whenever the consolidation or compression characteristic is needed, the use of triaxial test data is applicable. Brand and Phillipson (1985) noted that residual soils support many high-rise buildings and the foundations are usually taken down to below the residual soils, so that consolidation tests are regularly used. Trimming the residual soil samples is truly challenging because of the gravel content in some soils. There are various consolidation tests that have been carried out and the recognised features are described as follows:

- An oedometer can be used to assess the swelling characteristics of residual soil.
- Suction-controlled consolidometers can be used to control stress, soil suction and equilibrium electrolyte solution.

Geoguide 3 (1996) suggests the use of the more adaptable Rowe Cell which can accommodate larger samples. This is due to substantial evidence that larger, good quality undisturbed samples provide a better model of in-situ behaviour. Rowe cells also provide much greater versatility in terms of drainage conditions.

3.4 IN-SITU TESTS

Some of the most common in-situ tests available for routine investigation are listed in Table 3.8 (Schnaid *et al.*, 2004a; 2004b). In-situ testing equipment and testing procedures have spread worldwide for a wide range of geotechnical applications, and engineers can rely on a variety of commercial tools, Standards, guidelines and specifications supported by International Reference Test Procedures and well established national codes of practice.

The in-situ test techniques listed in Table 3.8 are all practical for the investigation of tropical residual soils, providing the necessary information for determining the sub-surface stratigraphy and for assessing the geotechnical parameters. Since most interpretation methods for estimating soil parameters are related to either sand (fully

Table 3.8 Commercial in-situ testing techniques (*After* Schnaid *et al.*, 2004a).

Test	*Designation*	*Measurements*	*Common Applications*
Geophysical Tests:			
Seismic refraction	SR	P-waves from surface	Ground characterisation
Surface waves	SASW	R-waves from surface	Small strain stiffness, G_o
Crosshole test	CHT	P & S waves in boreholes	
Downhole test	DHT	P & S waves with depth	
Standard Penetration Test	SPT	Penetration (N value)	Soil profiling Internal friction angle, ϕ'
Cone Penetration Test			
Electric	CPT	q_c, f_s	Soil profiling
Piezocone	CPTU	q_c, f_s, u	Undrained shear strength, s_u
Seismic	SCPT	$q_c, f_s, V_p, V_s, (+u)$	Relative density/friction angle, ϕ'
Resistivity	RCPT	q_c, f_s, ρ	Consolidation properties Stiffness (seismic cone)
Pressuremeter Test			
Pre-bored	PMT	$G, (\psi \times \varepsilon)$ curve	Shear modulus, G
Self-boring	SBPMT	$G, (\psi \times \varepsilon)$ curve	Undrained shear strength, s_u
Push-in	PIPPMT	$G, (\psi \times \varepsilon)$ curve	Internal friction angle, ϕ'
Full-displacement	FDPMT	$G, (\psi \times \varepsilon)$ curve	In-situ horizontal stress Consolidation properties
Flat Dilatometer Test			Stiffness
Pneumatic	DMT	p_o, p_l	Shear strength
Seismic	SDMT	p_o, p_l, V_p, V_s	
Vane Shear Test	VST	Torque	Undrained shear strength, s_u
Plate loading test	PLT	$(L \times \delta)$ curve	Stiffness and strength
Combined Test			Shear modulus, G
Cone pressuremeter	CPMT	$q_c, f_s, (+u), G, (\psi \times \varepsilon)$	Shear strength

drained conditions) or clay (fully undrained conditions), this chapter attempts to extend this existing background by summarizing the available experience on residual soils where the bonded structure and the cohesive-frictional nature of the soil has to be accounted for. It is here worth recalling that although most geomaterials are recognised as being 'structured', the natural structure of residual soils has a dominant effect on their mechanical response (Vaughan, 1985; 1997). In structured soils exhibiting a cohesive-frictional behaviour, a limited ground investigation based on penetration tests (CPT, CPTU or SPT) will not produce the necessary database for any rational assessment of soil properties, for the simple reason that two strength parameters (ϕ', c') cannot be derived from a single measurement (q_c or N_{60}). In practice, however, limited investigations are often the preferred option. In such cases involving cohesive frictional soil, engineers tend to (conservatively) ignore the c' component of strength and correlate the in-situ test parameters with the internal friction angle ϕ'. Average c' values may be later assessed from previous experience and backanalyses of field performance. More consistent approaches for characterizing the mechanical properties of cohesive-frictional materials are given by pressuremeter and plate load tests, despite the fact that analyses are complicated by a number of factors such as the influence of bonding on the stress-dilatancy response of soils and the effects of destructuration. These aspects are explored here and alternatives for testing interpretation are given.

Seismic tests

A geophysical survey is regarded as a powerful technique for subsurface exploration. Tests are generally non-destructive in nature and can be performed from the ground surface. Despite due recognition of its risks and limitations, there has been a steady increase in the perceived value of geophysics in representing complicated subsurface conditions involving large spatial variability and stratified soils. In addition, cross and downhole methods have been extensively used in geotechnical engineering, including the adaptation of sensors in the seismic cone.

The theoretical bases upon which seismic and other geophysical measurements are found are not within the scope of this chapter. For that purpose, there are a number of reference textbooks that extensively cover this subject area such as Richard *et al.* (1970), Sharma (1997) and Santamarina *et al.* (2001). For us, it is important to recall that geophysical methods rely on a significant contrast in physical properties of materials under investigation. Intrinsic properties such as density, resistivity or electrical conductivity, magnetic susceptibility and velocity of shock waves of the subsurface materials should be considered when evaluating the suitability of a given technique. Frequently used geophysical techniques are seismic refraction, high resolution surface wave reflection, vibration, down-hole and cross-hole, electrical resistivity, magnetic and gravity tests (e.g. Santamarina *et al.*, 2001; Becker, 2001; Stokoe *et al.*, 2004).

The primary applications in the use of geophysical methods in geotechnical engineering are (Becker, 2001): to map stratigraphy, determine thickness of strata, depth of bedrock and define major anomalies such as channels and cavities; to locate deposits of aggregates and other construction materials; and to determine engineering properties of strata and their spatial variation. Geo-environmental projects complement the list of applications. It is always necessary to bear in mind that geophysical techniques are intended to supplement ground investigation methods. To enhance its consistency, a site investigation campaign should always encompass a combination of geophysical surveys with a mesh of boreholes and/or penetration tests.

In cross-hole (CHT) and down-hole (DHT) tests, attention is given to the measurement of shear wave velocities from which it is possible to obtain the small-strain stiffness of the soil at induced strain levels of less than 0.001%:

$$G_o = \rho V_s^2 \tag{3.4}$$

where, G_o is the shear modulus, ρ the mass density and V_s the velocity of shear waves for a linear, elastic, isotropic medium. The CHT and DHT enable the velocity of horizontally propagating, vertically polarised (S_{hv}), vertically propagating, horizontally polarised (S_{vh}) and horizontally propagating, horizontally polarised (S_{hh}) shear waves to be measured.

Standard Penetration Test (SPT)

The SPT is the most widely used in-situ testing technique, primarily because of its simplicity, robustness and its ability to cope with difficult ground conditions in addition to providing disturbed soil samples. In many residual deposits the SPT, combined with continuous rotational coring, is the only investigation tool that is able to explore the stiff and hard layers of residual deposits.

Comprehensive reviews of the procedures and applications of the SPT are given by Decourt *et al.* (1988) and Clayton (1993). There is a range of types of SPT apparatus in use around the world (e.g. those employing manual and automatic trip hammers) and, consequently, variable energy losses cannot be avoided. Variability due to unknown values of energy delivered to the SPT rod system can now be accounted for properly by standardizing the measured N value to a reference value of 60% of the potential energy of the SPT hammer (N_{60}), as suggested by Skempton (1986). In many countries, however, this recommendation has not been incorporated into engineering practice. Moreover, even an SPT N value normalised to a given reference energy is not 'standard' because of the presently contentious issue of the influence of the length of the rod string.

The energy transferred to the composition of SPT rods was recently investigated by Odebrecht *et al.* (2004) and Schnaid (2005). This study has prompted a number of recommendations outlined to interpret the test in a more rational way on the basis of wave propagation theory. Recommendations are summarised as follows:

a. The energy transferred to the rod and to the sampler due a hammer impact should be obtained through integration of equation (3.5), calculated by the F-V method, and known as the Enthru energy:

$$E = \int_0^\infty F(t)V(t)\,dt \tag{3.5}$$

with an upper limit of integration equal to infinity (1/10 s is practical but may require longer time intervals of integration (1/5 s) in soft soils or long composition of rods).

b. the sampler energy can be conveniently expressed as a function of nominal potential energy E^*, sampler final penetration and weight of both hammer and rods.

The influence of rod length produces two opposite effects: wave energy losses increase with increasing rod length and in a long composition of rods, the gain in potential energy from rod weight is significant and may partially compensate measured energy losses.

c. efficiency is accounted for by three coefficients η_1, η_2 and η_3 that should be obtained from calibration. The hammer efficiency η_1 is obtained from measurement at the top of the rod stem. Efficiency factor η_2 can be assumed as unit. The energy efficiency η_3 is negatively correlated to the length of rods.

The *maximum potential energy*, PE^* delivered to the soil, should therefore be expressed as a function of the *nominal potential energy* E^*, and an additional energy related to the sampler penetration and the weight of both hammer and rods.

$$PE^* = E^* + (M_h + M_r)g\Delta\rho \tag{3.6}$$

where: M_h = hammer weight;
M_r = rod weight;
g = gravity acceleration;
$\Delta\rho$ = sample penetration under one blow;
E^* = nominal potential energy = 0.76 m 63.5 kg 9.801 m/s = 474 J

The *nominal potential energy* E^* = 474 J (ASTM, 1986) represents a part of the hammer potential energy to be transmitted to the soil. An *additional hammer potential energy* is given by $M_h, g, \Delta\rho$ which cannot be disregarded for tests carried out at great depths in soft soils, i.e. conditions in which $\Delta\rho$ and M_r are significant. For convenience, equation (3.6) can be written in two parts where the first represents the *hammer potential energy* (*nominal* + *additional*) and the second the *rod potential energy*:

$$PE^* = (0.76 + \Delta\rho)M_h g + \Delta\rho M_r g \tag{3.7}$$

Equation (3.7) deals with an ideal condition, where energy losses during the energy transference process are not taken into account. However, it is well known in engineering practice that these losses occur; they should not be disregarded and can be considered by efficiency coefficients. Equation (3.7) becomes:

$$PE_{h+r} = \eta_3[\eta_1(0.76 + \Delta\rho)M_h g + \eta_2 \Delta\rho M_r g] \tag{3.8}$$

where:

$$\eta_1 = \text{hammer efficiency} = \frac{\int_0^\infty F(t)V(t)\,dt}{(0.76 + \Delta\rho)M_h g}$$

$$\eta_2 = \beta_2 + \alpha_2 \ell \approx 1$$

$$\eta_3 = \text{energy efficiency} = 1 - 0.0042\ell$$

It follows from the foregoing that normalisation to a reference value of N_{60} is no longer sufficient to fully explain the mechanism of energy transfer to the soil. Furthermore,

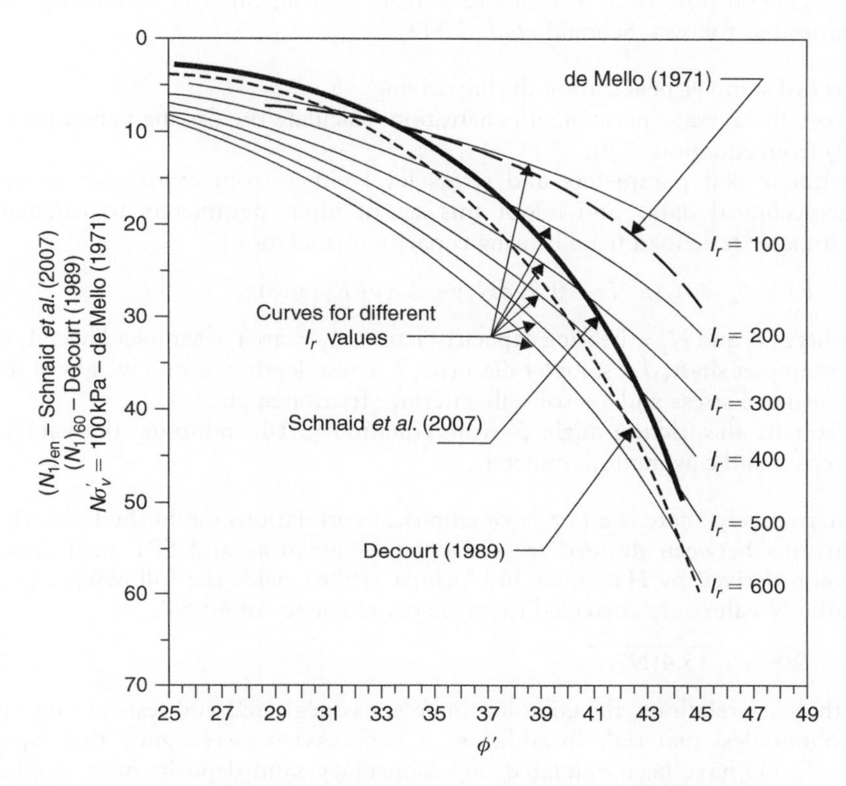

Figure 3.2 **Prediction of friction angle ϕ' from the SPT (*After* Schnaid et *al.*, 2009)**

it interesting to recall that maximum potential energy can be transformed into work by the non-conservative forces (W_{nc}) acting on the sampler during penetration, and since the work is proportional to the measured permanent penetration of the sampler, it is possible to calculate the dynamic force transmitted to the soil during driving:

$$PE_{h+r} = W_{nc} = F_d \Delta\rho \quad \text{or} \quad F_d = PE_{h+r}/\Delta\rho \tag{3.9}$$

The dynamic force F_d can be considered as a fundamental measurement for the prediction of soil parameters from SPT results.

The dynamic force applied to drive the sampler into the soil combined with bearing capacity and cavity expansion theories allows the internal friction angle ϕ' to be estimated directly from ordinary SPT sampler penetration (Schnaid *et al.*, 2009). The blow count number is defined as $(N_1)_{en}$ because the interpretation relies on correcting the measured value both to the energy transferred to the soil and the applied mean stress. The relationship between $(N_1)_{en}$ and ϕ' shown in Figure 3.2 is fairly sensitive to the variations in the rigidity index, an effect that is more significant for dense materials than for loose.

The general procedure to estimate ϕ' from bearing capacity formulation can be summarised as follows (Schnaid *et al.*, 2009):

a. record sampler penetration during driving;
b. from the average permanent penetration, calculate the driving penetration force F_d from equation (3.9);
c. estimate soil parameters and stress level either from experience or existing geotechnical data, and adopt this set of input parameters to calculate the ultimate static load from bearing capacity formulation:

$$F_d \approx F_e = A_p(p'N_q + 0.5\gamma d N_\gamma) + A_f(\gamma L K_s \tan\delta) \tag{3.10}$$

where N_q and N_γ = bearing capacity factors, A_p = area of sampler base, A_f = area of sampler shaft, d = sampler diameter, L = test depth, γ = unit weight of the soil, p' = mean stress and δ = soil-pile interface friction angle.
d. Estimate the friction angle ϕ' from equation (3.10), adopting the values from steps b and c as input parameters.

Alternatively, there is a family of empirical correlations established directly from comparisons between drained triaxial compression tests and SPT field data. The expression derived by Hatanaka and Uchida (1996) yields the following expression where the N values are corrected to an energy efficiency of 60(%):

$$\phi' \approx 20° + \sqrt{15.4(N_1)_{60}} \tag{3.11}$$

In all these correlations, the cohesion intercept is neglected and tests are interpreted as a cohesionless material. In addition, it is necessary to recognise that equations such as (3.11) have been validated on sedimentary sand deposits only. Application to an undisturbed sand database is presented in Figure 3.3, in which strength values quoted from these equations are shown to yield values that would not be unreasonable for granular materials. Its applicability for residual soils might depend on the degree of cementation of the deposit and may have to be locally validated.

The degree of cementation can be accessed from the ratio of penetration resistance and soil stiffness, since in principle, a material that is stiffer in deformation may be stronger in strength (e.g. Tatsuoka and Shybuya, 1991). Based on this concept, a methodology has been developed to identify the existence of distinctive behaviour emerging from ageing or cementation of residual soils based on the ratio of the elastic stiffness to ultimate strength, G_o/N_{60}. A guideline formulation to compute G_o from SPT tests is given by the following equations:

$$\frac{(G_0/p_a)}{N_{60}} = \alpha N_{60}\sqrt{\frac{p_a}{\sigma'_{vo}}} \quad \text{or} \quad \frac{(G_0/p_a)}{N_{60}} = \alpha(N_1)_{60} \tag{3.12}$$

where:

$$(N_1)_{60} = N_{60}\left(\frac{p_a}{\sigma'_{vo}}\right)^{0.5}$$

where σ'_{vo} is the mean vertical effective stress and p_a the atmospheric pressure. The α value is a dimensionless number that depends on the level of cementation and age as

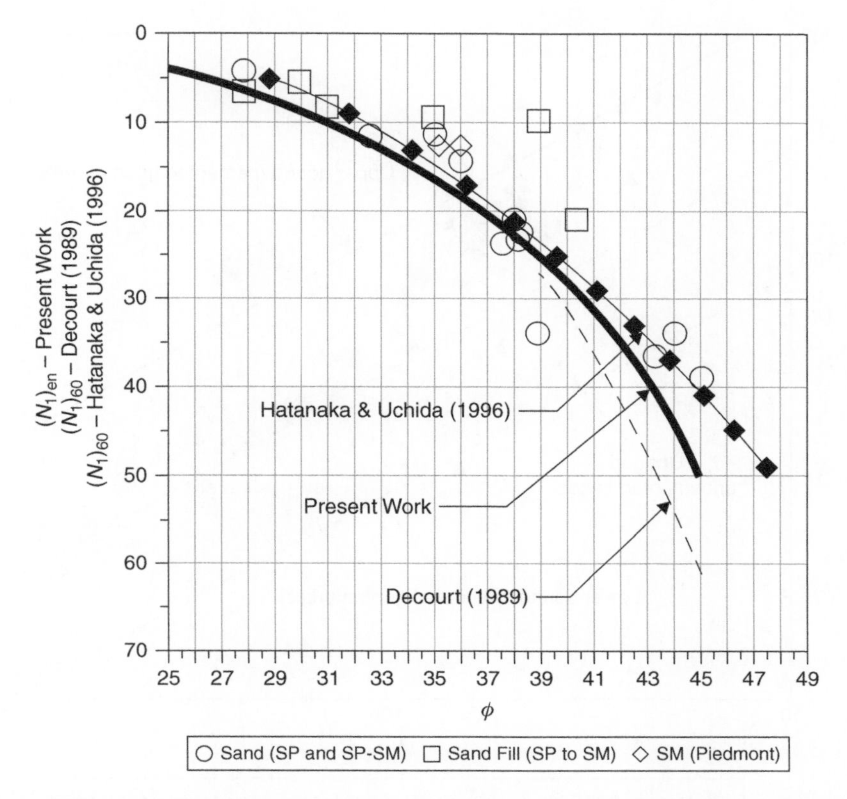

Figure 3.3 Friction angle of cohesionless soils from SPT resistance (*After* Schnaid *et al.*, 2009; data from Hatanaka and Uchida, 1996)

well as the soil compressibility and suction. The small strain stiffness to strength ratio embodied within the G_o/N_{60} term is seen on Figure 3.4, at a given $(N_1)_{60}$ (or relative density), to be generally appreciably higher for lateritic soils than that of the saprolites, primarily because the latter generally exhibit higher N_{60} (or strength) values.

The database comprises soils from Brazil and Portugal (Barros, 1998; Schnaid, 1997; Schnaid *et al.*, 2004a). The bond structure is seen to have a marked effect on the behaviour of residual soils, producing values of normalised stiffness (G_o/N_{60}) that are considerably higher than those observed in fresh cohesionless materials.

This database can be also useful for assessing the stiffness of natural deposits. Considering the variation observed in tropical residual soils, it is preferable to express correlations in terms of lower and upper boundaries designed to match the range of recorded G_o values (Schnaid *et al.*, 2004a):

$$\left.\begin{array}{ll} G_o = 1200\sqrt[3]{N_{60}\sigma'_{vo}p_a^2} & \text{upper bound: cemented} \\ G_o = 450\sqrt[3]{N_{60}\sigma'_{vo}p_a^2} & \text{lower bound: cemented} \\ & \text{upper bound: uncemented} \\ G_o = 200\sqrt[3]{N_{60}\sigma'_{vo}p_a^2} & \text{lower bound uncemented} \end{array}\right\} \tag{3.13}$$

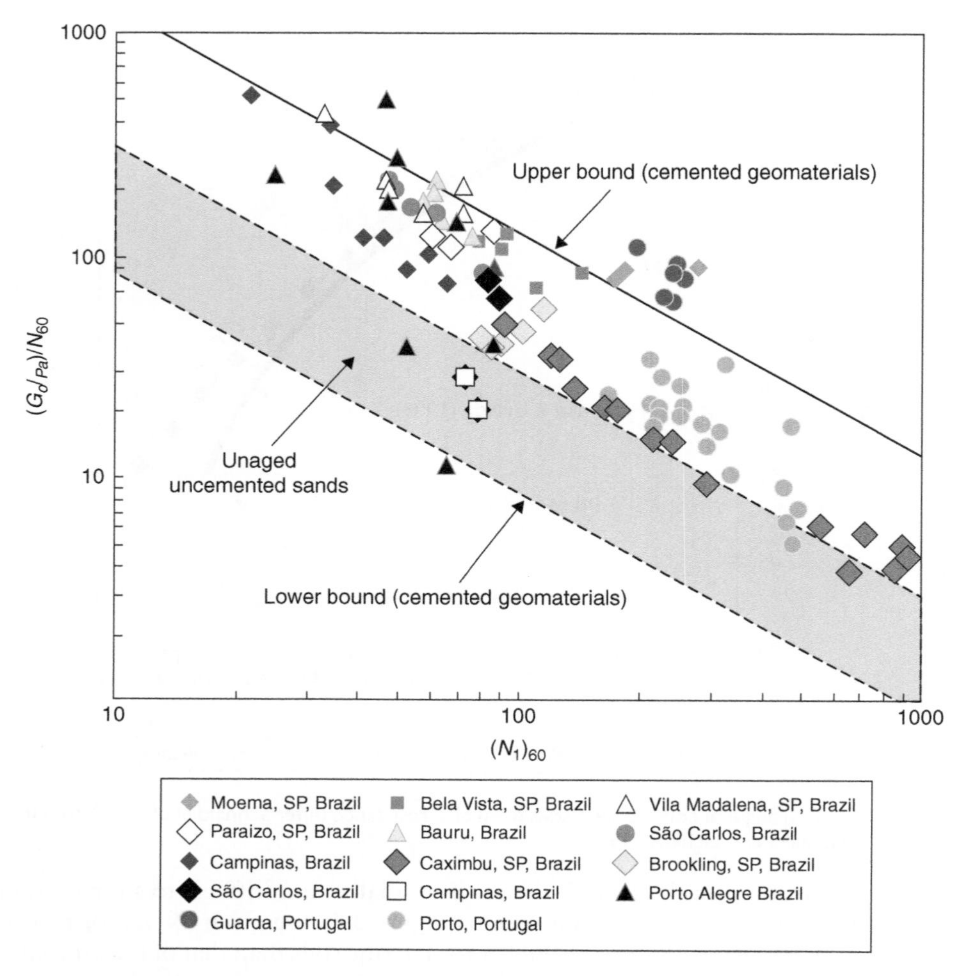

Figure 3.4 G_o/N_{60} vs $(N_1)_{60}$ space to identify cementation on residual soils (*After* Schnaid, 2005)

It is here important to emphasise that given the considerable scatter observed for different soils, correlations such as given in equation (3.13) are only approximate indicators of G_0 and do not replace the need for in-situ shear wave velocity measurements. The reduction in the ratio of G/G_o with shear stress and shear strain is known to be sensitive to degradation of cementation and structure, among several other factors (e.g. Tatsuoka *et al.*, 1997). The moduli degradation can be measured in the laboratory with high resolution sensors provided that high-quality undisturbed samples can be obtained.

Piezocone penetration test (CPTU)

The CPT, with the possible inclusion of pore water pressure, shear wave velocity and resistivity measurements, is recognised worldwide as an established, routine and

Figure 3.5 Range of CPT tools available for site investigation (courtesy of A.P. van den Berg)

cost-effective tool for site characterisation and stratigraphic profiling, and a means by which the mechanical properties of the subsurface strata may by assessed. CPTs were particularly popular in sands and in marine and lacustrine sediments in coastal regions, but are now also commonly used in residual soils provided that penetration is achieved. For a general review on the subject, the reader is encouraged to refer to Lunne *et al.* (1997) – *CPT in Geotechnical Practice*, and the *Proceedings of the Symposia on Penetration Testing* (1981, 1988, 1995, 1998, 2004).

Routine penetrometers have employed either one midface element for pore water pressure measurement (designated as u_1) or an element positioned just behind the cone tip (shoulder, u_2). The ability to measure pore pressure during penetration greatly enhances the profiling capability of the CPTU, but in unsaturated soil conditions, this measurement is of very little application. Geotechnical site characterisation can be further improved by independent seismic measurements, adding the downhole shear wave velocity (V_s) to the measured tip cone resistance (q_t), sleeve friction (f_s) and pore water pressure (u). The combination of different measurements into a single sounding provides a particularly powerful means of assessing the characteristics of unusual materials (Schnaid, 2005). Various penetrometers are available, for example a miniature electric element with a cone tip area of 2 cm^2, a 5 cm^2 minicone, a standard 10 cm^2 piezocone, a 10 cm^2 pore-water sensors piezocone, a seismic piezocone and a resistivity cone, some of which are shown in Figure 3.5.

Evaluation of the peak friction angle of cohesionless soils from CPT can be based on analytical, numerical or empirical approaches (e.g. Yu and Mitchell, 1998; Yu, 2004). There are a number of correlations often adopted in engineering practice such as (Lunne *et al.*, 1997):

$$\phi'_p = \arctan\left[0.1 + 0.38\log\left(\frac{q_t}{\sigma'_{vo}}\right)\right] \qquad (3.14)$$

More recently, an expression was derived by Mayne (2006) where the cone is normalised by the square root of the effective overburden stress following recommendations from Jamiolkowski *et al.* (1985):

$$\phi'_p = 17.6^\circ + 11.0 \log \left[\frac{\left(q_t/\sigma_{atm} \right)}{\left(\sigma'_{vo}/\sigma_{atm} \right)^{0.5}} \right] \tag{3.15}$$

While these correlations appear to work well for clean sands, there is very little systematic experience in intermixed and bonded soils and local validation is therefore recommended.

In bonded soils the contribution of the cohesive component linking particles is disregarded in equations (3.14) and (3.15), which may impact predictions of shear strength. The important feature of characterizing the presence of the bonding structure can be achieved by combination of strength and stiffness measurements. It follows from the foregoing on the SPT that a bonded/cemented structure produces G_o/q_c and G_o/N_{60} ratios that are systematically higher than those measured in cohesionless soils. These ratios provide a useful means of assisting site characterisation. Typical results from residual profiles are presented in Figure 3.6, in which the G_o/q_c ratios are plotted against the normalised parameter q_{c1} for CPT data (*After* Schnaid, 1999; Schnaid *et al.*, 2004a), where:

$$q_{c1} = \left(\frac{q_c}{p_a} \right) \sqrt{\frac{p_a}{\sigma'_v}} \tag{3.16}$$

p_a being the atmospheric pressure. The bond structure generates normalised stiffness values that are considerably higher than those for uncemented soils and as a result, the datapoints for residual soils fall outside and above the band proposed for sands by Eslaamizaad and Robertson (1997) and theoretically determined by Schnaid and Yu (2007).

The upper and lower bounds designed to match the range of recorded G_o values in residual soils can be used for a direct assessment of the variation of G_o with q_c (Schnaid *et al.*, 2004a):

$$\left.\begin{array}{ll} G_o = 800\sqrt[3]{q_c\sigma'_v p_a} & \text{upper bound: cemented} \\ G_o = 280\sqrt[3]{q_c\sigma'_v p_a} & \text{lower bound: cemented} \\ & \text{upper bound: uncemented} \\ G_o = 110\sqrt[3]{q_c\sigma'_v p_a} & \text{lower bound uncemented} \end{array}\right\} \tag{3.17}$$

Since considerable scatter is observed in the residual soils database, correlations such as given in equation (3.17) are only approximate indicators of G_o and do not replace the need for in-situ shear wave velocity measurements.

The magnitude of the small strain stiffness in bonded soils is better understood in comparison with values determined from natural sands. A reference equation adopted in the comparison is:

$$\frac{G_o(\text{MPa})}{F(e)} = S[p'(\text{MPa})]^n \tag{3.18}$$

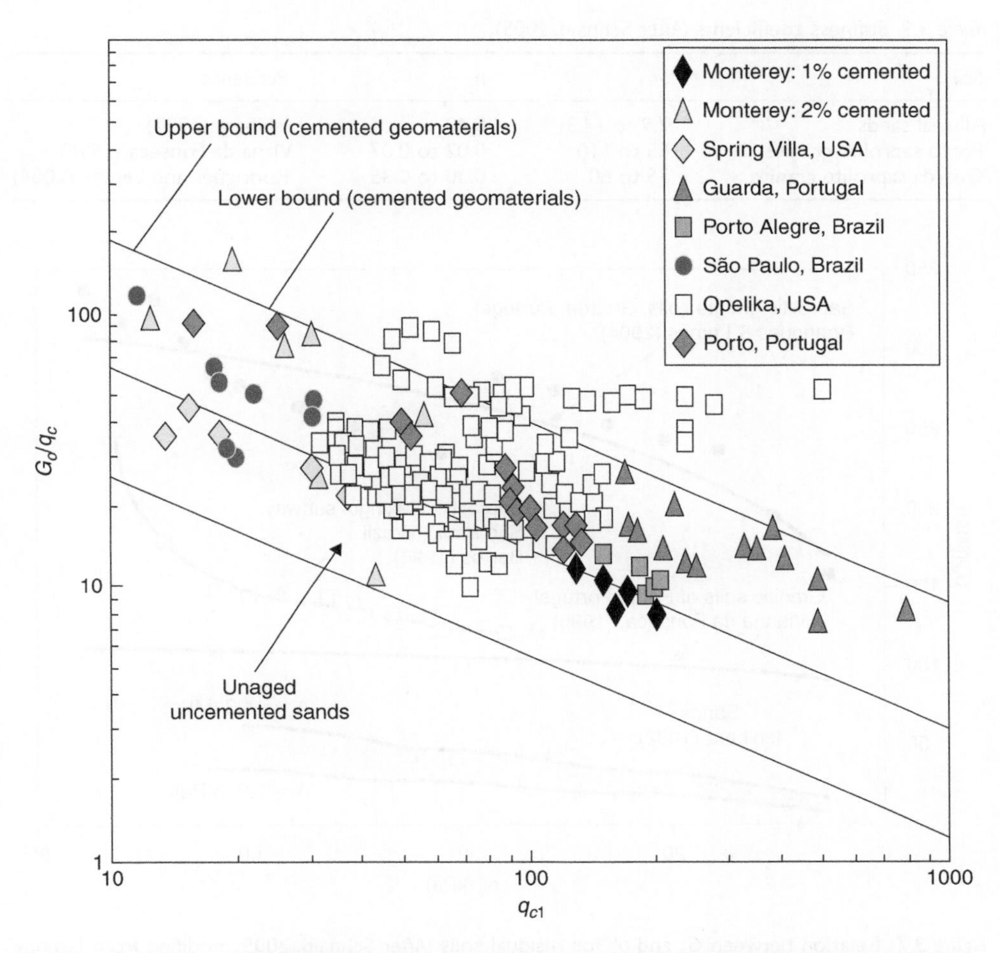

Figure 3.6 G_o/q_c ratios plotted against normalised parameter q_{c1} for CPT data (Schnaid, 2005)

with a void ratio function:

$$F(e) = \frac{(2.17 - e)^2}{1 + e} \tag{3.19}$$

Values of parameters S and n are given in Table 3.9 and a direct comparison is shown in Figure 3.7, having the data for alluvial sands from Ishihara (1982) as reference. Values of G_o diverge significantly from those established for transported soils when they exhibit the same granulometry but are uncemented. Parameter S is much higher than the value adopted for cohesionless soils, whereas n varies significantly as a result of local weathering conditions (e.g. Gomes *et al.*, 2004). Given the variations in both S and n, the need for site-specific correlations is demonstrated.

Table 3.9 Stiffness coefficients (*After* Schnaid, 2005)

Soil	*S*	*n*	*Reference*
Alluvial sands	7.9 to 14.3	0.40	Ishihara (1982)
Porto saprolite granite	65 to 110	0.02 to 0.07	Viana da Fonseca (1996)
Guarda saprolite granite	35 to 60	0.30 to 0.35	Rodrigues and Lemos (2004)

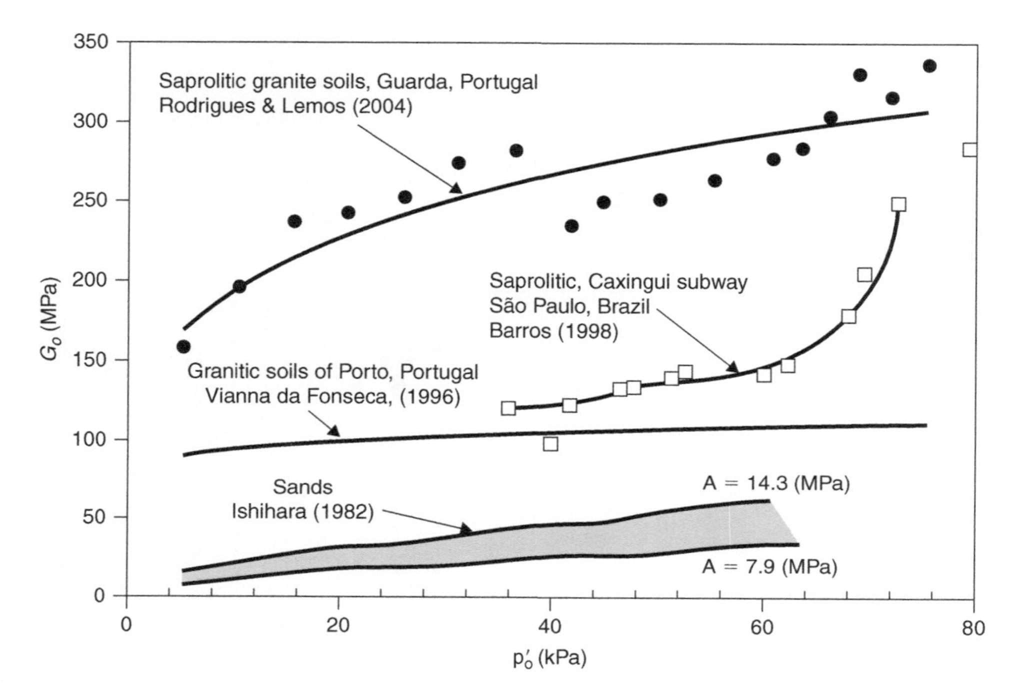

Figure 3.7 Relation between G_o and p'_o for residual soils (*After* Schnaid, 2005; modified from Gomes *et al.*, 2004)

Pressuremeter

Pressuremeters are cylindrical devices designed to apply uniform pressure to the wall of a borehole by means of a flexible membrane. Both pressure and deformation at the cavity wall are recorded and interpretation is provided by cavity expansion theories under the assumption that the probe is expanded in a linear, isotropic, elastic, perfectly plastic soil. Under this assumption, the soil surrounding the probe is subjected to pure shear only. Acknowledging that the greatest potential of the pressuremeter lies in the measurement of modulus, it is a common practice to carry out a few unloading-reloading cycles during the test. If the soil is perfectly elastic in unloading, then the unloading-reloading cycle will have a gradient of $2G_{ur}$, where G_{ur} is the unload-reload shear modulus. Numerous papers have been published on this theme and there are important textbooks such as Baguelin *et al.* (1978), Mair and Wood (1987), Briaud (1992), Clarke (1993) and Yu (2000).

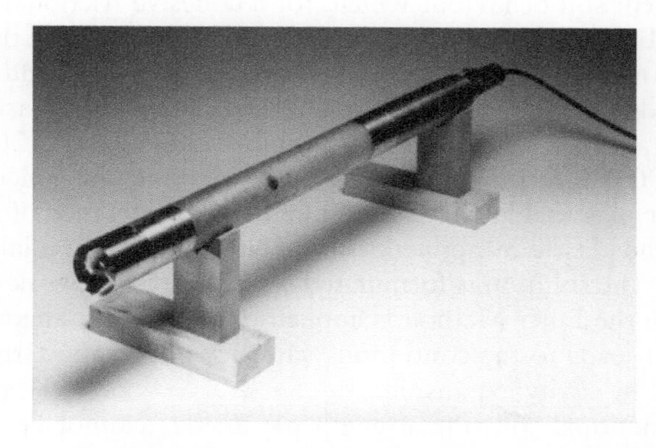

Figure 3.8 Cambridge self-boring pressuremeter (courtesy of Cambridge Insitu)

Pressuremeters are generally classified in three groups according to the method of installation into the ground. Pre-bored pressuremeter, self-boring pressuremeter and push-in pressuremeter are the three broad categories. The Menard pressuremeter is the most well known example of a pre-bored probe in which the device is lowered into a pre-formed hole. In a self-boring probe the device bores its own way into the ground with minimal disturbance (Figure 3.8), whereas in a push-in device the pressuremeter is pushed into the ground attached to a cone tip. The method of interpretation should take account of the installation process. Theoretical interpretation methods developed for pressuremeters involve axially symmetric expansion and contraction of an infinitely long cylindrical cavity. Under this fundamental assumption the cavity-expansion/contraction curve can be analytically modelled to obtain soil properties. The symmetry of the well defined boundary conditions of a pressuremeter is the main advantage of this technique over other in-situ tests.

The pressuremeter offers the possibility of characterizing the mechanical properties of residual soils, particularly as a means of inspecting the accurateness of a given set of design parameters. All the theories for the interpretation of the pressuremeter in bonded soils make use of the in-situ horizontal stress, soil stiffness and strength parameters: angle of internal friction, angle of dilation and cohesion intercept (which reduces with destructuration at high shear strains). The pressure expansion curve represents therefore a combination of all these parameters that cannot be assessed independently. In residual soils, instead of attempting to derive a set of parameters from a single test, the pressuremeter should be viewed as a "trial" boundary value problem against which a theoretical pressure-expansion curve, predicted using a set of independently measured parameters, is compared to a field pressuremeter tests. A good comparison between the measured and predicted curves gives reassurance to the process of selecting design parameters, whereas a poor comparison indicates that one or more of the constitutive parameters are unrealistic.

The solution for cavity expansion and contraction problems is extensive, and hence only a brief review is presented here to illustrate potential applications. Adequacy of interpretation methods largely depends upon the constitutive model adopted to represent soil behaviour which, for a cohesive-frictional material, is complicated by a number of factors such as the influence of bonding on the stress-dilatancy response of soils and the effects of destructuration. Ideally the c' and ϕ' should be coupled to stiffness, dilatancy and mean stress level, and for that reason the cylindrical cavity expansion analysis developed by Mantaras and Schnaid (2002) and Schnaid and Mantaras (2003) is recommended. In these solutions, the concept introduced by Rowe (1962) that plastic dilatancy is inhibited by the presence of soil bonding was used to describe the plastic components of the tangential and radial increments in an expanding cavity. The solution is formulated within the framework of non-associated plasticity in which the Euler Method is applied to solve simultaneously two differential equations that leads to the continuous variations of strains, stresses and volume changes produced by cavity expansion.

In cohesive-frictional soils the strength parameters cannot be extracted directly from the experimental curve. The method to assess parameters recognises that strength, stiffness and in-situ stresses interact to produce a particular pressuremeter expansion curve through a procedure that can be summarised as follows:

a. Record a field pressuremeter curve;
b. Compute values of stiffness and in-situ stresses directly from the average pressuremeter measurements of the three strain arms;
c. Input shear parameters should be from independent test data;
d. Calculate a pressure-expansion curve and compare to the experimental curve measured in-situ.
e. Re-evaluate the set of input parameters if a good comparison between measured and calculated data is not obtained and produce new comparisons.

By reproducing this process for tests carried out at several different depths, it is possible to estimate average constitutive parameters from a given deposit, bearing in mind that engineering judgment is required to avoid the selection of a set of *doubtful* parameter values that may produce a good fit to the data.

Case studies in the Hong Kong gneiss (Schnaid *et al.*, 2000), Porto granite (Mantaras, 2002) and São Paulo gneiss (Schnaid and Mantaras, 2003) have validated the application of the described methodology. Take the case study summarised by Schnaid and Mantaras (2004) which reports an extensive site investigation programme comprising laboratory triaxial tests, SPT and a large number of high quality pressuremeter tests in a residual gneiss soil profile. A typical example of the fit provided by the analytical solution is shown in Figure 3.9 and a summary of the shear strength data obtained from the interpretation of 15 such tests is presented in Figure 3.10, which also plots SPT N_{60} and the pressuremeter limit pressures (ψ_L). The SBPM yielded ϕ'_{ps} from 27 to 31° with considerable data scatter but within the range measured from laboratory testing data. The curve fitting applied to the loading portion of the SBPM tests gave results which are rather consistent, being slightly above the assumed critical state values and compatible with laboratory data. The presence of mica at given locations has yielded a lower boundary for predicted ϕ'_{ps} values, compatible with evidence provided by N values.

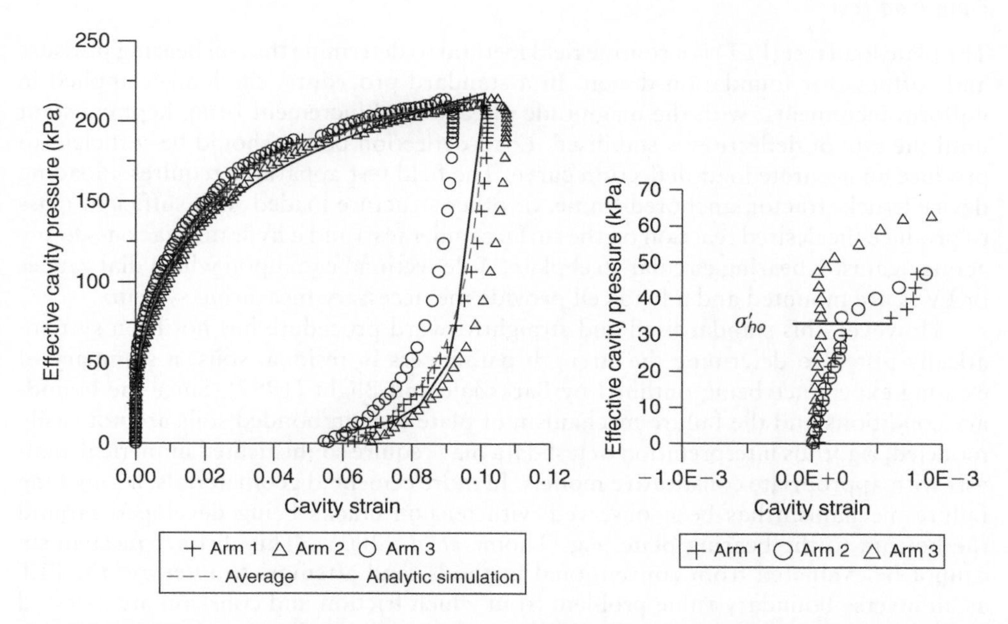

Figure 3.9 Typical example of a pressuremeter test carried out in the saprolite gneiss residual soil of Sao Paulo (*After* Schnaid and Mantaras, 2004)

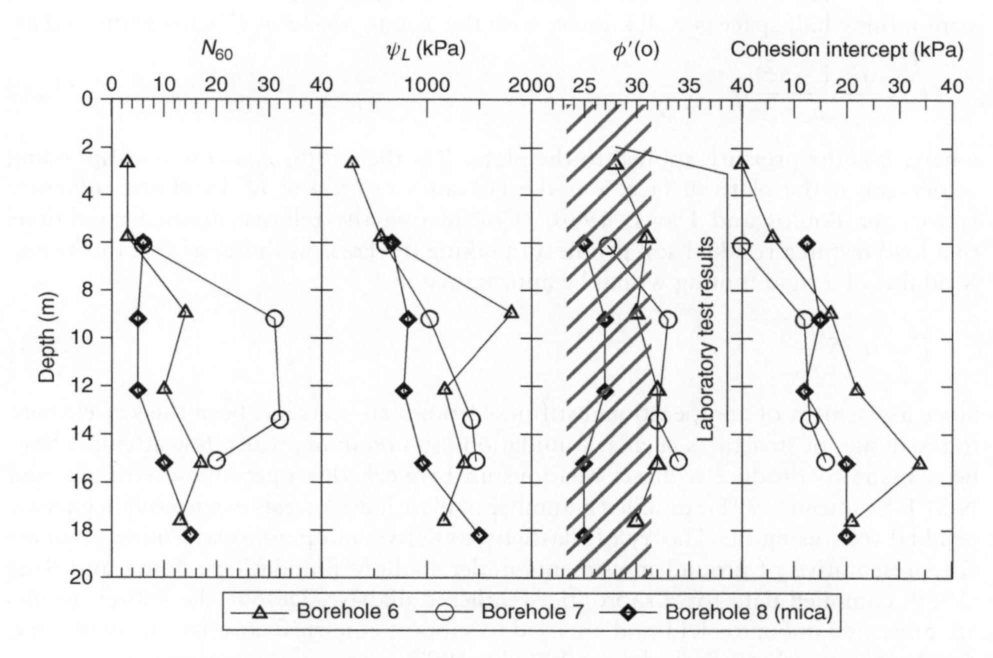

Figure 3.10 Prediction of soil properties for the Sao Paulo gneiss residual soil (*After* Schnaid and Mantaras, 2004)

Plate load test

The plate load test (PLT) is a routine field method to determine the soil bearing pressure and stiffness for foundation design. In a standard procedure, the load is applied in uniform increments, with the magnitude of each load increment being kept constant until the rate of deflection is stabilised. Load-deflection points should be sufficient to produce an accurate load-deflection curve. The field test apparatus requires a loading device (truck, tractor, anchored frame, or other structure loaded with sufficient mass to produce the desired reaction on the surface under test) and a hydraulic jack assembly acting against a bearing circular steel plate. A deflection beam upon which dial gauges or LVTs are mounted and a load cell provide the necessary measuring system.

However, this standardised and straightforward procedure has not been systematically tested to determine the strength parameters in residual soils, a summary of existing experience being outlined by Barksdale and Blight (1997). Since the boundary conditions and the failure mechanism of plate tests in bonded soils are not easily modeled, rigorous interpretation of test data may require sophisticated numerical analysis with appropriate constitutive models. In more cemented geomaterials, a punching failure mechanism has been observed with tension cracks being developed around the bottom to the bearing plate (e.g. Thomé *et al.*, 2005). This distinct mechanism cannot be evaluated from conventional methods, and attempts to interpret the PLT as an inverse boundary value problem from which friction and cohesion are assessed have proved unrealistic.

On the other hand, plate-load tests are a common field method to estimate soil stiffness. Solution for footings resting on a homogeneous, isotropic, linear-elastic, semi-infinite half space is well known, with the Young Modulus E being expressed as:

$$E = \frac{qB(1 - \nu^2)}{\rho_o} I_f \tag{3.20}$$

where, q is the pressure applied to the plate, B is the width, ρ_o is the corresponding settlement at the plate surface, ν is the Poisson's ratio, and I_f the elastic influence factor (see Poulos and Davis, 1980). Considering the relative drained conditions of a load test in a residual soil profile and taking the Poisson's ratio as 0.2, the Young Modulus of a rigid footing would be approximately:

$$E = 0.75 \frac{qB}{\rho_o} \tag{3.21}$$

Since assessment of an operational stiffness from plate tests has been the key element to the design of structures such as foundations and retaining walls, few attempts have been made to produce a direct relationship between this operational stiffness and N-SPT. Sandroni (1991) compiled a number of plate loading tests carried out in gneissic residual soils using the Theory of Elasticity to derive an operational Young modulus E representative of ground movements under shallow foundations. Jones and Rust (1989) compiled data for a saprolitic weathered diabase. Despite the scatter, results are presented in Figure 3.11 and are used to support empirical correlations between E and N values such as (Barksdale and Blight, 1997):

$$E = aN_{60}(\text{MPa}) \tag{3.22}$$

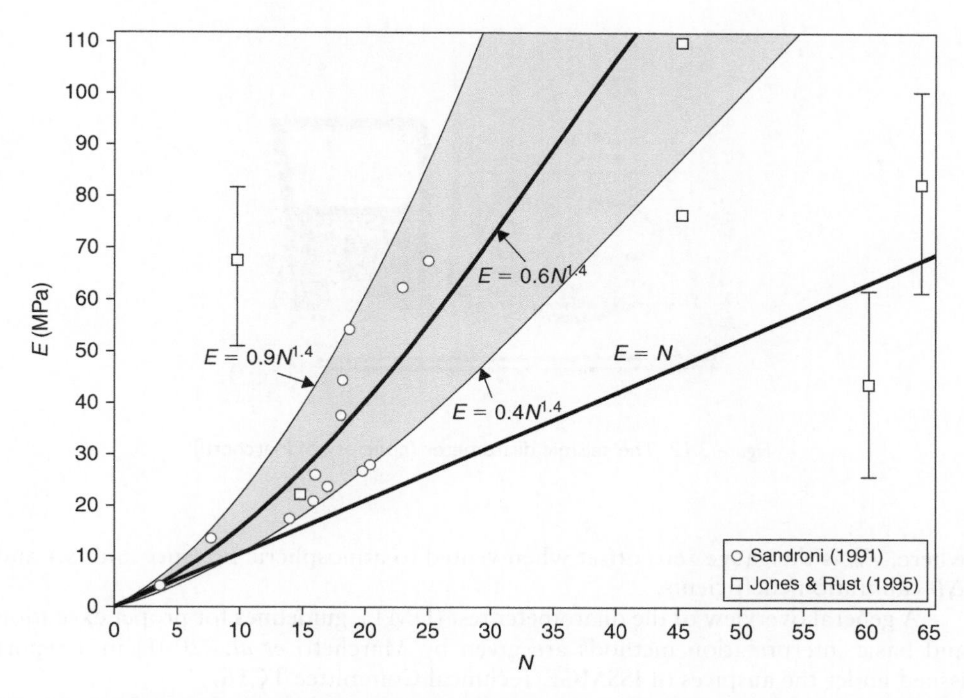

Figure 3.11 Relationships between N and Young modulus for residual soils N_{72} for Sandroni (1991) and N_{60} for Jones and Rust (1995) (*After* Schnaid, 2009)

with a ranging from 1 to 1.6 and (Sandroni, 1991):

$$E = bN_{72}^{c} \text{ (MPa)} \tag{3.23}$$

with coefficients $b = 0.6$ and $c = 1.4$.

Dilatometer

The dilatometer consists of a stainless steel blade having a flat, circular steel membrane mounted flush on one side (Figure 3.12).

The test starts by driving the blade into the soil at a constant rate, generally between 10 and 20 mm/s. After penetration, the membrane is inflated through a control unit and a sequence of pressure readings are made at prescribed displacements, corresponding to (a) the A-pressure at which the membrane starts to expand ("lift-off") and (b) the B-pressure required to move the centre of the membrane by 1.1 mm against the soil. Values of A and B readings are used to determine the pressures p_0 and p_1 from the following expressions:

$$p_0 = 1.05(A - Z_m + \Delta A) - 0.05(B - Z_m - \Delta B) \tag{3.24}$$

$$p_1 = (B - Z_m - \Delta B) \tag{3.25}$$

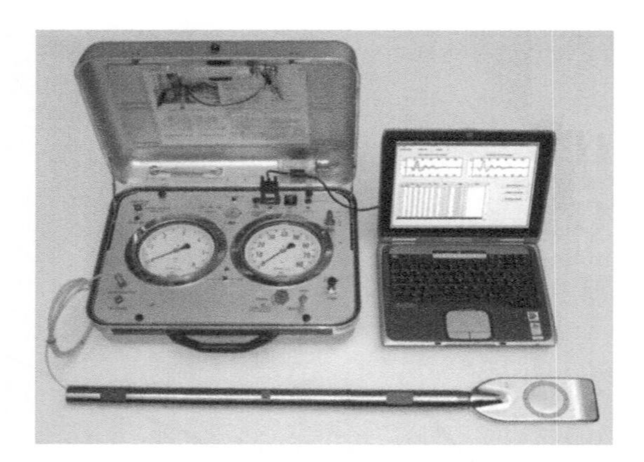

Figure 3.12 The seismic dilatometer (courtesy of Marchetti)

where, Z_m is the gauge zero offset when vented to atmospheric pressure and ΔA and ΔB calibration coefficients.

A general overview of the dilatometer test (DMT), guidelines for proper execution and basic interpretation methods are given by Marchetti *et al.* (2001) in a report issued under the auspices of ISSMGE Technical Committee TC16.

Interpretation methods are essentially based on correlations obtained by calibrating DMT pressure readings against high quality parameters (e.g. Lutenegger, 1988; Lunne *et al.*, 1989; Marchetti, 1997). These correlations are essentially empirically based, established against clay and sand material and are supported by a limited number of numerical studies (e.g. Baligh and Scott, 1975; Finno, 1993; Yu *et al.*, 1993; Smith and Houlsby, 1995; Yu, 2004).

There is no systematic experience of DMT in residual soils and it is not clear how the bonded structure of cohesive-frictional materials impacts parameters derived from the test. One potential application is the determination of the soil stiffness for settlement calculations. The dilatometer modulus is obtained by relating the displacement s_0 to p_0 and p_1 by the theory of elasticity (Gravesen, 1960). The solution assumes that (a) the space surrounding the dilatometer is formed by two elastic half-spaces in contact along the plane of symmetry of the blade and (b) zero settlement is computed externally to the loaded area:

$$s_0 = \frac{2D(p_1 - p_0)}{\pi} \frac{(1 - \nu^2)}{E} \tag{3.26}$$

where, E is the Young modulus and ν the Poisson's ratio.

For a diameter membrane D equal to 60 mm and a displacement s_o equal to 1.1 mm, equation (3.26) approaches:

$$E_D = 34.7(p_1 - p_0) \tag{3.27}$$

Although E_D is inherently an operational Young Modulus (both E and E_D are calculated from Elastic Theory), it is recognised that the expansion of the membrane from p_0 and p_1 reflects the disturbed soil properties around the blade produced by DMT penetration.

In-situ permeability

Permeability of undisturbed samples can be obtained by in-situ testing (falling head method) at various depths as the drilling of the borehole proceeds. The falling head, rising-head and constant head tests can be used in boreholes and employing packers to isolate a particular hole length. It is important that the inside surface of the hole used for permeability testing is free of loose or smeared material which can make the results of testing imprecise.

Permeabilities of residual soils are affected by the variations in grain size, void ratio, mineralogy, degree of fissuring and the characteristics of the fissures. Garga and Blight (1987) explained that the permeability of some residual soils is strongly controlled by the relict structure of the material, where the flow takes place along relict joints, quartz veins, termite and other biochannels. A permeability test is needed when seepage problems are involved or expected.

There are two methods concerning permeability, which are feed water and extract water. General guidelines for in-situ permeability tests are described as follows:

- Ponding test or infiltration test can be conducted when the water table is low.
- Pumping test from test pits or holes can be used when the water table is near to surface.

Tests for soil in unsaturated condition

Standard in-situ test interpretation does not consider the matric suction effect that emerges from the unsaturated conditions of various tropical residual soil deposits. In this case, the role of matrix suction and its effect on soil permeability has to be acknowledged and accounted for. The influence of partial saturation imparts a very distinct behaviour to a soil and should be considered in test interpretation. Few aspects of significance deserve attention: (a) suction measurement and its practical significance, (b) suction control in field tests and (c) soil collapsibility.

An important contribution in the analysis of unsaturated soils has been the extension of the elastic-plastic critical state concepts to unsaturated soil conditions by Alonso *et al.* (1990). In this method, the frame of reference is described by four variables – net mean stress $(p - u_a)$, deviator stress q, suction s $(u_a - u_w)$ and specific volume v, where u_a is the air pressure and u_w the pore water pressure. Several constitutive models have subsequently been proposed following these same concepts (Josa *et al.*, 1992; Wheeler and Sivakumar, 1995). These constitutive models allow derivation of the yield locus in the (p, q, s) space, an analysis that requires nine soil parameters. Model parameters are assessed from laboratory suction-controlled testing such as isotropic compression tests and drained shear strength tests. For isotropic conditions, the model is characterised by the loading-collapse (LC) yield curve whose hardening laws are controlled by the total plastic volumetric deformation. A third state parameter has to be incorporated to include the effect of the shear stress q. The yield curve for a sample at a constant

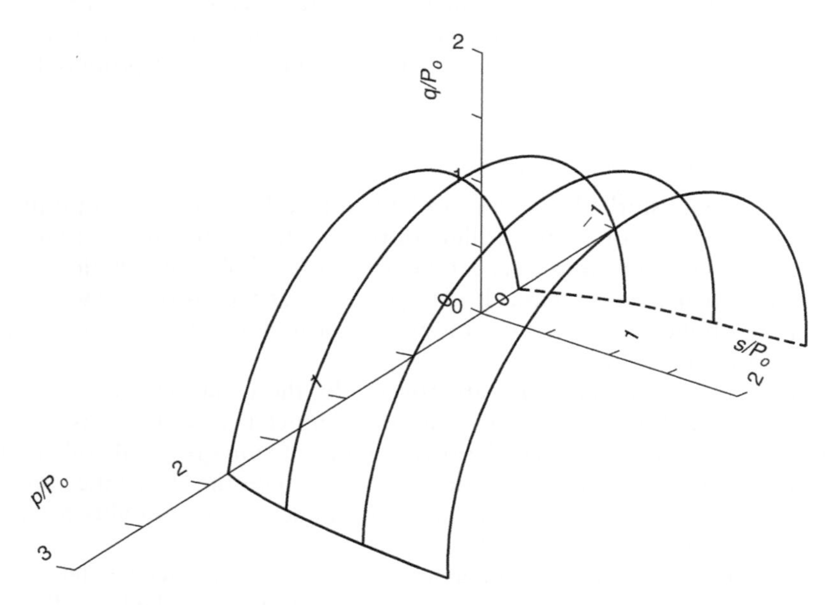

Figure 3.13 Three-dimensional view of yield surfaces in $(p/P_o, q/P_o, s/P_o)$ space

suction s is described by an ellipse, in which the isotropic preconsolidation stress is given by the previously defined p_o value that lies on the loading-collapse yield curve. The critical state line (CSL) for non-zero suction is assumed to result from an increase in (apparent) cohesion, maintaining the slope M of the CSL for saturated conditions, as illustrated in Figure 3.13.

For incorporating these concepts into the analysis of in-situ tests in unsaturated soil conditions, the first necessary step is to measure the in-situ matric suction. Several techniques have been developed recently to measure the matric suction, such as the non-flushable vacuum tensiometer, the flushable piezometer and the miniature non-flushable tensiometer (e.g. Ridley and Burland, 1995, Marinho, 2000). However, these techniques are normally used in the laboratory in compacted soils or in the field in sedimentary clay deposits.

Suction measurements in a granular granite residual soil site in southern Brazil, in which experience of such measurements is scarce, have been presented by Schnaid *et al.* (2004). At this site, the measured suction ranged from 25 to 70 kPa with a trend of increasing suction during any dry period, which can be attributed to evaporation processes. A typical result is illustrated in Figure 3.14 for measurements of up to 50 kPa. There is no marked difference between the readings recorded from the different instruments tested in the site, which implies that any technique can be used with some confidence in engineering practice for suctions less than 100 kPa in coarse grained soils. The relationship between matric suction and gravimetric water content for in-situ and laboratory specimens is shown in Figure 3.15, over the range of suctions being considered. In general, the data agree well with the general equation proposed by Fredlund and Xing (1994). This relationship is the so-called soil

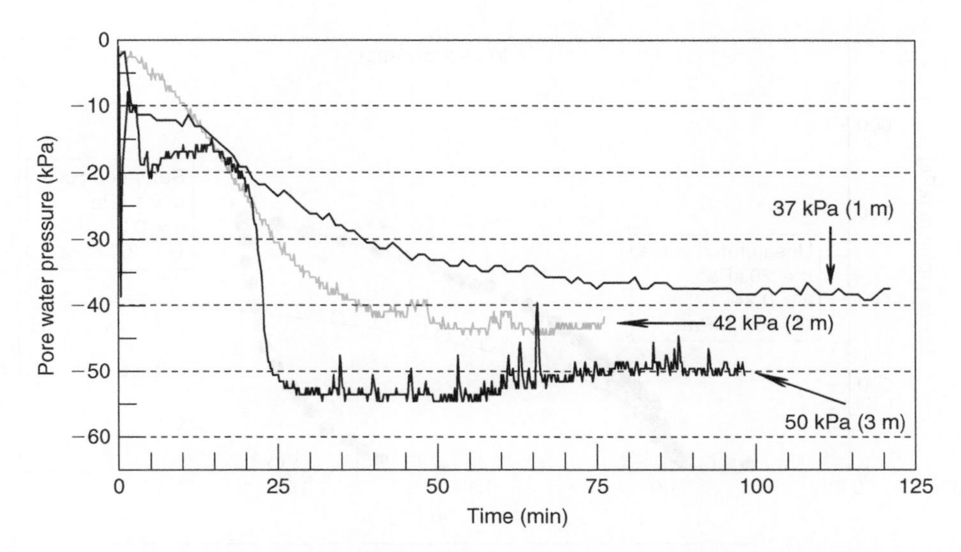

Figure 3.14 Typical in-situ suction measurements using the miniature tensiometer for the granite residual soil (*After* Schnaid *et al.*, 2004)

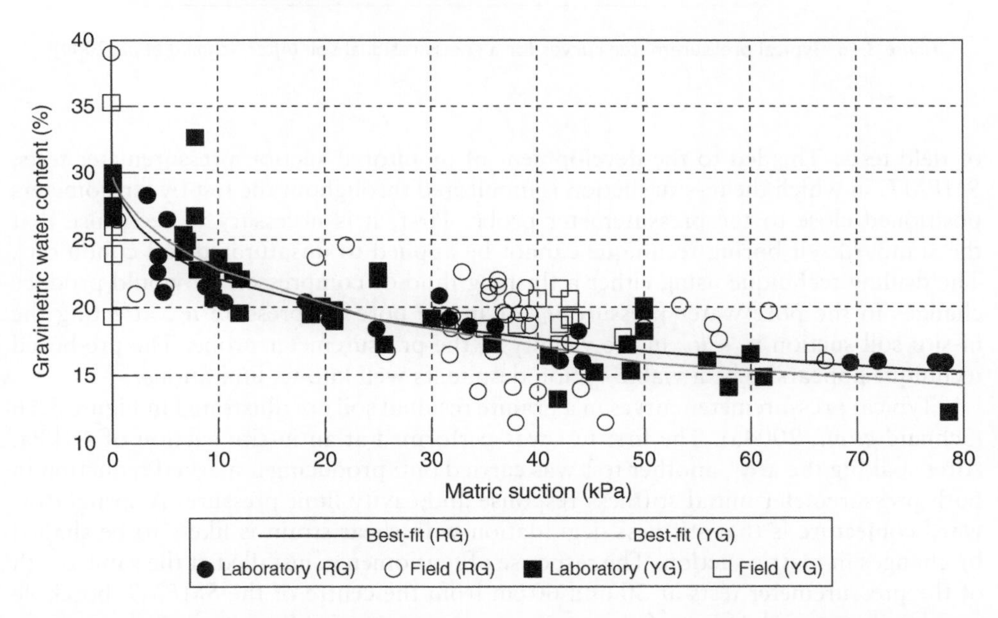

Figure 3.15 Relationship between matric suction and gravimetric water content for two decomposed granites

water characteristic curve and provides vital information concerning the hydraulic and mechanical behaviour of partially saturated soils.

The recognition that matric suction produces an additional component of effective stress suggests the need to link the magnitude of in-situ suction to the observed response

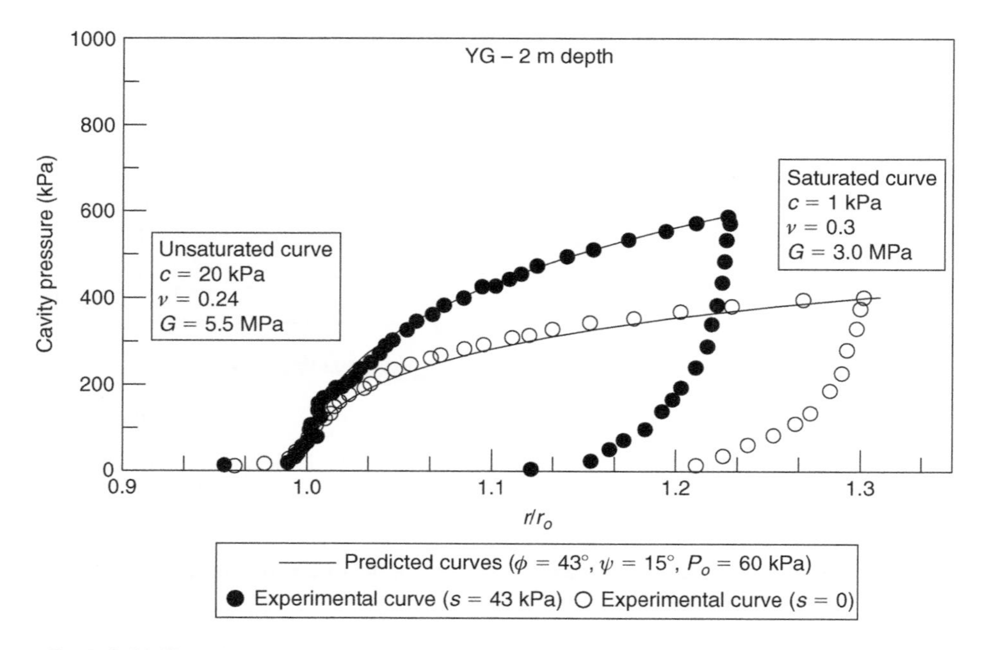

Figure 3.16 Typical pressuremeter curves for a gneiss residual soil (*After* Schnaid *et al.*, 2004)

of field tests. This led to the development of monitored suction pressuremeter tests, *SMPMT*, in which the in-situ suction is monitored throughout the test by tensiometers positioned close to the pressuremeter probe. First, it is necessary to recognise that the standard self-boring technique cannot be applied to unsaturated soil conditions. The drilling technique using either a flushing fluid or compressed air would produce changes in the pore water pressure u_w or in the pore air pressure u_a, affecting the in-situ soil suction $u_a - u_w$ in the vicinity of the pressuremeter probe. The pre-bored technique appears to be a viable option despite its well known limitations.

Typical pressuremeter curves in a granite residual soil are illustrated in Figure 3.16 (Schnaid *et al.*, 2004a). The first test was performed at an in-situ suction of 40 kPa. After soaking the area, another test was carried out; producing a marked reduction in both pressuremeter initial stiffness response and cavity limit pressure. A straightforward conjecture is that stiffness degradation with shear strain is likely to be shaped by changes in matric suction. The response of tensiometers installed at the same depth of the pressuremeter tests at 30 and 60 cm from the centre of the *SMPMT* borehole are also shown in the figure. Suction measurements remained approximately constant throughout the expansion phase in the tests carried out both in soaked and unsaturated soil conditions.

Similar patterns are observed during wetting-induced collapse investigated using both conventional suction-controlled oedometer tests and plate loading tests. Data suggest that shear strains induced by loading do not produce significant changes in matric suction and this enables cavity expansion theory to be extended to accommodate the framework of unsaturated soil behaviour in the interpretation of pressuremeter tests. As a consequence, it is possible to demonstrate that the pressuremeter system is not only suitable for estimating the potential collapse of soils but also for assessing the constitutive parameters that are necessary to describe the 3D-yield surfaces in a (p, q, s) space in unsaturated soils (Schnaid *et al.*, 2004a). The same cavity expansion theoretical background discussed for saturated drained materials remains valid (Schnaid *et al.*, 2004a; Gallipoli *et al.*, 2001).

Schnaid *et al.* (2004a) adapted the theoretical framework from Alonso *et al.* (1990) showing that the mean and deviator stresses on the elasto-plastic boundary becomes

$$p = \frac{P_o}{3}\left(2 + \frac{1}{K_o}\right) - u_a \tag{3.28}$$

and

$$q = P_o\left[\frac{(4K_o^2 - 2K_o + 1)}{K_o^2} - 3\cos\phi'_p\left(\frac{P_o^2\cos\phi'_p - 2P_o c\sin\phi'_p - c^2\cos\phi'_p}{P_o^2}\right)\right]^{1/2} \tag{3.29}$$

where, ϕ'_p is the peak frictional angle, c the cohesion intercept, P_o the in-situ horizontal stress or hydrostatic pressure and K_o the coefficient of earth pressure. The yield function for Cam clay, the yield pressure at the isotropic stress state to each given suction level and critical state slope can be calculated to describe the three-dimensional yield surfaces of the soil in a (p, q, s) space.

Model predictions from field and laboratory tests carried out in a gneiss, unsaturated residual soil site are shown in Figure 3.17 (Schnaid *et al.*, 2004) and are considered to be valuable in reproducing features of behaviour of these unsaturated soils. In this figure two- and three-dimensional plots correlating suction and mean net stress are presented in order to compare the load-collapse (*LC*) curve derived from pressuremeter testing data with the *LC* curve assessed from oedometer data. A discrepancy is observed between these two curves with the oedometer data yielding a collapsible response at much lower mean stress, for any given suction value. There is no reason to assume that these two tests would yield the same *LC* curve since (a) they reproduce rather distinctive stress paths with the oedometer giving rise to vertical displacements in a constrained ring whereas in the pressuremeter displacements are predominantly radial and (b) the oedometer is carried out in small samples whereas the pressuremeter tests a large body of soil that reflects its macro-structure including fissures and discontinuities. However, both oedometer and pressuremeter delivered the same qualitative information by indicating the collapse potential of the soil even at small suctions within the measured range up to about 100 kPa. The pressuremeter produces this evidence at a much lower cost and faster time, which makes it an attractive tool for practical applications.

In addition, a straightforward approach can be used to assess the collapse of residual soils. The term collapse is applied to unsaturated soils that exhibit a drastic rearrangement of particles and great loss of volume upon wetting with or without additional loading (Jennings and Knight, 1957). The collapse is likely to occur in a soil that has an open fabric with large void spaces giving rise to a metastable structure that emerges from a temporary strength due to capillary tensions in the pore water.

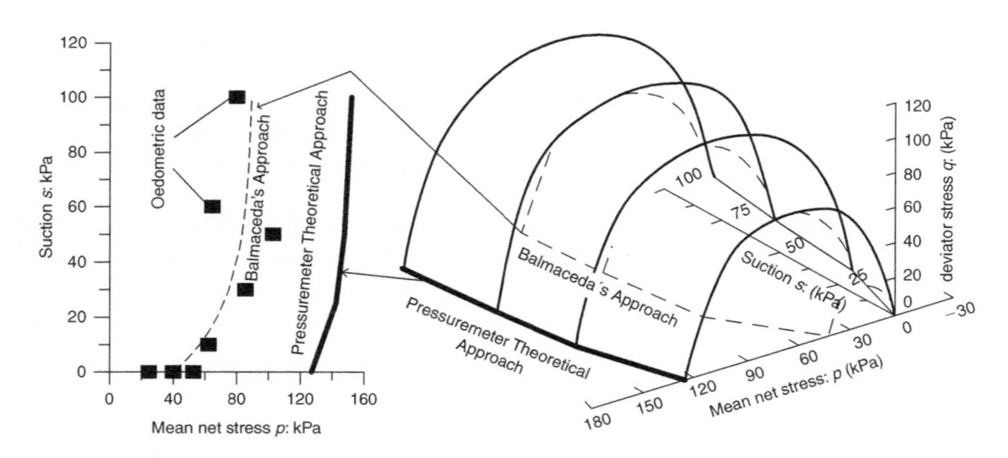

Figure 3.17 Yield surface prediction for the unsaturated granite residual soil (*After* Schnaid *et al.*, 2004a)

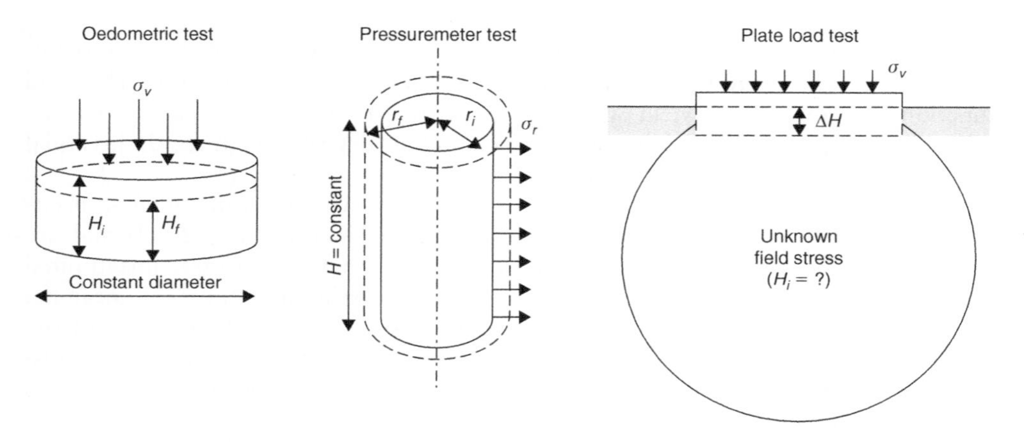

Figure 3.18 Boundary conditions at different collapsibility tests (*After* Schnaid *et al.*, 2003)

Several engineering problems are often associated with the collapse phenomenon, which can cause extensive damage to buildings, embankments, tunnels and many other engineered structures.

Results from both laboratory and in-situ tests in unsaturated soils can be used directly to assess the collapse potential by means of empirical approaches, or they can be used to determine the parameters associated with specific constitutive models. The three main testing techniques are the oedometer, the plate loading and the pressuremeter tests. Boundary conditions of the three tests during a wetting path are identified in Figure 3.18.

Jennings and Knight (1957) proposed the first method to predict the collapse potential using results from the double oedometer test. The method consists of running two oedometer tests, being one sample at constant natural water content and another

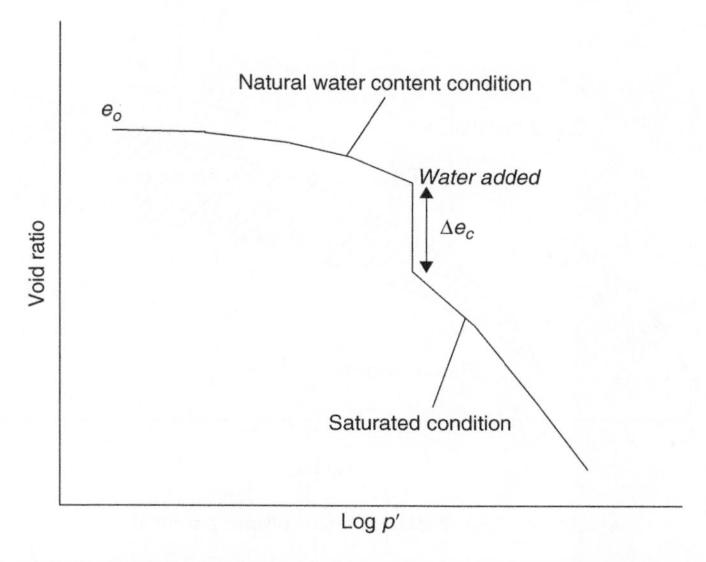

Figure 3.19 Single oedometer collapse test (*After* Jennings and Knight, 1957)

sample at soaked conditions. The collapse potential can be predicted by quantifying the volume decrease difference, at any given stress level, between the two compression curves. The collapse potential (CP) is then defined as

$$CP = \frac{\Delta e}{1 + e_o} \quad (3.30)$$

where, Δe is the difference in void ratio between saturated and natural water content conditions, at any stress level, and e_o is initial void ratio. Since the early stages, the authors recognised the difficulty in evaluating the collapse potential in natural soils based on this technique due to differences in the initial void ratio, e_o, of different laboratory specimens. But it was just in 1975 that Jennings and Knight presented an alternative testing procedure to assess the collapse potential in the laboratory. An ordinary oedometer test at natural water content is conducted at any load level, and then, the sample is flooded with water, left for 24 hours and the test is then carried on to its normal maximum loading limit. The collapse potential is still determined by equation (3.30), but Δe is conveniently replaced by Δe_c that signifies the change in void ratio upon wetting. Figure 3.19 shows an idealised view of this single oedometer collapse test. Later in the 1980s, constant suction oedometer tests became a routine in research enabling a more rational interpretation of test results by combining each oedometer curve to a single suction level.

As an alternative procedure, the collapse of soils can be evaluated from results of in-situ plate loading tests. The test is conducted by applying loading to a rigid circular plate at the base of a borehole; at a given stress level the water is introduced and the load displacement response of the soil is monitored (Ferreira and Lacerda,

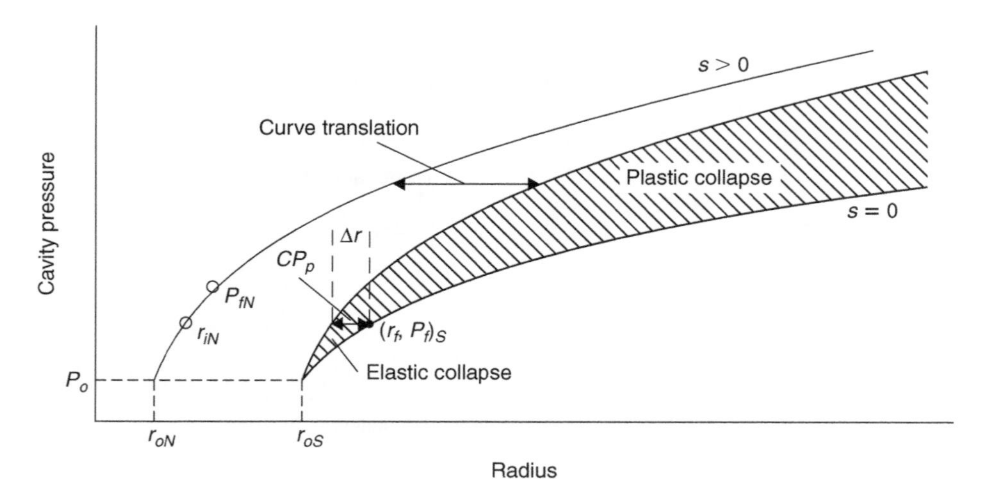

Figure 3.20 Pressuremeter collapse potential

1993; Houston *et al.*, 1995; Mahmoud *et al.*, 1995; Fonseca and Coutinho, 2008). Interpretation would require assessment to the depth of influence of the stress field, the change in stresses due to wetting and the depth of wetting throughout the test. Because evaluation of the stress field requires numerical analysis where coupling of flow and deformation takes place, a more empirical route has been taken.

A final remark is made to highlight the possibility of obtaining some direct and straightforward assessment of the collapse potential from pressuremeter data (Rollins *et al.*, 1994; Smith *et al.*, 1995; Smith and Rollins 1997; Schnaid *et al.*, 2004a; 2009). Different pressuremeter curves obtained both at the saturated and unsaturated states can give a direct assessment of soil collapsibility, being the collapse potential defined as the distance between a saturated and an unsaturated pressuremeter test (at a given in-situ matric suction), taking the yield stress at the saturated state as reference. This definition is similar to that adopted for the interpretation of the double oedometer test (equation (3.30)) and relies on the assumption that once wetted, the collapse sample follows approximately the stress-strain path of the initially saturated sample in compression or in shear. Since the initial cavity radius varies from one test to another, it is necessary to translate the pressuremeter curves for the same initial reference value. Figure 3.20 illustrates the method of empirically interpreting the collapse potential from the pressuremeter (CP_p) which can be calculated as:

$$CP_p = \frac{\Delta V}{V} = \frac{r_{fS}^2 - r_{iN}^2}{r_{iN}^2} - \frac{r_{oS}^2 - r_{oN}^2}{r_{oN}^2} \tag{3.31}$$

where ΔV and V are the volume change and total volume, r_{fS} and r_{iN} the radius at saturated and unsaturated states (taking the yield saturated pressure P_{fS} as reference) and r_{oS} and r_{oN} the initial radius of the saturated and unsaturated states, respectively.

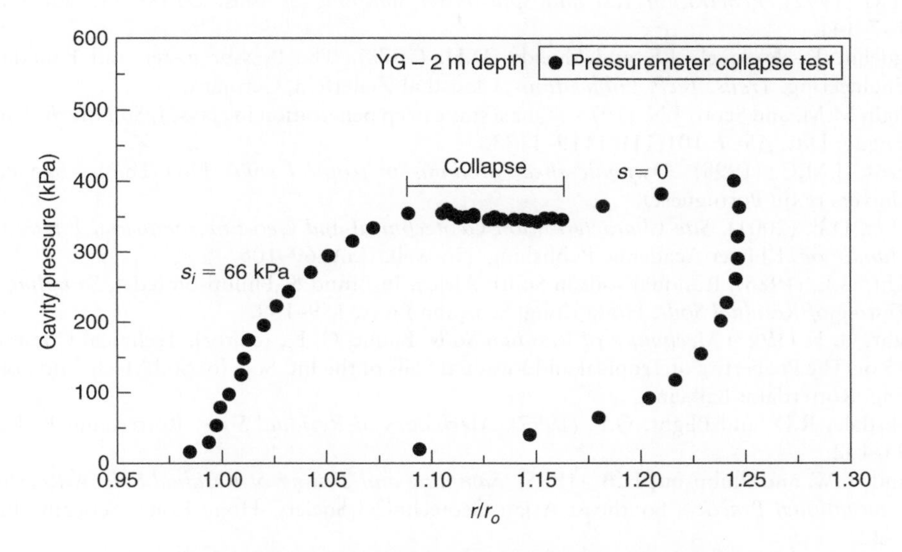

Figure 3.21 Typical collapse pressuremeter test in residual soil

An alternative testing procedure to assess the collapse potential from pressuremeter tests is presented in Figure 3.21, which is in general terms similar to the concepts adopted for the single oedometer test. An ordinary pressuremeter test at natural water content is performed up to a pressure close to the yielding pressure P_f, and then the soil is flooded with water until suction measurements at the tensiometers around the probe drops to values close to zero; the test is then carried on to its normal maximum loading limit. Observation of the increasing radial displacements at a constant radial pressure that results from reducing suction gives a direct assessment of the collapse potential of soils.

3.5 SUMMARY AND CONCLUSIONS

This chapter has briefly summarised the main laboratory and in-situ techniques, procedures and interpretation methods applied to residual soils. The unusual stress-strain-strength behaviour that deviates from classical textbook materials has been highlighted and has prompted the development of interpretation methods required in the detailed design of saturated and unsaturated soil conditions, bounded and structured organisation of particles and metastable collapsible response, among other characteristic features.

REFERENCES

Alonso, E.E., Gens, A. and Josa, A. (1990). A constitutive model for partially saturated soils. *Géotechnique*, 40(3): 405–430.

ASTM (1992). *Penetration Test and Split-barrel Sampling of Soils*. D1586-84, Vol. 04.08, 137–141.

Baguelin, F., Jezequel, J.F. and Shields, D.H. (1978). The Pressuremeter and Foundation Engineering. *Trans. Tech. Publications*, Clausthal-Zellerfeld, Germany.

Baligh, M.M. and Scott, J.N. (1975). Quasi static deep penetration in clays. *J. Soil Mech. Found. Engng. Div., ASCE* 101(11): 1119–1133.

Barros, J.M.C. (1998). *Dynamic shear modulus in tropical soils*. PhD Thesis, São Paulo University (in Portuguese).

Becker, D.E. (2001). *Site Characterisation Geotechnical and Geo-Environmental Engineering Handbook*. Kluwer Academic Publishing. Norwell, USA, 69–105.

Blight, G.E. (1985). Residual soils in South Africa. In Brand & Phillipson (eds), *Sampling and Testing of Residual Soils*. Hong Kong: Scorpion Press, 159–168.

Blight, G. E. (1997) *Mechanics of Residual Soils*. Blight, G. E., (editor). Technical Committee 25 on The Properties of Tropical and Residual Soils of the Int. Soc. for Soil Mech. and Found. Eng. Rotterdam: Balkema.

Barksdale, R.D. and Blight, G.E. (1997). *Mechanics of Residual Soils*. Rotterdam: Balkema, 95–152.

Brand, E.W. and Philipson, H.B. (1985). *Sampling and Testing of Residual Soil. A Review of International Practice*. Southeast Asian Geotechnical Society. Hong Kong: Scorpion Press. 7–22.

Brenner, R.P., Garga, V.K. and Blight, G.E. (1997). Shear strength behavior and the measurement of shear strength in residual soils. In Blight (eds) *Mechanics of Residual Soils*. Rotterdam: Balkema, 155–217.

Briaud, J.L. (1992). *The Pressuremeter*. Rotterdam: Balkema.

BSI, (1990). BS1377. *Soil for Civil Engineering Purposes*. British Standards Institution.

Chandler, R.J., Crilly, M.S. and Montgomery-Smith, G. (1992). A low-cost method of assessing clay dessication for low-rise buildings. *Civil Eng., Proc. ICE*: 92(2): 82–89.

Clarke, B.G. (1993). *Pressuremeter in Geotechnical Design*. Chapman & Hall, London, UK.

Clayton, C.R.I. (1993). *The Standard Penetration Test (SPT) – Methods and Use*. Construction Industry Research and Information Association, Funder Report/ CP/7, CIRIA, London.

Decourt, L., Muromachi, T., Nixon, I.K., Schmertmann, J.H., Thorbum, C.S. and Zolkov, E. (1988). Standard Penetration Test (SPT): International reference test procedure. *Proc. Int. Symp. on Penetration Testing*, ISOPT-1, USA, 1: 37–40.

Eslaamizaad, S. and Robertson, P.K. (1997). A Framework for In-situ determination of Sand Compressibility. *49th Can. Geotech. Conf.*; St John's Newfoundland.

Ferreira, S.R.M., and Lacerda, W.A. (1993). Soil collapsibility from laboratory and field tests. *Soils Rocks*, 16(4): 245–253 (in Portuguese).

Finno, R.J. (1993). Analytical interpretation of dilatometer penetration through saturated cohesive soils. *Géotechnique*, 43(2): 241–254.

Fonseca, A.V. and Coutinho, R.Q. (2008). Characterization of residual soils. *Proc. of the Third Int. Conf. on Site Charact.* – ISC'3, Taipei, Taiwan. Taylor & Francis Group, London, 195–249.

Fookes, P.G. (1997) (Ed). *Tropical residual soils*. A Geological Society engineering group working party revised report. The Geological Society. London, UK.

Fredlund, D.G. (1987). Slope stability analysis incorporating the effect of soil suction. In Anderson & Kemp (eds). *Slope Stability*. John Wiley, Chichester, UK.

Fredlund, D.G. and Xing, A. (1994). Equations for the soil-water characteristic curve. *Can. Geotech. J.*, 31(4): 521–532.

Gallipoli, D., Karstunen, M. and Wheeler, S.J. (2001). Numerical modelling of pressuremeter tests in unsaturated soil. *Proc. 10th Int. Conf. on Computer Methods and Advances in Geomech.*, Tucson, Arizona, USA: Balkema, 807–812.

Garga, V.K. and Blight, G.E. (1997). Permeability. In Blight (ed.) *Mechanics of Residual Soils*. ISSMFE (TC 25): Rotterdam: Balkema, 79–93.

Gasmo, J., Hritzuk, K.J., Rahardjo, H. and Leong, E.C. (1999). Instrumentation of an unsaturated residual soil slope. *Geotechnical Testing Journal, GTJODJ*. 22(2): 128–137.

Geoguide 3. (1996). *Tropical Weathered in situ Materials – Laboratory Testing*. JKR (Public Works Department, Malaysia).

Gomes, C., A., Viana da Fonseca, A. and Gambim, M. (2004). Routine and advanced analysis of mechanical in situ tests. Keynote Lecture. *2nd Int. Conf. Geotech. Site Characterization*, Millpress, Amsterdam, The Netherlands, 1: 75–96.

Gravesen, S. (1960). *Elastic Semi-Infinite Medium Bounded by a Rigid Wall with a Circular Hole*. Laboratoriet for Bygninsteknik, Danmarks Tekniske Højskole, Meddelelse No. 10, Copenhagen.

Hatanaka, M. and Uchida, A. (1996). Empirical correlation between penetration resistance and effective friction angle for sandy soils. *Soils Found*., 36(4): 1–9.

Head, K.H. (1992). *Manual of Soil Laboratory Testing*. Volume 1: Soil classification and compaction tests. London: Pentech Press.

Houston, S.L., Mahmound, H.H. and Houston, W.N. (1995). Down hole collapse test system. *J. Geotech. Eng*., 121(4): 341–349.

Ishihara, K. (1982). Evaluation of soil properties for use in earthquake response analysis. *Int. Symp. Num. Models in Geomech*., Zurich, 237–259.

Jennings, J.E. and Knight, K. (1957). The additional settlement of foundations due to a collapse of structure of sandy subsoils on wetting. *Proc., 4th Int. Conf. on Soil Mechanics and Foundation Engineering*, Butterworths Scientific, London, 3, 316–319.

Jones, D.L. and Rust, E. (1989). Foundations on residual soil using a pressuremeter moduli. *12th Int. Conf. Soil Mech. Found. Engng*., Rio de Janeiro, 1: 519–524.

Josa, A., Balmaceda, A., Gens, A. and Alonso, E.E. (1992). An elastoplastic model for partially saturated soils exhibiting a maximum of collapse. *Proc. 3rd Int. Conf. Comput Plasticity*, Barcelona, 1:815–826.

Lunne, T., Lacasse, S. and Rad, N.S. (1989). SPT, CPT, pressuremeter testing and recent developments on in-situ testing of soils. *12th Int. Conf. Soil Mech. Found. Engng*., Rio de Janeiro, Brazil.

Lunne, T., Robertson, P.K. and Powell, J.J.M. (1997). *Cone Penetration Testing in Geotechnical Practice*. Blackie Academic & Professional.

Lutenegger, A.J. (1988). Current status of the Marchetti dilatometer test. *Proc. 1st Int. Symp. on Penetration Testing, ISOPT-1*, Orlando, Florida 1: 137–155.

Mahmound, H.H., Houston, W.N. and Houston, S.L. (1995). Apparatus and procedure for an *in situ* collapse test. *Geotech. Test. J*., 18(4): 431–440.

Mair, R.J. and Wood, D.E. (1987). *Pressuremeter Testing: Methods and Interpretation*. CIRIA Report, Buterworths, UK.

Mantaras, F.M. and Schnaid, F. (2002). Cavity expansion in dilatant cohesive-frictional soils. *Geotechnique*, 52(5), 337–348.

Marchetti, S. (1997). The flat dilatometer: design applications. Keynote Lecture. *Proc. 3rd Int. Geotech. Engng Conf*., Cairo, 421–448.

Marchetti, S., Monaco, P., Totani, G. and Calabrese, M. (2001). The flat dilatometer test (DMT) in soil investigation. *Report by the ISSMGE Committee TC 16. Proc. Int. Conf. on In Situ Measurement of Soil Properties*. Baligh, 95: 132.

Marinho, F.A.M. (2000). Discussion – The use of miniature pressure transducer in measuring matric suction in unsaturated soils. ASTM. *Geotechnical Testing Journal*, USA, v. 23, n. 12, pp. 532–534.

Mayne, P.W. (2006). Undisturbed sand strength from seismic cone tests. The 2nd James K. Mitchell Lecture. *J. Geomech. & Geoengng*., 1(4): 239–258.

Mutaya, P. and Huat, B.B.K. (1993). Effect of pretest treatment on engineering properties of local residual soils. *Journal, Institution of Engineers Malaysia* 55: 29–41.

Nicholls, R.A. (1990). Sampling techniques in soft ground and residual soils. *Proc. of seminar on geotechnical aspects of the north – south expressway. Kuala Lumpur* : 1–7.

Novais Ferreira, H. (1985). Characterisation, identification and classification of tropical lateritic and saprolitic soils for geotechnical purposes. *General Report, Int. Conf. Geomech. in Tropical Lateritic and Saprolitic Soils*, Brasília, 3: 139–170.

Odebrecht, E., Schnaid, F., Rocha, M.M. and Bernardes, G.P. (2004). Energy measurements for standard penetration tests and the effects of the length of rods. *2nd Int. Conf. on Site Charact.*, Milpress, Porto, 1: 351–358.

Poulos, H.G. and Davis, E.H. (1980). *Pile Foundation Analysis and Design*. Wiley, USA.

Richards, B.G. (1985). Geotechnical aspects of residual soils in Australia. In Brand & Phillipson (eds). *Sampling and Testing of Residual Soils*. Hong Kong: Scorpion Press, 23–30.

Richard, F.E. Jr., Hall, J.R. and Woods, R.B. (1970). *Vibrations of Soils and Foundations*. Englewood Cliffs, New Jersey, Prentice Hall.

Ridley, A.M. and Burland, J.B. (1995). A pore water pressure probe for the in situ measurement of a wide range of soil suctions. *Proc. 1st Int. Conf. on Advances in Site Investigation Practice*, ICE, Thomas Telford, London, 510–520.

Rodrigues, C.M.G. and Lemos, L.J.L. (2004). SPT, CPT and CH tests results on saprolitic soils from Guarda, Portugal. *Proc. of the 2nd Int. Conf. on Site Charact.*, Milpress, Porto, 2: 1345–1352.

Rollins, K.M., Rollins, R.L., Smith, T.D. and Beckwith, G.H. (1994). Identification and characterization of collapsible gravels. *J. Geotech. Eng.*, 120(3): 528–542.

Rowe, P.W. (1962). The stress-dilatancy relation for static equilibrium of an assembly of particles in contact. *Proc. Royal Soc.*, A 269, 500–527.

Sandroni, S.S. (1991). Young metamorphic residual soils. *9th Panamerican Conf. Soil Mech. Found, Engng.*, Viña del Mar, Argentina.

Santamarina, J.C., Klein, K.A. and Fam, M.A. (2001). *Soils and waves*. John Wiley & Sons, Ontario, Canada.

Shibuya, S., Yamashita, S., Watabe, Y. and Lo Presti, D.C.F. (2004). In situ seismic servey in characterising engineering properties of natural ground. *Proc. 2nd Int. Conf. on Site Charact.*, Milpress, Porto, 1: 167–185.

Schnaid, F. (1997). Panel Discussion: Evaluation of in situ tests in cohesive frictional materials. *Proc. 14th Int. Conf. Soil Mech. Found. Engng*, Hamburg, 4: 2189–2190.

Schnaid, F. (1999). On the interpretation of in situ tests in unusual soil conditions. *Pre-failure Deformation Charac. of Geomaterial*, Torino, Italy, Balkema, 2: 1339–1348.

Schnaid, F. (2005). Geocharacterisation and properties of natural soils by *in situ* tests. State-of-the-Art Report. *Proc. 16th International Conference on Soil Mechanics and Geotechnical Engineering (ICSMGE)*, Osaka, 1: 3–46.

Schnaid, F. (2009). *In Situ Testing in Geomechanics*. Taylor & Francis, London.

Schnaid, F. and Mantaras, F.M. (2003). Cavity expansion in cemented materials: structure degradation effects. *Géotechnique*, 53(9): 797–807.

Schnaid, F. and Mantaras, F.M. (2004). Interpretation of pressuremeter tests in a gneiss residual soil from Sao Paulo, Brazil. *Proc. 2nd Int. Conf. on Site Charact.*, Milpress, Porto, 2: 1353–1360.

Schnaid, F. and Yu, H.S. (2007). Theoretical interpretation of the seismic cone test in granular soils. *Géotechnique*, 57(3), 265–272.

Schnaid, F., Oliveira, L.A.K. and Gehling, W.Y.Y. (2004a). Unsaturated constitutive surfaces from pressuremeter Tests. *Journal of Geotechnical and Geoenvironmental Engineering*, 130(2): 1–12.

Schnaid, F., Lehane, B.M. and Fahey. M. (2004b). *In situ* test characterisation of unusual geomaterials. *Proc. 2nd Int. Conf. on Site Charact.*, Milpress, Porto, 1: 49–74.

Schnaid, F., Odebrecht, E. and Rocha, M. (2007a). On the mechanics of dynamic penetration tests. *Geomechanics and Geoengineering: an Int. J.*, 3: 12–16.

Schnaid, F., Odebrecht, E., Rocha, M.M. and Bernandes, G.P. (2007b). Prediction of soil properties from the concepts of energy transfer in dynamic penetration tests. *J. Geotech. and Geoenvironm. Engng, ASCE.*

Schnaid, F., Odebrecht, E., Rocha, M.M. and Bernardes, G. de P. (2009). Prediction of Soil Properties from the Concepts of Energy Transfer in Dynamic Penetration Tests. *Journal of Geotechnical and Geoenvironmental Engineering*, v. 135, pp. 1092–1100.

Sharma, P.V. (1997). *Environmental and Engineering Geophysics*. Cambridge University Press, Cambridge, UK.

Skempton, A.W. (1986). Standard penetration test procedures and effects in sands of overburden pressure, relative density, particle size, aging and over consolidation. *Géotechnique*, 36(3), 425–447.

Smith, M.G. and Houlsby, G.T. (1995). Interpretation of the Marchetti dilatometer in clay. *Proc. 11th Int. Conf. Soil Mech. and Found. Engng.*, 1: 247–252.

Smith, T.D. and Rollins, K.M. (1997). Pressuremeter testing in arid collapsible soils. *Geotech. Test. J.*, 20(1): 12–16.

Smith, T.D., Duquette, J. and Deal, C. (1995). A database of moisture induced soil collapse from the pressuremeter. *Proc. 4th Int. Symp. on Pressuremeters*, Sherbrooke, Quebec, Canada, 81–85.

Souza, Neto, J.B., Coutinho, R.Q. and Lacerda, W.A. (2005). Evaluation of the collapsibility of a sandy soil by in situ collapse tests. *Proc. XVI International Conference on Soil Mechanics and Geotechnical Engineering*, Osaka.

Stokoe II, K.H., Joh, S.H. and Woods, R.D. (2004). The contributions of in situ geophysical measurements to solving geotechnical engineering problems. *Proc. 2nd Int. Conf. on Site Charact.*, Milpress, Porto, 1: 97–132.

Tatsuoka, F. and Shibuya, S. (1991). Deformation characteristics of soils and rocks from field and laboratory tests. *Proc. 9th Asian Regional Conf. on Soil Mech. and Found. Engng.*, Bangkok, 2: 101–170.

Tatsuoka, F., Jardine, R.J., Lo Presti, D., Di Benedetto, H. and Kodaka, T. (1997). Theme Lecture: Characterising the pre-failure deformation properties of geomaterials. *Proc. 14th Int. Conf. Soil Mech. Found. Engng*, Hamburg, 4: 2129–2164.

Vargas, M. (1974). Engineering properties of residual soils from southern-central region of Brazl. *Proc. 2nd Int. Cong. IAEG*, São Paulo, 1:5.1–5.26.

Vaughan, P.R. (1985). Mechanical and hydraulic properties of tropical lateritic and saprolitic soil, General Report, *Proc. Int. Conf. Geomech. in Tropical Lateritic and Saprolitic Soils*, Brasilia, 3: 231–263.

Vaughan, P.R. (1997). Engineering behaviour of weak rock: Some answers and some questions. *Proc. 1st Int. Conf. on Hard Soils and Soft Rock*s, Athens, Vol. 3: 1741–1765.

Viana da Fonseca, A. (1996). *Geomechanics of residual soils from Porto Granite: design criteria for shallow foundations*. PhD Thesis, Porto University, Porto (in Portuguese).

Wheeler, S.J. and Sivakumar, V. (1995). An elasto-plastic critical state framework for unsaturated soil. *Géotechnique*, 45(1):35–53.

Yu, H.S. (2000). *Cavity Expansion Methods in Geomechanics*. Kluwer Academic Publishers, Dordrecht, The Netherlands.

Yu, H.S. (2004). The James K. Mitchell Lecture: In situ testing: from mechanics to prediction. *Proc. 2nd Int. Conf. on Site Charact.*, Milpress, Porto, 1: 3–38.

Yu, H.S., Carter, J.P. and Booker, J.R. (1993). Analysis of the dilatometer test in undrained clay. *Predictive Soil Mech.*, London, 783–795.

Yu, H.S. and Mitchell, J.K. (1998). Analysis of cone resistance: a brief review of methods. *J. Geotech. and Geoenvironm. Engng., ASCE*, 124(2):140–149.

4

The behaviour of unsaturated soil

David G. Toll
Durham University, Durham, United Kingdom

This chapter provides an overview of the important aspects of unsaturated soil behaviour that are of relevance for tropical residual soils. Many tropical soils exist in an unsaturated state due to soil water deficits induced by the tropical climate. The concepts of suction are introduced and how suction can be measured. The role of suction in determining shear strength is described, together with test results, showing the variations of ϕ^a and ϕ^b for two tropical soils. The volume change behaviour of unsaturated soils (shrinkage, swelling and collapse) is also discussed as this can be of relevance to tropical clay soils.

4.1 INTRODUCTION

Unsaturated soils are those where the pores are filled with both water and gas (usually air), unlike saturated soils where the voids are filled entirely with water. Many tropical soils exist in an unsaturated state due to soil water deficits induced by the tropical climate. While many tropical regions can have high rainfall, this can be offset by even greater evaporation and transpiration which removes water from the soil. For this scenario, the ground water table can be at significant depths (perhaps greater than 10 m) which means that the zone of soil involved in engineering and construction operations will be above the water table and potentially unsaturated. This unsaturated zone above the water table is known as the Vadose zone.

In unsaturated conditions, the water phase is held in the soil by a negative pressure (or suction). The effect of suction is very important in understanding how the soil will behave in an engineering context. Suction affects the shear behaviour and also controls volume changes in response to wetting and drying. The fact that a soil is unsaturated also has a significant effect on the water permeability (hydraulic conductivity).

4.2 SUCTION

4.2.1 Components of suction

Soil suction is made up of two components: *matric suction* and *osmotic suction* (also called *solute suction*). The sum is known as the *total suction*. Matric suction is due to surface tension forces at the interfaces (menisci) between the water and the gas (usually air) phases present in unsaturated soils (the surface tension effect is sometimes referred to as capillarity). Osmotic suctions are due to the presence of dissolved salts within the pore water.

Total suction is defined as the negative pressure which must be applied to a pool of *pure* water at the same elevation and temperature in order for it to be in equilibrium with the soil water (Review Panel Statement, 1965). This negative pressure is needed to balance the suction forces acting within the soil due to capillarity (matric suction) and the suction induced by different concentration of salts in the pore water in the soil

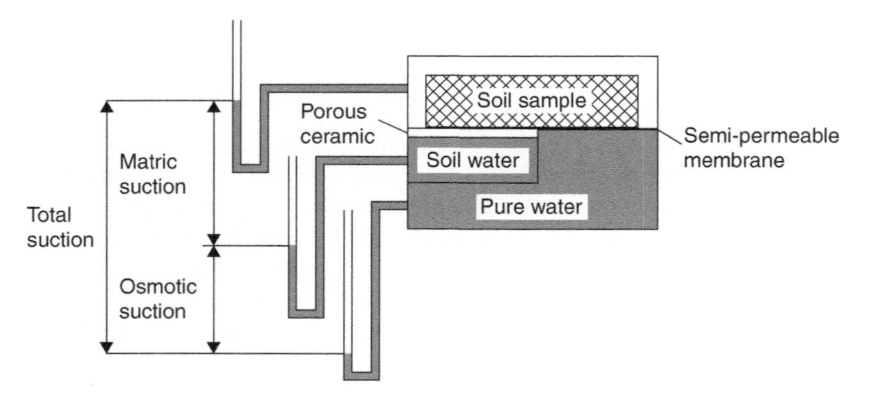

Figure 4.1 Components of Total Suction: Matric and Osmotic Suction

and the pure water outside (osmotic suction). Water will try to flow from pure water to soil water in order to equalise the differences in salt concentration, thus trying to draw water into the soil and inducing suction.

If the soil was put in contact with a pool of water identical in composition to the soil water then the negative pressure needed to achieve equilibrium would be equal to the matric suction. The difference between total suction and matric suction is made up by the osmotic suction.

Imagine a soil sample that is separated from water by a porous ceramic (which allows salts to pass through freely) (Figure 4.1). In this case, salts will move through the ceramic until equilibrium is established between the salt concentrations in the sample and the water beneath the ceramic. The pressure difference across the ceramic will then be equal to the matric suction, since the water on the other side of the ceramic will be of the same composition as the soil water and there will be no osmotic suction component.

If the water beneath the porous ceramic is separated from pure water by a semi-permeable membrane (which allows water molecules to pass through, but prevents the passage of salts), the difference in pressure across the membrane will be equal to the osmotic suction (Figure 4.1). Therefore, where the soil sample is separated from pure water by the semi-permeable membrane, the pressure difference across the membrane between the soil and pure water will be equal to the total suction as both and osmotic suction components are present.

4.2.2 Axis translation

Matric suction is controlled by the pressure difference across the water/air interface $(u_a - u_w)$, where u_a is the pore air pressure and u_w is the pore water pressure. Generally, in the field, the pore air pressure (u_a) will be at atmospheric pressure $(u_a = 0)$ so the suction is defined by the negative pore water pressure $(-u_w)$. However, highly negative pore water pressures can be difficult to measure directly, as cavitation can occur in conventional measuring systems if the water pressures become too negative (below -100 kPa). Cavitation is the process by which bubbles form when water is

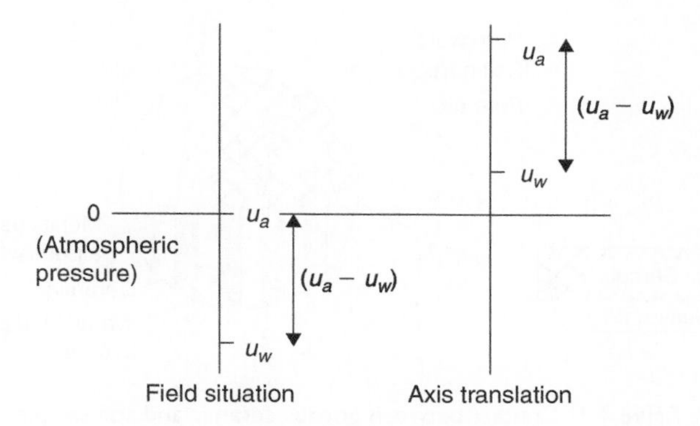

Figure 4.2 The concept of axis translation

subjected to tensile stresses. Therefore, suctions are often measured (or controlled) in the laboratory by elevating the pore air pressure, u_a, within the sample, a technique known as 'axis translation' (See Figure 4.2).

It has been shown by Hilf (1956) that the suction $(u_a - u_w)$ remains unchanged by shifting the axis of pressure in this way. The advantage of using an elevated air pressure is that the pore water pressure can be made to become positive, and so cavitation is prevented. It can then be measured using high air entry discs (ceramics or membranes).

High air entry discs can sustain a pressure difference between air pressure on one side of the disk and water pressure on the other side, when they are saturated. This is because the pores in the disc are fine enough to allow menisci to form that prevent air entry (Figure 4.3). By placing a sample on the high air entry disc, where the pore water in the sample is in contact with the water within the disc, allows the water pressures to be measured or controlled independently of the applied air pressure.

Using the axis translation technique, suction can be controlled or measured by applying a known air pressure to the sample. If the water reservoir below the high air entry disc is closed then the water pressure can be measured and the suction is given by the difference between the air pressure and water pressure $(u_a - u_w)$. Alternatively, the water pressure below the high air entry ceramic can be maintained at a known value and the water pressure in the sample can be allowed to equilibrate with the water pressure below the plate. In this way, a known value of suction is imposed in the sample. Contact is required between the pore water within the soil and the water below the plate to ensure equalisation (Figure 4.3).

4.2.3 The suction scale

In much of the soil science literature, suction is expressed in pF units i.e. the logarithm (to base 10) of the suction expressed in centimetres of water (Schofield, 1935). For engineering applications, it is generally more convenient to use conventional stress units. The relationship to convert from pF units to kPa is given by:

$$\text{Suction (kPa)} = 9.81 \times 10^{[\text{pF}-2]} \qquad (4.1)$$

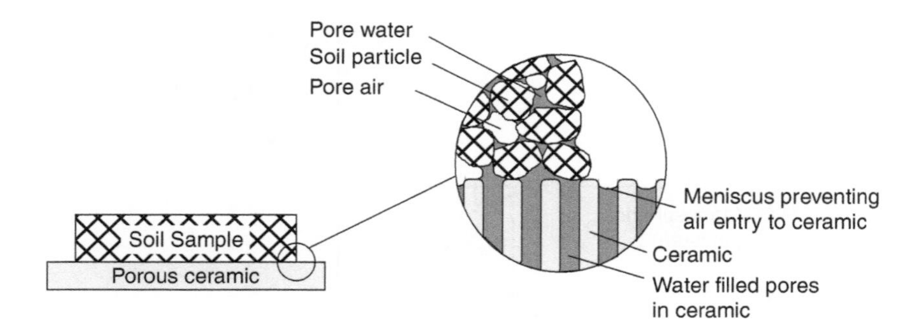

Figure 4.3 Contact between porous ceramic and soil sample

Table 4.1 The Suction Scale

Suction (kPa)	*Suction (pF)*	*Reference points*	*Moisture Condition*
1,000,000	7	Oven dry	Dry
100,000	6		
10,000	5		
1,000	4	Wilting point for plants Plastic Limit	Moist
100	3		
10	2		
1	1	Liquid Limit	Wet
0.1	0	Saturated	

The suction scale (showing both kPa and pF units) with some points of reference and indications of the moisture condition of a soil are shown in Table 4.1.

4.2.4 Limiting suctions

The maximum suction that can be sustained within the pores of a soil will depend on the pore size. If we consider the pores to be like capillary tubes, we can relate the maximum suction to the diameter of a pore using a simple capillary model. From capillary theory the pressure difference across a gas/water interface (assuming a hemispherical meniscus with a contact angle of zero) is given by:

$$u_a - u_w = \frac{2T}{R} \tag{4.2}$$

where T is the surface tension ($0.073\,\text{Nm}^{-1}$ for water) and R is the radius of the meniscus.

A meniscus requiring a radius less than the minimum pore radius cannot be sustained. Therefore the maximum sustainable suction (or pressure difference, $u_a - u_w$) will be reached when the radius of the meniscus equals the minimum radius of the pore.

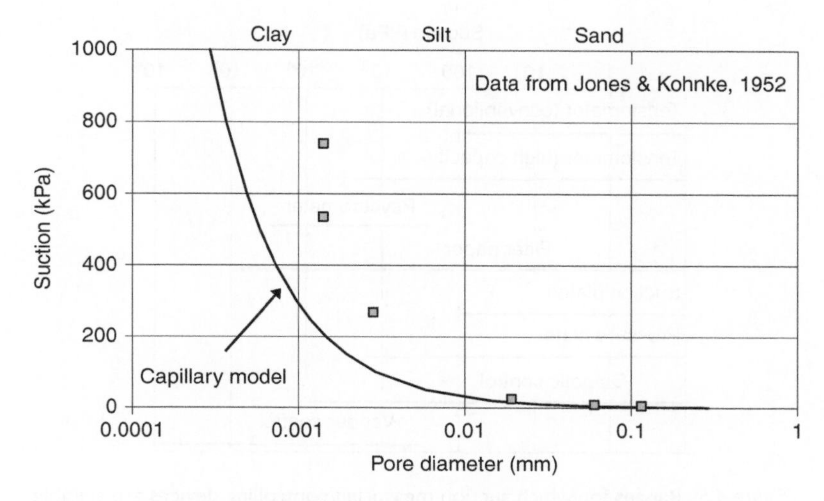

Figure 4.4 Limiting suction as a function of pore size

This limiting value of suction for the case of a soil having uniformly shaped pores is shown as a line in Figure 4.4. Experimental observations by Jones and Kohnke (1952), also shown in the figure, suggests that the simple capillary model underestimates the limiting suctions, as pores within a real soil will not be smooth and tubular but will vary in diameter and be tortuous in nature. Nevertheless, it is clear from Figure 4.4 that limiting suction increases with decreasing pore size.

This means that in a clean sandy soil, where pore sizes will be of the order of 0.1 mm or larger, that the maximum suctions will be very small (typically <5 kPa). In clean silty materials, where pore sizes might be of the order of 0.01 mm, the maximum suctions are likely to be less than 100 kPa. However, in clayey soils, where pore sizes can be less than 0.001 mm, then high suctions greater than 1000 kPa can be sustained. This explains why clean sandy soils have no strength when they dry out (they lose cohesion) as they cannot sustain high suctions. However, clayey soils can become very strong when they dry out due to the high suctions that are maintained in the fine pores of the soil. The suctions pull the soil particles together and give the soil considerable cohesive strength in a dry state.

4.2.5 Suction measurement

There are a number of different techniques for suction measurement and control. Their suitability varies according to the range of suctions operating. An indication of appropriate ranges is shown in Figure 4.5. It is generally necessary to use a variety of techniques in order to cover the entire suction scale.

In the laboratory, direct measurement techniques (tensiometers) can be used, and high capacity devices now allow measurements up to about 2 MPa. Psychrometers can be used for a higher range of suction (200 kPa to 20 MPa). Filter paper methods

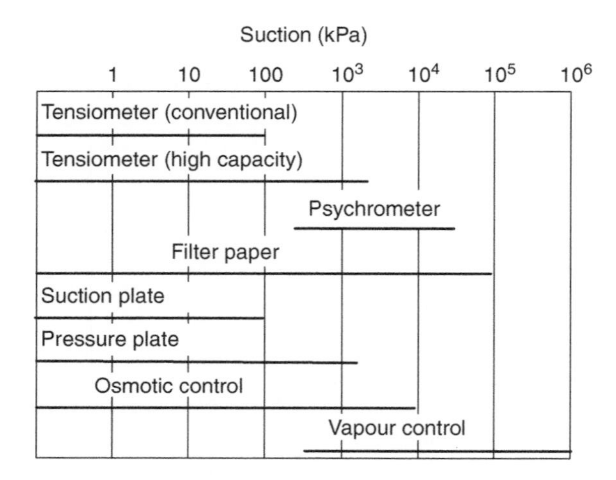

Figure 4.5 Ranges for which suction measuring/controlling devices are suitable

can also be used to give an indication of suction in the range 1 kPa to 100 MPa. The technique can be used to obtain both total and matrix suctions.

Pressure plate techniques can be used to control suctions up to 1.5 MPa (based on the axis translation technique) depending on the air entry value of the ceramic used (some devices are limited to below 500 kPa). Suction plates can be used to control low suctions (<100 kPa) but have limited usefulness because of the narrow range. Osmotic control can be used up to 10 MPa. For the higher suction range, vapour control can be used for imposing suctions from 400 kPa to 400 MPa (or even higher).

Pressure membrane techniques also use axis translation, but use a semi-permeable membrane in place of the high air entry porous disk. This can extend the range to 150 MPa (Croney *et al.*, 1958), although the range can be limited by the strength of the cell (not the air entry limit of the membrane) to about 3 MPa. Although pressure membrane devices were used successfully in the early days in the development of unsaturated soil testing, they have fallen out of favour and are not usually found in modern laboratories and hence will not be covered in the following.

Osmotic control techniques use a semi-permeable membrane with a polyethylene glycol (PEG) solution. The value of suction is controlled by the concentration of the PEG solution. The semi-permeable membranes used are permeable to salts and so allow control of matric suction.

It should be noted that suction plate, pressure plate and osmotic control techniques allow the control of matric suction. On the other hand, the vapour control techniques provide control of total suction. The chilled mirror dew-point technique is another alternative for total suction measurement in the laboratory (Leong *et al.*, 2003).

In the field, tensiometers can be used for direct measurement, and high capacity devices allow measurements to 2 MPa. Above this psychrometers can be used, but are highly sensitive to temperature variations. Porous block sensors are available that use indirect methods to determine suction, but the response time can be reduced compared to direct measurements.

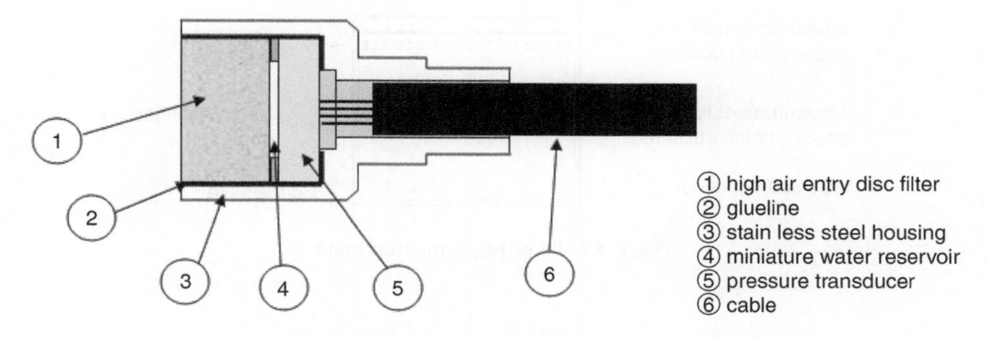

Figure 4.6 High Capacity Tensiometer (*After* Lourenço *et al.*, 2006)

High capacity tensiometers

High capacity tensiometers can exceed the conventional cavitation limit of 100 kPa (negative pressure) by designing the water reservoir to have a small volume (Figure 4.6) and by ensuring a high standard of saturation by pre-pressurisation (between 1–4 MPa). The first device was reported by Ridley and Burland (1993) who demonstrated a cavitation pressure of 1250 kPa. Further devices are now available with higher cavitation pressures and Tarantino and Mongiovi (2001) and Lourenço *et al.* (2006) have reported cavitation limits in excess of 2 MPa. The use of such devices in the field is described by Toll *et al.* (2011).

Filter paper

Filter paper is a highly controlled material that has a closely defined suction-water content relationship. Therefore, by allowing suction to equalise between the soil and a piece of filter paper, the water content of the filter paper can be determined and hence related to suction. If the filter paper is placed in contact with the soil, the equalisation can take place through liquid water flow, and hence matric suction is measured. If the filter paper is placed inside a sealed container with the soil, but not in contact, then equalisation takes place by vapour transfer and hence total suction is measured.

Figure 4.7 shows a typical setup for measuring suction using filter paper. The soil sample is cut in two parts and three sheets of dry filter paper are sandwiched between them and pressed together to ensure good contact. The two filter papers in immediate contact with the soil are sacrificial and are used to protect the inner measurement filter paper, to avoid soil particles sticking to that filter paper and inducing errors in its weight. A further dry filter paper is placed on top of the sample, separated by a mesh to prevent direct contact. The arrangement is sealed and left to achieve equilibrium (usually between 7 and 14 days).

The measurement filter papers are weighed (using a balance with an accuracy of 0.0001 g) to determine the filter paper water content. Calibrations are then applied to determine the suction. A set of widely used calibrations are reported by Leong *et al.* (2002). Separate curves are defined for matric and total suction (for suctions below 1000 kPa). This is because the normal equilibration time (even 14 days) is insufficient

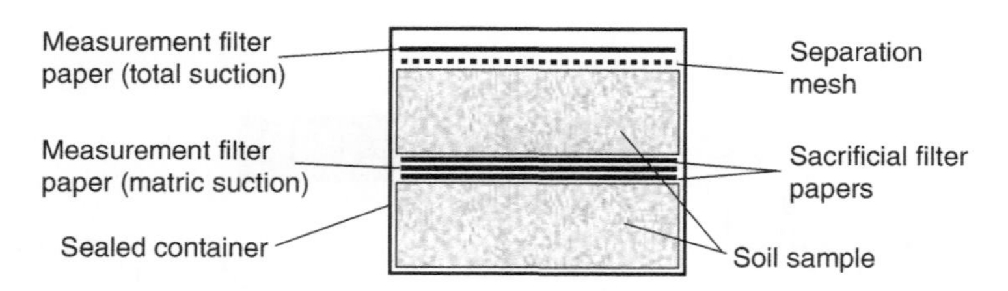

Figure 4.7 Filter paper measurement

to ensure full equalisation due to vapour transfer, so a separate calibration curve is used for total suction.

Psychrometer

A psychrometer is a device for measuring relative humidity. This is done by measuring the difference in temperature between wet and dry junctions. There are two main forms of psychrometer, the thermocouple psychrometer and the transistor psychrometer. With the thermocouple psychrometer, the 'wet' junction is cooled to induce condensation. The temperature of the condensed water is then measured, which can be related to the ambient relative humidity. In the transistor psychrometer, a drop of water is placed on the 'wet' transistor. Evaporation will take place from the water drop until equilibrium is established with the ambient relative humidity, and the difference in temperature between the wet and dry transistors can be related to the relative humidity. The total suction can be calculated from the relative humidity using Kelvin's equation:

$$\psi = \gamma_w \frac{RT}{Mg} \log_e \left(\frac{H}{100} \right) \tag{4.3}$$

where ψ is total suction
γ_w is the unit weight of water (9.81 kN m^{-3})
R is the universal gas constant (8.317 J mole^{-1} deg^{-1})
T is absolute temperature (°K)
M is the molecular weight of water (0.018 kg mole^{-1})
g is the acceleration due to gravity (9.81 m s^{-2})
H is relative humidity.

The psychrometer is sealed into a chamber that contains the soil sample and a small air space. The relative humidity of the air surrounding the sample is allowed to equilibrate with the total suction in the sample, and measurements are taken with the psychrometer until an equilibrium value is obtained. This might take several hours or even days to achieve depending on the size of the air space. Psychrometers are highly sensitive to very small temperature variations and special measures are needed (e.g. water bath, insulation) so that temperature can be controlled to ±0.1°C.

Porous block sensors

Porous block sensors provide an indirect measurement of suction by determining the water content of a porous block when buried in the soil. Since porous block sensors have to reach suction equilibrium with the surrounding soil there can be a time lag in establishing a suction reading. This lag can be 2–3 days, or with some sensors as much as 2–3 weeks. This makes them unsuitable for use in a rapidly changing moisture environment. The water content of the porous block can be determined by measuring electrical conductivity, thermal conductivity or dielectric permittivity.

For electrical conductivity devices, the porous element is either of gypsum or a granular matrix of silt size (Watermark^{TM} sensors). Such devices are generally inexpensive, but are more suitable for trend measurement rather than accurate determination of suction. Gypsum blocks can be used over the range 20 kPa to 2 MPa, whereas Watermark^{TM} sensors are limited to 200 kPa. Gypsum blocks have the advantage that they tend to buffer the soil salinity, but they do this by dissolution and so have a limited life (2–5 years).

Thermal conductivity sensors rely on the measurement of the water content of the block by measuring the temperature rise induced by a heater embedded in the centre of the porous block. Shuai and Fredlund (2000) show that full equalisation time can be about 2 weeks although a dry sensor may only take 4 days. In addition, these devices need corrections for temperature changes and hysteresis. The suction range is 10 kPa to 10 MPa.

The Equitensiometer is a device that uses a ThetaProbe sensor (that determines water content by measuring the apparent dielectric constant) embedded in a "specially formulated porous matric material" (Delta-T 2005) which is intended to eliminate hysteretic behaviour between drying and wetting. However, Ireson *et al.* (2005) suggest there may be some hysteresis in the sensor. The suction range is 10 kPa to 2.5 MPa.

Pressure plate

A pressure plate apparatus consists of an enclosing cell that contains a high air entry ceramic disc (Figure 4.8). The cell top is normally held down with quick release bolts and has a rubber seal. The sample is placed on the plate (ceramic disc), and an air pressure is applied in the cell, thus controlling the sample pore air pressure. The pore water pressure is controlled by the pressure of water beneath the ceramic.

Suction is imposed by applying a known air pressure in the air chamber and allowing the water pressure in the sample to equilibrate with the water pressure below the plate. Normally the water pressure is maintained at atmospheric pressure, so the imposed suction is given by the applied air pressure. To check that equilibrium has been achieved, a graduated burette may be used to observe when there is no further flow of water from the sample. The sample can also be removed from the apparatus and weighed to check when a stable mass has been achieved.

In commercial pressure plate apparatuses (e.g. those supplied by Soil Moisture Equipment Corp.) the water reservoir below the ceramic plate may be formed from a neoprene membrane sealed around the sides of the ceramic. The plate and reservoir are then placed inside the air chamber, with a tube connected to the water reservoir leading out of the apparatus.

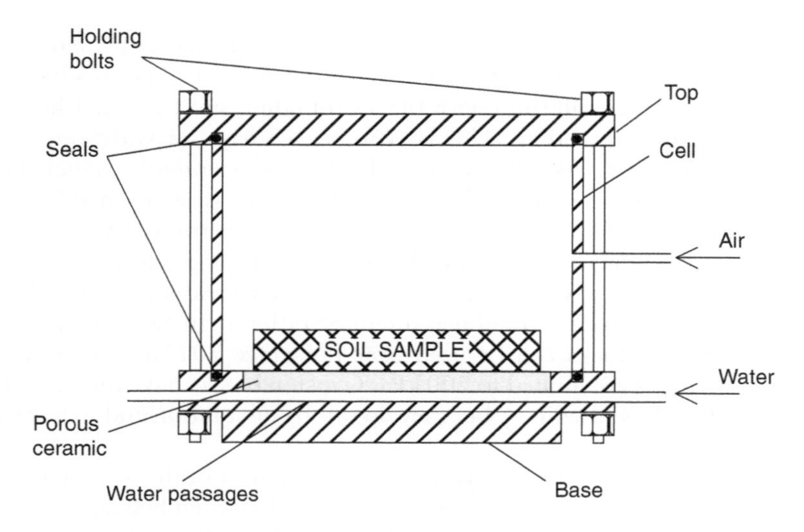

Figure 4.8 The pressure plate apparatus

Vapour control

The vapour control technique is based on controlling the relative humidity surrounding the soil sample. The sample reaches equilibrium with its surroundings by vapour transfer. The suction imposed in this way is controlled by using aqueous solutions of salts or acids of known concentration. The partial vapour pressure imposed by a solution of known concentration is used to determine the relative humidity and hence total suction. Alternatively, a tensiometer or psychrometer can be used to check the suction that has been imposed.

The relative humidity is controlled by suspending the sample over carefully controlled aqueous solutions within a sealed chamber (Figure 4.9). Accurate control of temperature is essential. Even so large errors can arise due to temperature effects at suctions less than 3 MPa. Examples of solutions which have been used for this purpose are NaCl (common salt) and KCl which can impose suctions up to about 40 MPa. Higher suctions (up to 1 GPa) can be imposed using acids. Blatz *et al.* (2008) provide a recent review of the technique.

Osmotic control

In the osmotic technique (Delage and Cui, 2008) a semi-permeable membrane is used to separate the soil from an aqueous solution of polyethylene glycol (PEG). The PEG molecules are too large to pass through the membrane, but water can, so an osmotic suction is applied to the sample through the membrane. The value of the suction depends on the concentration of the PEG solution. Since the membrane is permeable to salts, the osmotic technique controls the matrix suction, not total suction.

The technique can be used simply by immersing a soil sample enclosed in the semi-permeable membrane into the PEG solution (Figure 4.10), which needs to be stirred to

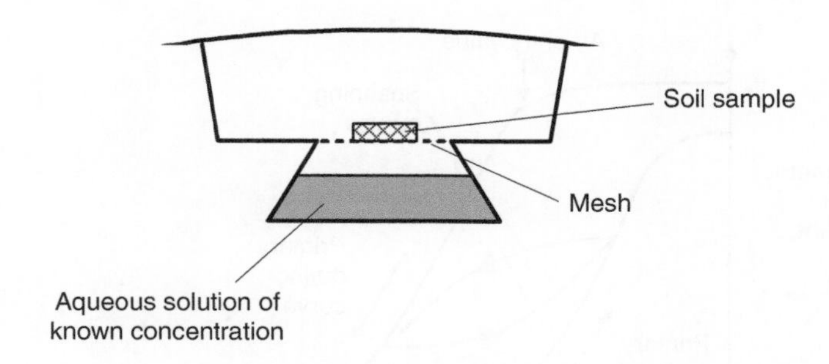

Figure 4.9 The vapour desiccator

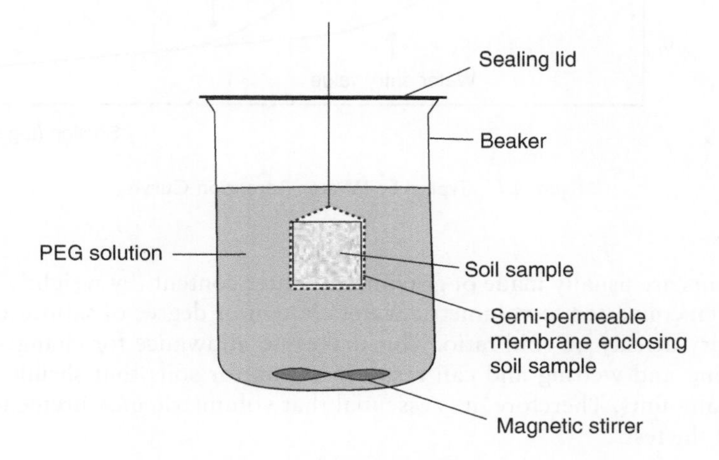

Figure 4.10 Osmotic control

maintain uniformity of concentration. The sample can be weighed to check when the soil has equilibrated with the solution. The technique can also be used within other test equipment (e.g. triaxial, oedometer) by fitting the semi-permeable membrane to the base pedestal and circulating the PEG solution beneath the membrane.

4.3 WATER RETENTION BEHAVIOUR

As a soil dries out, or wets up, the suction within the soil will change. The relationship between water content and suction is known as the Soil Water Retention Curve (SWRC). This relationship has also been called the Soil Water Characteristic Curve (SWCC). However, since the relationship is highly dependent on other factors, such as void ratio, it cannot be properly referred to as a 'characteristic' and hence the term Soil Water Retention Curve is preferred.

Soil water retention curves are often expressed in terms of volumetric water content, θ, or degree of saturation, S_r versus suction. However, in geotechnical practice,

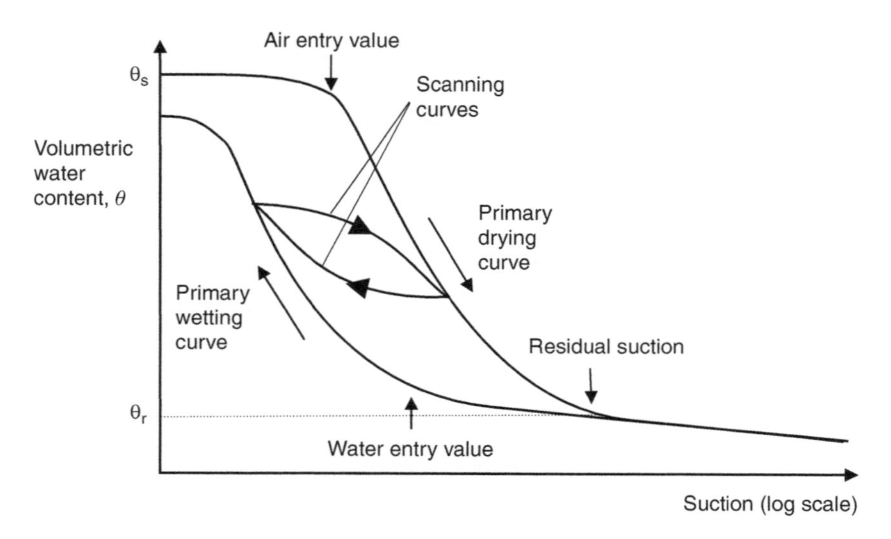

Figure 4.11 Typical Soil Water Retention Curve

measurements are usually made of gravimetric water content (by weight). These readings are often converted to volumetric water content or degree of saturation based on the initial dry density or void ratio. This makes no allowance for changes in volume due to drying and wetting and can result in errors for soils that shrink or swell by significant amounts. Therefore, it is essential that volumetric measurements are made throughout the test.

$$\theta = w\frac{\rho_d}{\rho_w} \tag{4.4}$$

$$\theta = S_r\frac{e}{1+e} \tag{4.5}$$

where θ is volumetric water content
w is water content (gravimetric)
ρ_d is dry density
ρ_w is the density of water ($1.0\,\mathrm{Mg}\cdot\mathrm{m}^{-3}$)
S_r is degree of saturation
e is void ratio.

Figure 4.11 shows a typical SWRC. If the soil starts from a saturated state and is then subject to drying, it will follow the *Primary Drying Curve*. As suction increases, the largest pores will reach their limiting value of suction, and they will start to desaturate. The suction at which this starts to occur, when air first enters the soil, is known as the *Air Entry Value*. Beyond the air entry value, the soil will reduce in water content as finer and finer pores progressively desaturate. At a value of suction known as the *Residual Suction* (with a corresponding residual water content, θ_r) the SWRC may flatten and much smaller changes in volumetric water content result from an increase in suction. It is suggested that within the residual zone, beyond the residual suction, water is held as adsorbed water on clay particles (McQueen and Miller, 1974) rather than as capillary water held by menisci. To achieve zero water content (equivalent to an oven dried condition) requires a suction of the order of 170 MPa (pF 6.25) (McKeen, 1992) or 1 GPa (pF 7) (Fredlund and Xing, 1994).

On wetting from an oven dried state, the soil will follow the *Primary Wetting Curve* (Figure 4.11). At a point known as the *Water Entry Value*, a significant increase in water content occurs as more pores start to fill. When the suction is reduced to zero, the final volumetric water content may be lower than the initial saturated value, θ_s, either due to air bubbles remaining trapped within the soil, or as a result of irrecoverable shrinkage of the soil.

The primary drying and wetting curves define an envelope of possible states within which the soil can exist. If drying is halted partway down the primary drying curve and wetting is started, the soil will follow an intermediate *Scanning Curve* that is flatter than the primary wetting curve, until the primary wetting curve is reached. Similarly, if wetting is halted partway up the primary wetting curve and drying is started, the soil will follow another Scanning Curve, that is flatter than the primary drying curve, until the primary drying curve is reached.

Soil water retention curves can be determined by measuring suctions during drying and wetting using tensiometers, psychrometers or filter paper techniques. Alternatively pressure plates or humidity control can be used to impose known suctions. Normally the drying curve is determined, starting from an initially saturated state. However, it should be appreciated that it is the wetting process that is of most concern in many engineering applications, and the appropriate SWRC should be determined.

4.4 SHEAR BEHAVIOUR

4.4.1 Stress state variables

The first attempts to explain the engineering behaviour of unsaturated soils used an equivalent effective stress approach, attempting to extend the successful use of Terzaghi's effective stress principle for saturated soils. To do this, the stress variables of total stress and suction were combined into a single effective stress. Bishop (1959) adopted an effective stress approach which was expressed as:

$$\sigma' = \sigma - u_a + \chi(u_a - u_w) \tag{4.6}$$

where σ' is effective stress
σ is total stress
u_a is pore air pressure
u_w is pore water pressure
χ is a factor related to the degree of saturation.

The χ variable was an empirical factor that varied between 0 and 1 as a function of degree of saturation, with $\chi = 1$ coinciding with full saturation. If $\chi = 1$ the equation reduces to the effective stress equation for saturated soils, so this provided a simple transition between saturated and unsaturated conditions.

The use of an effective stress law for unsaturated soils was criticised by Jennings and Burland (1962) and Burland (1965). They pointed out that the proposed effective stress law, while appearing to explain shear strength behaviour, could not fully explain volume change behaviour (particularly collapse). In addition, Burland explained that suction did not only increase the contact stresses between particles but that the surface tension at the menisci also provided a stabilising force that contributed to shear resistance. The importance of separating the stress state variables $(\sigma - u_a)$ and $u_a - u_w$ (and treating them differently) was emphasised. Therefore, the effective stress approach for unsaturated soils will not be used further in this chapter.

In an unsaturated soil there are three stress components, the total stress, σ, the pore water pressure, u_w, and pore air pressure, u_a. Fredlund and Morgenstern (1977) suggest that the best pair of stress variables to adopt are the net stress $(\sigma - u_a)$ and the matric suction $(u_a - u_w)$. This provides a pair of independent variables, where net stress represents the effect of the total stress applied (referenced to the applied air pressure) and the matric suction represents the pressure difference between the water and air phases. The pressure difference across the air-water interfaces within the soil controls the shape of the menisci which in turn influences the degree to which soil particles are held together by surface tension. Therefore, the matric suction term represents these attractive forces between particles independently of the total stress applied to the soil.

Other stress variables have been used to explain soil behaviour. Attempts have been made to combine the stress contributions from the pore air and pore water phases, usually using degree of saturation as a weighting factor to represent the proportions of each phase in the soil pores. This stress component (variously described as *Bishop's stress* (Bolzon *et al.*, 1996) or *average soil skeleton stress* (Jommi, 2000)) is similar to Bishop's original formulation but uses degree of saturation, S_r, in place of the empirical factor χ and is defined as:

$$\sigma^* = \sigma - [S_r u_w + (1 - S_r)u_a] \tag{4.7}$$

Bishop's stress can be used in combination with suction, so the criticisms of not representing the surface tension component of stress are overcome. There seems to be some advantage in using such a combined stress approach. However, there are also good reasons for using the variables of net stress $(\sigma - u_a)$ and suction $(u_a - u_w)$. The first reason is that this simple approach requires only the measured stress components for calculation (not a volumetric term that may be unknown or subject to measurement error). The second reason is that these variables of net stress and suction have been widely used in the literature. Therefore, the stress state variables of net stress $(\sigma - u_a)$ and the matric suction $(u_a - u_w)$ will be used throughout this chapter. However, the important role of degree of saturation will also be recognised.

4.4.2 The extended Mohr-Coulomb failure criteria

The most commonly used approach to interpreting shear strength behaviour in unsaturated soils is to adopt an extended version of the traditional Mohr-Coulomb approach. This extension to unsaturated soils was put forward by Fredlund *et al.* (1978).

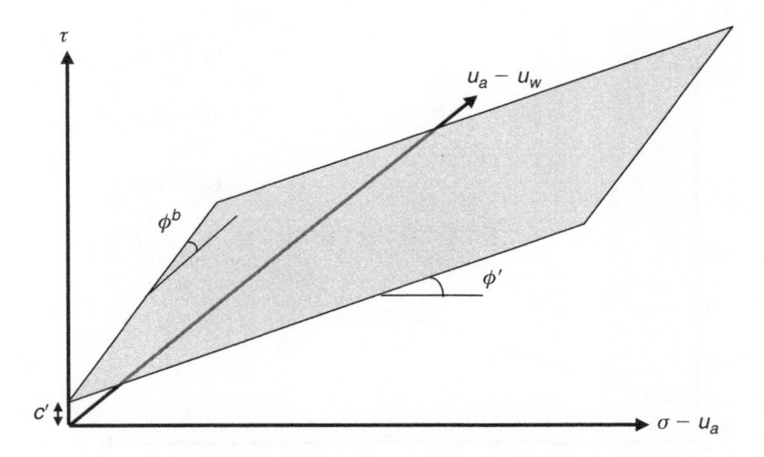

Figure 4.12 The extended Mohr-Coulomb failure envelope

It involves two separate friction angles to represent the contribution to strength from the net stress and matric suction, giving the shear strength equation as:

$$\tau = c' + (\sigma - u_a)\tan\phi^a + (u_a - u_w)\tan\phi^b \qquad (4.8)$$

where τ is shear strength
c' is the effective cohesion
ϕ^a is the angle of friction for changes in net stress $(\sigma - u_a)$
ϕ^b is the angle of friction for changes in matrix suction $(u_a - u_w)$.

This separates the effects of net stress $(\sigma - u_a)$ and $(u_a - u_w)$ and treats them differently, as suggested by Burland (1965). The differences are recognised by having two angles of friction relating to the two components of stress. The extended Mohr-Coulomb failure surface is shown in three dimensions in Figure 4.12. The surface is shown by views in the net stress plane in Figure 4.13 and in the suction plane in Figure 4.14.

Figure 4.13 shows that the strength envelope increases as the suction increases. This can be represented as an increase in the total cohesion, c where:

$$c = c' + (u_a - u_w)\tan\phi^b \qquad (4.9)$$

Figure 4.14 shows the increase in total cohesion, c, as the suction increases. The slope of the graph is defined by ϕ^b. The relationship between τ and $(u_a - u_w)$ has been found to be non-linear by Escario and Saez (1986) and Fredlund *et al.* (1987). Fredlund *et al.* (1987) took ϕ^b to vary as a function of suction. Below the air entry value (when the soil remains saturated) ϕ^b is equal to ϕ', but at higher suctions the value of ϕ^b reduces (Figure 4.14). The tangent value may fall to zero at high suctions, implying no further increase in strength at higher suctions.

Fredlund *et al.* (1978) suggested that the slope of the failure envelope in net stress space, ϕ^a, could be assumed to be equal to the effective stress angle of friction measured

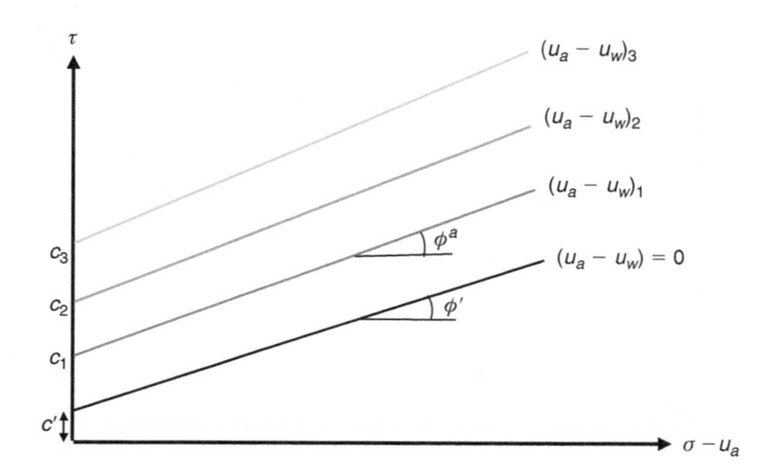

Figure 4.13 The extended Mohr-Coulomb failure envelope in net stress space

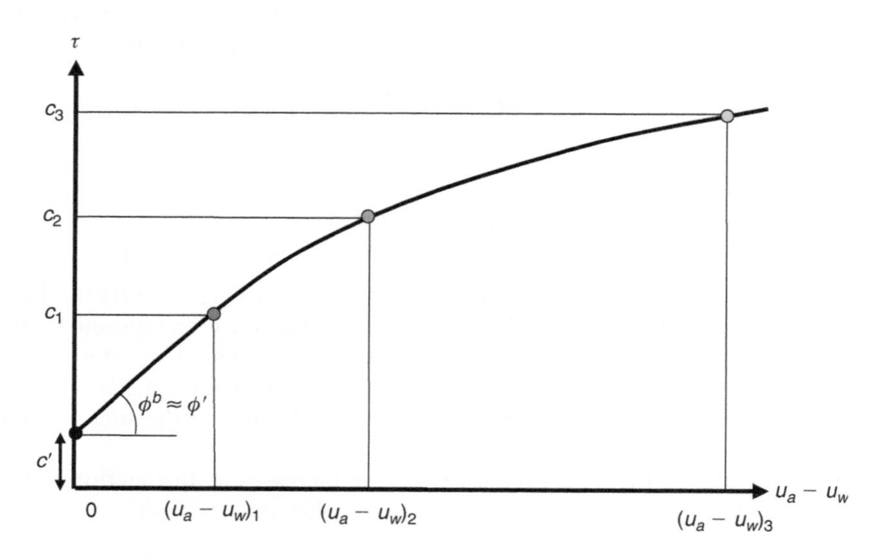

Figure 4.14 The extended Mohr-Coulomb failure envelope in matric suction space

in saturated conditions (ϕ'). This would suggest that ϕ^a was constant for all values of matrix suction. However, Delage *et al.* (1987), Toll (2000) and Toll *et al.* (2008) have shown that ϕ^a cannot always be assumed to be equal to ϕ'.

Triaxial testing of unsaturated soils can be carried out under constant suction conditions (analogous to drained tests in saturated soils). Such tests can be time consuming, as drainage of water can be slow in unsaturated soils due to low water permeabilities at low degrees of saturation. Alternatively, tests can be performed under constant water content conditions, where changes of suction are measured. This is analogous

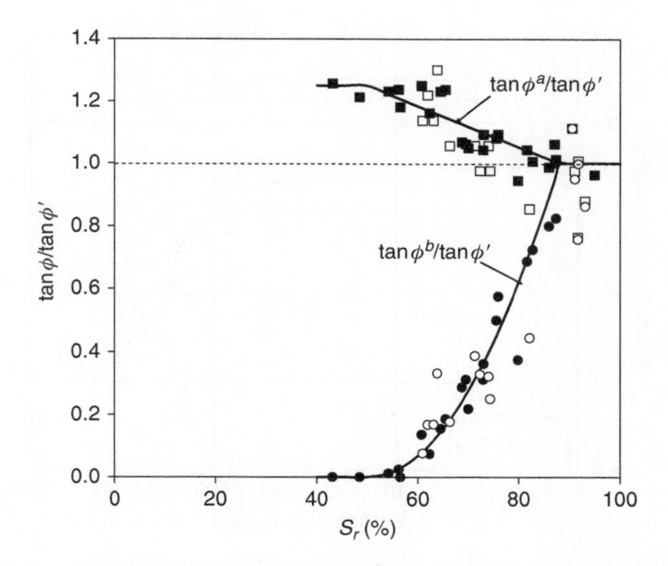

Figure 4.15 Values for ϕ^a and ϕ^b related to degree of saturation for two tropical soils [Black symbols for Kiunyu Gravel (*After* Toll, 1990): Open symbols for Jurong soil (*After* Toll and Ong, 2003)]

to undrained testing in saturated soils. However, it has to be appreciated that constant water content does not imply constant volume, as samples can still change in volume due to movement of air in or out of the sample during shear.

Toll (1990) and Toll and Ong (2003) have reported results of constant water content triaxial tests on unsaturated samples of tropical soils, a lateritic gravel from Kiunyu, Kenya and a residual sandy clay from Jurong, Singapore. The results are plotted in Figure 4.15 showing the variation of ϕ^a and ϕ^b with degree of saturation, S_r.

At lower degrees of saturation it can be seen that ϕ^a has greater values than ϕ' (the value for saturated conditions). This was explained by Toll (1990, 2000) as being due to the presence of aggregations at lower degrees of saturation, causing the soil to behave in a coarser fashion than would be justified by the grading. In unsaturated conditions the aggregated fabric can be maintained during shear because the suction gives strength to the aggregations. In a saturated soil, the aggregations would be broken down during shear and would not be expected to affect the strength parameters.

Figure 4.15 also shows that at low degrees of saturation ϕ^b becomes significantly lower than ϕ' and eventually drops to zero. This represents a similar effect to the reduction in ϕ^b as suction increases, as observed by Escario and Saez (1986) and Fredlund *et al.* (1987), as degree of saturation will reduce as suction increases.

Fredlund *et al.* (1995) and Vanapalli *et al.* (1996) have suggested that ϕ^b can be related to normalised volumetric water content or degree of saturation, determined from the SWRC. The functions used were:

$$\frac{\tan\phi^b}{\tan\phi'} = \Theta^k = \left(\frac{\theta - \theta_r}{\theta_s - \theta_r}\right)^k = \left(\frac{S_r - S_{r(res)}}{100 - S_{r(res)}}\right)^k \tag{4.10}$$

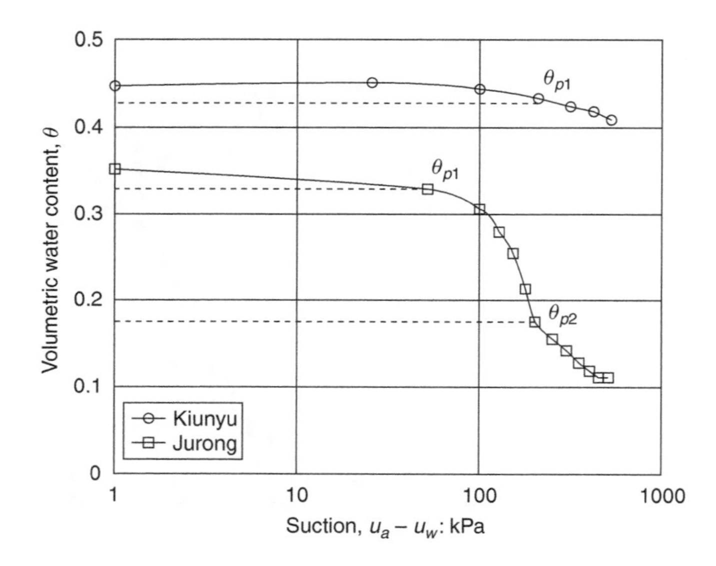

Figure 4.16 Transition points on the SWRC for Kiunyu gravel and Jurong soil (*After* Toll and Ong, 2003)

where ϕ^b is the angle of friction for changes in matrix suction $(u_a - u_w)$
ϕ' is the effective angle of friction for saturated conditions
Θ is normalised water content
k is a fitting constant
θ is volumetric water content
θ_s is the saturated volumetric water content
θ_r is the residual volumetric water content
S_r is degree of saturation
$S_{r(res)}$ is the residual degree of saturation.

Toll and Ong (2003) found that a similar approach could be used to fit the data from Kiunyu gravel and the Jurong soil. A value of the fitting parameter $k = 2.0$ was used, similar to that used by Vanapalli *et al.* (1996) where typically values of 2.2–2.5 were found to be appropriate for a glacial till. However, the reference points of θ_s and θ_r were replaced by θ_{p1} and θ_{p2} which seemed to represent the beginning and end of the primary transition zone of the SWRC (Figure 4.16).

Care must be taken in using this type of assumption for selecting appropriate values of ϕ^b. Toll *et al.* (2008) found that for bonded sand, with uniform grading and an SWRC with a very steep desaturation zone, that the Vanapalli *et al.* (1996) approach could not explain the variation in ϕ^b observed. In this case, ϕ^b values remained higher than would be predicted within the desaturation zone of the SWRC, and only started to drop significantly within the residual zone.

4.5 VOLUME CHANGE

Shrinkage and swelling are two sides of the same process whereby soils change in volume when water is removed from the soil (shrinkage) or added to the soil (swelling).

Soils that have a large intrinsic expansiveness, and hence show large swelling strains, will also show large shrinkage strains. The intrinsic expansiveness of a soil will be controlled by its mineralogy, particularly by the presence of highly active clay minerals, such as smectites. However, the changes of volume for a particular soil will result from changes in suction.

Collapse is a different phenomenon where soils in a loose state, with an open fabric, will reduce in volume when wetted. However, the process can again be explained by a consideration of suction and fabric changes in the soil during wetting.

4.5.1 Shrinkage

As soils dry out they reduce in volume (shrink). A typical shrinkage curve is shown in Figure 4.17. With progressive drying, the shrinkage limit is reached, where little further volume change takes place.

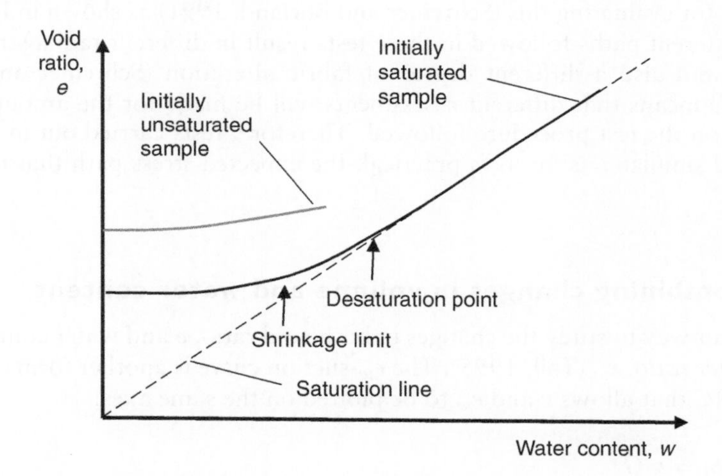

Figure 4.17 Typical shrinkage curve

The relationship between void ratio and water content is given by:

$$e = \frac{wG_s}{S_r} \qquad (4.11)$$

where e is void ratio

w is water content (gravimetric)

G_s is the specific gravity (ρ_s/ρ_w)

ρ_s is the particle density

ρ_w is the density of water ($1.0\,\text{Mg}\cdot\text{m}^{-3}$)

S_r is degree of saturation.

If the soil is initially saturated, it will follow a saturation line defined by $e = wG_s$ (since $S_r = 1$). The point at which the curve deviates from this line indicates air entering the soil, indicating the soil has reached a suction equal to the Air Entry Value. An initially unsaturated sample will tend to follow a flatter shrinkage curve and show less shrinkage (Marinho and Chandler, 1993).

Shrinkage can often lead to cracking of a soil horizon. This can have major implications for engineering behaviour, in terms of both strength and permeability of the soil.

4.5.2 Swelling

There are three important aspects to swelling behaviour (i) the amount of swell that might take place if unrestrained (free swell) (ii) the amount of swell that might take place under a defined load and (iii) the swelling pressure that can be generated if the soil is prevented from swelling (constant volume). Swell under a defined load is usually of most interest for engineering applications, and is usually evaluated in one-dimensional conditions in an oedometer. However, it should be noted that there are different test procedures for evaluating this (Schreiner and Burland, 1991) as shown in Figure 4.18.

The different paths followed in these tests result in different radial stresses being developed and also a different degree of fabric alteration (Schreiner and Burland, 1991). This means that different assessments will be made for the amount of swell, depending on the test procedure followed. Therefore, tests carried out in the laboratory should simulate, as far as is practical, the expected stress path that the soil will experience.

4.5.3 Combining changes in volume and water content

A convenient way to study the changes in both void ratio, e and water content, w is to use the *water ratio*, e_w (Toll, 1995). The e_w-suction curve is another form of portrayal of the SWRC that allows e and e_w to be plotted on the same axes.

$$e_w = eS_r = wG_s \tag{4.12}$$

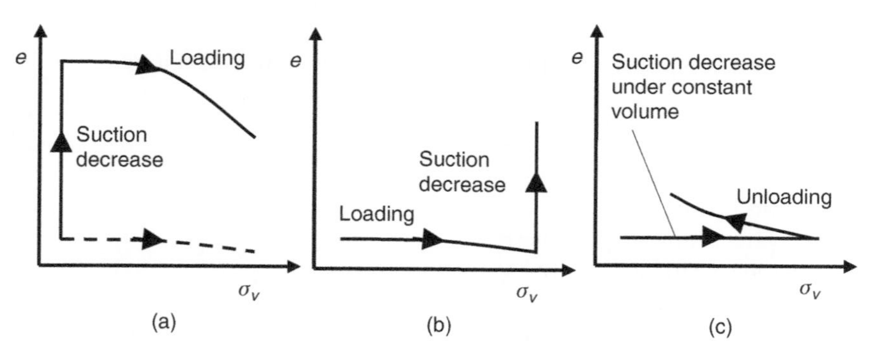

Figure 4.18 Swell test procedures (*After* Schreiner and Burland, 1991) [e – void ratio, σ_v – vertical stress]. (a) initial wetting followed by loading (dotted line shows comparable loading at initial water content) (b) initial loading followed by wetting (c) wetting (with vertical stress increase to prevent swelling) followed by unloading

Figure 4.19(a) shows the path (in terms of both e and e_w against log suction) followed by a normally consolidated soil as it is dried. From the initial point A to point B, the soil remains saturated, ($e_w = e$) and follows the Virgin Consolidation Line (VCL).

At point B the suction reaches the desaturation level of the largest pore (air entry value) and air starts to enter the soil. The finer pores remain saturated and will continue to decrease in volume as the suction increases. However the desaturated pores will be much less affected by further changes in suction and will not change significantly in volume. The overall volume change will therefore be less than in a mechanically compressed saturated soil, and the void ratio-suction line will become less steep than the VCL – points B to C in Figure 4.19(a). The volume of water removed by desaturating a pore will be close to the pore volume. This will be greater than the volume of water which would be expelled from the soil by reducing the size of the pore. Therefore the e_w-suction line (called the Virgin Drying Line, VDL) will become steeper than the VCL as shown by B–D in Figure 4.19(a). From point B to C there are sufficient saturated pores for increasing suction to produce significant consolidation. As more and more pores desaturate, a point will be reached beyond which the amount of consolidation produced by increasing the suction reduces considerably. This point represents the shrinkage limit (Point C). Continued water content change is produced by desaturation of the pores, but little change in void ratio results. With respect to the e_w curve, point D is eventually reached (the residual suction) and the e_w line flattens.

It is possible for a soil to exist in a saturated state with the same initial suction as a normally consolidated sample but with a lower void ratio due to overconsolidation. Figure 4.19(b) shows the desaturation curves for two samples at different overconsolidation ratios. The lightly overconsolidated soil at point A will follow the saturated reload path A–B until the VCL is reached and will then follow the same desaturation path B–C–D as for the normally consolidated soil. For a more heavily overconsolidated soil at point E, the desaturation point may be reached at point F before the VCL is reached. The e_w path would then follow a steeper path than the saturated reloading path, but shallower than the path followed by the normally consolidated soil (F–H in Figure 4.19(b)) since the pores in the overconsolidated soil are smaller and less water will be expelled when they desaturate. At the point where the e_w path reaches the path followed by the normally consolidated sample at point H, a change in the water content-suction relationship might be expected.

Figure 4.19(c) shows a cycle of wetting and drying for an initially saturated normally consolidated soil. Both e and e_w follow the virgin drying paths for the initial drying (A–B) and (A-E). On wetting an increase in void ratio (B–C) is produced and also an increase in e_w (E–F). If the soil is again subjected to drying there may be a decrease in void ratio (C–D) and e_w will follow the path F–H with a kink in the path at G where the virgin drying line is reached.

4.5.4 Collapse

Collapse can occur if a soil is in a meta-stable state and is wetted. A soil exists in a meta-stable state if the void ratio is above the saturated virgin consolidation line. If the soil exists in this state (with a loose, open fabric), because of suction (rather than chemical bonding), then wetting will induce collapse.

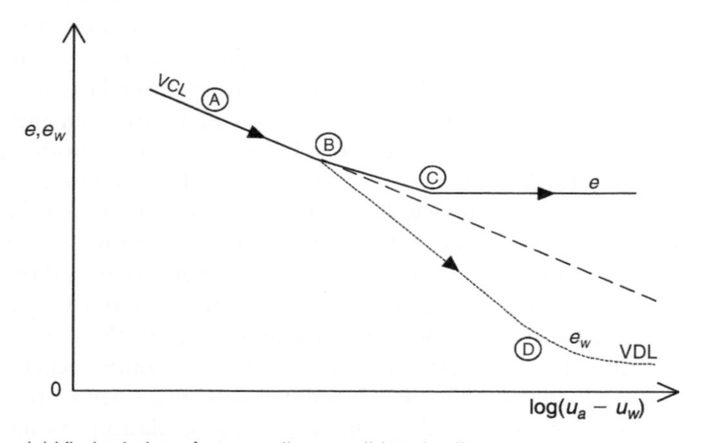

(a) Virgin drying of a normally consolidated soil

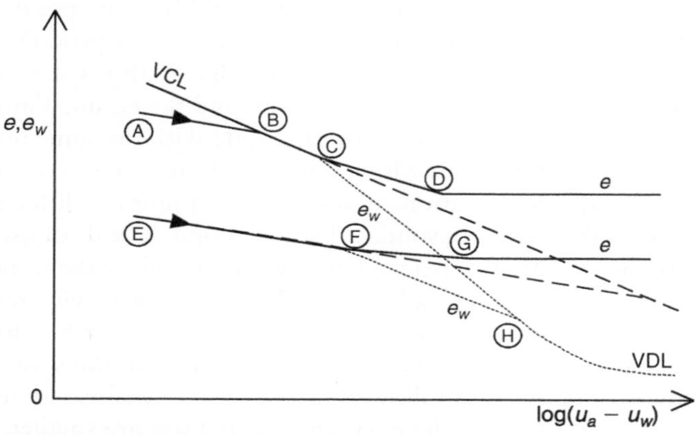

(b) Virgin drying of an overconsolidated soil

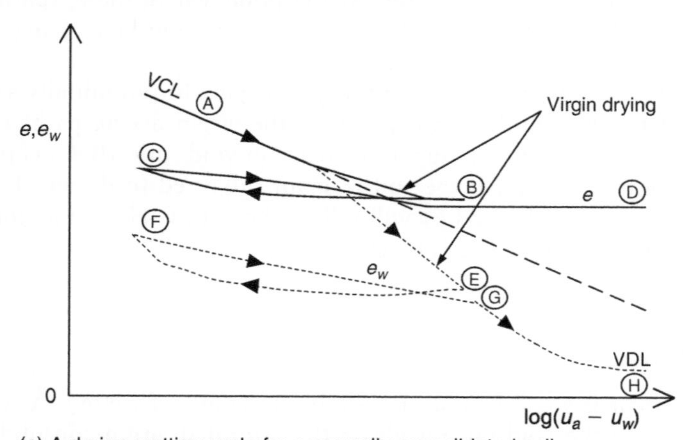

(c) A drying-wetting cycle for a normally consolidated soil

Figure 4.19 A simple conceptual picture of drying and wetting (*After* Toll, 1995)

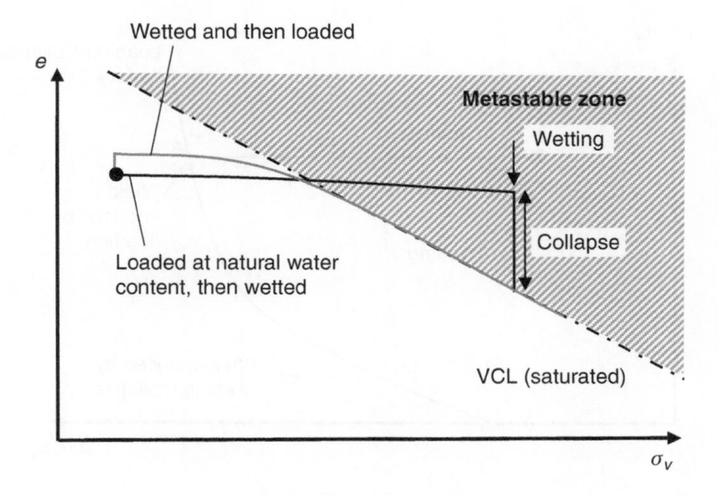

Figure 4.20 Collapse on wetting

Figure 4.20 shows that a sample that is initially wetted, so that suction becomes zero before loading, will join the Virgin Consolidation Line (VCL) for the saturated soil as it is loaded. However, a soil that is loaded at natural water content, where suction exists, can cross the saturated VCL to reach the metastable zone. This is because the suction can support a very loose, open fabric as the capillary forces link particles together. If this sample is wetted while in the metastable zone, then the suction supporting the open fabric is removed and the loose state can no longer be sustained. On loss of suction, the void ratio would drop to that of saturated (or zero suction) state, thus causing a sudden and possibly large collapse settlement.

This concept can be further explained by the Loading-Collapse (LC) surface put forward by Alonso *et al.* (1990). This surface defines a yield condition in terms of mean net stress $(p - u_a)$ and suction $(u_a - u_w)$ (Figure 4.21). Providing the soil remains within the LC surface, then strains will be elastic. If the LC surface is reached then large plastic strains can be induced, either by loading (increase in net mean stress) or by wetting (reduction in suction).

It can be seen in Figure 4.21 that if a soil was loaded along path A–B (without wetting) it would yield at a value of mean net stress defined by point B. However, if the soil was loaded from A–C and then wetted, then the soil would yield at a lower value of mean net stress defined by point D, producing a collapse settlement.

4.6 PERMEABILITY

4.6.1 Water permeability (hydraulic conductivity)

The permeability to water (hydraulic conductivity) of an unsaturated soil can be considerably lower than the soil in a saturated state. This is because water finds it easier to flow through pores that are already full of water. If the soil has a lower degree of

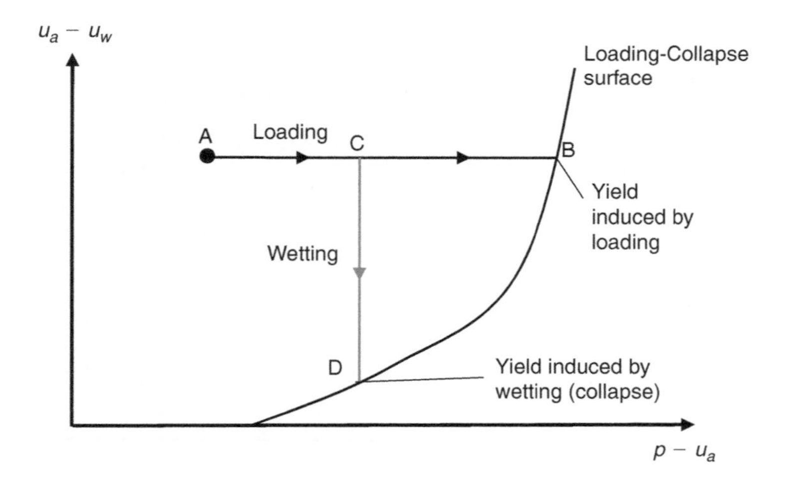

Figure 4.21 The Loading-Collapse surface proposed by Alonso *et al.* (1990)

saturation, there are fewer water-filled pores through which flow can take place. At very low degrees of saturation (within the residual zone of the SWRC), water movement may have to take place in the vapour phase, which can be extremely slow.

The relative water permeability (k_{rw}) is defined as the ratio of the water permeability (k_w) at any suction or degree of saturation, to that in the saturated state (k_s):

$$k_{rw} = \frac{k_w}{k_s} \tag{4.13}$$

The unsaturated water permeability can be expressed as a function of suction or degree of saturation. There are a number of models that are used to estimate the unsaturated permeability function from the SWRC, such as Brooks and Corey (1964), van Genuchten (1980) and Fredlund *et al.* (1994). However, great care should be taken in using such estimated functions without any experimental validation, as there can be major discrepancies (e.g. Karthikeyan *et al.*, 2008).

Figure 4.22 shows the relative water permeability, k_{rw}, as a function of degree of saturation, as predicted by the Brooks and Corey model. If water permeability is expressed as a function of suction, then the function will show significant hysteresis between drying and wetting, mirroring the hysteretic behaviour of the SWRC.

The unsaturated water permeability will only be relevant when considering flow under a suction gradient, where there is no positive pore water pressure involved. The unsaturated water permeability will control the process of moisture equalisation and redistribution, but not that of infiltration. In the case of infiltration, a wetting front moves through the soil, and behind the front the soil approaches saturation. Flow into the soil will, therefore, be governed by the saturated permeability, although the rate of progress of the wetting front will depend on the suction in the soil ahead of the front.

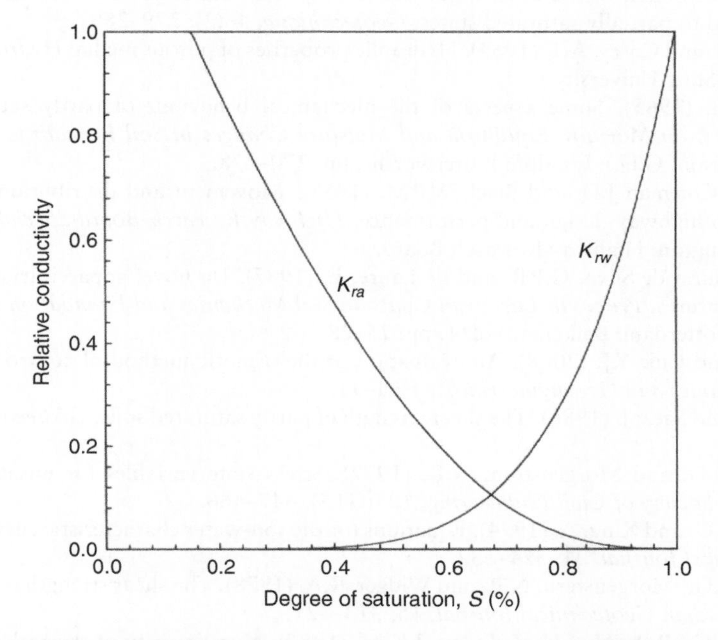

Figure 4.22 Relative air and water permeability as predicted by the Brook and Corey model (*After* Lu and Likos, 2004)

4.6.2 Air permeability (air conductivity)

The permeability to air (air conductivity) of an unsaturated soil can be high at low degrees of saturation when the air phase is continuous. However, as the degree of saturation increases, the air permeability will reduce. At the point that air becomes occluded (present as bubbles), typically at a degree of saturation of 85–90%, the air permeability becomes very low.

The relative air permeability (k_{ra}) is defined as the ratio of air permeability (k_a) at any suction or degree of saturation to that in the dry state (k_d). The variation with degree of saturation, based on the Brooks and Corey model, is shown in Figure 4.22.

$$k_{ra} = \frac{k_a}{k_d} \tag{4.14}$$

REFERENCES

Alonso, E.E., Gens, A. and Josa, A. (1990). A constituitive model for partially saturated soils, *Géotechnique*, 40(3): 405–430.

Bishop A.W. (1959). The principle of effective stress, *Tecknisk Ukeblad*, 106(39): 859–863.

Blatz, J.A., Cui, Y.J. and Oldecop, L. (2008). Vapour equilibrium and osmotic technique for suction control, *Geotechnical and Geological Engineering*, 26(6): 661–673.

Bolzon, G., Schrefler, B.A. and Zienkiewicz, O.C. (1996). Elasto-plastic soil constitutive laws generalised to partially saturated states, *Géotechnique*, 46(2): 279–289.
Brooks, R.H. and Corey, A.T. (1964). Hydraulic properties of porous media: *Hydrology Papers*, Colorado State University.
Burland, J.B. (1965). Some aspects of the mechanical behaviour of partly saturated soils, *Proc. Conf. on Moisture Equilibria and Moisture Changes in Soil Beneath Covered Areas* (ed. Aitchison, G.D.), London: Butterworths, pp. 270–278.
Croney D., Coleman J.D. and Black W.P.M. (1958). Movement and distribution of water in relation to highway design and performance, *Highway Research Board Special Report No. 40*, Washington: Highway Research Board.
Delage, P., Suraj de Silva, G.P.R. and de Laure, E. (1987). Un novel appareil triaxial pour les sols non saturés, *Proc. 9th European Conf. on Soil Mechanics and Foundation Engineering, Dublin,* Rotterdam: Balkema, Vol. 1, pp. 25–28.
Delage, P. and Cui, Y.J. (2008). An evaluation of the osmotic method of controlling suction. *Geomechanics and Geoengineering*, 3(1): 1–11.
Escario, V. and Saez, J. (1986). The shear strength of partly saturated soils, *Géotechnique*, 36(3): 453–456.
Fredlund, D.G. and Morgenstern, N.R. (1977). Stress state variables for unsaturated soils, *American Society of Civil Engineering*, 103(GT5): 447–466.
Fredlund, D.G. and Xing, A. (1994). Equations for the soil-water characteristic curve, *Canadian Geotechnical Journal*, 31: 521–532.
Fredlund, D.G., Morgenstern, N.R. and Widger, R.A. (1978). The shear strength of unsaturated soils, *Canadian Geotechnical Journal*, 15: 313–321.
Fredlund, D.G., Rahardjo, H. and Gan, J.K.M. (1987). Non-linearity of strength envelope for unsaturated soils, *Proc. 6th Int. Conf. Expansive Soils, New Delhi*, Rotterdam: Balkema, pp. 49–54.
Fredlund, D.G., Xing, A. and Huang S.Y. (1994). Predicting the permeability function for unsaturated soils using the soil-water characteristic curve, *Canadian Geotechnical Journal*, 31: 533–546.
Fredlund, D.G., Xing, A., Fredlund, M. D. and Barbour, S. L. (1995). The relationship of the unsaturated soil shear strength to the soil-water characteristic curve. *Canadian Geotechnical Journal*, 32: 440–448.
Hilf, J.W. (1956). An investigation of pore water pressure in compacted cohesive soils, *US Bureau of Reclamation, Tech. Mem. 654*, Denver: US Bureau of Reclamation.
Ireson, A.M., Wheater, H.S., Butler, *et al.* (2005). Field monitoring of matric potential and soil water content in the chalk unsaturated zone. *Proc. Int. Symp. Advanced Experimental Unsaturated Soil Mechanics, Trento, Italy*, pp. 511–518.
Jennings J.E.B. and Burland J.B. (1962). Limitations to the use of effective stresses in partly saturated soils, *Géotechnique*, 12: 125–144.
Jommi, C. (2000). Remarks on the constitutive modelling of unsaturated soils, in *Experimental Evidence and Theoretical Approaches in Unsaturated Soils* (eds. Tarantino, A. and Mancuso, C.) Rotterdam, Balkema, pp. 139–153.
Jones, H.E. and Kohnke, H. (1952). The influence of soil moisture tension on vapour movement of soil water, *Proc. Soil Science Soc. America*, 16: 245–248.
Karthikeyan, M., Toll, D.G. and Phoon, K.K. (2008). Prediction of changes in pore-water pressure response due to rainfall events, *Unsaturated Soils: Advances in Geo-Engineering, Proc. 1st European Conf. Unsaturated Soils, Durham* (eds. Toll, D.G., Augarde, C.E., Gallipoli, D. and Wheeler, S.J.) Leiden: CRC Press/Balkema, pp. 829–834.
Leong, E.C., He, L. and Rahardjo, H. (2002). Factors affecting the filter paper method for total and matric suction measurements, *Geotechnical Testing Journal*, 25(3): 322–333.

Leong, E.-C., Tripathy, S., and Rahardjo, H. (2003). Total suction measurement of unsaturated soils with a device using the chilled-mirror dew-point technique, *Géotechnique*, 53(2): 173–182.

Lourenço, S.D.N., Gallipoli, D., Toll, D.G. and Evans, F.D. (2006). Development of a commercial tensiometer for triaxial testing of unsaturated soils, *Proc. 4th International Conference on Unsaturated Soils, Phoenix, USA, Geotechnical Special Publication No. 14*, Reston: ASCE, Vol. 2, pp. 1875–1886.

Lu, N. and Likos, W.J. (2004). *Unsaturated Soil Mechanics*, Wiley, New York.

Marinho, F.A.M. and Chandler, R.J. (1993). Aspects of the behaviour of clays on drying, in *Unsaturated Soils. Geotechnical Special Publication No. 39* (eds. Houston, S.L. and Wray, W.K.), New York: ASCE, pp. 77–90.

McKeen, R.G. (1992). A model for predicting expansive soil behaviour, *Proc. 7th Int. Conf. on Expansive Soils*, Dallas, Vol. 1, pp. 1–6.

McQueen, I.S. and Miller, R.F. (1974). Approximating soil moisture characteristics from limited data: empirical evidence and tentative model, *Water Resources Research*, 10: 521–527.

Review Panel Statement (1965). Engineering concepts of moisture equilibria and moisture changes in soils, *Proc. Conf. on Moisture Equilibria and Moisture Changes in Soil Beneath Covered Areas* (ed. Aitchison, G.D.), London: Butterworths, pp. 7–21.

Ridley, A.M. and Burland, W.K. (1993). A new instrument for the measurement of soil moisture suction, *Géotechnique*, 43(2): 321–324.

Schofield, R.K. (1935). The pF of water in soil. *Trans. 3rd Int. Cong. Soil Science*, 2: 37–48.

Schreiner, H.D. and Burland, J.B. (1991). A comparison of the three swell test procedures, Geotechnics in the African Environment, *Proc. 10th Regional Conf. on Soil Mechanics and Foundation Engineering/3rd Int. Conf. on Tropical and Residual Soils*, Vol. 1, pp. 259–266.

Shuai, F. and Fredlund, D.G. (2000). Use of a new thermal conductivity sensor to measure soil suction, in *Advances in Unsaturated Soils, Geotechnical Special Publication No. 99* (eds. Shackleford, C., Houston, S.L. and Chang, N-Y.), Reston: American Society of Civil Engineers, pp. 1–12.

Tarantino, A. and Mongiovi, L. (2001). Experimental procedures and cavitation mechanisms in tensiometer measurements, in *Unsaturated Soil Concepts and their Application in Geotechnical Practice* (ed. Toll, D.G.), Dordrecht: Kluwer Academic, pp. 189–210.

Toll, D.G. (1990). A framework for unsaturated soil behaviour, *Géotechnique*, 40(1): 31–44.

Toll, D.G. (1995). A conceptual model for the drying and wetting of soil, *Unsaturated Soils, Proc. 1st Int. Conf. Unsaturated Soils, Paris* (eds. Alonso, E.E. and Delage, P.), Rotterdam: Balkema, Vol. 2, pp. 805–810.

Toll, D.G. (2000). The influence of fabric on the shear behaviour of unsaturated compacted soils, *Advances in Unsaturated Soils, Geotechnical Special Publication No. 99* (eds. Shackleford, C., Houston, S.L. and Chang, N.Y.), Reston: American Society of Civil Engineers, pp. 222–234.

Toll, D.G. and Ong, B.H. (2003). Critical state parameters for an unsaturated residual sandy clay, *Géotechnique*, 53(1): 93–103.

Toll, D.G., Ali Rahman, Z. and Gallipoli, D. (2008). Critical state conditions for an unsaturated artificially bonded soil, *Unsaturated Soils: Advances in Geo-Engineering, Proc. 1st European Conf. Unsaturated Soils, Durham* (eds. Toll, D.G., Augarde, C.E., Gallipoli, D. and Wheeler, S.J.) Leiden: CRC Press/Balkema, pp. 435–440.

Toll, D.G., Lourenço, S.D.N., Mendes, J., *et al.* (2011) Soil suction monitoring for landslides and slopes, *Quarterly Journal of Engineering Geology and Hydrology*, 44(1): 23–33.

van Genuchten, M.Th. (1980). A closed-form equation for predicting the hydraulic conductivity of unsaturated soils, *Soil Science Society of America Journal*, 44(5): 892–898.

Vanapalli, S.K., Fredlund, D.G., Pufahl, D.E. and Clifton, A.W. (1996). Model for the prediction of shear strength with respect to soil suction, *Canadian Geotechnical Journal*, 33: 379–392.

5

Volume change of tropical residual soils

Sudhakar M. Rao
Indian Institute of Science, Bangalore, India

5.1 INTRODUCTION

Residual soils are formed by the mechanical and chemical weathering of parent rocks at the original location. Poor drainage conditions favour formation of montmorillonite-rich, expansive soils (vertisols) in semi-arid zones, while good internal drainage favours the development of kaolinite-dominant ferruginous soils in the sub-tropical zones (FAO, 1989; Fookes, 1990). Occurrence of low groundwater table coupled with evapo-transpiration effects often render the residual soils unsaturated (degree of saturation, $S_r < 1$). A special feature of these unsaturated residual soils is their volumetric instability to changes in water content at constant net applied stress ($\sigma - u_a$). Moisture induced volumetric instability in residual soils manifest as swell (increase in void ratio upon water absorption), shrinkage (decrease in void ratio from reduction in water content) and collapse (decrease in void ratio upon water absorption) as illustrated in Figure 5.1. The rates of swell and shrink of unsaturated soils are much slower than rate of collapse. The swell process is essentially governed by moisture migration into the microstructure, while shrinkage depends on moisture evaporation from the microstructure. Microstructure-dependent moisture exchange makes the swelling and shrinkage a slow process. Collapse is a more rapid process as it essentially depends on the rate of moisture migration into the porous macrostructure. The magnitudes of volumetric deformations in response to changes in moisture content at given net applied stress

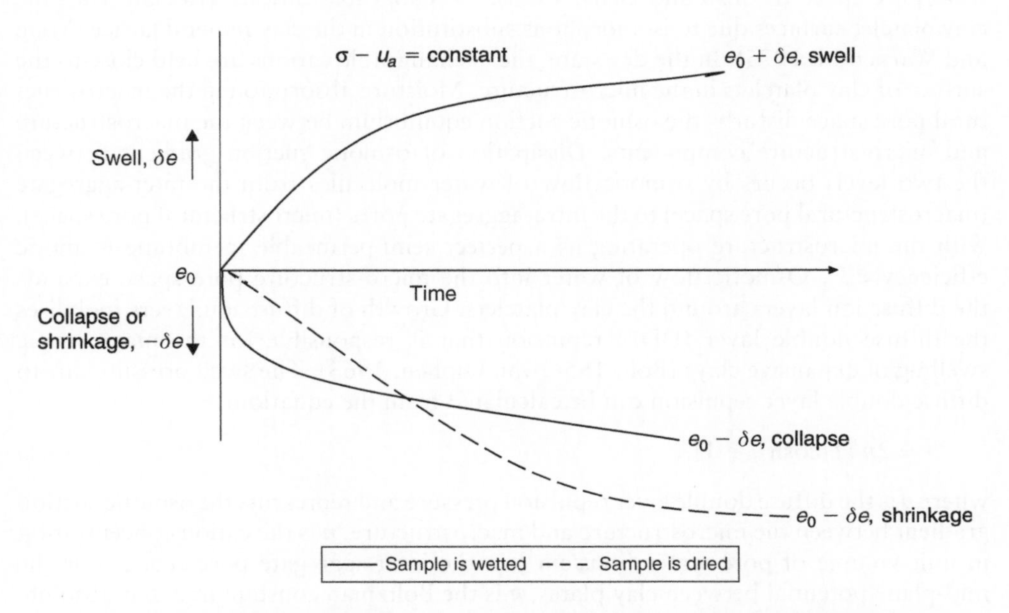

Figure 5.1 Conceptual illustration of swell-collapse-shrinkage in unsaturated soils

are influenced by the mineralogy of clay fraction, soil density and moisture content, surcharge pressure and soil microstructure.

5.2 SWELLING AND SHRINKAGE

Expansive soils tend to swell upon infiltration of water from sources such as (a) precipitation, (b) gardening and lawn irrigation and (c) leaks in sewers and water lines. The volume of these soils decreases with water loss as a function of the rate and completeness of drying and the volume change rearranges soil particles and aggregates into denser arrangement (Smiles, 2000). Shrinkage-drying of expansive soils is generally facilitated by (a) evapo-transpiration and (b) local desiccation due to heating systems such as boilers, furnaces, etc. The large swell-shrink potentials of expansive soils make them poor foundation soils and earth construction material.

The swelling tendencies of expansive soils are quantified by the parameters of swell potential and swelling pressure. The following sections consider the mechanisms of swell and shrinkage, their laboratory measurement procedures and factors influencing these soil properties.

Swell mechanism of expansive soils

Unsaturated expansive soils swell upon water absorption by the soil matrix. According to the framework postulated for the mechanical behaviour of unsaturated expansive clays, the microstructural pore space is saturated, which implies that the matrix suction responsible for moisture absorption mainly operates in the macrostructural pore space. The framework further postulates that the exchangeable cations responsible for diffuse ion layer formation around the clay platelets are encountered in the microstructural pore space (Alonso and Gens, 1994). Exchangeable cations associate with the clay platelet surfaces due to isomorphous substitution in the clay mineral lattice (Yong and Warkentin, 1975). In the dry state, the exchangeable cations are held close to the surface of clay platelets in the microstructure. Moisture absorption in the macrostructural pore space disturbs the osmotic suction equilibrium between the macrostructure and micro-structure components. Dissipation of osmotic suction gradient between the two levels occurs by osmotic flow of water molecules from the inter-aggregate (macrostructural pore space) to the intra-aggregate pores (microstructural pore space), with the microstructure operating as a perfect semi-permeable membrane (osmotic efficiency = 1). Osmotic flow of water into the micro-structure pore space expands the diffuse ion layers around the clay platelets. Growth of diffuse ion layers mobilises the diffuse double layer (DDL) repulsion that is responsible for the macroscopic swelling of expansive clays (Bolt, 1956; van Olphen, 1963). The swell pressure due to diffuse double layer repulsion can be calculated from the equation:

$$p = 2nkT(\cosh u - 1) \tag{5.1}$$

where p is the diffuse double layer repulsion pressure and represents the osmotic suction gradient between the microstructure and macrostructure, n is the cation concentration in unit volume of pore water (ions/cm^3) in the inter-aggregate pore space, u is the mid-plane potential between clay plates, k is the Boltzman constant and T is absolute temperature.

More recently, Wheeler (1994) proposed that the swelling pressure of unsaturated clay defines the critical value of net stress at which the behaviour of unsaturated clay changes from swelling to collapse. In other words, wetting unsaturated clay at external stresses lower than its swell pressure causes it to expand, while wetting the same clay at external stresses greater than its swell pressure causes it to collapse. However, unsaturated clays with large swell pressures (>100 kPa) will experience lower potential for collapse as their collapse tendency is offset by the strong electrical forces associated with the clay mineral surface (Rao and Reddy, 1998).

Laboratory measurements of swell pressure and swell potential

Swell pressure is defined as the pressure which an expansive soil exerts, if the soil is prevented from swelling upon moisture absorption (BIS 2720 Part 61). The swell potential of a soil is defined as the percentage increase in ratio of change in vertical height (Δh) to original height of a column of in-situ soil/compacted soil (h) upon absorption of moisture (ASTM D 4546-03). The swell potential and swelling pressure values of expansive clays depend on factors such as (a) type and amount of clay in the soil, (b) initial water content and dry density, (c) the nature of pore fluid, (d) stress history of the soil including the confining pressure and (e) drying and wetting cycles to which the soil has been subjected (BIS 2720 Part 61; Seed *et al.*, 1962; Chen, 1988). Specially designed oedometer tests have been found useful to determine the swell potential and swelling pressure magnitudes of expansive soils. Based on the oedometer test procedures, Holtz and Gibbs (1956) and Seed *et al.* (1962) have classified the relative expansion of swelling soils. The criterion of Holtz and Gibbs (1956) is based on the swell potential developed by undisturbed soil specimens upon expanding from dry to saturated state under vertical pressure of 1 psi (approximately 7 kPa). Comparatively, the classification of Seed *et al.* (1962) was based on swell potential measurements performed with compacted soil specimens (compacted at maximum dry density and optimum moisture content conditions corresponding to standard AASHTO compaction test) that were inundated under vertical pressure of 1 psi. Table 5.1 provides the expansivity categories proposed by the two groups of researchers. Chen (1988) has provided a guide for estimating the probable volume change of expansive soils based on the swell potential and swell pressure values (Table 5.2).

Oedometer test procedures

The initial water content and void ratio should be representative of the in-situ soil immediately prior to construction. In case where it is necessary to use disturbed soil samples, the soil sample should be compacted to the required field density and

Table 5.1 Classification of expansive soils

Degree of expansion	*Holtz and Gibbs (1956), % swell*	*Seed et al. (1962), % swell*
Low	<10	0 to 1.5
Medium	10 to 20	1.5 to 5
High	20 to 30	5 to 25
Very high	>30	>25

Table 5.2 Degree of soil expansion (After Chen, 1988)

Probable expansion, % total volume change	*Swell pressure, kPa*	*Degree of expansion*
>10	958	Very high
3–10	239–958	High
1–5	144–239	Medium
<1	48	Low

Table 5.3 Index and compaction characteristics of red soil (RS) and black cotton soil (BCS) specimens

Soil	*Sand*	*Silt*	*Clay*	*Liquid limit, %*	*Plastic limit, %*	*Standard Proctor* OMC, %	*Standard Proctor MDD, Mg/m³*
BCS1	11	28	61	77	31	30	1.41
BCS2	23	18	59	95	24	28	1.44
BCS3	15	30	55	75	31	29	1.35
BCS4	0	34	66	103	52	31	1.27
BCS5	16	29	55	84	23	28	1.42
RS1	59	16	24	39	19	17	1.73
RS2	42	26	32	37	19	18	1.77
RS3	43	34	23	45	25	17	1.76

OMC = Optimum moisture content, MDD = Maximum dry density

water content in a Proctor compaction mould. The swell potential and swell pressure of a soil specimen can be determined by any of the three methods specified by the ASTM Standard (D 4546-03) for determining one dimensional swell potential of cohesive soils following the procedure for specimen preparation outlined in this standard. The index properties and compaction characteristics of black cotton soil (BCS) specimens referred to as illustrative examples are provided in Table 5.3.

Free-swell and load

The compacted expansive soil specimen is inundated and allowed to swell vertically at seating pressure of at least 1 kPa (Rao, 2006). The resultant time-swell curve typically consists of three regions: an initial swell region, primary swell region and secondary swell region (Figure 5.2). The specimen is stepwise loaded after primary swell is complete using a load increment ratio of unity. The loading process is continued until the swollen specimen regains its initial void ratio/height. The external pressure needed to regain the initial void ratio e_0, defines the swell pressure (Figure 5.3).

The swell potential at the seating pressure relative to the initial void ratio e_0 is given as:

$$\text{Swell potential } (\%) = e_{se} - \frac{e_0}{1 + e_0} \times 100 \tag{5.2}$$

where e_{se} is the void ratio after stabilised swell at seating pressure and e_0 is the initial void ratio of the soil specimen at the seating pressure. This test method measures

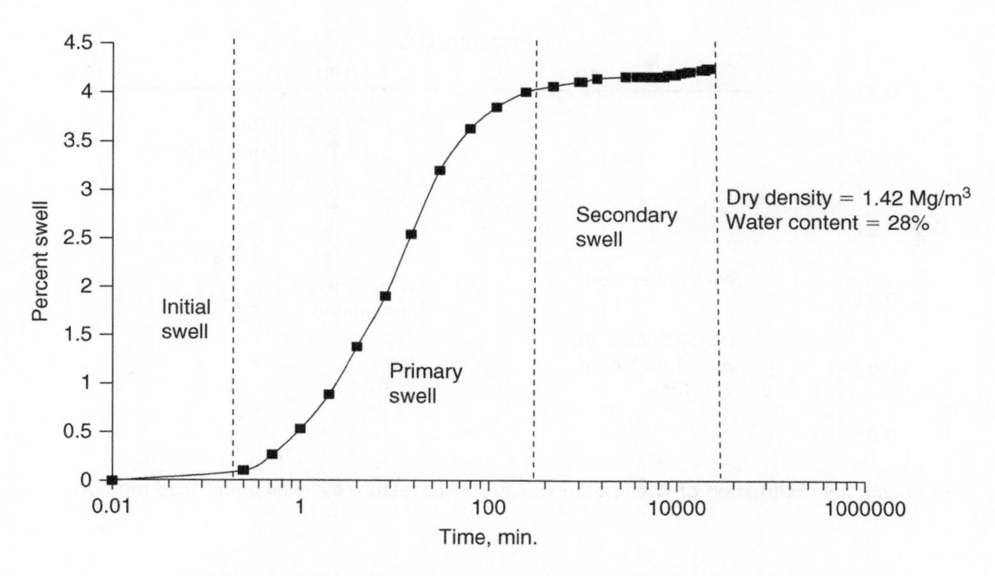

Figure 5.2 Time-swell behaviour of compacted BCS4 specimen

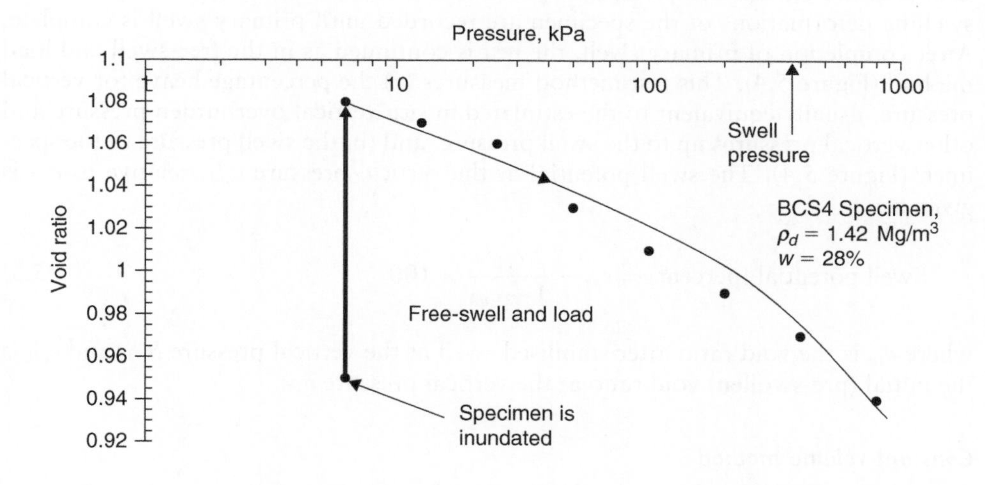

Figure 5.3 Estimation of swell potential and swell pressure by free swell and load method

(a) the percentage swell of the specimen at the seating pressure and (b) the swell pressure of the specimen (Figure 5.3).

Swell under load

A vertical pressure exceeding the seating pressure is applied to the soil specimen. The magnitude of vertical pressure (σ_{v0}) is usually equivalent to the in-situ overburden

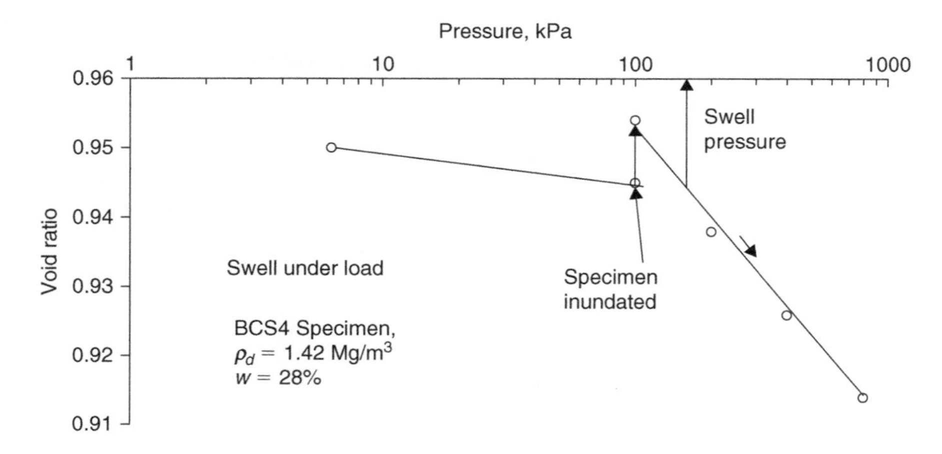

Figure 5.4 Estimation of swell potential and swell pressure by swell under load method

pressure, or structural loading or both. After the axial deformations under the vertical pressure σ_{v0} are complete, the specimen is given free access to water, and axial swelling deformations of the specimen are recorded until primary swell is complete. After completion of primary swell, the test is continued as in the free-swell and load method (Figure 5.4). This test method measures (a) the percentage heave for vertical pressure, usually equivalent to the estimated in situ vertical overburden pressure and other vertical pressures up to the swell pressure, and (b) the swell pressure of the specimen (Figure 5.4). The swell potential at the vertical pressure σ_{v0}, relative to e_{v0} is given as:

$$\text{Swell potential (percent)} = e_{se} - \frac{e_{v0}}{1 + e_{v0}} \times 100 \tag{5.3}$$

where e_{se} is the void ratio after stabilised swell at the vertical pressure σ_{v0}, and e_{v0} is the initial (pre-swollen) void ratio at the vertical pressure σ_{v0}.

Constant volume method

A vertical pressure (σ_1) equivalent to the estimated vertical in-situ pressure or swell pressure is applied to the specimen. After completion of axial deformations under the vertical pressure, σ_1, the soil specimen is inundated with water. Increments of vertical stress are applied to the wetted specimen to prevent any swell. Variations from the dial gage readings, at the time the specimen is inundated at stress σ_1, should preferably be kept within 0.005 mm and not more than 0.01 mm. The vertical pressure at which the wetted specimen shows no further tendency to swell defines the swell pressure of the specimen. The specimen is stepwise loaded after showing no further tendency to swell. The applied load increments should be sufficient to define the maximum curvature on the consolidation curve and to determine the slope of the virgin compression curve (Figure 5.5).

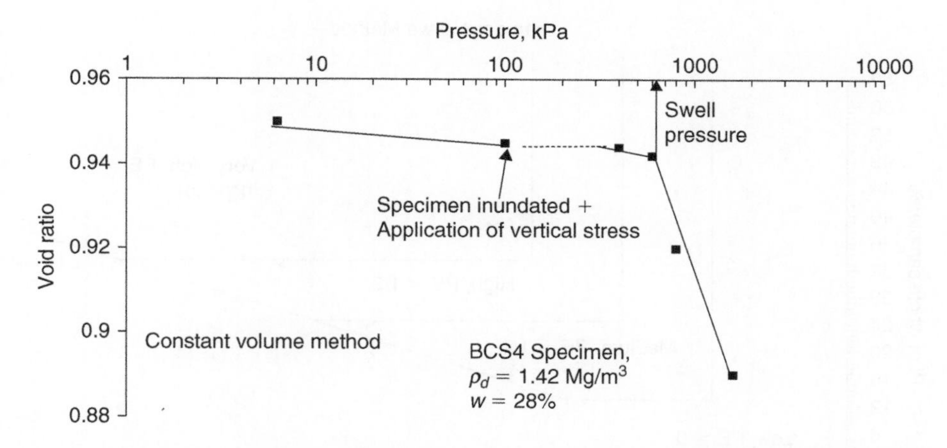

Figure 5.5 Estimation of swell pressure by constant volume method for compacted BCS4 specimen

Table 5.4 Identification criteria for expansive clays (*After* Holtz and Gibbs, 1956)

Colloid content, %	*Plasticity index, %*	*Shrinkage limit, %*	*Degree of expansion*	*Probable expansion (% total volume change)*
<15	<18	<10	Low	<10
13–23	15–28	10–20	Medium	10–20
20–31	25–41	20–30	High	20–30
>28	>35	>30	Very high	>30

Indirect evaluation of swell potential

Besides direct quantification of swell potential from the oedometer test, it is also possible to indirectly estimate the degree of expansivity of clay soils from their index properties or from the differential free swell test. The Atterberg limits and swell potentials of clays depend on the quantity of water that they can imbibe. These indirect classification systems work on the principle that the higher the plasticity index of the clay, the greater is the amount of water that can be imbibed by it and hence the swell potential would be larger. Likewise, a low shrinkage limit is indicative that a soil will begin to swell at low water content. The colloid content (<2 μm) constitutes the most active part of the soil contributing to swelling and a high colloid content naturally means a greater possibility of expansion. Together, the United States Bureau of Reclamation uses these three parameters to indicate the criterion for identification of expansive soils, which are reproduced in Table 5.4.

Similarly, the van der Merwe method (van der Merwe, 1964) evolved from empirical relationships between the degree of expansion, the plasticity index, the percentage clay fraction and the surcharge pressure. The total heave at the ground surface is found from the relation:

$$\Delta H = \sum_{D=1}^{D=n} F \times PE \tag{5.4}$$

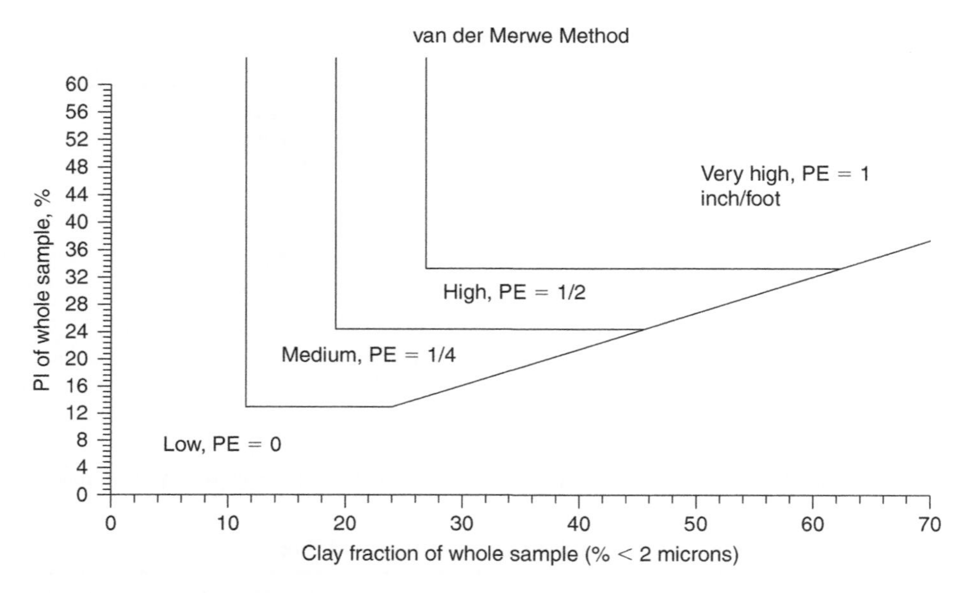

Figure 5.6 Expansivity classification by the van der Merwe method (PE = Potential Expansivity, PI = Plasticity Index)

where
ΔH = total heave, inches
D = depth of soil layer in increments of 1 foot increment at the deepest level
F = reduction factor for surcharge pressure and $F = 10^{-D/20}$
PE = potential expansiveness in inch/foot of depth.

The PE values is found by assumed values of $PE = 0$, ¼, ½ and 1 inch/foot for low, medium, high and very high levels, respectively of potential expansiveness as depicted in Figure 5.6.

Besides using the index properties, the swell potential of clay soils can be indirectly estimated from the differential free swell (DFS) test (BIS 2720 Part 60). In this method, two oven-dried samples weighing 10 g each and passed through a 425 μm sieve are taken. One sample is put in a 100 ml graduated glass cylinder containing kerosene (non-polar liquid) and the other sample is put in a similar cylinder containing distilled water. Both the samples are stirred and left undisturbed for 24 hours and then their volumes are noted. The DFS is expressed as:

$$\text{DFS} = \frac{\text{Soil volume in water} - \text{soil volume in kerosene}}{\text{Soil volume in kerosene}} \times 100\% \tag{5.5}$$

The degree of expansivity and possible damage to lightly loaded structures may be qualitatively assessed from Table 5.5. In areas where the soils have high or very high DFS values, conventional shallow foundations may not be adequate.

Table 5.6 presents swell potentials predicted based on index properties of black cotton soil specimens from Karnataka, India (BCS1 to BCS3, Table 5.3) using

Table 5.5 Degree of expansiveness and free swell

Degree of expansiveness	*DFS, %*
Low	<20
Moderate	20–35
High	35–50
Very high	>50

Table 5.6 Estimation of swell potential of black cotton soil specimens based on index properties (*After* Holtz and Gibbs, 1956)

Soil No.	*Clay content, %*	*Plasticity index, %*	*Shrinkage limit, %*	*Probable expansion %*
BCS1	61	46	10	>30
BCS2	59	71	8	>30
BCS3	55	61	10	>30

Table 5.7 Oedometer swell potentials of black cotton soils

Soil No.	*Dry density, Mg/m³*	*Moisture content, %*	*Swell potential, %*
BCS1	1.34	32	5.2
BCS2	1.37	30	4
BCS3	1.35	33	5
BCS3	1.35	28	10
BCS3	1.42	28	11
BCS3	1.42	23	16

the Holtz and Gibbs (1956) criteria for indirect estimation of swell potentials. Table 5.7 presents the oedometer swell potentials of these compacted black cotton soil specimens using the free swell and load method. The percent swell predicted by the index properties (Table 5.6) far exceeds the oedometer swell potentials (Table 5.7). Further, variations in compaction conditions impact the oedometer swell potentials of the compacted black cotton soil specimens (BCS3, Table 5.7). Indirect predictions based on index properties needs to be modified to reflect sensitivity of swell potentials to variations in initial soil compaction conditions.

Prediction of vertical heave of expansive soil

The potential vertical heave (Δh_{swell}) of a uniform layer of expansive soil of thickness z_i is calculated from the equation (TM 5-818-7):

$$\Delta h_{swell} = \frac{e_f - e_0}{1 + e_0} z_i N \tag{5.6}$$

where e_f and e_0 refer to the swollen and initial void ratios of the soil specimen determined in the oedometer test as per swell under the load method; in this method the soil

Table 5.8 Swell potential of compacted soils at vertical pressure of 6.25 kPa by free-swell and load method

Compaction condition	*RS1* specimens, % Swell*	*BCS4 specimens, % Swell*
MDD & OMC	0.6	8
95% of MDD & OMC	0.4	5
95% MDD & 5% wet of OMC	Nil	2

*Properties of red soil (RS) specimens are provided in Table 5.3

specimen is inundated at the magnitude of vertical pressure (σ_{v0}) equivalent to the in situ overburden pressure + structural loading (refer Figure 5.4). The value of N refers to the fraction of heave that occurs in the vertical direction, and the value of N depends on the soil fabric and anisotropy. Vertical heave of intact soil with few fissures may account for all of the volumetric swell such that $N = 1$, while vertical heave of heavily fissured isotropic soil may be as low as 1/3 of the volumetric swell (TM 5-818-7).

Swelling behaviour of compacted specimens

The swell potential of unsaturated soils are influenced by several factors. These factors can be divided into two groups. Factors in the first group depend on the nature of soil particles including type and amount of clay mineral present, while factors in the second group depend on features such as compaction dry density/water content, surcharge pressure and method of compaction (Seed *et al.*, 1962).

The black cotton soil (BCS4) specimens develop large swell potentials of 2 to 8% (Table 5.8), while the red soil (RS1) specimens swell marginally (0 to 0.6%) (Mishra *et al.*, 2008). Black cotton soil specimens swell more owing to the predominant presence of montmorillonite, while the red soil specimens swell marginally owing to the predominant presence of non-swelling kaolinite. The vastly different Atterberg limits of RS1 and BCS4 specimens (Table 5.3) are indicative of the different clay mineralogies of the two residual soil types.

The initial compaction conditions also have a definite bearing on the swell potentials of residual soils (Table 5.8). For example, at 95% MDD, the BCS or RS specimens develop larger swell potentials at OMC than at 5% wet of OMC (Table 5.8). Reduction in compaction water content increases the soil's affinity for water and thereby increases the swell magnitude. Similarly, the swell potentials of the RS or BCS specimens increase with dry density at constant compaction water content. Increase in dry density promotes closer spacing of clay particles in the microstructure that in turn favour stronger electrical repulsion forces and larger swell potentials.

Tables 5.8 and 5.9 bring out the role of clay mineralogy in the swelling behaviour of residual soils (Mishra *et al.*, 2008). Though red soil specimens are compacted at much higher density and lower water content than the black cotton soil specimens (Table 5.8), they develop a lower range of swell potentials as clay mineralogy (reflected by the Atterberg limits of the residual soil specimens, Table 5.9) dominates over compaction conditions in influencing swell behaviour of these residual soil specimens.

Table 5.9 Comparison of swell potentials of soil specimens

Soil type	*Liquid limit, w_L (%)*	*Plasticity index, I_P, %*	*Range of oedometer swell (%) at 6.25 kPa*
Red soil (RS1)	39	20	0–6
Black cotton soil (BCS4)	84	61	2–11

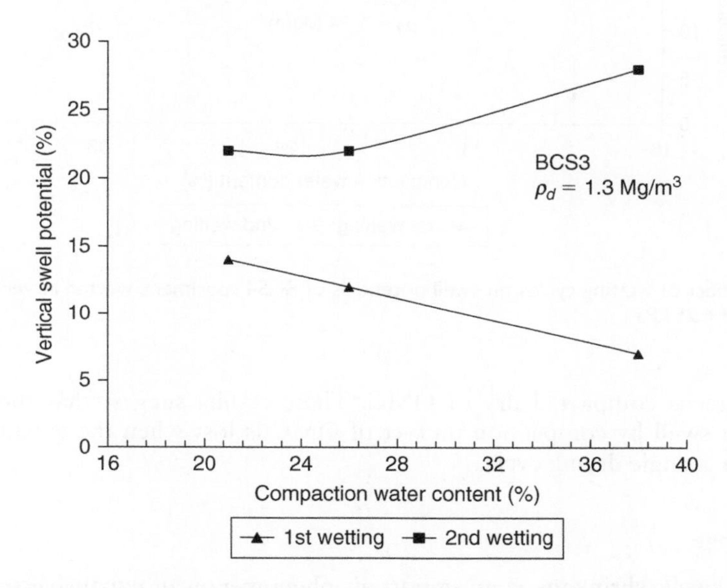

Figure 5.7 Effect of wetting cycles on swell potential of BCS3 specimens wetted at vertical pressure of 6.25 kPa

Influence of wetting-drying cycles on swell potential of compacted clay specimens

Clay soils in arid and semi-arid regions are subjected to cycles of wetting and drying in the field due to climate patterns. The data of Day (1994), Al-Homoud *et al.* (1995) and Subba Rao *et al.*, (2000) have illustrated that cyclic wetting and drying overrides the influence of initial compaction conditions on the swell potentials of expansive clays. Figures 5.7 and 5.8 (Rao, 2006) illustrate the impact of one drying-wetting cycle on the swell potentials (determined by the free-swell and load method at vertical pressure of 6.25 kPa) of compacted black cotton soils (BCS3 and BCS4 specimens). Table 5.3 provides their index properties and compaction characteristics.

The expansive soil specimen compacted wet of OMC (OMC of BCS3 and BCS4 specimens correspond to 29 and 31%) develops a smaller swell potential than specimen compacted dry of OMC, illustrating the role of compaction control on first wetting (Figures 5.7 and 5.8). After the swelling process was complete, the compacted specimens were dried at 45°C. The dried specimens were again wetted. On second wetting, the specimens compacted wet of OMC, swell similarly (Figure 5.8) or more (Figure 5.7)

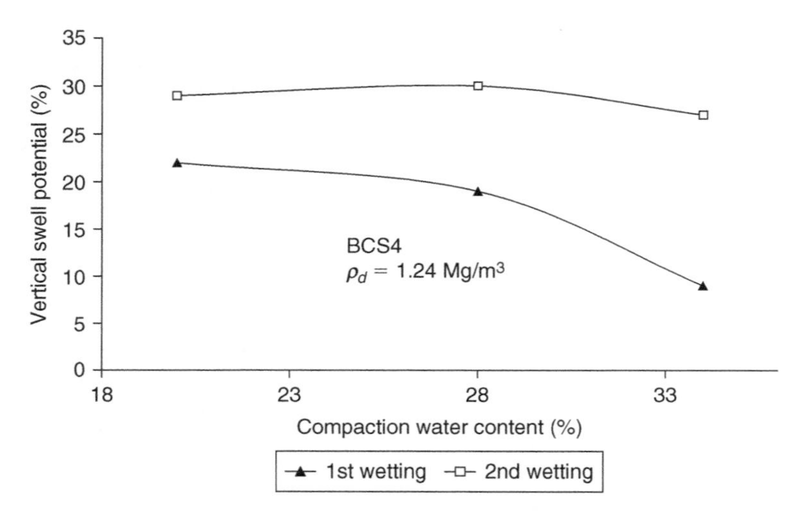

Figure 5.8 Effect of wetting cycles on swell potentials of BCS4 specimens wetted at vertical pressure of 6.25 kPa

than specimens compacted dry of OMC. These results suggest that the benefit of controlling swell by compaction on wet of OMC is lost when the expansive clay is exposed to a single drying cycle.

Soil shrinkage

Similar to swell, shrinkage is an important phenomenon in expansive soils. Drying expansive soil causes it to shrink both in the lateral and vertical directions. Factors that influence swell also influence the shrinkage, but to different degrees. The shrinkage magnitude is a major function of (a) type and amount of clay present in the soil, with the shrinkage magnitude increasing with amount of clay-sized particles, and (b) initial water content with the magnitude of shrinkage increasing with initial water content (Yong and Warkentin, 1975; Chen, 1988).

Shrinkage mechanism

Low humidity and high temperatures cause soil to experience shrinkage. Terzaghi (1925) viewed the shrinking process as a compaction process occurring under the action of capillary stresses. During shrinkage, at the air-water interface, capillary menisci form between adjacent soil particles with the pore water pressure (u_w) in the convex inner side being lower than the atmospheric pressure (u_a) on the concave outer side of the menisci (Yong and Warkentin, 1975). This pressure difference ($u_a - u_w$) constitutes the matric suction in unsaturated soils (Fredlund and Rahardjo, 1992) and acts to compress the soil skeleton through a suction stress, $\chi(u_a - u_w)$, where χ is a material property (Bishop *et al.*, 1963; Lu and Likos, 2004) that depends on the degree of saturation, varying between zero (perfectly dry soil) to unity (saturated soil).

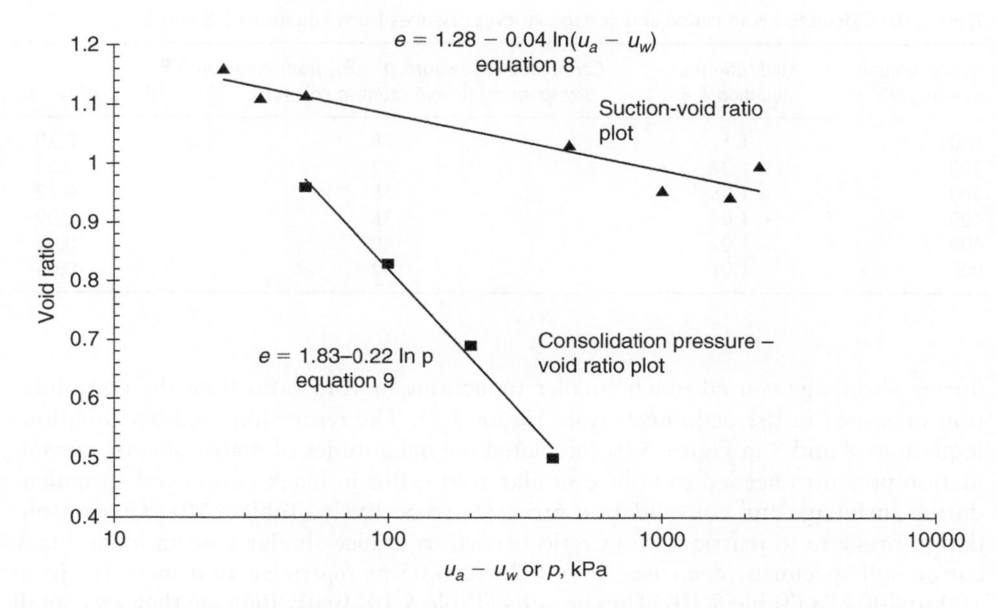

Figure 5.9 Void ratio-matric suction ($u_a - u_w$) and void ratio-consolidation pressure (p) plots for black cotton soil specimens

Continued evaporation increases the meniscus curvature in soil pores and the magnitude of matric suction, as the latter is inversely related to the capillary radius (r in m):

$$u_a - u_w = \frac{2T}{r} \cos\alpha \tag{5.7}$$

where T is the surface tension of water-air interface (72 mN/m at 25°C) and α is the contact angle and equals zero for contact between pure water and clean glass (Das, 1998). During shrinkage, volume change occurs so long as the compressive stresses exerted by matric suction exceed the particle resistance to closer approach. Eventually, a situation is realised when particle interaction resists the shrinkage stresses and further increments of water removed are partly replaced by air. Small residual shrinkage may further occur from fabric re-arrangement or bending of particles (Yong and Warkentin, 1975).

An insight into the relative contribution of suction stress, $\chi(u_a - u_w)$, to soil compressibility during shrinkage is provided by Figure 5.9 that re-plots the data (presented later) in Figure 5.11 in terms of void ratio-log matric suction. The matric suction data in Figure 5.9, obtained from Sarita (2004), were measured by the filter-paper method. The figure also includes the void ratio-consolidation pressure plot for the black cotton soil specimen (oedometer data from Murthy, 1988). The initial void ratios and degree of saturation of the black cotton soil specimens in shrinkage and oedometer experiments are similar (1.11 to 1.15 and 90 to 95%). Matric suction generated

Table 5.10 Calculated void ratios and consolidation pressures from equations 5.8 and 5.9

Matric suction, $u_a - u_w$, kPa	*Void ratio from equation 5.8*	*Consolidation pressure, p, (kPa) from equation 5.9 corresponding to void ratios in column 2*	$p/u_a - u_w$
100	1.1	28	0.28
200	1.07	32	0.32
300	1.05	35	0.12
400	1.04	36	0.09
600	1.02	40	0.07
800	1.01	42	0.05

during shrinkage caused much smaller reductions in void ratio than the consolidation pressures in the oedometer test (Figure 5.9). The regression analysis equations (equations 8 and 9 in Figure 5.9) calculated the magnitudes of matric suction consolidation pressures needed to induce similar void ratios in black cotton soil specimens during shrinkage and consolidation processes respectively (Table 5.10). The consolidation pressure to matric suction ratio needed to induce similar void ratios in black cotton soil specimens decreases from 0.28 to 0.05 as matric suction increases from 100 to 800 kPa (Table 5.10). These results (Table 5.10) firstly indicate that very small fractions of matric suction translate to suction stress [$\chi(u_a - u_w)$] that is responsible for soil compressibility during shrinkage. Secondly, the magnitudes of matric suction contributing to suction stress progressively decrease with an increase in matric suction.

Laboratory measurements of shrinkage

Shrinkage limit

Shrinkage limit is defined as the smallest water content at which the soil remains saturated (Taylor, 1948); it also represents the transition water content between semi-solid and solid states, and the water content below which no further change of soil mass occurs on drying (Das, 1998). The shrinkage limit of soil can be determined by ASTM D427-98 (2004), "Standard test method for shrinkage factors of soil by mercury method" or the more recent (and recommended) D 4943-02, 2004, "Standard test method for shrinkage factors of soil by wax method". Shrinkage limit water content is indicative of the shrinkage potential of residual soils; soils with higher Atterberg limits are characterised with lesser shrinkage limits and are expected to experience greater shrinkage movements in the field. For example, data of Sridharan *et al.*, (1992) show that montmorillonitic black cotton soil specimens with higher liquid limit (range from 72 to 104%) possess lower shrinkage limits of 10 to 12%, while the red soils with lower liquid limits (range from 47 to 75%) possess larger shrinkage limits of 16 to 20%. The shrinkage limits of soils have also been used as an indicator of their potential expansion (Table 5.11, Holtz and Gibbs, 1956).

Measuring water content – void ratio relations during shrinkage

The relationships between void ratio and water content during shrinkage are a useful tool to understand shrinkage process and to predict shrinkage induced settlement in

Table 5.11 Degree of soil expansion based on shrinkage limits

Shrinkage limit, %	*% probable expansion from air-dry to saturated state at 1 psi vertical load*	*Degree of expansion*
<11	>30	Very high
7–12	20–30	High
10–16	10–20	Medium
>15	<10	Low

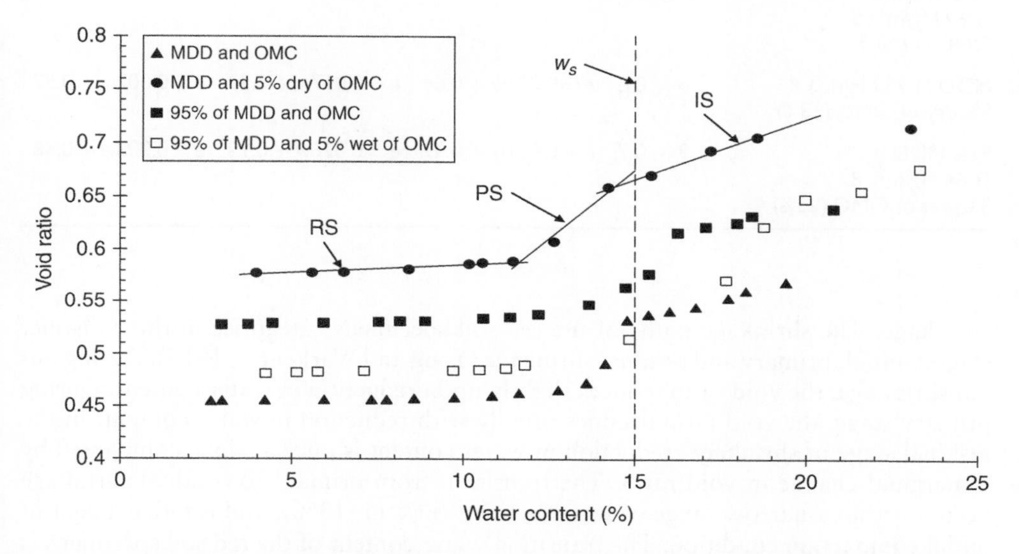

Figure 5.10 Shrinkage paths of swollen RS1 specimens: IS, PS, RS and w_s represent initial shrinkage, primary shrinkage, residual shrinkage and shrinkage limit; properties of red soil (RS1) specimens are given in Table 5.3

clays. The reduction in volume (Δv) calculates the void ratio and water content of the specimen at any stage of drying from the relations (Mishra *et al.*, 2008):

$$\frac{\Delta v}{V_0} = \frac{\Delta e}{1 + e_0} \tag{5.8}$$

$$w = \frac{w_0 - (\Delta v \times \rho_w)}{w_s} \times 100\% \tag{5.9}$$

where V_0, e_0 and w_0 are the volume, void ratio and water content of the specimen in the swollen state; Δv and Δe represent the reduction in soil volume and void ratio at any stage of drying, ρ_w is the density of water and w is the gravimetric water content of the specimen at any stage of drying. The drying process is terminated when the mass of the specimens becomes nearly constant.

Figure 5.10 (Mishra *et al.*, 2008) plots the void ratio-water content paths of swollen (at vertical pressure of 6.25 kPa in oedometer cells) red soil (RS1) specimens during

Table 5.12 Shrinkage magnitudes of red soil (RS1) specimens

	Shrinkage			Linear regression equation		
Compaction condition	*IS (%)*	*PS (%)*	*RS (%)*	*Initial shrinkage (IS)*	*Primary shrinkage (PS)*	*Residual shrinkage (RS)*
MDD & OMC (1.73 Mg/m^3) (17%)	2	5	0.6	$e = 0.015w + 0.41$	$e = 0.038w - 0.05$	$e = 0.001w + 0.45$
95% of MDD (1.64 Mg/m^3) & OMC (17%)	1	5	0.7	$e = 0.005w + 0.54$	$e = 0.028w + 0.19$	$e = 0.001w + 0.52$
MDD (1.73 Mg/m^3) & 5% dry of OMC (13%)	1	6	0.6	$e = 0.003w + 0.64$	$e = 0.017w + 0.39$	$e = 0.002w + 0.57$
95% MDD (1.64 Mg/m^3) & 5% wet of OMC (23%)	2	9	0.7	$e = 0.011w + 0.42$	$e = 0.026w + 0.12$	$e = 0.003w + 0.48$

shrinkage. The shrinkage paths of the red soil specimens categorise in three distinct stages: initial, primary and residual shrinkage (Yong and Warkentin, 1975). During initial shrinkage, the void ratio reduces slightly upon reduction in water content. During primary stage, the void ratio declines rapidly with reduction in water content. In the residual stage of shrinkage, reduction in water content is once again accompanied by a marginal change in void ratio. The transition from primary to residual shrinkage occurs within a narrow range of water content (i.e., 11–13%), and is independent of initial compaction condition. The transition water content of the red soil specimens is close to its shrinkage limit of 15%. The magnitudes of initial, primary and residual shrinkage (in percentage) are calculated from the equations (Mishra *et al.*, 2008):

$$\text{Initial Shrinkage (IS)} = \frac{(e_s - e_{is})}{1 + e_s} \times 100 \tag{5.10}$$

$$\text{Primary Shrinkage (PS)} = \frac{(e_{is} - e_{ps})}{1 + e_{is}} \times 100 \tag{5.11}$$

$$\text{Residual Shrinkage (RS)} = \frac{(e_{ps} - e_{rs})}{1 + e_{ps}} \times 100 \tag{5.12}$$

where e_s is swollen void ratio and e_{is}, e_{ps} and e_{rs} are the void ratios at the end of initial, primary and residual shrinkage of the specimens. The initial shrinkage magnitude ranges from 1 to 2%, primary shrinkage from 5 to 9% and residual shrinkage is <1% (Table 5.12). The red soil specimens with higher initial (swollen) water content also experience larger shrinkage (Yong and Warkentin, 1975).

Equations based on linear regression analysis are listed in Table 5.12 for the red soil specimens. The slopes of the linear segments for primary shrinkage region (0.017–0.038) are higher than slopes for initial (0.003–0.015) and residual (0.001–0.003) shrinkage regions.

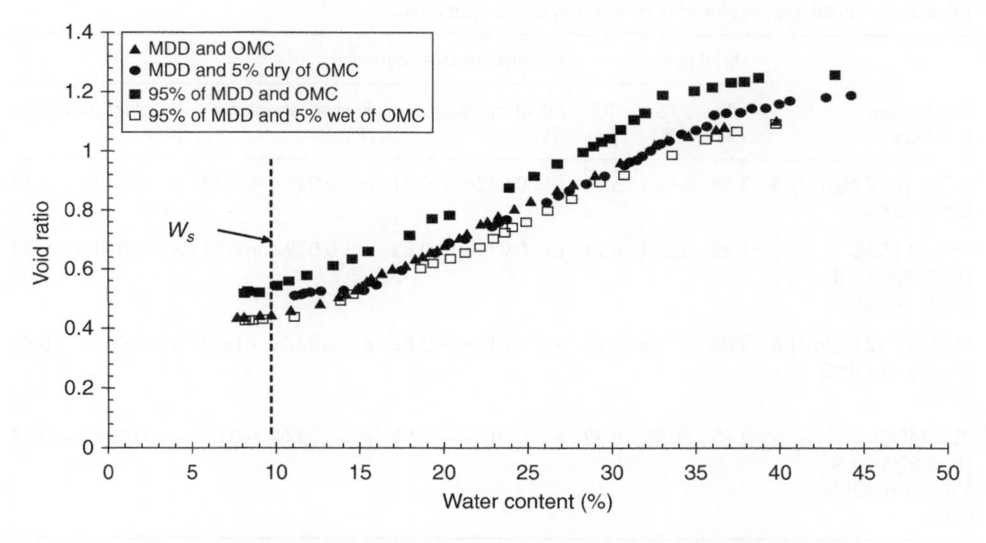

Figure 5.11 Shrinkage paths of swollen BCS5 specimens; w_s = shrinkage limit

Figure 5.11 (Mishra *et al.*, 2008) plots the void ratio-water content paths during shrinkage of swollen (at vertical pressure of 6.25 kPa in oedometer cells) black cotton soil (BCS5) specimens. The black cotton soil specimens also shrink in three stages; transition from primary to residual shrinkage occurs at water contents (13–15%), that is, slightly larger than the shrinkage limit of the soil ($w_s = 10\%$). Primary shrinkage magnitude varies between 24 and 32%, while the residual shrinkage magnitude ranges from 0.2 to 6% for the black cotton soil specimens. The slopes for the initial, primary and residual shrinkage regions for the black cotton soil specimens (Table 5.13) correspond to 0.011–0.013, 0.027–0.029 and 0.004–0.012 respectively (Mishra *et al.*, 2008).

Table 5.14 compares the shrinkage magnitudes of the black cotton soil (BCS5) and red soil (RS1) specimens (Mishra *et al.*, 2008). The montmorillonitic black cotton soil specimens are characterised by larger initial water contents (23–33%, Table 5.14) and therefore experience larger primary shrinkage magnitudes (24 to 32%).

Prediction of shrinkage settlement

Briaud *et al.* (2003) have suggested that the shrinkage settlement (Δh_{sh}) of uniform soil layer of thickness z_i can be calculated from the equation:

$$\Delta h_{sh} = \frac{z_i \cdot f \cdot \Delta w}{E_w} \tag{5.13}$$

where z_i is the thickness of the soil layer experiencing shrinkage settlement, f is the ratio of axial shrinkage strain to volumetric shrinkage strain and is approximated as 0.33 (Briaud *et al.*, 2003), Δw is the reduction in water content of the soil layer upon

Table 5.13 Shrinkage magnitudes of swollen BCS5 specimens

	Shrinkage			*Linear regression equation*		
Compaction condition	*IS (%)*	*PS (%)*	*RS (%)*	*Initial shrinkage (IS)*	*Primary shrinkage (PS)*	*Residual shrinkage (RS)*
MDD (1.42 Mg/m^3) & OMC (28%)	3.88	24.43	5.96	$e = 0.012w + 0.71$	$e = 0.027w + 0.13$	$e = 0.012w + 0.33$
95% of MDD (1.35 Mg/m^3) & OMC (28%)	1.43	32.14	0.21	$e = 0.011w + 0.82$	$e = 0.029w + 0.21$	$e = 0.018w + 0.34$
MDD (1.42 Mg/m^3) & 5% dry of OMC (23%)	3.08	27.23	2.33	$e = 0.011w + 0.82$	$e = 0.027w + 0.12$	$e = 0.004w + 0.47$
95% MDD (1.35 Mg/m^3) & 5% wet of OMC (33%)	2.65	30.95	0.32	$e = 0.013w + 0.59$	$e = 0.027w + 0.08$	$e = 0.004w + 0.37$

Table 5.14 Shrinkage magnitudes of residual soil specimens

	RS1 specimens			*BCS5 specimens*		
Compaction condition	*IS (%)*	*PS (%)*	*RS (%)*	*IS (%)*	*PS (%)*	*RS (%)*
MDD & OMC	2.34	4.56	0.61	3.88	24.43	5.96
95% of MDD & OMC	1.12	5.13	0.65	1.43	32.14	0.21
MDD & 5% dry of OMC	1.09	6.41	0.63	3.08	27.23	2.33
95% of MDD & 5% wet of OMC	2.37	8.91	0.67	2.65	30.95	0.32

shrinkage and E_w is the shrink-swell modulus and represents the slope of water content versus volumetric strain/void ratio plot and is defined as:

$$E_w = \frac{\Delta w}{\Delta v / v_0} \tag{5.14}$$

where $\Delta w = (w_0 - w)$; w_0 = initial water content corresponding to zero volume change; w = water content corresponding to the relative volume change $\Delta V/V_0$. The shrink-swell modulus E_W is the reciprocal of the slope of the void ratio versus water content (in percentage) plot generated during shrinkage (Figure 5.11). The slope of the void ratio versus water content plot (Figure 5.11) is defined as the index of volumetric compressibility, C_W (Nelson and Miller, 1992).

Bulk of the shrinkage of the expansive black cotton soil (BCS5) occurs in the primary shrinkage region (Table 5.14) and hence the shrink-swell modulus for primary shrinkage is considered in the example for estimating shrinkage settlements of a black cotton soil deposit. The index of volumetric compressibility for the black cotton soils

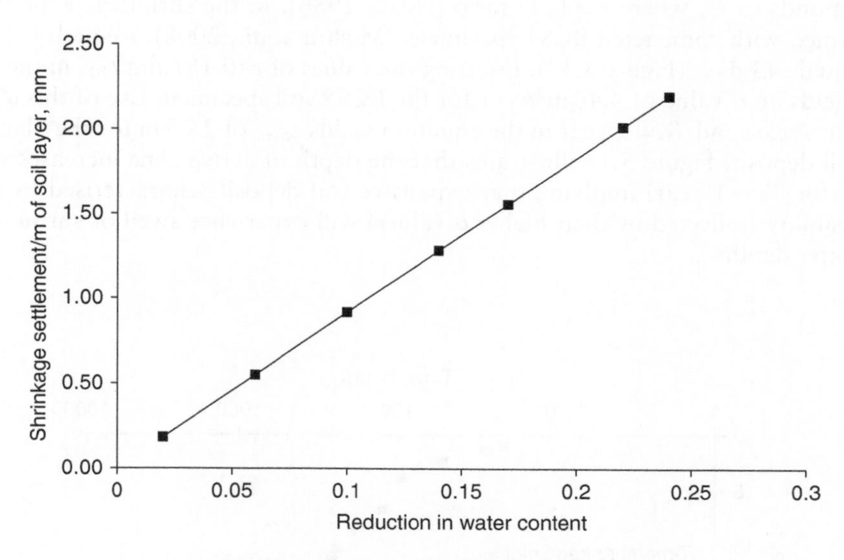

Figure 5.12 Shrinkage settlement as function of Δw

varies between narrow range of 2.7 and 2.9 (Table 5.13) and the average value 2.8 is considered in the example calculation. The swollen BCS5 specimens experience up to 24% decrease in water content during primary shrinkage (Figure 5.11). The shrinkage settlement experienced by 1 m thick black cotton soil layer calculated using equation 5.13 for Δw values ranging from 0 to 24% is plotted in Figure 5.12 ($f = 0.33$, $\Delta w = 0.02$ to 2.4 and $E_w = 0.36$ which is the reciprocal of 2.8). The 1 m thick black cotton soil layer experiences shrinkage settlements of 0.2 to 2.2 mm upon reduction in water contents of 2 to 24%.

The expansive soil deposit will experience swell or shrinkage in the soil zone that is subjected to moisture content fluctuations. This zone is generally termed as the zone of seasonal fluctuation or the active zone (Nelson and Miller, 1992). Briaud *et al.* (2003) suggest that the depth of active zone (z_{max}) can be calculated from the equation:

$$z_{\max} = 1.3(T_0\alpha)^{0.5} \tag{5.15}$$

where T_0 is related to duration of one wetting-drying cycle and may range from 0.74 to 2 years depending on the local climatic conditions, and α is a parameter similar to the time rate of consolidation (c_v, units m^2/year) and can be derived from an equation of the type (Head, 1986):

$$c_v = \alpha = \frac{\pi D^2}{\lambda t_{100}} \tag{5.16}$$

where D is diameter of sample, and λ is a parameter depending on the drainage boundary conditions. For the shrinkage test performed in the laboratory (Mishra *et al.*, 2008), water can evaporate from both the ends of the specimen and therefore the value of λ

corresponds to r^2, where $r = L/D$ ratio (Head, 1986). In the shrinkage experiments performed with compacted BCS4 specimens (Mishra *et al.*, 2008), r equals 0.19 and t_{100} equals 43 days (Figure 5.13). Inserting the values of r (0.19) and t_{100} in the equation yields an α value of 4.46 m^2/year for the BCS5 soil specimen. Use of this α value (4.46 m^2/year) and $T_0 = 1$ year in the equation yields z_{max} of 2.75 m for the black cotton soil deposit. Figure 5.14 illustrates that the depth of active zone increases with α value (for $T_0 = 1$ year) implying that expansive soil deposits characterised by larger permeability (reflected by their higher α values) will experience swell or shrinkage up to greater depths.

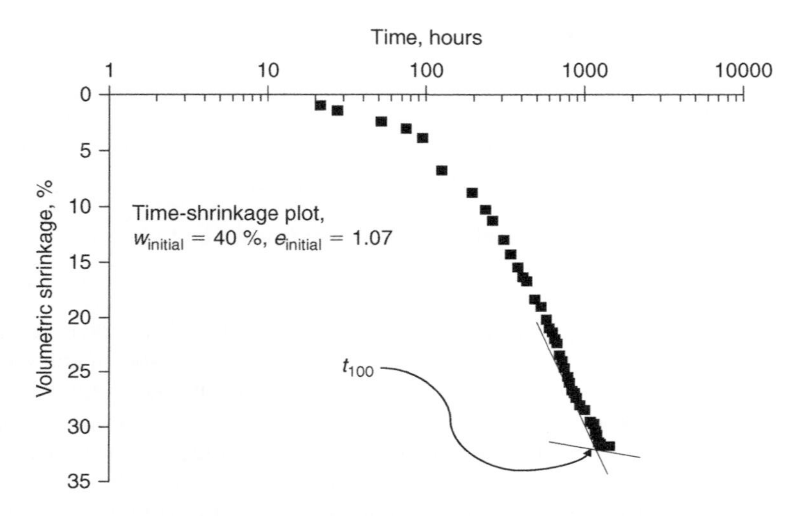

Figure 5.13 Time-shrinkage plot of BCS4 soil specimen

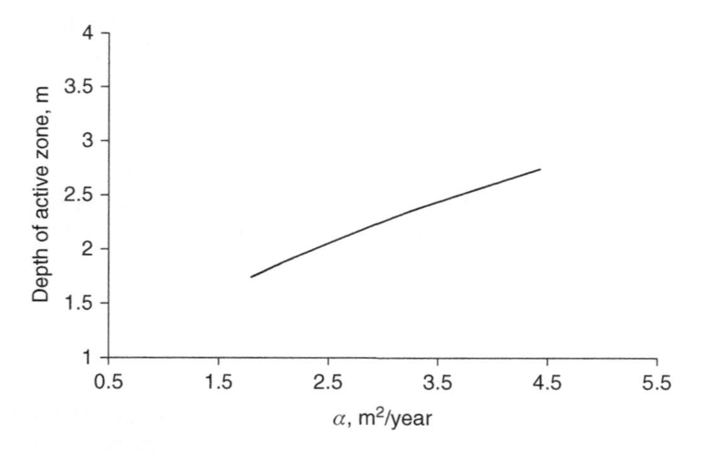

Figure 5.14 Variation of depth of active zone with α

5.3 COLLAPSIBLE RESIDUAL SOILS

Certain categories of residual soils exhibit collapse settlement (Fookes, 1990; Barksdale and Blight, 1997). The first category of collapsing residual soils is transported soils that have undergone post-depositional pedogenesis. An example of this soil type is the sandy soils occurring in Southern Africa (Jennings, 1957). The sands are believed to be either colluvial or windblown deposits of Quaternary age and contain a small percent of silt and clay because of partial in-situ weathering. The second category of collapsing residual soils is the unusually highly weathered and leached residual soils. An example of this category is the red soils of Brazil formed by the decomposition of gneiss and basalt (Vargas, 1953, 1973) and the partially saturated residual soils of Bangalore District (Siddappa, 2000). Figure 5.15 presents the frequency distribution of the in-situ void ratios of red soils of Bangalore District. The in-situ void ratios of these residual soils vary between 0.4 and 1.5 with bulk (87%) of the values ranging between 0.5–1.1 (Figure 5.15). Based on IAEG (International Association of Engineering Geologists) classification (Fookes, 1990) of void ratio and porosity (Table 5.15), the red soils of Bangalore mostly classify as moderately to highly porous. Examination of the S_r values of undisturbed residual soil samples showed their S_r values to range between 20 and 100% (Figure 5.16). The bulk of the samples (89%) have S_r values less than 70%, while only 11% of the samples have degree of saturation (S_r) in excess of 70%.

Table 5.15 IAEG* classification void ratio and porosity

Term	*Void ratio*	*Porosity (%)*
Very high	>1.00	>50
High	1.00–0.80	50–45
Medium	0.80–0.50	45–35
Low	0.50–0.43	35–30
Very low	<0.43	<30

*Fookes (1990)

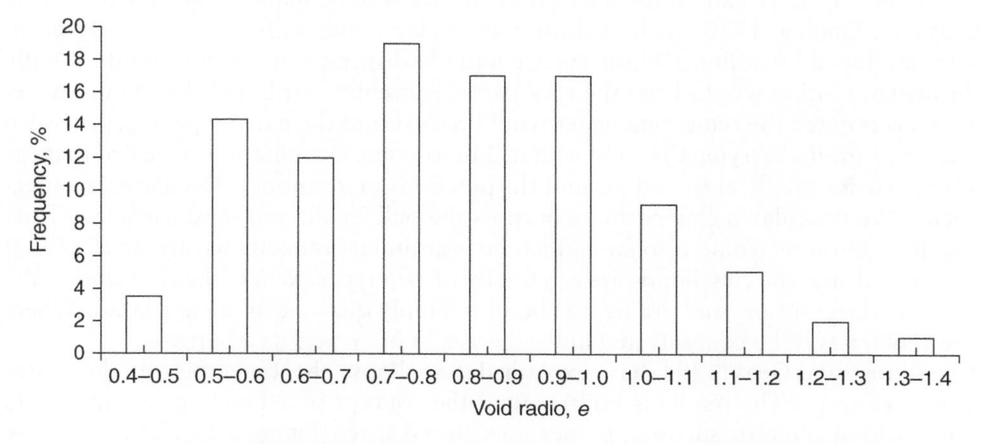

Figure 5.15 Frequency distribution of void ratios of undisturbed red soil specimens

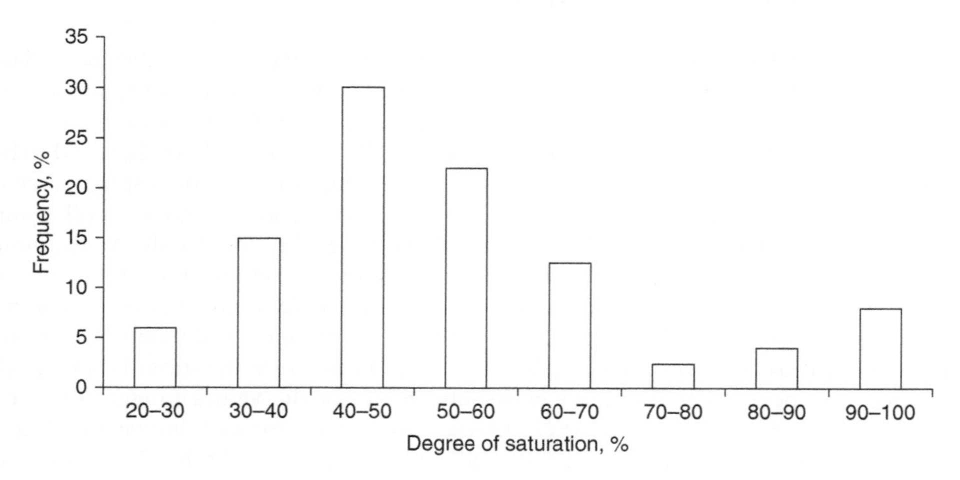

Figure 5.16 Frequency distribution of degree of saturation values of undisturbed red soil specimens

Mechanisms of collapse

Appreciable collapse is experienced by unsaturated soils when the following conditions are met (Barden *et al.*, 1973; Mitchell, 1993):

1 The soil has an open, potentially unstable, unsaturated structure.
2 A high enough value of external stress is applied to develop a metastable condition.
3 A high enough value of matric suction is available to stabilise the inter-granular contacts and whose reduction on wetting leads to collapse of the soil.

Dudley (1970) and Barden *et al.* (1973) have highlighted the role of soil structure in the collapse behaviour of natural soils in the context of inter-granular bonds. Both researchers have visualised the inter-granular bonds to be mainly contributed by clay particles. Dudley (1970) visualised the role of clay bonds as follows. The clay particles developed by authigenic process are initially dispersed in the pore fluid. Initially Brownian motion would keep the clay particles evenly distributed. But as the water later evaporated the remaining water would retreat into the narrow passages between the larger grains carrying the solid with it. Microscopic examinations revealed that the clay particles finally clustered around the junctions in a random flocculated arrangement. The flocculated clay particles buttress the bulkier silt and sand particles. Gross capillary tensions would also strengthen inter-granular contacts. Barden *et al.* (1973) postulated that the clay bonds are essentially of two types. Where the clay clothes the surface of the silt or sand grains, the bond is simply quartz-clay-quartz bond. Where the clay tends to be concentrated in local areas, it often acts as a buttress.

Alonso and Gens (1994) have explained the collapse phenomenon from the elasto-plastic concept. The model is built around the concept of a Loading-Collapse yield curve within s (matric suction), p (mean net stress) space (Figure 5.17). The L-C curve bounds an elastic zone and wetting inside this curve (decrease in s) is expected to induce

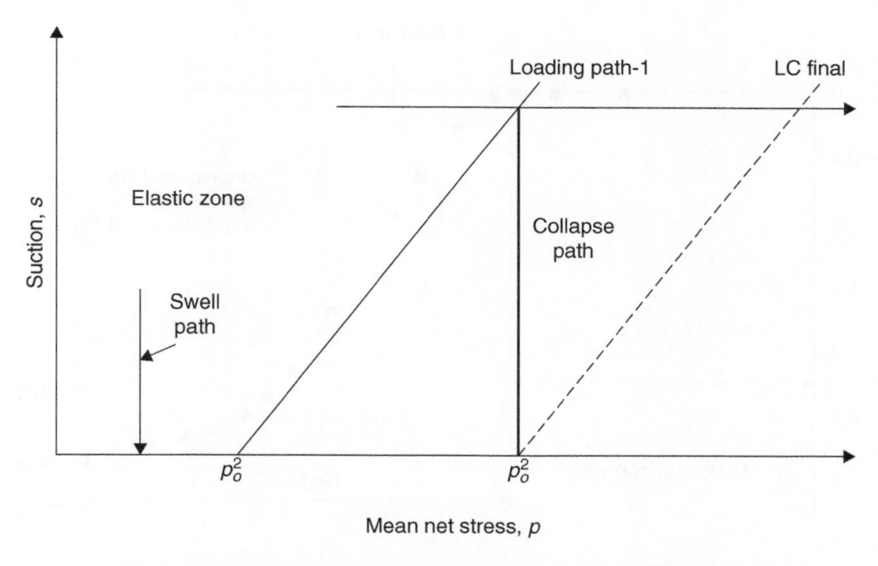

Figure 5.17 Yield curves for load-collapse behaviour of an unsaturated soil

elastic swelling of the unsaturated specimen. Wetting (decrease in s) the unsaturated specimen at pressures (p) beyond the yield locus shifts the yield curve to a new position (shown by the dotted line in Figure 5.17) and plastic compression or collapse occurs as the clay buttresses and cementation bonds are no longer able to resist the shear induced by contact forces, once the suction is reduced. Catastrophic collapse of the unsaturated bonded specimen occurs as the work-softening effect arising from damage to bonds exceeds the normal work-hardening effect of plastic strains (Wheeler, 1994). Degeneration of inter-granular cementation bonds may largely be a product of motion of granular particles resulting from the loss of two other sources of strength-matric suction and clay buttresses (Dudley, 1970). Two stress paths are shown in Figure 5.17; (1) loading path from increase in p, and (2) collapse path from reduction in s. Both stress paths take the LC curve to the same final position and therefore produce the same plastic deformations. Parameters $(p_o^*)_1$ and $(p_o^*)_2$ represent the pre-consolidation pressure of the soil specimen in the unsaturated and saturated conditions (Figure 5.17).

Laboratory determination of collapse potential

Two loading-wetting sequences are used in the laboratory to determine the collapse potential of unsaturated soils (Lawton *et al.*, 1992). The more common method consists of loading the unsaturated soil to a specific state of stress without soaking the specimen, allowing the sample to attain equilibrium under the applied stress, and then providing the soil with access to water. This soaked-after-loading method is also known as single oedometer collapse test (Fookes, 1990; Lawton *et al.*, 1992). The second load-wetting sequence is the double oedometer test method of Jennings and Knight (1957). This method is based on the assumption that the deformations induced by wetting are

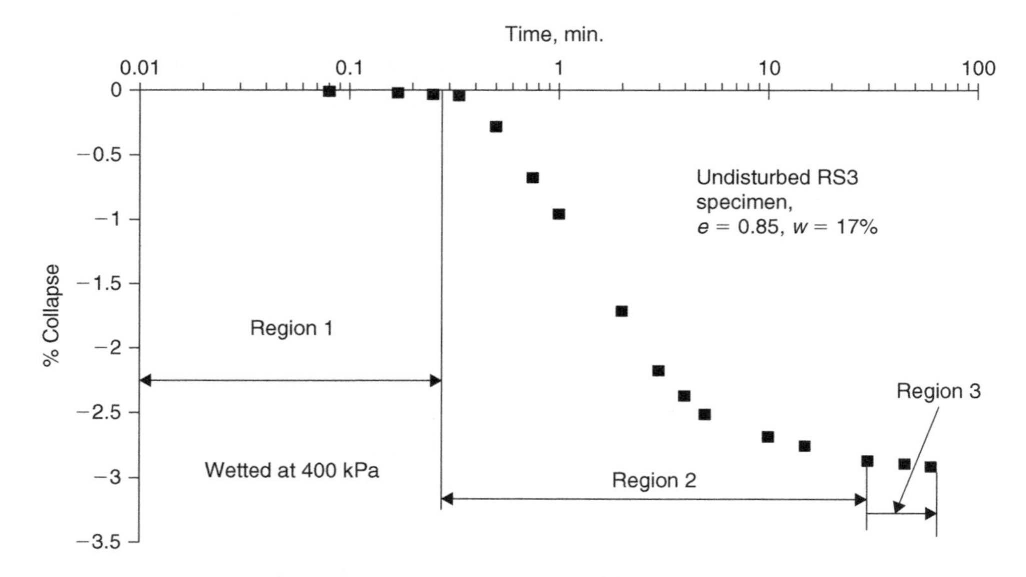

Figure 5.18 Time-collapse behaviour of undisturbed RS3 specimen

independent of the loading-wetting sequence (Fookes, 1990; Lawton *et al.*, 1992). More details of the two oedometer procedures are provided in the ensuing sections.

Soaked After Loading Test (SALT) method

In this test, the compacted/undisturbed soil specimen is incrementally loaded up to a specific state of stress without wetting the specimen. On attainment of equilibrium under the applied load, the soil specimen is inundated with water. Time-collapse observations are continuously noted during the collapse process (Figure 5.18). The time-collapse plot categorises into three distinct regions (Lawton *et al.*, 1992); region 1 where an initial time lapse occurs prior to water infiltrating the soil sample, region 2, where the bulk of collapse occurs as water infiltrates through the soil specimen, and region 3 where some residual collapse occurs. After residual collapse is complete (or region 3 is established), additional loadings are applied to the soil specimen. Collapse of sandy to clayey samples rapidly occurs ($\leq$60 min) in laboratory oedometer tests (Booth, 1977; Lawton *et al.*, 1992).

The soaked-after-loading test data is plotted as e (void ratio) versus log p (pressure) plot (Figure 5.19). The collapse potential of the soil specimen is determined from the equation:

$$\text{Collapse potential} = \frac{-\Delta e}{1 + e_{unsoaked}} \times 100\% \tag{5.17}$$

where $-\Delta e$ is the reduction in void ratio of the unsaturated specimen on wetting and $e_{unsoaked}$ is the void ratio of the unsoaked soil specimen. The compacted red soil specimen exhibits collapse potential of -12.1% on wetting at vertical stress of 200 kPa.

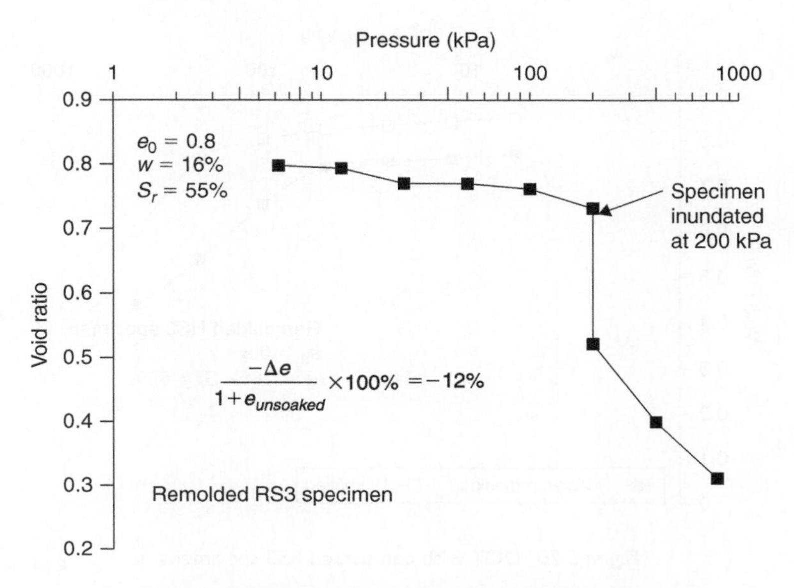

Figure 5.19 SALT method on compacted RS3 specimen

Table 5.16 Severity of soil collapse (*After* Fookes, 1990)

Percent collapse	*Severity of problem*
0–1	No problem
1–5	Moderate trouble
5–10	Trouble
10–20	Severe trouble
20	Very severe trouble

Table 5.16 classifies the severity of collapse of unsaturated soils based on their percent collapse values (Fookes, 1990). The compacted red specimen classifies as severe troublesome soil based on its collapse potential value of −12.1% at 200 kPa.

Double Oedometer Test (DOT) method

In this procedure, two identical specimens of a soil are tested in the conventional oedometer assemblies. One specimen is stepwise loaded in the unsoaked condition up to the desired vertical pressure (example, 400 kPa). The other specimen is inundated at a nominal pressure of 6.25 kPa. After heave or collapse of the soaked soil is completed, additional loads are applied to the soaked specimen. Duration and sequence of loading were identical for the unsoaked and soaked specimens (Figure 5.20). The difference between the void ratios (Δe) of the soaked and unsoaked soil specimens at any vertical stress divided by the value of $1 + e_{unsoaked}$ (equation 5.17) represents the collapse strain (Figure 5.20) that this specimen will experience on wetting at the specified load. It may be noted from Figure 5.20 that the compacted red soil specimen

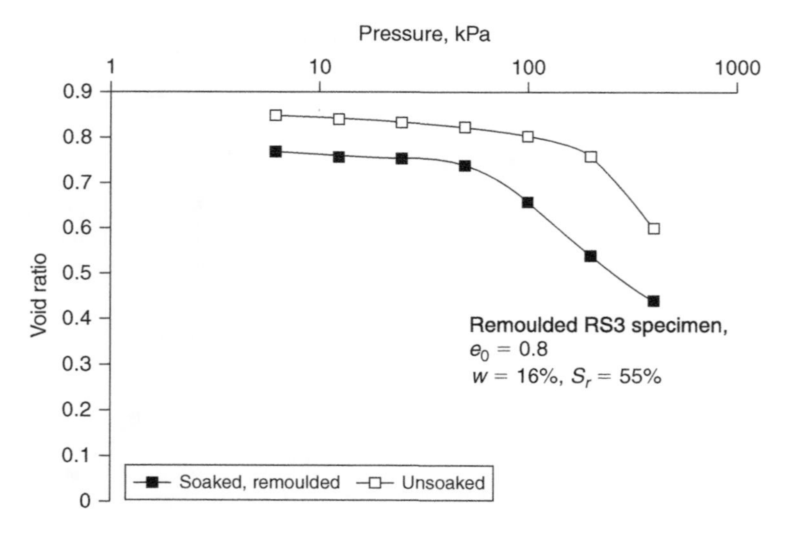

Figure 5.20 DOT with compacted RS3 specimens.

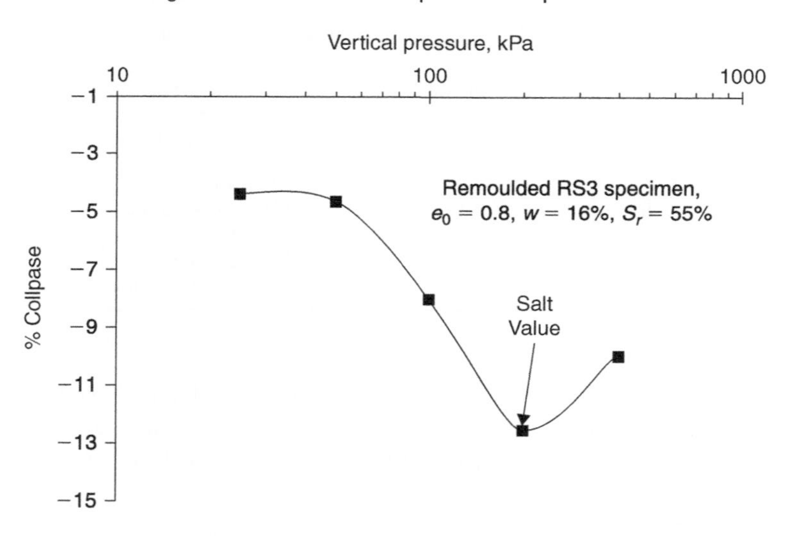

Figure 5.21 Prediction of collapse potential by DOT procedure for remoulded RS3 specimen

has a tendency to collapse on wetting at all experimental stresses. Figure 5.21 plots the collapse potential predicted by the DOT method for compacted red soil specimens for wetting at various vertical stresses. The DOT procedure predicts that the specimen will experience collapse potential of −12.5% on wetting at 200 kPa, that is, similar to the value measured in the SALT method (−12.1%).

Identification of collapsible soils using index tests

Rapid methods of identification of collapsible soils have been developed using index properties such as natural water content, degree of saturation, natural void ratio and

Table 5.17 Identification criteria for collapsing soils (*After* Fookes and Best, 1969)

Investigator	*Criteria*
Denisov (1951)	Coefficient of subsidence: $K = \frac{e_L}{e_0}$ where, e_L is liquid limit void ratio and e_0 is natural void ratio $K = 0.5$ to 0.75, highly collapsible $K = 1.0$, non-collapsible loam $K = 1.5$ to 2.0, non-collapsible soils
Soviet Building Code (1962)	$L = \frac{e_0 - e_L}{1 + e_0}$ For natural degree of saturation $<60\%$, if $L > -1$, it is a collapsible soil
Feda (1964)	$K_L = \frac{w_0/S_r - w_p}{I_p}$ where w_0 is natural water content, S_r = natural degree of saturation, w_p is plastic limit and I_P is plasticity index For $S_r < 100\%$, if $K_L > 0.85$ it is a collapsible soil
Prilonski (1952)	$K_D = \frac{w_0 - w_P}{I_P}$ $K_D < 0$ highly collapsible soils $K_D > 0.5$ non-collapsible soils $K_D > 1$ for swelling soils

Table 5.18 Index properties of compacted red soil (RS2) specimens

Series	ρ_d, *Mg/m*^{3}	e	*w*, %	S_r, %	w_L, %	I_p, %
I	1.4 (RC = 79%)	0.91	10.6 17.6	31 52	37	18
II	1.49 (RC = 84%)	0.80	10.6 17.6	36 59	37	18
IV	1.77 (RC = 100%)	0.51	10.6 17.6	56 92	37	18

Atterberg limit water content (Denisov, 1952; Prilonski, 1952; Soviet Building Code, 1962; Feda, 1964) as referenced by Fookes and Best (1969). These tests, primarily developed for collapsing aeolian soils, are summarised in Table 5.17. Table 5.18 (Rao, 2002) provides the index properties of compacted red soil (RS2, Table 5.3) specimens whose compaction details are provided in Table 5.20. Table 5.19 (Rao, 2002) classifies the collapse behaviour of the compacted red soil specimens.

The Denisov criteria, Soviet Building Code and Feda Criteria classify the compacted red soil specimens as non-collapsing (Table 5.19). These compacted red soil specimens are observed to significantly collapse at most vertical stresses in oedometer tests (Figures 5.22, 5.23, 5.24). The Prilonski criterion based on natural water content

Table 5.19 Collapse classification of compacted red soil (RS2) specimens

Series	*w, %*	*Denisov*	*Soviet building code*	*Prilonski*	*Feda*
I	10.6	1.12 NC	−0.06 NC	−0.47 C	0.84NC
	17.6	1.12 NC	−0.06 NC	−0.08 C	0.82 NC
II	10.6	1.28 NC	−0.12 NC	−0.47 C	0.58 NC
	17.6	1.28 NC	−0.12 NC	−0.08 C	0.60 NC
IV	10.6	2.0 NC	−1.0 C	−0.47 C	−0.004 NC
	17.6	2.0 NC	−1.0 NC	−0.08 C	−0.007 NC

NC = non-collapsing, C = collapsing

Table 5.20 Compaction details of red soil (RS2) specimens

Series	*RC (%)*	ρ_d *(Mg/m³)*	*w (%)*
I	79	1.4	10.6
			14.6
			17.6
II	84	1.49	10.6
			14.6
			17.6
III	94	1.66	10.6
			14.6
			17.6
IV	100	1.77	10.6
			14.6
			17.6

and Atterberg limits water contents classifies the compacted red soil (RS2) specimens as collapsing soils. According to these criteria, soils possessing coefficient of subsidence (K_D) < 0 are collapsing in nature (Table 5.17). As most residual soils (exceptions being allophanes and halloysitic residual soils) have natural water contents lower than their plastic limit they would classify as collapsible soils according to these criteria. Use of index properties for identification of collapse tendency of unsaturated soils hence tend to be unsatisfactory needs refinement.

Role of compaction conditions in collapse

Residual soils are extensively used in construction of compacted fills and embankments. The roles of compaction dry density, water content and vertical pressure in determining the collapse behaviour of compacted residual soils are essential to help identify the critical compaction parameters and applied stress level that need to be controlled by the fill designer in order to minimise their wetting-induced collapse.

The influence of compaction water content on the swell/collapse potentials of loosely (RC = 79%, $\rho_d = 1.40\,\text{Mg/m}^3$) and densely (RC = 100%, $\rho_d = 1.77\,\text{Mg/m}^3$) compacted red soil (RS2) specimens at different consolidation pressures is illustrated

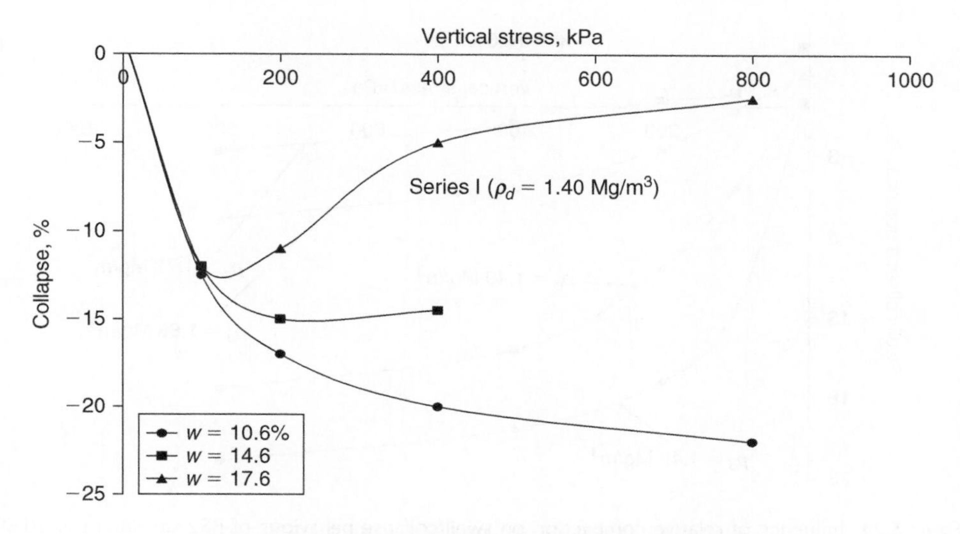

Figure 5.22 Collapse behaviour of series I-RS2 specimens

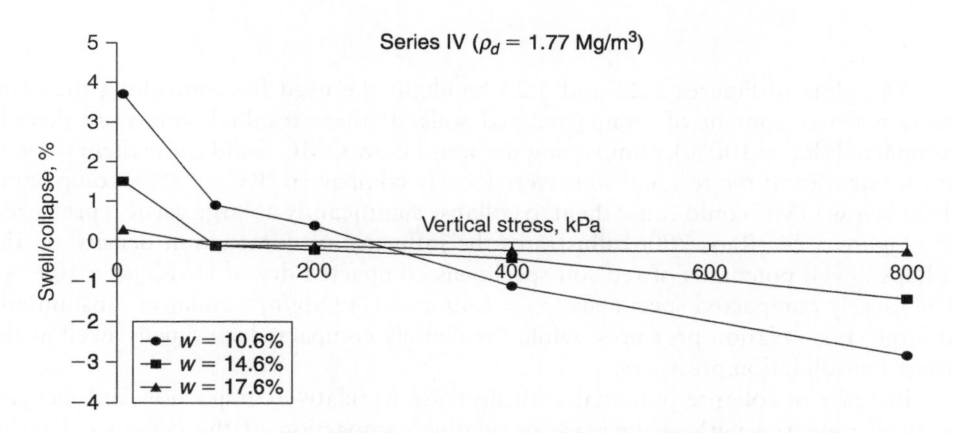

Figure 5.23 Swell/collapse behaviour of series IV-RS2 specimens

in Figures 5.22 and 5.23 (Rao, 2006). Plots in Figure 5.22 show that the variations in initial water content have no effect on the collapse behaviour of the loosely compacted (RC = 79%) specimens up to vertical pressures of 100 kPa. Interestingly, the densely compacted specimens (Figure 5.23) that were compacted at 7% dry of OMC ($w = 10.6\%$) exhibit notable swelling tendency at vertical stresses <100 kPa. In general, for any series, at large consolidation pressures, a compacted specimen with lower initial water content exhibits a higher collapse potential (Cox, 1978; Lawton *et al.*, 1989).

Comparatively, at lower scale of consolidation pressures (less than 100 kPa), the densely compacted residual soil specimens at lower initial water content, exhibit higher swell potentials.

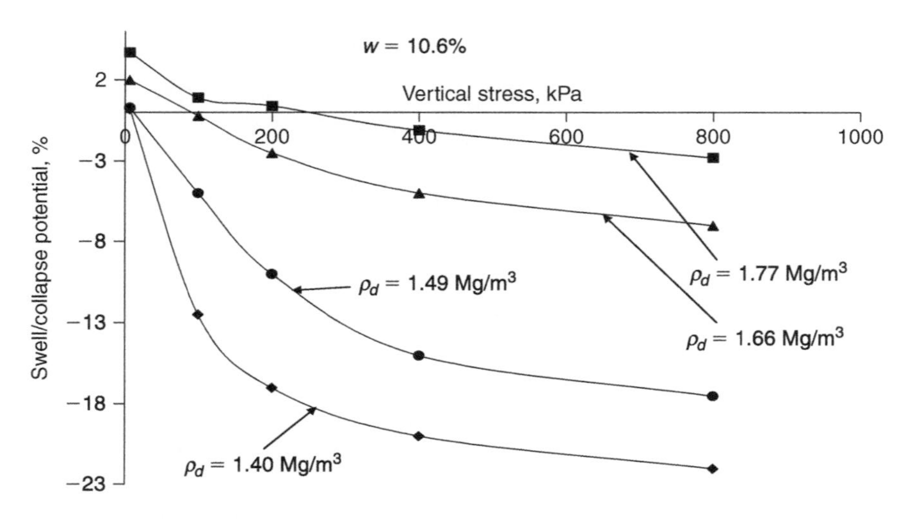

Figure 5.24 Influence of relative compaction on swell/collapse behaviour of RS2 specimens at 10.6% water content

The plots in Figures 5.22 and 5.23 highlight the need for controlling the compaction water content of compacted red soils. If these residual soils were densely compacted (RC = 100%), compacting the soils below OMC could cause them to swell at low stresses. If the residual soils were loosely compacted (RC = 79%), compacting them below OMC could cause them to collapse significantly at large vertical pressures.

Figure 5.24 (Rao, 2006) illustrates the influence of compaction density on the collapse/swell potentials of red soil specimens compacted dry of OMC ($w = 10.6\%$). The loosely compacted specimens ($\rho_d = 1.40$ and $1.49\ \text{Mg/m}^3$) collapse substantially at large consolidation pressures, while the densely compacted specimens swell at the lower consolidation pressures.

Increase in collapse potential with decrease in relative compaction and increase in swell potential with an increase in relative compaction of the compacted residual soil specimens is an established trend (Dudley, 1970; Foss, 1973, Lawton *et al.*, 1992). Figure 5.24 underlines the importance of controlling relative compaction of the red soils. Densifying ($\rho_d = 1.66$ to $1.77\ \text{Mg/m}^3$) red soil specimens at OMC imparts to them stability towards wetting-induced volume changes as opposed to the loosely compacted ($\rho_d = 1.40$ to $1.49\ \text{Mg/m}^3$) residual soil specimens (Figure 5.25). It is therefore beneficial to compact residual soil fills at OMC and at relative compactions $\geq 94\%$, to minimise wetting-induced volume changes at a range of overburden pressures (Figure 5.25).

The above results hence bring out the importance of properly controlling the compaction of red soil fills. Poor compaction leading to low dry density and water content of compacted fills and embankments must be avoided to minimise wetting-induced volume changes. Failure to do so can result in substantial collapse settlements of structures founded on them.

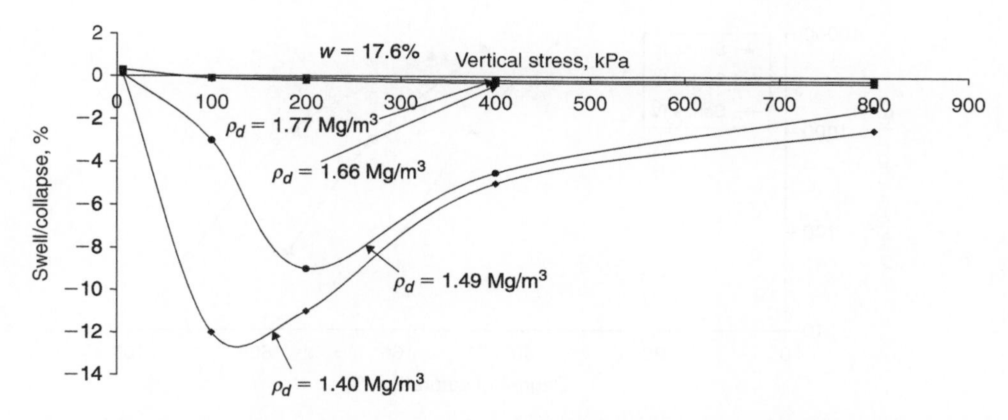

Figure 5.25 Influence of relative compaction on swell/collapse behaviour of RS2 specimens at OMC (17.6% water content)

Matric suction and collapse

The importance of matric suction in the collapse behaviour of unsaturated soils is well recognised. Structural stability is imparted to a collapsible soil by matric suction that stabilises the inter-granular contacts in the partially saturated condition. Addition of water destroys the capillary bonds and causes the inter-granular contacts to fail in shear, resulting in a reduction in total volume (Burland, 1961). This section examines the influence of compaction conditions on the matric suction of compacted red soil (RS2) specimens discussed in Table 5.20. It additionally examines the impact of variations in matric suction on the collapse behaviour of these compacted residual soils. The ASTM Filter Paper Method (ASTM, 1994) was used to evaluate the matric suction of the compacted soil specimens. The filter paper method is a simple, inexpensive method and can be reliably used to measure suctions from 80 kPa to values in excess of 1000 kPa (Hamblin, 1981; Fredlund and Rahardjo, 1993; Houston *et al.*, 1994; Ridley, 1995; Chandler and Gutierrez, 1986).

Influence of compaction conditions on matric suction

Variations in matric suction as function of degree of saturation of compacted red soil specimens (RC = 84 to 100%) are illustrated in Figure 5.26 (Rao, 2006). An increase in degree of saturation markedly decreases the matric suction of the compacted soil specimens. Variations in relative compaction at constant degree of saturation also impacts the matric suction (Figure 5.26), as smaller pores of the denser specimen (larger RC) are able to support larger matric suction.

Role of matric suction and collapse

Variations in collapse potential as function of matric suction are illustrated in Figures 5.27 and 5.28 (Rao, 2006) for compacted red soil specimens flooded at consolidation pressures of 200 and 400 kPa respectively. Matric suction of the specimens

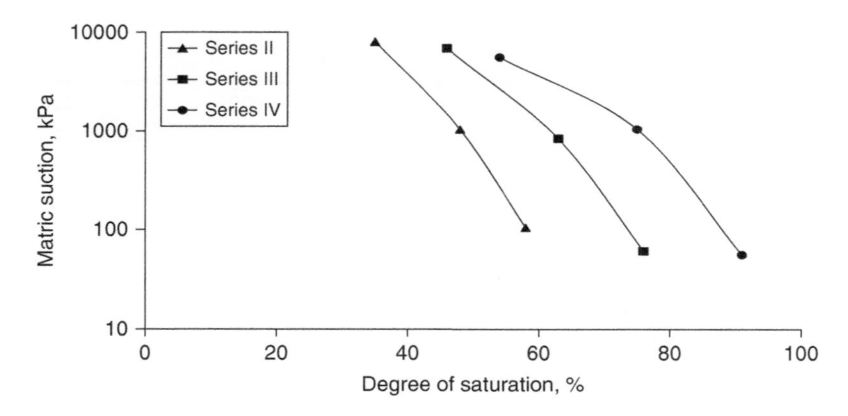

Figure 5.26 Matric suction of compacted RS2 specimens

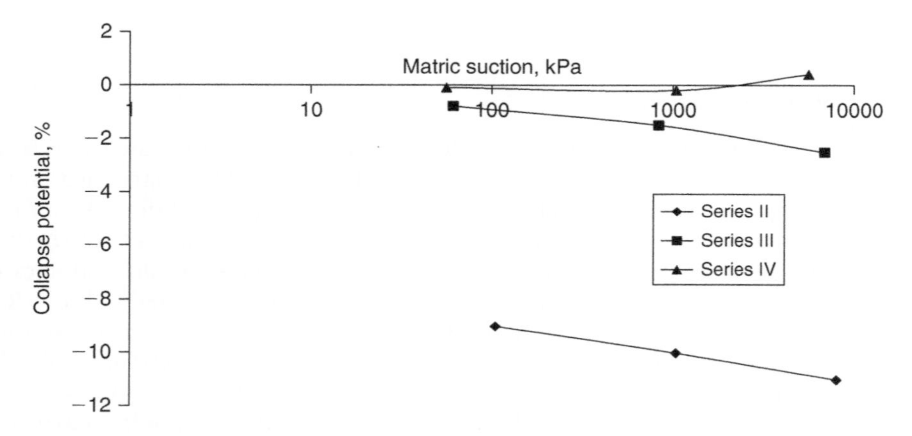

Figure 5.27 Influence of matric suction on collapse of RS2 specimens at 200 kPa

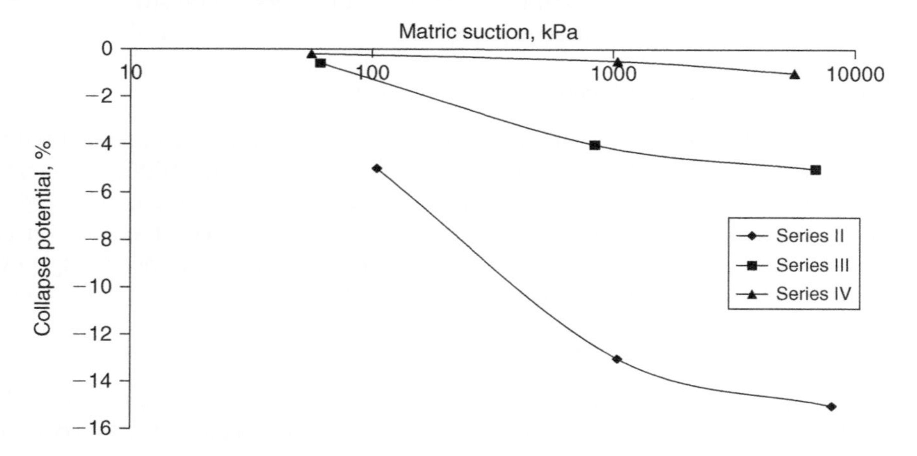

Figure 5.28 Influence of matric suction on collapse of RS2 specimens at 400 kPa

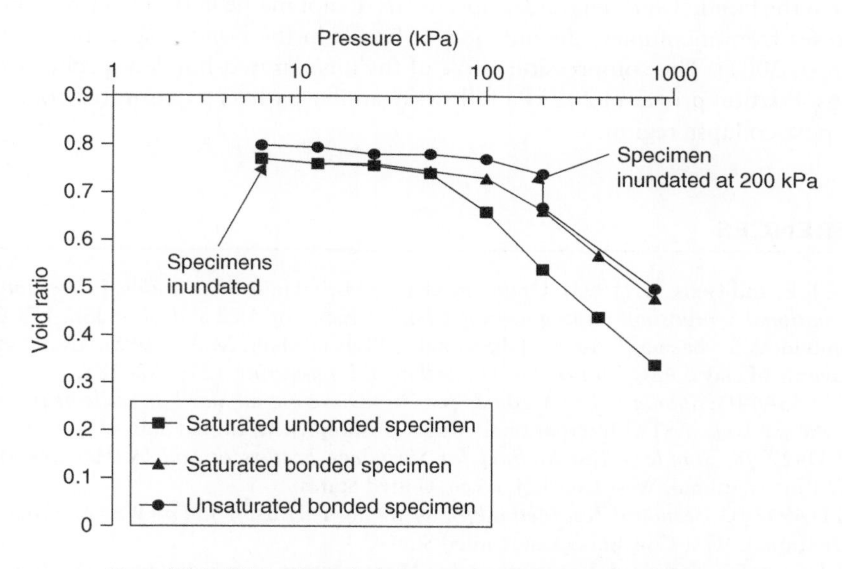

Figure 5.29 Effect of bonded structure on compression behaviour

varies from changes in compaction water content (10.6, 14.6, and 17.6%) at given relative compaction.

Variations in matric suction marginally impact the collapse tendency (collapse potential $<-1\%$) of densely compacted (RC = 100%, $\rho_d = 1.77\,\mathrm{Mg/m^3}$) red soil specimens despite being characterised by matric suctions of 56 and 5619 kPa (compaction water contents 14.6 and 10.6%). Loosely compacted specimens (RC = 84%, $\rho_d = 1.49\,\mathrm{Mg/m^3}$) characterised by matric suctions of 105 and 8042 kPa collapse by −4.4 and −15% on wetting at 400 kPa. Further, the collapse potentials of the soil specimens are more susceptible to changes in matric suction on wetting at 400 kPa than at 200 kPa (Figures 5.26 and 5.27). Influence of matric suction on the collapse behaviour of the compacted red soils depends on the relative compaction and the consolidation pressure at which they were inundated.

Role of bonding in collapse

Partially saturated residual soils are characterised by a bonded structure that plays an important role in their compression and strength behaviour (Vaughan, 1988; Fookes, 1990; Alonso and Gens, 1994; Barksdale and Blight, 1994; Wheeler, 1994). In moderately weathered soils, some of the bonding may be inherited from the parent rock. However, in a fully developed residual soil, it is more likely to be due to the effects of crystallisation during weathering and mineral alteration, and the precipitation of cementing material (Terzaghi, 1958; Newill, 1961; Vargas, 1973; Wallace, 1973; Mitchell and Sitar, 1982; Wesley, 1990).

The role of the bonded structure in the compression behaviour of red soil specimens is illustrated in Figure 5.29. Remoulded and bonded red soil (RS3) soil specimens were wetted at 6.25 kPa and stepwise loaded to 800 kPa in oedometer tests. The compression curves of the bonded and remoulded specimens do not merge in the experimental range of stresses from incomplete degradation of bonds in the bonded specimen (Rao and Siddappa, 2001). The compression curve of the unsaturated-bonded specimen wetted at (consolidation pressure) 200 kPa follows a similar path as its saturated counterpart in the post-collapse region.

REFERENCES

Alonso, E.E. and Gens. A. (1994). On the mechanical behaviour of arid soils. *Proceedings First International Symposium on Engineering Characteristics of Arid Soils*, London, 173–205.

Al-Homoud, A.S., Basma, Husein Malkavi and Al-Bashabshah, M.A. (1995). Cyclic swelling behaviour of clays. *ASCE Journal of Geotechnical Engineering* 121: 582–585.

ASTM D 4546-03: *Standard Test Methods for One-Dimensional Swell or Settlement Potential of Cohesive Soils*. ASTM International, West Conshohocken, United States.

ASTM D427-98. *Standard Test Method for Shrinkage Factors of Soil by Mercury Method.* ASTM International, West Conshohocken, United States.

ASTM D 4943-02. *Standard Test Method for Shrinkage Factors of Soil by Wax Method.* ASTM International, West Conshohocken, United States.

ASTM D5298-03. *Standard Test Method for Measurement of Soil Potential (Suction) Using Filter Paper*. ASTM International, West Conshohocken, United States.

Barksdale, R.D. and Blight, G.E. (1997). Compressibility and settlement of residual Soils, in *Mechanics of Residual Soils*, G. E. Blight, (ed.), Balkema, Rotterdam

Barden, L. Mcgown, A. and Collins, K. (1973). The collapse mechanism in partly saturated soils. *Engineering Geology*, 7: 49–60.

Bishop, A.W. and Blight, G.E. and Donald, B. (1960). Factors controlling the strength of partly saturated cohesive soils. *Proc. ASCE Conf. Shear Strength of Cohesive Soils*, Colorado, 503–532.

BIS 2720 Part 60. *Methods of Test for Soils: Determination of Free Swell Index of Soils*. Bureau of Indian Standards, New Delhi.

BIS 2720 Part 61. *Method of Test for Soils: Measurement of Swelling Pressure of Soils*. Bureau of Indian Standards, New Delhi.

Bolt, G. (1956). Physico-chemical analysis of the compressibility of pure clays. *Géotechnique* 6: 920–934.

Booth, A.R. (1977). Collapse settlement in compacted soils. *CSIR Research Report* 324, NITRR Bulletin 13, Pretoria, South Africa.

Briaud, J.L., Zhang, X. and Moon, S. (2003). Shrink test-water content methods for shrink and swell predictions. *ASCE Journal of Geotechnical and Geoenvironmental Engineering* 129: 590–600.

Burland, J.B. (1961). Discussion on Collapsible Soils. *Proceedings 5th International Conference on Soil Mechanics and Foundation Engineering*, Paris, Vol. 3, 219–220.

Chandler, R.J. and Gutierrez, C.I. (1986). The filter paper method of suction measurement. *Géotechnique*, 36: 265–268.

Chen, F.H. (1988). *Foundations on Expansive Soils*, Elsevier, New York.

Cox, D.W. (1978). Volume change of compacted clay fill. *Proceedings of Institution of Civil Engineers Conference on Clay Fills*, London, 79–86.

Das, B.M. (1998). *Principles of Geotechnical Engineering*, PWS Publishing Company, New York.

Day, R.W. (1994). Swell-shrink behaviour of compacted clay. *ASCE Journal of Geotechnical Engineering*, 120: 618–623.

Dudley, J.H. (1970). Review of collapsing soils. *ASCE Journal of Soil Mechanics and Foundation Division* 96, 925–947.

El-Sohby, M.A. and Rabbaa, S.A. (1984). Deformation behaviour of unsaturated soils upon wetting. *Proceedings, 8th Regional Conference for Africa*, Harare, 129–137.

Fookes, P.G. (1990). *Report on Tropical Residual Soils, The Quarterly Journal of Engineering Geology* 23:103–108.

Fookes, P.G. and Best, R. (1969). Consolidation characteristics of some late pleistocene periglacial metal stable soils of East Kent. *Quarterly Journal of Engineering Geology* 2: 103–128.

Foss, I. (1973). Red soil from Kenya as a foundation material. *Proceedings 8th International Conference on Soil Mechanics and Foundation Engineering*, Vol. 2, Moscow, 73–80.

Fredlund, D.G. and Rahardjo, H. (1993). *Soil Mechanics for Unsaturated Soils*, Wiley, New York.

Hamblin, A.P. (1981). Filter paper method for routine measurement of field water potential. *Journal of Hydrology* 53: 355–360.

Houston, S.L., Houston, W.N. and Wagner, A.M. (1994). Laboratory filter paper suction measurements. *ASTM Geotechnical Testing Journal*, 17: 185–194.

Head, K.H. (1986). *Manual of Soil Laboratory Testing: Volume 3, Effective Stress Tests*. Pentech press, London.

Holtz, W.G. and Gibbs, H.J. (1956). Engineering properties of expansive clays. *Trans. American Soc. Civil Engineers* 121: 641–663.

Jennings, J.K. and Knight, K. (1957). The additional settlement of foundations due to collapse structure of sandy subsoils on wetting. *Proceedings, 4th International conference on Soil Mechanics and Foundation Engineering*, London, 1: 316–319.

Lawton, E.C., Fragaszy, R.J. and Hotherington, M.D. (1992). Review of wetting induced collapse in compacted soil. *ASCE Journal of Geotechnical Engineering* 118: 1376–1394.

Lu, N. and Likos, W.J. (2004). *Unsaturated Soil Mechanics*. Wiley, New York.

Mishra, A.K., Dhawan, S. and Rao, S.M. (2008). Analysis of swelling and shrinkage behaviour of compacted clays. *Geotechnical and Geological Engineering* 26: 289–298.

Mitchell, J.K. (1993). *Fundamentals of Soil Behaviour*. Wiley, New York.

Mitchell, J.K. and Sitar, N. (1982). Engineering properties of tropical residual soils. *ASCE Geotechnical Engineering Specialty, Conference on Engineering and Construction in Tropical Residual Soils*, Hawaii, 30–57.

Murthy, N.S. (1988). *Factors and mechanisms controlling volume change behaviour of fine grained soils*. Ph.D. Dissertation, Indian Institute of Science, Bangalore 560012, India.

Nelson, J.D. and Miller, D.J. (1992). *Expansive Soils Problems and Practices in Foundation and Pavement Engineering*. Wiley, New York.

Newill, D.A. (1961). Laboratory investigation of two red clays from Kenya. *Géotechnique*, 11: 302–318.

Rao, S.M. (2002). Collapse behavior of red soils. *Proc. Second International Conference on Geotechnical and Geoenvironmental Engineering in Arid Lands*, Riyadh, Saudi Arabia, (H. A. Alwaji, (ed.), Balkema, Tokyo, 79–85.

Rao, S.M. (2006). *Identification and Classification of Expansive Soils. Expansive Soils; Recent Advances in Characterization and Treatment*, Al-Rawas and Goosen (ed.), Taylor and Francis, New York, 15–24.

Rao, S.M. and Venkataswamy, B. (2002). Lime pile treatment of black cotton soils. *Ground Improvement*, 6(2): 85–93.

Rao, S.M. and Reddy, P.M.R. (1998). Physico-chemical behaviour of dry silty clays, *ASCE Journal of Geotechnical and Geoenvironmental Engineering* 124: 451–453.

Ramaiah, B.K. and Sheshagiri Rao, K. (1969). *Soil distribution and engineering problems in Bangalore area*. Research Bulletin, Golden Jubilee Volume, Bangalore University, Bangalore.

Radhakrishna, B.P. and Vaidyanadhan, R. (1997). *Geology of Karnataka*. Geological Society of India, Bangalore.

Reddy, D.V. (1995). *Engineering Geology for Civil Engineers*. Oxford & IBH Publishing Company, New Delhi.

Ridley, A.M. (1995). Discussion-Laboratory filter paper suction measurements. *ASTM Geotechnical Testing Journal*, 18: 391–396.

Sarita, D. (2004). *An experimental study on shrinkage behaviour and SWCC relations of clays*. MSc. (Engg.) Dissertation, Indian Institute of Science, Bangalore, India.

Seed, M.B., Woodward, R.J. and Lundgren, R. (1962). Prediction of swelling potential of compacted soils. *ASCE Journal of Soil Mechanics and Foundation Engineering* 85: 86–128.

Siddappa, R.K. (2000). *Collapse behaviour of red soils of Bangalore District*. Ph.D. Thesis, Indian Institute of Science, Bangalore.

Sridharan, A., Rao, S.M. and Murthy, N.S. (1992). Physico chemical effects on compressibility of tropical soils. *Soil and Foundations* 32: 156–163.

Subba Rao, K.S., Rao, S.M. and Gangadhara, S. (2000). Swelling behaviour of desiccated clay. *ASTM Geotechnical Testing Journal* 23: 193–198.

Taylor, D.W. (1959). *Fundamentals of Soil Mechanics*. Asia Publishing House, New Delhi.

Terzaghi, K. (1925). *Erdbaumchanik* (in German). Vienna: Franz Deuticke.

Terzaghi, K. (1958). Design and performance of Sasumua dam. *ASCE Proceedings Institution of Civil Engineers*, 9: 369–394.

Toll, D.G. (1990). A framework for unsaturated soil behaviour. *Géotechnique* 40: 31–44.

Wallace, K.B. (1973). Structural behaviour of residual soils of the continually wet high lands of Papua New Guinea. *Géotechnique* 23: 203–218.

Wesley, L.D. (1990). Influence of structure and compaction on residual soils. *ASCE, Journal of Geotechnical Engineering* 116: 589–603.

Yong, R.N. and Warkentin, B.P. (1975). *Soil Properties and Behaviour*. Elsevier, New York.

van der Merwe, H. (1964). The prediction of heave from plasticity index and the percentage clay fraction of the soils. *The Civil Engineer, South African Institute of Civil Engineers* 6, 103–107.

Vaughan, P.R. (1988). Characterizing the mechanical properties of in-situ residual soil. *Proceedings of Second International Conference on Geomechanics in Tropical Soils*, Singapore, Vol. 2, 469–487.

van Olphen, H. (1963). *An Introduction to Clay Colloid Chemistry*. Wiley, New York.

Vargas, M. (1953). Some engineering properties of residual clay soils occurring in Southern Brazil. *Proceedings, 3rd International Conference on Soil Mechanics and Foundation Engineering Zurich*, Vol. 1, 67–71.

Vargas, M. (1973). Structurally unstable soil Southern Brazil. *Proceedings, 8th International Conference on Soil Mechanics and Foundation Engineering*, Moscow, Vol. 2, 239–246.

Wheeler, S.J. (1994). General report – Engineering behaviour and properties of arid soils. *Proceedings First International Symposium on Engineering Characteristics of Arid Soils*, London, 161–172.

6

Shear strength model for tropical residual soil

Mohd Jamaludin Md. Noor
Universiti Teknologi Mara, Shah Alam, Malaysia

Bujang B.K. Huat
Universiti Putra Malaysia, Serdang, Malaysia

Faisal Ali
Universiti Pertahanan Nasional Malaysia, Kuala Lumpur, Malaysia

6.1 INTRODUCTION

Shear strength is a soil property which is its ability to resist movement between grains along a plane. When there is shearing along a plane in a soil mass such as a slope failure or movement of backfill behind a retaining wall, stress is developed to resist particle-to-particle sliding along the failure plane. This stress is referred as soil shear strength. When the stress is lower than the maximum value, this stress is known as ***mobilised shear strength***. As this stress increases with denser particle packing, a maximum value is achieved at some point before the stress starts to decrease. The maximum stress attained is referred as the ***shear strength at failure***. Since these stresses are produced when there is particle movement within a soil mass, they must have a strong influence on the soil mechanical behaviour. In other words, shear strengths must be the controlling factor for slope failures, settlement, bearing capacity and movement of retaining walls. Normally, the term ***shear strength*** is used to refer to the shear strength at failure. Therefore, shear strength can be defined as the maximum shear stress that a soil can mobilise along a plane to resist movement along the plane. The SI unit for shear strength is kN/m^2.

Soil mechanical behaviours are, for example, landslides, settlement, pile shaft friction and the bearing capacity of shallow foundations. These behaviours are usually viewed from the viewpoint of the shear strength at saturation which is characterised purely based on the effective stress, $(\sigma - u_w)$. However, there are behaviours which are categorised as complex when the effective stress concept, which is the existing fundamental concept in soil mechanics cannot explicitly explain. These are like the shallow rainfall-induced landslides and wetting settlement. These behaviours are taking place under effective stress decrease which is in contrary to the existing concept of effective stress. Surface water infiltration and inundation of ground water increase the magnitude of pore water pressure and thereby makes the effective stress decrease; moreover these are triggering the shallow landslide and the wetting settlement respectively. Then geotechnical researchers realised that shear strength is not solely governed by effective stress. It is something associated with the soil degree of wetness. As partially saturated soils become wetter, their shear strength diminishes to the minimum value when fully saturated. The stress state variable associated with the effect of wetting began to be incorporated, which is suction, $(u_a - u_w)$. From this point, soil shear strength is characterised based on the two independent stress state variables which are net stress, $(\sigma - u_a)$ and suction. Effective stress is applied when the condition is saturated, and once the soil becomes partially saturated the relevant independent stress state reverts to net stress and suction. The form of soil shear strength model which defines the shear strength behaviour, mathematically, has been expanding since its first introduction by the concept of the Coulomb frictional law (1776). This indicates the continuous advancement of the knowledge in shear strength behaviour. This chapter starts by presenting the beginning and the advancement of soil shear strength models thereafter. The common testing methods for obtaining shear strength parameters both in the laboratory as well as in the field are described in Chapter 3.

Before various soil shear strength models can be developed, there has to be development of shear strength testing methods in the laboratory. Basically, there are two types of laboratory shear strength tests, which are direct shear and triaxial tests. The equipment involved may be upgraded from testing only saturated soil to the testing of partially saturated soil. This is achieved through some modification which allows for the control of pore water and pore air pressure applied to the specimen. The former is a relatively simple test that involves only two soil phases, which are solid and water, whereas the latter is more complex as it involves four soil phases which are solid, water, air and air-water inter-phase. The testing procedures for the unsaturated soils will be presented. This is followed by the interpretation of the shear strength at saturation and the elevated shear strength due to partial saturation which is known as apparent shear strength.

6.2 DEVELOPMENT OF SOIL SHEAR STRENGTH MODELS

The knowledge of soil shear strength behaviour has been continuously expanding since its first introduction. This is because researchers are continuously seeking for a model that can explicitly explain the observed soil mechanical behaviour from the standpoint of shear strength. Soil shear strength and volume change behaviour are two very complex behaviours and difficult to characterise. There are basically two types of approach, which are the Mohr-Coulomb failure criterion and the critical state method. However, there have been many shortcomings in the latter and thus it has become less popular. For example, the modified Cam clay critical state model of Roscoe and Burland (1968) was quoted by Woods (1990) as perverse when the shape of the yield locus bears little resemblance to those obtained experimentally for many natural soils. Then the critical state model of Alonso *et al.* (1990), developed for saturated and partially saturated soil conditions, was quoted by Wheeler *et al.* (2003) as failing to explain the soil volume change behaviour of unsaturated kaolin clay due to alternate wetting and drying. Despite having a simple linear equation to define the critical state line, the models failed to explain those complex soil behaviours from the standpoint of critical state strength. Perhaps it is not the critical state strength that governed the soil mechanical behaviour.

Coulomb frictional law in soil mechanics

Coulomb (1776) introduced the law of mechanics for a rigid body sliding on a plane. The concept has been extended to the mechanics of soils.

Consider a rigid body being pulled along a horizontal plane by a horizontal force P as shown in Figure 6.1. W is the weight of the body acting vertically downwards from its centre of mass while N is the normal reaction acting by the sliding plane on to the body. F is the frictional force developed between the body and the plane. R is the resultant force from the normal force N and the frictional force F. The angle, ϕ' between the normal force N and the resultant force R is called the friction angle. Coulomb's (1776) law states that,

$$F = N \tan \phi' \tag{6.1}$$

where, $\mu = \tan \phi' =$ coefficient of friction.

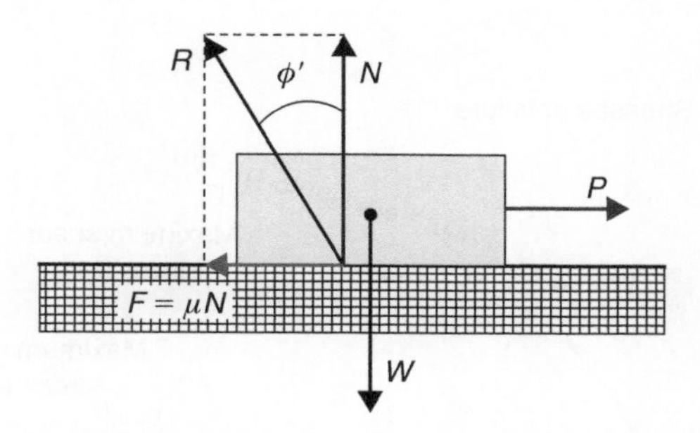

Figure 6.1 The application of Coulomb's (1776) law to the mechanics of a sliding body

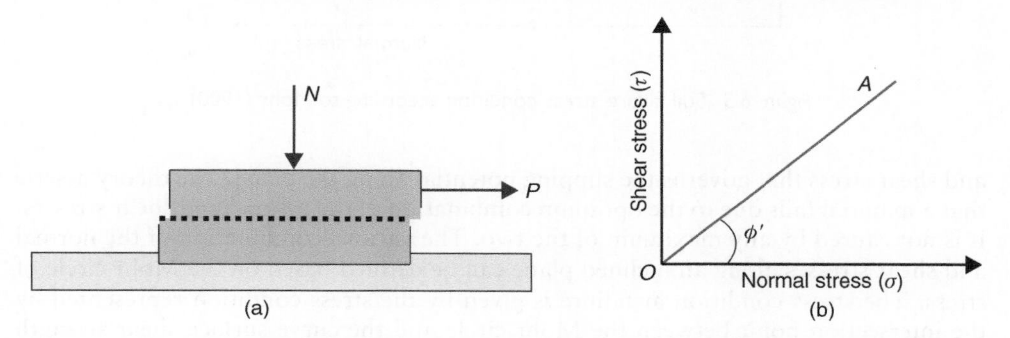

Figure 6.2 Application of Coulomb's (1776) frictional law to the shear strength behaviour of soil (a) shearing of two soil segments (b) graph of normal stress and the maximum mobilised shear stress

Now consider two segments of soil encased in a split rectangular container as shown in Figure 6.2(a). The bottom segment is fixed while the top segment is pulled horizontally. N is the normal force applied on the top segment.

Assume "A" as the cross sectional area of the sliding plane between the two soil segments. The normal stress, σ along the sliding plane is given by:

$$\sigma = N/A \tag{6.2}$$

The shear stress, τ, is

$$\tau = P/A \tag{6.3}$$

If the maximum mobilised shear stress, τ is plotted against the normal stress, σ, a straight line graph, OA is obtained as shown in Figure 6.2(b). The angle, ϕ' between the normal stress axis and the variation of shear stress, OA, is called the internal friction angle of the soil. Therefore Coulomb's (1776) law can be applied to model soil shear strength behaviour.

Mohr-Coulomb failure criterion

A theory for the rupture of materials was introduced by Mohr (1900). Along any incline plane that cuts through a material, there is a certain combination of normal

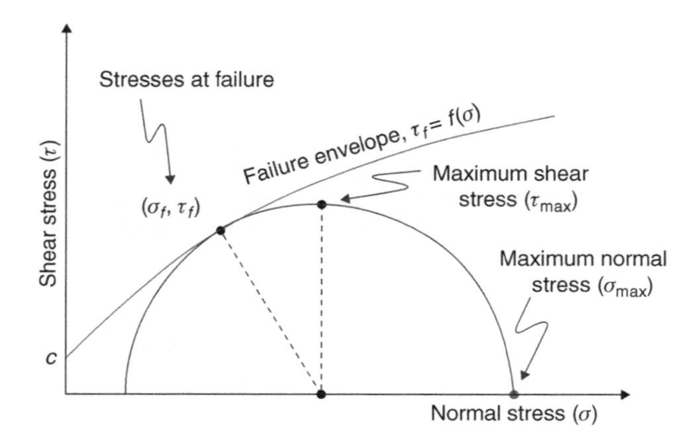

Figure 6.3 Soil failure stress condition according to Mohr (1900)

and shear stress that governs the slipping potential along the plane. The theory asserts that a material fails due to the optimum combination of the normal and shear stresses. It is not caused by any maximum of the two. The various combination of the normal and shear stresses along an inclined plane can be defined based on the Mohr circle of stress. The stress condition at failure is given by the stress condition represented by the intersection point between the Mohr circle and the curve surface shear strength envelope as shown in Figure 6.3.

The coupling between the Mohr (1900) stress circle and the Coulomb's (1776) frictional law is able to define the soil shear failure based on the subjected stress state and the limiting stress condition respectively. This is recognised as the Mohr-Coulomb failure criterion. If the Mohr stress circle which represents the subjected stress condition grows to indicate stress increase and reaches the failure envelope which is the limiting stress condition, failure will be triggered. At this point the soil mass will develop a shearing plane and the corresponding normal and shear stresses developed represent the stress state at failure. Thence, the stress state at failure is defined by the coordinates of the intersection point between the Mohr stress circle and failure envelope in the space of shear stress versus normal stress as shown in Figure 6.3. By this concept, it is impossible for the stresses in the soil mass to lie beyond the failure envelope since it is a condition that the soil cannot withstand. The concept of Mohr-Coulomb failure criterion has become the fundamental basis for the developments of most soil shear strength models. The curved type functional relationship between the shear stress at failure and the normal stress suggested by Mohr is given as Equation 6.4 and its shape is as sketched in Figure 6.3. Note that this is based on the total stress concept.

$$\tau_f = f(\sigma) \tag{6.4}$$

where

τ_f = shear stress on failure envelope
σ = normal stress on the failure envelope.

At this point, the knowledge of effective stress is not being revealed yet. The concept put forward is based purely on the total stress. For materials other than soils where pore water pressure is non-existent, the concept of the Mohr-Coulomb failure criterion can be applied directly. However, in soil, the stress between the soil solids is affected by the pressure of water in the voids. This is where the effective stress concept becomes relevant.

Total stress Mohr-Coulomb shear strength model

The application of Coulomb's frictional law to soil mechanics suggested that for most soil mechanics problems, it is sufficient to approximate the shear stress on a failure plane as a linear function of the normal stress. The equation of the linear type of soil shear strength envelope is suggested as Equation 6.5. This equation is based on the Mohr-Coulomb failure criterion.

$$\tau_f = c + \sigma \tan \phi \tag{6.5}$$

where,

c = cohesion
ϕ = angle of internal friction.

Therefore there are two assumptions made in the concept of the Mohr-Coulomb failure criterion:

a. It is not necessary that the maximum shear stress or the maximum normal stress is causing the failure, but it is the combination of the two in which the coordinate intersects with the failure envelope.
b. It is thought that for most practical problems involving soils, it is sufficient to assume a linear type of shear strength envelope.

Effective stress Mohr-Coulomb shear strength model

Terzaghi (1936) introduced the concept of effective stress. Effective stress, σ' is the stress carried by the soil solids. When a saturated soil is subjected to a total stress, σ, the stress is taken up by the effective stress and the pore water pressure, u_w. This concept is represented by Equation 6.6.

$$\sigma = \sigma' + u_w \tag{6.6}$$

Rearranging the equation,

$$\sigma' = \sigma - u_w \tag{6.7}$$

Equation 6.7, is in fact the effective stress equation of Terzaghi. By the concept of effective stress, he contended that:

> "All the measurable effects of a change in stress, such as compression, distortion, and a change in shearing resistance, are exclusively due to changes in effective stress $\sigma_1', \sigma_2', \sigma_3'$."

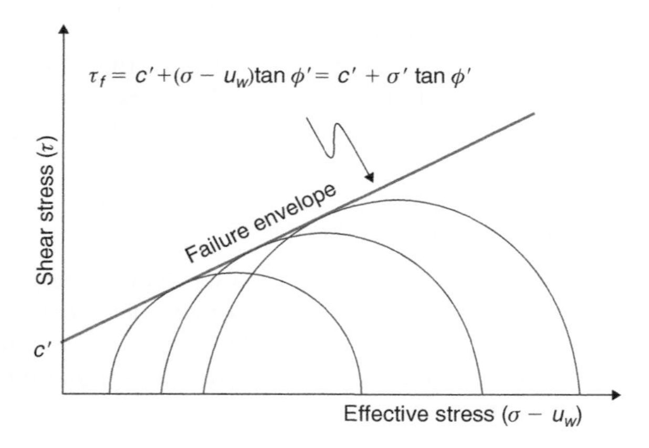

Figure 6.4 Mohr-Coulomb failure envelopes by the effective stress concept (Terzaghi, 1936)

Then, Terzaghi applied the concept of effective stress to the Mohr-Coulomb failure criterion and the shear strength equation becomes Equation 6.8. The envelope is shown in Figure 6.4.

$$\tau_f = c' + (\sigma - u_w)\tan\phi' = c' + \sigma'\tan\phi' \tag{6.8}$$

This is again considering a linear type of shear strength envelope. Note that this equation is valid for saturated soil conditions only. Nevertheless, the equation has been applied to evaluate slope failures, force on retaining structures and bearing capacity of foundations, despite the in-situ soil condition being partially saturated in most cases. This is because the saturated soil condition is giving the lowest shear strength and thus its application is thought to make the design very conservative. However, there are many soil mechanical behaviours that cannot be explained by the equation such as the shallow rainfall-induced slope failures. The application of the shear strength equation (i.e. Equation 6.8) into slope stability analysis would indicate failure whereas the slope is still standing safely (Othman, 1989). This is because of the application of the shear strength at saturation, whereas the in-situ soil condition is partially saturated. There is extra shear strength known as apparent shear strength when the soil is partially saturated due to the presence of negative pore water pressure or suction in the soil. Thence, to incorporate the increase in shear strength when the soil is partially saturated, the existing shear strength equation needs to be extended to encompass unsaturated soil conditions.

Extended Mohr-Coulomb shear strength model

The occurrence of shallow type of rainfall-induced slope failures is very rampant in tropical residual soils. The application of Terzaghi's effective stress shear strength equation cannot explicitly explain the behaviour. The equation tends to produce a deep type of slope failure. Researchers realised that the rainfall-induced slope failures are triggered by the softening of the soil fabric when it is wetted. Thence, the degree of wetness has some influence on the shear strength. In another words, it not just the effective

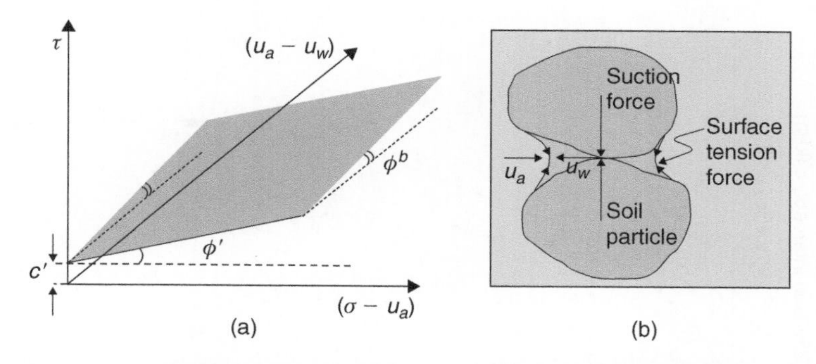

Figure 6.5 Extended Mohr-Coulomb shear strength model of Fredlund *et al.* (1978)

stress that governs the magnitude of the shear strength. This is where two independent stress state variables, which are net stress $(\sigma - u_a)$ and suction $(u_a - u_w)$, were taken to characterise the shear strength behaviour of partially saturated soils. The magnitude of suction signifies the soil degree of wetness. This is where Fredlund *et al.* (1978) introduced an extended Mohr-Coulomb failure envelope soil shear strength model as shown in Figure 6.5(a). It takes the form of an inclined plane which indicates indefinite linear increase in shear strength with respect to both effective stress and suction. The planar failure envelope intersects the shear stress axis giving the cohesion intersect, c'. The angles, ϕ' and ϕ^b are the inclination of the envelope from net stress and suction axis respectively. The cohesion intersect, c' and the angles, ϕ' and ϕ^b are the three shear strength parameters required to define the planar surface envelope. The shear strength equation that represents the planar envelope is as in Equation 6.9.

$$\tau = c' + (\sigma - u_a)\tan\phi' + (u_a - u_w)\tan\phi^b \quad (6.9)$$

Shear strength is derived from inter-particle shearing stress, in turn derived from particle inter-locking. When the condition becomes partially saturated, there is an enhancement in the interlocking effect. This is due to the extra inter-particle attraction or suction force derived from the surface tension force that exists on the surface of the water meniscus as shown in Figure 6.5(b). Suction force is derived from suction, $(u_a - u_w)$ which a stress state variable defined as the difference between pore air, u_a and pore water, u_w pressures. However, this extra inter-particle attraction will completely vanish when the soil is saturated. Since this extra shear strength is not permanent, the shear strength produced by the existence of suction is referred to as ***apparent shear strength***. The variation of the apparent shear strength relative to the moisture content has a strong link to the soil mechanical behaviours like the rainfall-induced slope failure and the wetting collapse, also known as inundation settlement. The shear strength model of Fredlund *et al.* (1978) is the first to incorporate apparent shear strength.

However, there is much evidence that shear strength varies non-linearly relative to the suction. The first evidence is the laboratory results from direct shear tests conducted by Donald (1956) on unsaturated fine sand and coarse silt using a modified direct shear box. The pore-air and the pore-water pressures were controlled during shearing. This is followed by the reports from Escario and Saez (1986) for Madrid grey clay.

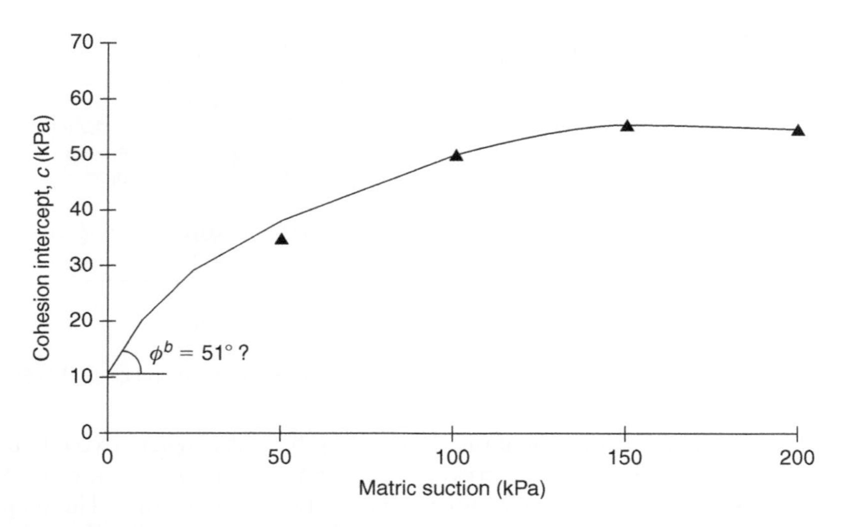

Figure 6.6 Non-linear variation of shear strength relative to suction for granitic residual soil from Bukit Timah formation, Singapore (*After* Toll *et al.*, 2000)

This was further confirmed by authors such as Gan *et al.* (1988), Escario and Juca (1989) and Toll *et al.* (2000). Figure 6.6 shows a typical non-linear variation of the apparent shear strength relative to suction for granitic residual soil from Bukit Timah formation, Singapore, reported by Toll *et al.* (2000). The apparent shear strength increases to a maximum value and the corresponding suction is called residual suction (Vanapalli *et al.*, 1996). Beyond residual suction, the apparent shear strength starts to decrease as the soil desaturates.

With regards to the shear strength behaviour relative to effective stress, there have been numerous reports that the behaviour is curvilinear with zero strength at zero effective stress. It is the combination of non-linear behaviour at low stress levels, becoming linear at high stress levels. The first evidence was provided by Bishop (1966) which encompassed soils of various grain sizes including clay, silt and sand. This behaviour is further reinforced by reports from Charles and Watts (1980) for gravels, Fukushima and Tatsuoka (1984) for sands and Indraratna *et al.* (1993) for greywacke rockfill. The curvilinear shear strength envelope for grade VI granitic residual soil from Rawang, Malaysia has been reported by Md. Noor *et al.* (2008) and the envelope is shown in Figure 6.7.

The reinterpretation of the linear shear strength envelope for sedimentary residual soil from Jurong formation, Singapore, as reported by Rahardjo *et al.* (1995), has showed that there is a steep non-linear drop of shear strength as effective stress approaches zero, as shown in Figure 6.8. The selected data were taken from specimens having SPT N values ranging between 10 and 12 only. This again reinforced the curvilinear shape of the shear strength behaviour relative to effective stress for the tropical residual soils.

With this evidence for the soil shear strength behaviour relative to suction and net or effective stress, further improvement must be made on the existing extended

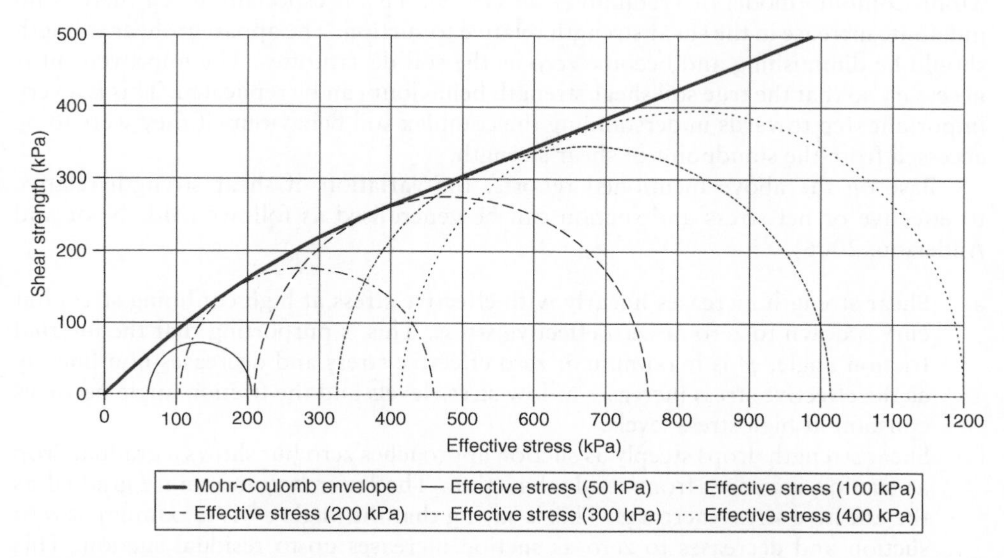

Figure 6.7 Curvilinear shear strength envelope of grade VI granitic residual soil from Rawang, Malaysia (*After* Md. Noor *et al.*, 2008)

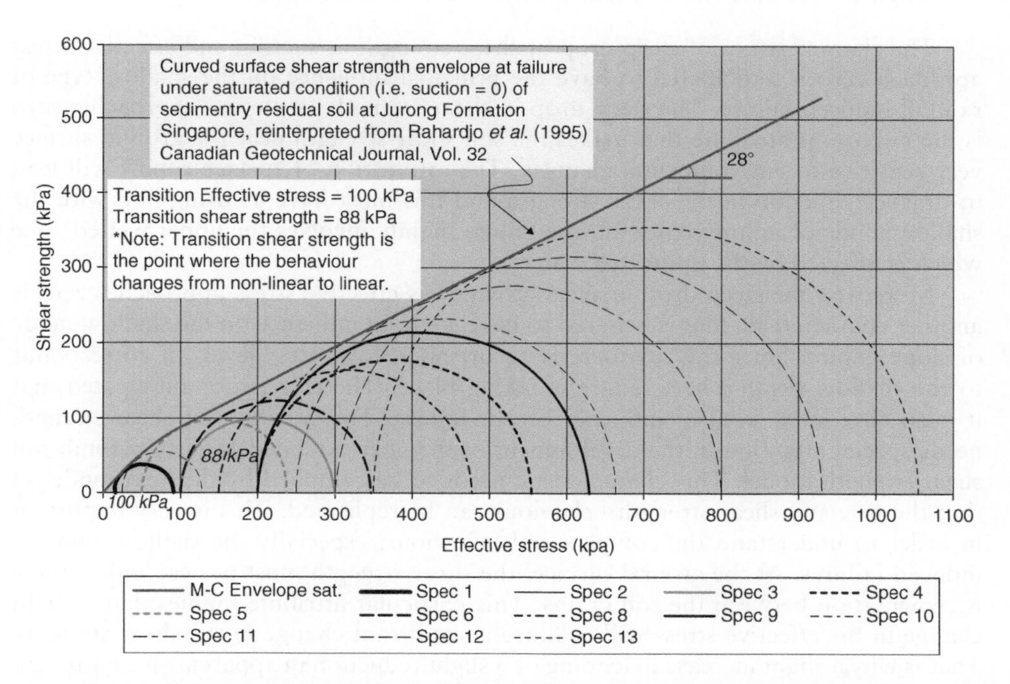

Figure 6.8 Steep drop in shear strength as effective stress approaches zero for sedimentary residual soil from Jurong formation, Singapore (*After* Rahardjo *et al.*, 1995)

Mohr-Coulomb model of Fredlund *et al.* (1978). This is especially when there is no indefinite increase in the shear strength relative to suction. The apparent shear strength should be diminishing and become zero as the soil de-saturates. The improvement is necessary so that the true soil shear strength behaviour can be replicated. This is a very important step towards understanding the complex soil behaviours if they were to be accessed from the standpoint of shear strength.

Base on the above mentioned reports, the variation of shear strength relative to effective or net stress and suction can be generalised as follows (Md. Noor and Anderson, 2006).

a. Shear strength increases linearly with effective stress at high confining stress but curves down to zero at zero effective stress. This is purporting that the internal friction angle, ϕ' is maximum at zero effective stress and decreases non-linearly as the effective stress increases at low stress levels and the friction angle becomes constant at high stress levels.
b. Shear strength drops steeply as suction approaches zero but shows a gradual drop as suction increases from residual suction. The latter becomes more gradual as the effective stress increases. This is saying that the angle ϕ^b is maximum at zero suction and decreases to zero as suction increases up to residual suction. This behaviour, up to residual suction, is uniform irrespective of the effective stress. At residual suction, ϕ^b is zero since that is the turning point where it starts to decrease. Beyond that, ϕ^b continues to increase negatively. The reduction in the angle ϕ^b becomes more gradual as the effective stress increases.

The characteristics of steep drop in shear strength as suction and effective stress approach zero is anticipated to have the principal influence on the shallow type of rainfall-induced failure. The steep drop in shear strength as suction approaches zero is the important attribute that makes the soil shear strength near the ground surface very sensitive to environmental changes. The infiltration of surface runoff will lead to drastic reduction in the shear strength and this must have a strong link with the shallow mode of failure. After all, the failure mainly involves the upper wetted zone which is affected by the infiltrated water.

Moreover, the steep drop in shear strength as effective stress approaches zero is another characteristic that is believed to have a strong influence on the shallow mode of slope failure. Since this attribute is occurring at low stress levels, it corresponds to the shallow depth where failure is taking place. Therefore, it is anticipated that it must have some kind of influence on the failure. This attribute of shear strength needs special attention in the development of the next extended Mohr-Coulomb soil shear strength model. This characteristic needs to be complied by the next model so that the true soil shear strength behaviour can be replicated. This is very important in order to understand the complex soil behaviour; especially the shallow rainfall-induced failures. At the ground surface, the shear strength must be zero unless there is cementation between the soil grains. This particular attribute implies that a slight change in the effective stress will produce a significant change in the shear strength. That is why a slight increase in loading or a slight reduction in apparent shear strength when the soil is wetted would trigger a small failure usually referred as "local failure".

There have been many reports that the angle ϕ^b is normally equal or less than ϕ'. Now, when the envelope is non-linear relative to suction, then ϕ^b is not a constant any

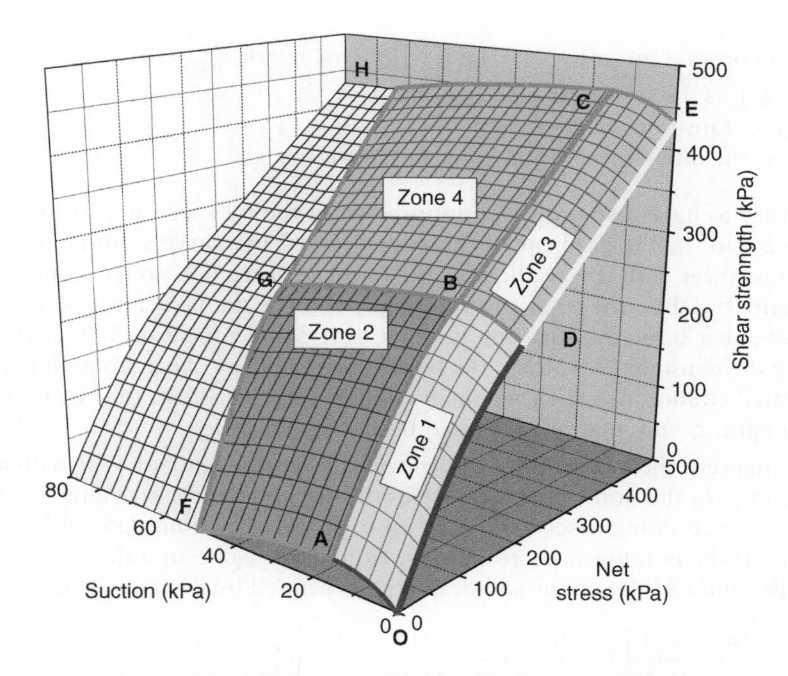

Figure 6.9 Curved-surface envelope soil shear strength model

more. Maximum ϕ^b occurs at zero suction. Moreover, reports that maximum ϕ^b is greater than ϕ' are also emerging now. The curved-surface envelope soil shear strength model has been developed to comply with the mentioned complex soil shear strength characteristics.

Curved-surface envelope Mohr-Coulomb soil shear strength model

The curved-surface envelope soil shear strength model is able to explicitly comply with the description of the shear strength behaviour relative to effective or net stress and suction as described in the preceding section. The shape of the model is as shown in Figure 6.9. The surface envelope is divided into four zones to ease the mathematical representation. It would be very difficult to define a complex surface by just a single equation. For each zone, the shear strength is defined by adding the apparent shear strength to the shear strength at saturation. The latter is the lower limit of the shear strength, while the apparent shear strength varies with suction.

The shear strength at saturation, τ_{sat} is divided into two sections which are the non-linear section, OD on the lower levels of effective stress defined by Equation 6.10, and the linear section, DE for the higher stress levels defined by Equation 6.11.

$$\tau_{sat} = \frac{(\sigma - u_a)}{(\sigma - u_w)_t}\left[1 + \frac{(\sigma - u_w)_t - (\sigma - u_a)}{N\,(\sigma - u_w)_t}\right]\tau_t \tag{6.10}$$

$$\tau_{sat} = (\sigma - u_a)\tan\phi'_{\min_f} + [\tau_t - (\sigma - u_w)_t \tan\phi'_{\min_f}] \tag{6.11}$$

where,

N is a constant given by; $N = 1/(1 - [(\sigma - u_w)_t (\tan \phi'_{\min_f}/\tau_t)])$
$(\sigma - u_w)_t$ is the transition effective stress
τ_t is the transition shear strength
$\phi'_{\min_f}$ is the minimum friction angle at failure

In order to have a smooth transition between the two sections (i.e. zones 1 and 3; zones 2 and 4), the gradient of the upper end of the curve sections (i.e. zones 1 and 2) must meet with the gradient of the linear sections (i.e. zones 3 and 4). Along their boundary DBG, the total shear strength and the gradient must be equal when calculated either using the equation for zone 1 or 3 and zone 2 or 4. This is achieved by having the constant N which is the function of the three shear strength parameters for saturated conditions which are the transition effective stress, $(\sigma - u_w)_t$, transition shear strength, τ_t and minimum internal friction angle, $\phi'_{\min_f}$.

The apparent shear strength, τ_{app} behaviour relative to suction is also divided into two sections, (1) the non-linear increase in the apparent shear strength (i.e. OA, DB and EC), as suction increases from zero up to residual suction; defined by Equation 6.12, and (2) the non-linear decrease as suction increases beyond the residual suction (i.e. AF, BG and CH) when the soil desaturates; defined by Equation 6.13.

$$\tau_{app} = \frac{(u_a - u_w)}{(u_a - u_w)_r}\left[1 + \frac{(u_a - u_w)_r - (u_a - u_w)}{(u_a - u_w)_r}\right]c_s^{\max} \tag{6.12}$$

$$\tau_{app} = \left[\frac{(u_a - u_w)_u - (u_a - u_w)}{(u_a - u_w)_u - (u_a - u_w)_r}\right] \times \left[1 - \frac{(u_a - u_w)_r - (u_a - u_w)}{(u_a - u_w)_u - (u_a - u_w)_r}\right]c_s^{\max} \tag{6.13}$$

where

$(u_a - u_w)_r$ is the residual suction
$c_s^{\max}$ is the maximum apparent shear strength
$(u_a - u_w)_u$ is the ultimate suction where the apparent shear strength vanishes.

The non-linear increase in the apparent shear strength from zero suction to residual suction is considered uniform irrespective of the value of net stress. This is demonstrated by the uniform curvature of OA, DB and EC. However, the non-linear decrease in apparent shear strength beyond residual suction is getting more gradual as the net stress increases. This is well demonstrated as the curve becomes more gradual from AF to BG and to CH.

The value of ultimate suction (i.e. along FGH) increases linearly with net stress according to Equation 6.14 when seen from plan view; which is in the plane of net stress versus suction. Nevertheless, the line FGH is seen as a curve in the perspective view as shown in Figure 6.9.

$$(u_a - u_w)_u = \zeta(\sigma - u_a) + (u_a - u_w)_u^{\sigma'=0} \tag{6.14}$$

where,

ζ is the rate of increase of ultimate suction relative to net stress
$(u_a - u_w)_u^{\sigma'=0}$ is the ultimate suction at zero net stress.

Table 6.1 Sum of shear strength equations at saturation and partial saturation to represent total shear strength for each zone

	Components of shear strength		
Zone	*Saturation*	*Partial saturation*	*Resulting representing equation*
1	Equation 6.10	Equation 6.12	Equation 6.15
2	Equation 6.10	Equation 6.13	Equation 6.16
3	Equation 6.11	Equation 6.12	Equation 6.17
4	Equation 6.11	Equation 6.13	Equation 6.18

The value of ultimate suction obtained from Equation 6.14 is then substituted into Equation 6.13 in order to determine the value of apparent shear strength at the respective value of net stress for zone 2 or 4 respectively. This is because the value of ultimate suction is not a constant but instead there is a linear increase in ultimate suction as net stress increases. The summit of the curvature in the direction of suction axis is at the residual suction (i.e. along ABC) where the apparent shear strength is a maximum. For a smooth transition of the shear strength behaviour along the suction axis between the two sections (i.e. left and right of ABC), the upper end of the right section and the beginning of the left section must have zero gradient (i.e. along ABC).

The state of shear strength at failure, τ_f at any point on the envelope is taken as the sum of the shear strength at saturation, τ_{sat} and the apparent shear strength, τ_{app} when the soil is partially saturated. The components of shear strength that make up the total shear strength at failure for each zone is summarised in Table 6.1. The resulting equations to represent zones 1, 2, 3 and 4 are Equations 6.15, 6.16, 6.17 and 6.18 respectively.

$$\tau_f = \frac{(\sigma - u_a)}{(\sigma - u_w)_t}\left[1 + \frac{(\sigma - u_w)_t - (\sigma - u_a)}{N(\sigma - u_w)_t}\right]\tau_t + \frac{(u_a - u_w)}{(u_a - u_w)_r}\left[1 + \frac{(u_a - u_w)_r - (u_a - u_w)}{(u_a - u_w)_r}\right]c_s^{\max} \tag{6.15}$$

$$\tau_f = \frac{(\sigma - u_a)}{(\sigma - u_w)_t}\left[1 + \frac{(\sigma - u_w)_t - (\sigma - u_a)}{N(\sigma - u_w)_t}\right]\tau_t + c_s^{\max}\left[\frac{(u_a - u_w)_u - (u_a - u_w)}{(u_a - u_w)_u - (u_a - u_w)_r}\right] \times \left[1 - \frac{(u_a - u_w)_r - (u_a - u_w)}{(u_a - u_w)_u - (u_a - u_w)_r}\right] \tag{6.16}$$

$$\tau_f = (\sigma - u_a)\tan\phi'_{\min_f} + \left[\tau_t - (\sigma - u_w)_t \tan\phi'_{\min_f}\right] + \frac{(u_a - u_w)}{(u_a - u_w)_r}\left[1 + \frac{(u_a - u_w)_r - (u_a - u_w)}{(u_a - u_w)_r}\right]c_s^{\max} \tag{6.17}$$

$$\tau_f = (\sigma - u_a)\tan\phi'_{\min_f} + \left[\tau_t - (\sigma - u_w)_t \tan\phi'_{\min_f}\right] + c_s^{\max}\left[\frac{(u_a - u_w)_u - (u_a - u_w)}{(u_a - u_w)_u - (u_a - u_w)_r}\right]\left[1 - \frac{(u_a - u_w)_r - (u_a - u_w)}{(u_a - u_w)_u - (u_a - u_w)_r}\right] \tag{6.18}$$

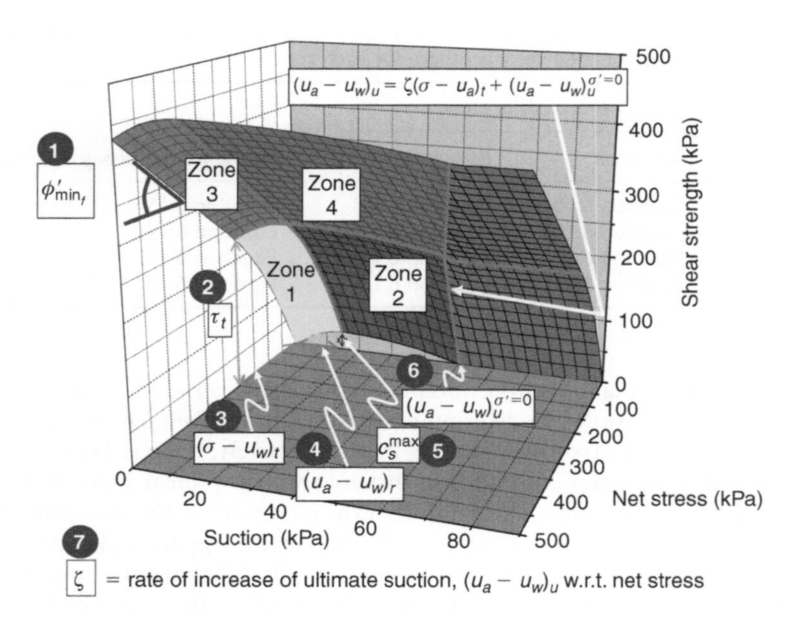

Figure 6.10 Seven soil shear strength parameters required to define the curved-surface envelope

Table 6.2 Seven shear strength parameters in the curved-surface envelope extended Mohr-Coulomb model

No.	*Shear strength parameter*	*Parameter name*	*Soil condition*
1	τ_t	Transition shear strength	Saturated
2	$(\sigma - u_w)_t$	Transition effective stress	
3	ϕ'_{min_f}	Minimum internal friction angle	
4	$(u_a - u_w)_r$	Residual suction	Partially saturated
5	$(u_a - u_w)_u^{\sigma'=0}$	Ultimate suction	
6	c_s^{max}	Maximum apparent shear strength	
7	ζ	Rate of increase in ultimate suction relative to net stress	

There are seven soil shear strength parameters required to define the curved-surface envelope of the model and the dimensions that they represent are illustrated in Figure 6.10. Three shear strength parameters are required to define the behaviour at saturation and a further four shear strength parameters are required to characterise the behaviour at partial saturation. The shear strength parameters are summarised in Table 6.2.

6.3 LABORATORY MEASUREMENT OF SHEAR STRENGTH USING TRIAXIAL APPARATUS

Two common types of equipment used to test soil shear strength in the laboratory are the shear box and triaxial tests apparatuses. However, of these two types of equipment,

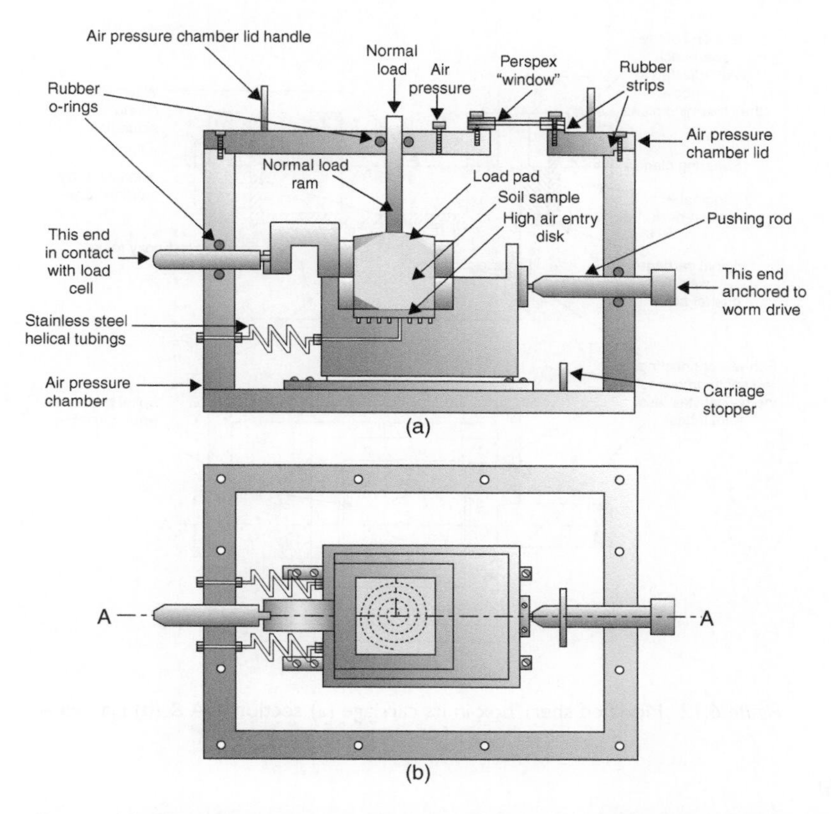

Figure 6.11 Modified direct shear box assembly (a) section A-A & (b) plan view

the triaxial test apparatus is more widely used. This is because the actual field stress states can be well replicated in this equipment by controlling the cell pressure, pore water pressure, pore air pressure, deviator stress and the drainage conditions.

Shear box

To enable soil samples to be tested in the unsaturated state (under suction), the conventional shear box apparatus may be modified as shown in Figures 6.11–6.13. The modification required involves the following:

- The air pressure chamber,
- The normal loading system,
- The shear box assembly inside the air pressure chamber, and
- Addition of data acquisition devices to enhance the performance and simplify the usage of the modified shear box test apparatus to make it suitable for testing unsaturated residual soils.

A detailed description of the modified shear box can be found in Huat *et al.* (2005a).

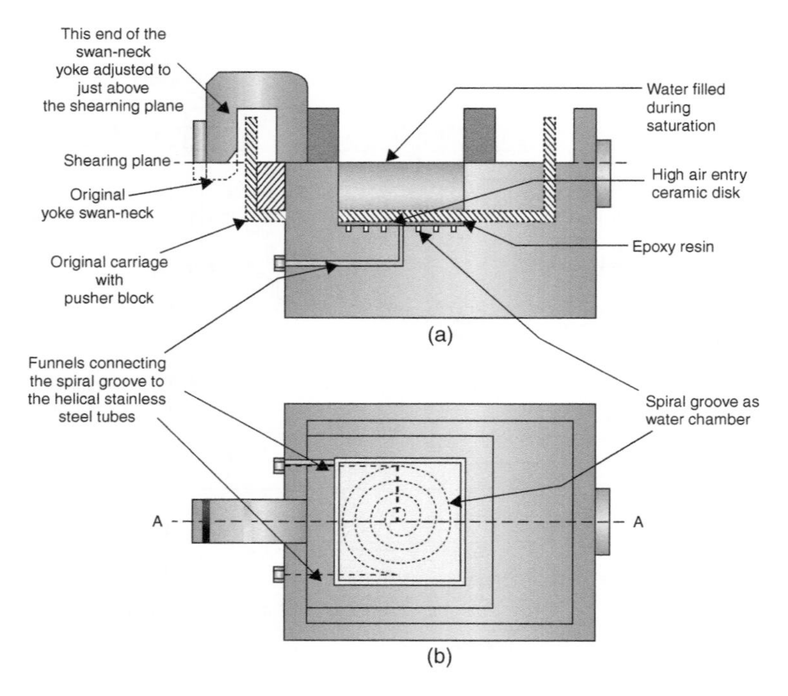

Figure 6.12 Modified shear box in its carriage (a) section A-A & (b) plan view

Triaxial

Soil shear strength behaviour is characterised by two types of laboratory test. They are the shear strength tests on saturated and partially saturated specimens. For the case of the triaxial test, the equipment used in the former is referred to as conventional triaxial apparatus and for the latter it is called double-walled triaxial apparatus.

Triaxial tests on saturated specimens

This test is to determine the soil shear strength behaviour when the condition is fully saturated. The results from the test will determine the shape of the envelope along the line ODE in Figure 6.9 if it is interpreted according to the curved-surface envelope extended Mohr-Coulomb model. It will give three saturated shear strength parameters which are transition shear strength, τ_t, transition effective stress, $(\sigma - u_w)_t$, and minimum friction angle at failure, $\phi'_{\min_f}$. However, if it is interpreted according to the extended Mohr-Coulomb envelope, it will give the friction angle, ϕ' and cohesion, c'. Conventional triaxial apparatus is used for this test. The typical plumbing of the apparatus is shown in Figure 6.14.

The instruments required for the test are as follows:

- Axial strain transducer using linear variable differential transformer (LVDT). It is a type of electrical transformer used for measuring linear displacement.

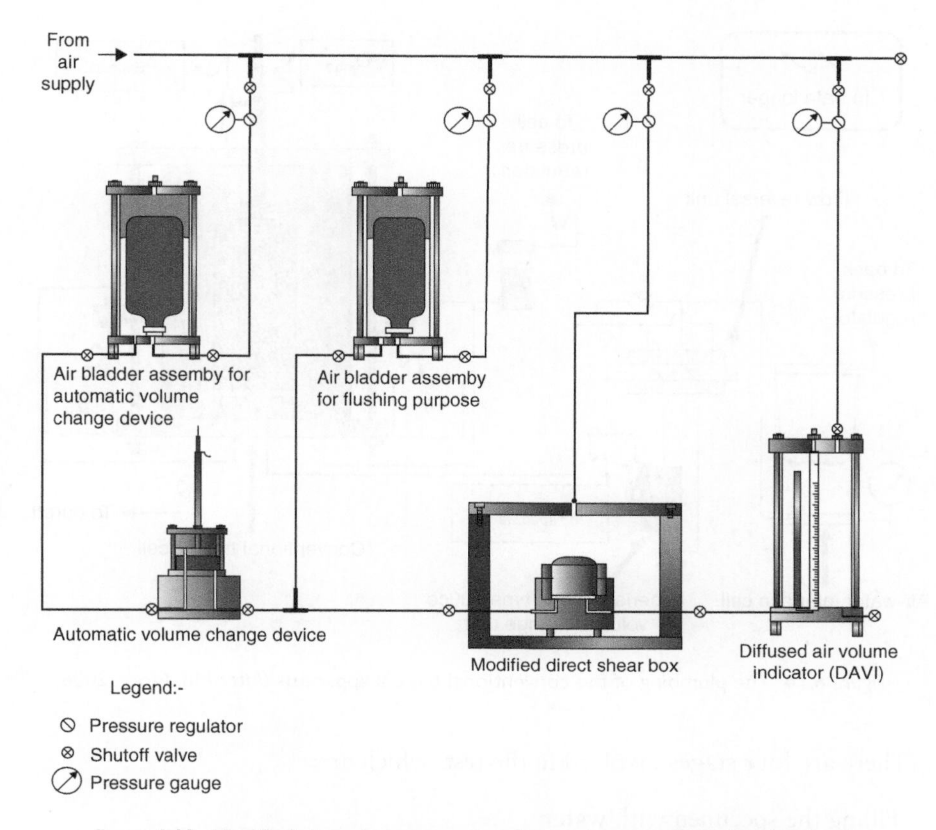

Figure 6.13 Modified shear box test set-up for testing unsaturated residual soil

This is an indirect method of measuring displacement. The transducer is to measure the overall change in the specimen's height during consolidation and shearing.

- Specimen water volume change unit.
- Cell pressure transducer.
- Pore water pressure transducer.
- Load cell.

The change in the specimen's volume during the test is monitored from the change in the volume of water that is expelled or drawn into the specimen. The movement of water in and out of the specimen is monitored through a Volume Change Unit (VCU) connected to the bottom of the specimen as shown in Figure 6.14. Any expansion of the cell wall due to the applied cell pressure does not affect the specimen's volume change measurement. The water line connected to the top of the specimen is opened during the filling stage, when water is pushed into the specimen through the VCU and lets the displaced air out. Thereafter, the outlet valve is always closed. The pore water pressure transducer is used to monitor the change in the pore water during the saturation, consolidation and shearing stage. The data on pore water pressure variation are needed to interpret shear strength based on the effective stress analysis.

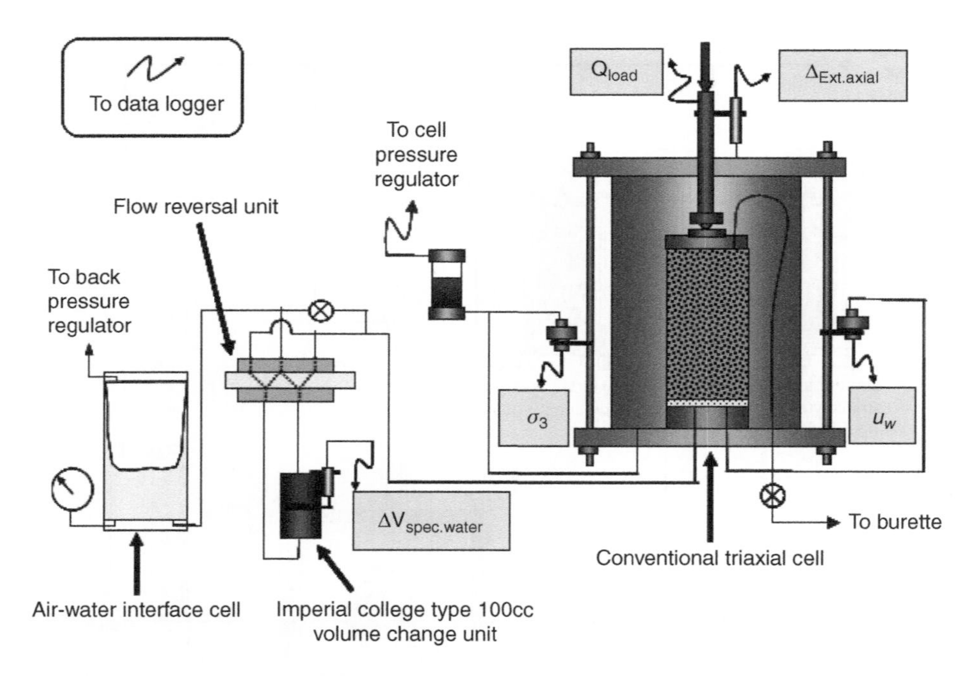

Figure 6.14 The plumbing of the conventional triaxial apparatus (*After* Md. Noor, 2006)

There are four stages involved in the test, which are:

a. Filling the specimen with water
b. Saturation
c. Consolidation
d. Shearing

a. Saturation stage

Saturation is a process of forcing water into the specimen until the trapped air in the specimen is fully dissolved in it. This filling process is to be carried out under pressure difference (i.e. between cell pressure and back pressure) of not more than 10 kPa. It involves two alternate processes which are, (1) forcing of water into the specimen or filling and (2) Skempton's 'B' value (i.e. $\Delta u/\Delta\sigma_3$) checks. It is very important to note that the back pressure line is only open for filling process and other than that it should be closed; especially when the cell pressure is to be elevated for 'B' value check. If it is open during the elevation of the cell pressure, then the specimen will be accidentally consolidated before conducting the real consolidation stage. During the filling process, the back pressure is always maintained at 10 kPa less than the cell pressure to avoid consolidating the specimen. This is the pressure difference that is pushing the water via the VCU into the specimen. During the filling stage, as water is pushed into the specimen, the pore water pressure will rise towards the value of the applied back pressure. However, the maximum value that the pore water pressure will attain will be less than the applied back pressure. This is when the specimen is not fully saturated. The pore-water pressure at the end of the filling stage is noted and the rise in the pore-water pressure, when the cell pressure is elevated, is recorded. Ultimately, the rise in the pore-water pressure will be equal to the increase in the cell pressure when full (100%) saturation is achieved. A minimum back pressure of 200 kPa is required (Head, 1981) to fully dissolve the trapped air. An example of the saturation stage for coarse grained soil is shown in Table 6.3.

b. Consolidation and shearing stage

The consolidation stage follows immediately after the saturation process is complete. With the back pressure line closed, the cell pressure is elevated higher than the back pressure by the magnitude of the required effective stress for the shearing stage. In most cases, the back pressure achieved at the end of the saturation stage is maintained. However, in certain cases, the back pressure needs to be reduced. If this is the case, it should not be reduced to less than 150 kPa and it must be greater than 200 kPa to avoid air coming out of the solution and forming air bubbles in the specimen (Head, 1981). Then the consolidation stage is started by opening the back pressure line. The water in the specimen will be squeezed out through the VCU. If this is carried out manually, the volume of water expelled is noted at 0 s, 15 s, 30 s, 1 min, 2 min, 4 min, 8 min, 15 min, 30 min, 1 h, 2 h, 4 h, or until the water stops flowing out from the specimen (i.e. to mark that constant specimen volume is achieved), or the constant pore water pressure equals the back pressure. There are two parameters that need to be determined from the consolidation stage. The first one is the reduction in height of the specimen and second is the amount of water expelled from the specimen. Once the consolidation stage has completed, the deviatoric stress can be applied to begin the shearing stage. The appropriate shearing rate needs to be calculated based on the consolidation curve.

Triaxial tests on partially saturated specimen

Triaxial tests on partially saturated specimen will give four shear strength parameters according to the curved-surface envelope of the extended Mohr-Coulomb model. The parameters are residual suction, $(u_a - u_w)_r$, maximum apparent shear strength, $c_s^{\max}$, ultimate suction, $(u_a - u_w)_u^{\sigma'=0}$ and the rate of increase in ultimate suction relative to net stress, ζ. If the data are interpreted according to the extended Mohr-Coulomb envelope, the test will give the value of ϕ^b. It requires the use of a double-walled triaxial cell. For the triaxial test on saturated specimens, the specimen volume change is determined from the amount of water that is expelled from or drawn into the specimen. However, this technique cannot be applied for the partially saturated specimen. This is because when the specimen is partially saturated, the change in the specimen volume comes from the change in the volume of air and water in the void spaces; and measuring each of them is quite difficult. However, measuring the combined effect of the two can be much easier. This is achieved by measuring the amount of water that goes in or out of the inner cell which is in contact with the specimen. That is why a double-walled triaxial cell is required for the partially saturated specimen where the pressure in the inner and outer cells is maintained equal to avoid the expansion of the inner cell wall. This is because any expansion of the inner cell will affect the specimen volume change measurement.

Table 6.3 Specimen saturation recording form

Process type	Clock time	Elapsed time	Cell pressure (kPa)	Cell pressure difference (kPa) $\Delta\sigma_3$	Back Pressure (BP) (kPa)	Pore Pressure (kPa)	PWP diff. (kPa) Δu	B Value $\left(=\frac{\Delta u}{\Delta\sigma_3}\right)$	Back pressure volume change (cc)	
									Before	After
Filling	8:45 am	8:50 am	50	–	45	30.77	–	–	65.62	48.128
B value check			Elevate cell pressure while BP line closed	This is the increase in the cell pressure	Back pressure line closed to stop consolidation	Record for the increase in the pore pressure	Increase in the PWP is 80.45 minus 30.77	$\frac{49.68}{50}$	No VCU reading since BP line closed	
	–	–	100	50	–	80.45	=49.68	=0.994	–	–
Filling					Increase the back pressure to 95 kPa and open the BP line to let water into the specimen	Record the pore water pressure at the end of filling	Increase in the PWP is 131.58 minus 82.15		Record the initial VCU reading before BP line is open	Record the reading of VCU when it is constant and close
	8:58 am	9:15 am	100	–	95	82.15	–	–	48.13	44.94
B value check	–	–	150	50	Close	131.58	49.43	$\frac{49.43}{50}=0.99$	No VCU reading since BP line closed	

Note: 1. In this example, the process of saturation needs to be continued until a minimum of back pressure 200 kPa is attained despite the B value of 0.99.
2. BP = Back Pressure.

For the test on shear strength at partial saturation, the pore air pressure and the pore water pressure are controlled separately in order to achieve the magnitude of suction required. In other words, suction is imposed on the specimen and not measured. The process of subjecting the specimen to certain suction by elevating the pore air pressure is called axis-translation. This is because the pore-air pressure applied to the specimen is higher than the field pore-air pressure, which is atmospheric. This is to avoid the formation of bubbles, usually known as cavitations, in the pore water pressure measuring system when the pore water pressure is less than −1 atm (i.e. −101 kPa). Cavitations will affect the pore water pressure measurement. The suction is the difference between the applied pore air and the pore water pressure. By elevating the pore-air pressure, the pore-water pressure will also need to be elevated to achieve certain suction. In this manner, the working pore-water pressure during the test can be made higher than −101 kPa and thereby cavitations can be avoided. The high air-entry ceramic disk prevents air from flowing past the saturated disk, but will only allow the passage of water. This is provided so that the application of the air pressure to the specimen does not exceed the air-entry value of the ceramic disk. Nevertheless, even though the air pressure is kept lower than the air-entry value of the ceramic disk, there is still a problem with air diffusion where air in the specimen can still diffuse into the contacting water body in the pores of the high-air entry ceramic disk and reappear in the water chamber underneath the disk in the form of air bubbles. This, thereby, will affect the pore water pressure and the specimen water volume change measurement. However, it can be resolved by conducting periodic flushing of the air bubbles in the pore water pressure system and measuring the volume of the flushed air bubbles using a twin burette apparatus.

The stages involved in the triaxial test on partially saturated specimen are as follows:

a. Saturation of ceramic disk
b. Equalisation
c. Consolidation
d. Shearing

a. Saturation of ceramic disk

The following procedure has been applied for disk saturation. It has proved to be effective (Md. Noor, 2006). Saturation of the ceramic disk is carried out by applying a pressure of 240 and 500 kPa to force de-aired water through the ceramic disk at 1 and 3 bar high air-entry value respectively. Every day, for three consecutive days, water is allowed to flow through the ceramic disk for 5 minutes. This technique will saturate the disk and any air bubbles present in the disk will be dissolved. The saturation of the disk is very important because it will stop the air phase in the specimen from passing through throughout the test. In this way, the pore air and pore water pressure can be controlled independently. The disk can be fixed on the pedestal of the triaxial cell, the cell filled with de-aired water and the de-aired water forced through. The water in contact with the disk must de-aired and prevented from coming in contact with air. This is because air can be readily dissolved in the water if in contact, and when it comes out of the solution, air bubbles will be formed in the void spaces of the ceramic disk. Right after the disk saturation process, it is very important for the disk to be kept

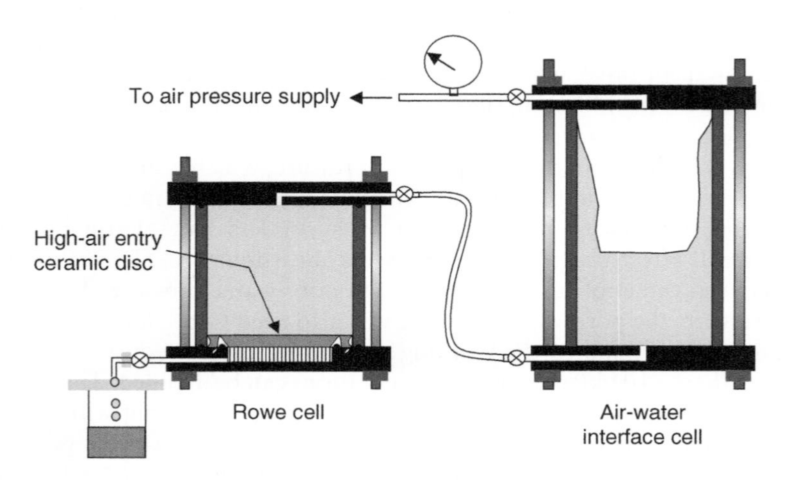

Figure 6.15 A typical set up for disk saturation process conducted outside the triaxial cell

covered with water before the soil specimen is put in place. A typical set up for disk saturation, conducted outside the triaxial cell, is shown in Figure 6.15. Note that the application of the air-water inter-phase cell prevents the water from coming in contact with air. The same system can be applied for disk saturation in the triaxial cell also.

b. Equalisation

Equalisation is a process whereby the moisture content of the soil is allowed to equilibrate with the applied pore water and pore air pressure, or in other words, to equilibrate with the suction applied to the soil. Excess water in the specimen will seep to the pore water pressure measuring system underneath the ceramic disk. Conversely, if the specimen is too dry for the applied suction, water in the compartment will seep to the specimen via the ceramic disk. This process of water migration will continue until a state of equilibrium is achieved. The volume of water that is drawn in or expelled from the specimen is measured by a VCU. However, the reading has to be corrected for the volume of diffused air collected underneath the high air-entry ceramic disk. The volume is determined by flushing the air through a twin burette system. The reading indicated by the specimen water volume change unit is the volume of specimen water that seeps through the ceramic disk plus the volume of diffused air. When the volume of the diffused air is determined, the actual specimen water volume change can be calculated. The equalisation can be considered complete when the flow has decreased to $0.05\,cm^3$ in 12 hours.

c. Consolidation and shearing

During the equalisation, the cell pressure is set at 10 kPa higher than the pore-air pressure. At the end of the equalisation stage, the specimen is consolidated to a certain net stress by elevating the cell pressure. The required cell pressure is raised while the inner and the outer cell pressure valve are in the closed position. The consolidation is

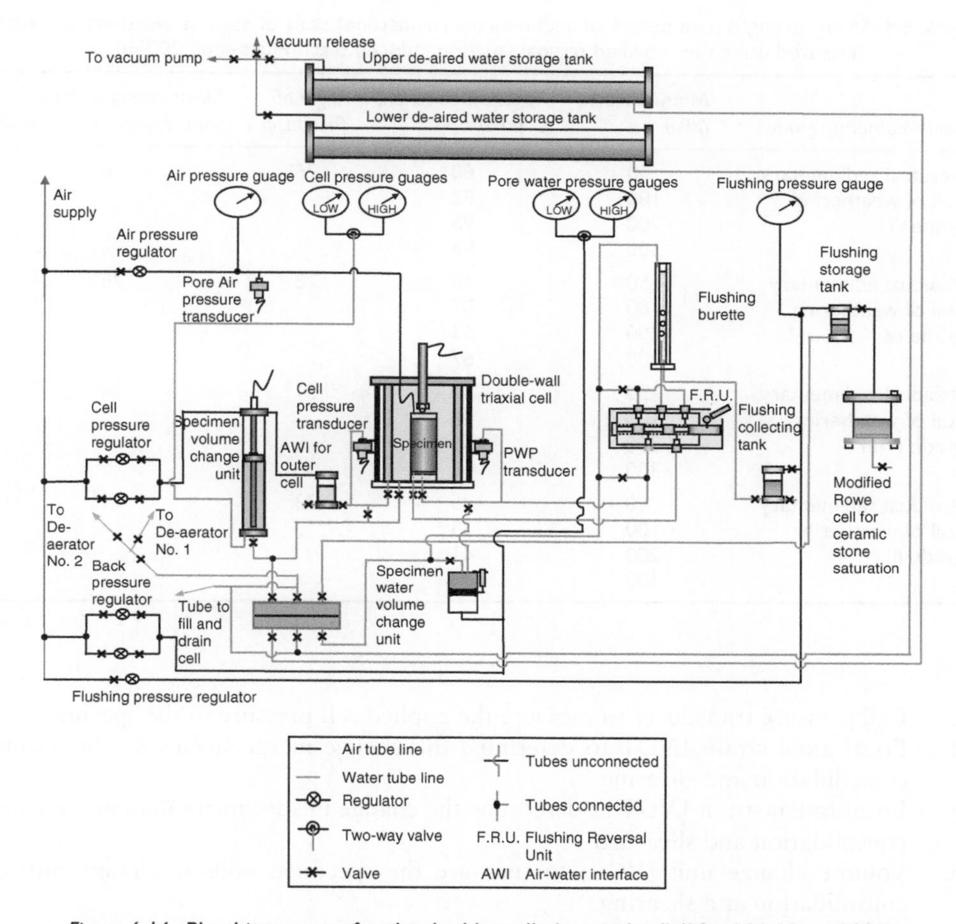

Figure 6.16 Plumbing system for the double-walled triaxial cell (*After* Md. Noor, 2006)

started by opening both valves simultaneously. The consolidation is considered complete when the specimen compression has ceased and when there is no more volume change indicated by the specimen volume change unit.

The shearing stage is carried out by applying the deviatoric stress until the specimen fails or a specified strain is achieved. The strain rate needs to be slow enough so that the pore water and pore air pressures are not affected during the shearing. The shearing rate applied for partially saturated gravel of 6.0 mm nominal size is 0.0256 mm/min.

The instrumentations required with the double-walled triaxial cell are as follows:

a. Axial strain transducer using linear variable differential transformer (LVDT).
b. Load cell to measure the applied deviatoric stress during the shearing stage.
c. Pore-air pressure transducer to measure the pore-air pressure applied to the specimen.
d. Pore-water pressure transducer to measure the applied pore-water pressure to the specimen through the high air-entry ceramic disk.

Table 6.4 Shear strength parameters of sedimentary (sandstone) soils of various weathering grades, measured using the modified triaxial (suction induced test) (Huat *et al.*, 2005b)

Soil/Weathering grade	*Matric suction, (kPa)*	*Apparent cohesion, c′ (kPa)*	*Angle of friction, ϕ'*	*Shear strength change with change in suction, ϕ^b*
Residual sedimentary soil of weathering grade V	50	60	26°	26°
	100	82		
	200	95		
	300	97		
Residual edimentary soil of weathering grade IV	50	48	28°	26°
	100	57		
	200	63		
	300	67		
Residual sedimentary soil of weathering grade IV-III	50	37	31°	24°
	100	48		
	200	55		
	300	61		
Residual sedimentary soil of weathering grade III	50	25	33°	19°
	100	35		
	200	41		
	300	43		

e. Cell pressure transducer to measure the applied cell pressure to the specimen.
f. Local axial strain LVDT to determine the change in specimen's height during consolidation and shearing.
g. Local radial strain LVDT to determine the change in specimen's diameter during consolidation and shearing.
h. Volume change unit (VCU) to measure the specimen volume change during consolidation and shearing.
i. Volume change unit (VCU) to measure the specimen's water volume change especially during equalisation, consolidation and shearing.

A typical plumbing for double-walled triaxial cell is shown in Figure 6.16.

Table 6.4 shows the values of shear strength parameters of some tropical residual sedimentary (sandstone) soils obtained from the double wall (modified) triaxial test. The apparent cohesion, c' increases with an increase in the matric suction. For a given level of matric suction, the cohesion increases with an increase in the soil weathering grade, i.e. as the soils/rocks are weathered more. The angle of friction, ϕ' apparently decreases with an increase in the soil's weathering grade.

6.4 CONCLUSIONS

The real soil shear strength behaviour in saturated and partially saturated conditions is very complex. It is very difficult to define it by a single equation. Nevertheless, it has been defined by the curved-surface envelope of the Mohr-Coulomb model. The application of the true soil shear strength behaviour will help to explain complex soil mechanical behaviour. The model exhibits the steep drop in shear strength near saturation (i.e. suction approaching zero) and the steep drop in shear strength when approaching zero effective stress. These are the shear strength behaviour at low stress levels and they are very much related to the shallow rainfall-induced failure. The steep drop in shear strength near saturation is very much related to the occurrence of wetting collapse or inundation settlement. However, the equipment and the test procedures required to comprehensively characterise soil shear strength behaviour are quite extensive. Furthermore, it is a very challenging task to incorporate the true soil shear strength behaviour in a theory or a framework that can characterise this complex soil behaviour.

REFERENCES

Alonso, E.E., Gens, A. and Josa, A. (1990). A constitutive model for partially saturated soil. *Géotechnique*, 40(3), 405–430.

Bishop, A.W. (1966). The strength of soils as engineering materials. *Géotechnique*, 16(2), 91–130.

Charles, J.A. and Watts, K.S. (1980). The influence of confining pressure on the shear strength of compacted rockfill. *Géotechnique*, 30(4), 353–367.

Coulomb, C.A. (1776). Essai sur une application des regles de maximia et minimis a quelques problèmes de statique relatifs a l'architecture. *Mémoires de la Mathématique et de Physique, présentes a l'Académique Royale des Sciences, par divers savants, et lus dans ces Assemblées*. L'Imprimerie Royale, Paris, 3–8.

Donald, I.B. (1956). Shear strength measurements in unsaturated non-cohesive soils with negative pore water pressures. *Proc. 2nd Australia-New Zealand Conf. Soil Mechanics Foundation Eng.* (Christchurch, New Zealand), 200–205.

Escario, V. and Juca, J. (1989). Strength and deformation of partly saturated soils. *Proceedings of the 12th International Conference on Soil Mechanics and Foundation Engineering*, Rio de Janeiro: 3, 43–46.

Escario, V. and Saez, J. (1986). The shear strength of partly saturated soils. *Géotechnique*, 36(3), 453–456.

Fredlund, D.G., Morgenstern, N.R. and Widger, R.A. (1978). Shear strength of unsaturated soils. *Canadian Geotechnical Journal*, 15(3), 313–321.

Fukushima, S. and Tatsuoka, F. (1984). Strength and deformation characteristics of saturated sand at extremely low pressures. *Soils and Foundations, JSSMFE*, 24(4), 30–48.

Gan, J.K.M., Fredlund, D.G. and Rahardjo, H. (1988). Determination of the shear strength parameters of an unsaturated soil using the direct shear test. *Canadian Geotechnical Journal*, 25(3), 500–510.

Head, K.H. (1981). *Manual of Soil Laboratory Testing*, Pentech Press, London.

Huat, B.B.K., Ali, F.H. and Hashim, S. (2005a). Modified shear box Test apparatus for measuring shear strength of unsaturated residual soil. *American Journal of Applied Sciences*. USA: New York. 2(9), 1283–1289.

Huat, B.B.K., Ali, F.H. and Abdullah, A. (2005b). Shear strength parameters of unsaturated tropical residual soils of various weathering grades. *Electronic Journal of Geotechnical Engineering*, Vol. 10 (Bundle D):12.

Indraratna, B., Wijewardena, L.S.S. and Balasubramaniam, A.S. (1993). Large-scale triaxial testing of greywacke rockfill. *Géotechnique*, 43(1), 37–51.

Md. Noor, M.J. (2006). *Shear strength and volume change behaviour of unsaturated soils*. Ph.D. Thesis, University of Sheffield, UK.

Md. Noor, M.J. and Anderson, W.F. (2006). A comprehensive shear strength model for saturated and unsaturated soil. *Proc. 4th Int. Conf. on Unsaturated Soils*. ASCE Geotechnical Special Publication No. 147, Carefree, Arizona, Vol. 2, pp. 1992–2003.

Md. Noor, M.J., Zulkipli, M.H. and Fauzi, M.F. (2008). Shear strength behaviour relative to effective stress for granitic residual soil grade VI. *International Seminar on Civil and Infrastructure Engineering (ISCHE) 2008 in UiTM*, Malaysia.

Mohr, O. (1900). Welche Umstande bedingen die elastizitatsgrenze und den bruch eines materiales? *Zeitschrift des Vereines Deutscher Ingenieure*, Vol. 44, 1524–1530, 1572–1577.

Othman, M.A. (1989). *Highway Cut Slope Instability Problems in West Malaysia*. Ph.D. Thesis, Department of Geography, University of Bristol, UK.

Rahardjo, H., Lim, T.T., Chang, M.F. and Fredlund, D.G. (1995). Shear strength characteristics of a residual soil in Singapore. *Canadian Geotechnical Journal*, 32, 60–77.

Roscoe, K.H. and Burland, J.B. (1968). On the generalised stress-strain behaviour of 'wet' clay. In: J. Heyman and F.A. Leckie (eds), *Engineering Plasticity*. Cambridge UK, Cambridge University Press, pp. 535–609.

Terzaghi, K. (1936). The shear resistance of saturated soils. *Proc. 1st. International Conference on Soil Mechanics and Foundation Engineering*. Cambridge MA, 1, 54–56.

Toll, D.G., Ong, B.H. and Rahardjo, H. (2000). Triaxial testing of unsaturated samples of undisturbed residual soil from Singapore. *Proc. Unsaturated Soil for Asia.*, Singapore, 581–586.

Vanapalli, S.K., Fredlund, D.G., Pufahl, D.E. and Clifton, A.W. (1996). Model for the prediction of shear strength with respect to soil suction. *Canadian Geotechnical Journal*, 33(3), 379–392.

Wheeler, S.J., Sharma, R.S. and Buisson, M.S.R. (2003). Coupling of hydraulic hysteresis and stress-strain behaviour in unsaturated soils. *Géotechnique*, 53(1), 41–54.

Wood, D.M. (1990). *Soil Behaviour and Critical State Soil Mechanics*. Cambridge UK, Cambridge University Press.

7

Slopes

H. Rahardjo & E.C. Leong
Nanyang Technological University, Singapore

J.A.R. Ortigao
Terratek, Rio de Janeiro, Brazil

R.B. Rezaur
Golder Associates Ltd., Calgary, Canada

7.1 INTRODUCTION

This chapter focuses on the geological factors and their effect on slope stability, landslip type, effect of relict structures, rainfall and suction and finally addresses the stabilisation and remedial measures to soil slopes. The authors' experience covers a wide range of tropical regions including the Far East and Central and South America.

Many tropical residual soil slopes have a deep groundwater table with a significant unsaturated zone above the water table. Rainfall-induced slope failures are a common problem in tropical residual soil slopes. The infiltration rate of rainwater through the unsaturated zone and the changes in shear strength of unsaturated soils due to rainwater infiltration are essential parameters for the stability assessment of slopes against rainfall. This chapter illustrates the significance of rainfall intensity, rainfall amount, antecedent rainfall, infiltration, and soil properties towards evaluating slope instability. Different methods of slope stability analyses that incorporate unsaturated soil mechanics principles are described. Slope stabilisation methods using surface cover, horizontal drains and capillary barriers are also described. The significance of different parameters related to the slope stability assessment and slope stabilisation methods described are illustrated through results and knowledge gained from parametric studies, laboratory experiments and field observations made in residual soil slopes in Singapore.

7.2 GEOLOGICAL FACTORS OF SLOPE BEHAVIOUR

Kanji (2004) reviewed geological factors that have impact on slope stability. This section presents a summary of these factors.

Significant geological features

There are some geological features that are more significant than others as related to the geotechnical behaviour of slopes and other engineering works.

Faults: The rupture of the rock mass by geological (tectonic) action, followed by some displacement, causes the formation of a continuous plane of discontinuity, often accompanied by the fragmentation of the rock along this plane. This allows water percolation, and more weathering than in the adjoining rock mass. A fault plane can have a much lower friction coefficient with respect to the rest of the rock mass. An unfavourable orientation of this plane with respect to the cut is when it has a direction parallel to, or a low angle to, the slope surface and dips towards the slope. If the plane is exposed in the excavation and if the friction angle of this plane is equal to or smaller than the dip angle, it will cause the sliding of the rock mass above it.

Joints, bedding or schistosity planes, and shear zones: These planes of discontinuity act in the same manner as faults, representing planes of weakness within the soil-rock

Figure 7.1 Gneiss saprolites relict structures (Rio de Janeiro)

mass (Figure 7.1). When the joints are oriented in preferential spatial attitudes and conform to jointing systems along which the rock mass is weaker with respect to shearing stresses, they become directions of weakness. These discontinuities can be critical when adversely oriented with respect to the slope. Figure 7.2 shows profiles in residual soils of sedimentary Jurong formation.

Stress relief joints: The slow natural removal of topsoil by erosion promotes the decrease of stresses acting on the rock by the self weight of the material. The rock then tends to decompress, giving rise to tension in planes parallel to the topographic surface. As the rock has a low tensile strength (about one tenth of the compressive strength), it fractures, forming a family of joints, which become more tightly spaced close to the rock surface. They are often weathered, as they constitute easier ways for the water infiltrating into the ground, besides the fact that after heavy rains, they are saturated; allowing the building up of high hydrostatic pressures.

Tension cracks: In clefts, where the slope is very steep, sometimes even vertical, and when the strength of the rock is low, vertical tension cracks may be formed. In rainy periods, the water infiltrates very easily within these fissures and, since their drainage is usually much slower than the infiltration rate, high hydrostatic pressures are built up ("cleft water pressures"), exerting hydrostatic lateral thrust on the thin vertical slice of rock, causing failures by toppling.

Sheared planes: Planes that underwent displacement in the geological past, like bedding planes in folded strata and planar discontinuities, had their friction angle decreased due to the destruction of roughness during shear, and had their friction coefficient decreased, eventually attaining their residual strength.

Figure 7.2 Excavated slope in Singapore, showing profiles of residual soil of the sedimentary Jurong formation

Figure 7.3 Slope cut in phyllites, Belo Horizonte, Brazil

Weak layers: The rock mass may contain layers of weaker material, which may represent potential zones of failure. This is commonly the case of stratified rock, with the presence of shales within harder rocks, and of schists rich in mica within quartzites and other rocks (Figures 7.3 and 7.4), or porous and poorly cemented sandstone within hard sandstone, and so on. Their adverse effect will also depend on the orientation of these layers with respect to the slope surface.

Figure 7.4 Weak layers in phyllites

Weathered zones: The rock mass may be more weathered in certain zones or layers than others; similarly to weak layers, they represent weaker zones.

Permeable layers: The existence of more permeable layers covered by impermeable layers, such as the saprolite horizon covered by clayey soil, may help in developing uplift pressures that decrease the strength of the soils. Another undesirable condition may occur when, within the slope or at the foundation of high fills or dumps, there are permeable layers intercalated with less permeable layers; in this situation, an artesian condition may occur if the piezometric level is high, decreasing the stability of the slope.

Colluvium deposits: These deposits result from the accumulation of loose material transported by gravity or by erosion from the upper parts of the hill. These deposits occur at the toe of the hill or at the hill slope, in portions of lower inclination. Due to their mode of formation, they stayed practically at their natural angle of repose and also are quite irregular and under-consolidated. In tropical regions, where chemical weathering predominates, they have an earthy matrix and rounded rock fragments and boulders. They commonly show movements and instabilities in rainy seasons due to the rise of water level and decrease in its suction tension. In arid or semiarid regions, under prevailing physical weathering, they may have a sandy matrix and angular rock fragments and are more commonly called talus. In any case, they are very unstable deposits (mainly the colluvium) when subjected to cuts.

7.3 GEOLOGY AND MODE OF SLOPE FAILURE

The geological characteristics of a slope play a major role on its mode of failure and determine how it should be analysed. Typical situations are outlined below for 2 and 3 dimensional analyses.

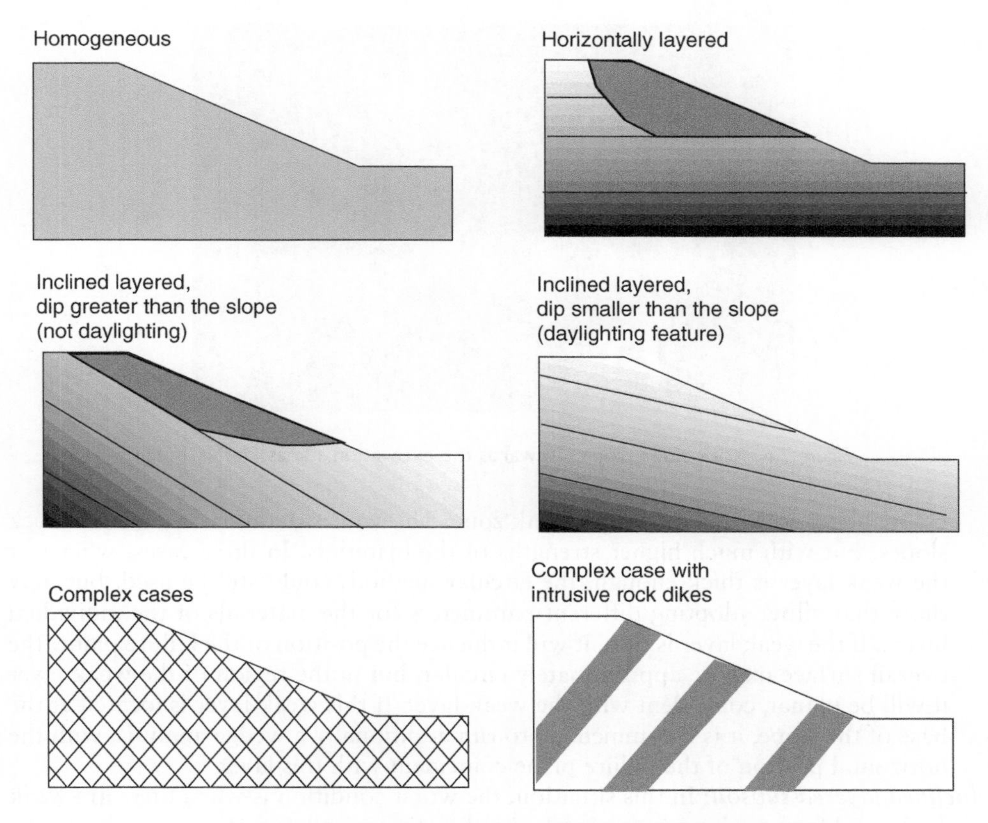

Figure 7.5 Modes of failure resulting from different geological conditions

Planar or two-dimensional failure mode

The planar or two-dimensional failure mode is used when the kinetics of the landslide leads to a plane state deformation case. Figure 7.5 shows typical cases and these are outlined below.

Weak and homogeneous materials: This is the case of soil slopes or very weak rocks without conditioning weak planes. The failure surface will be determined by the concentration of shear stress in the soil or rock mass, which from limit-state analysis, is shown to be approximately circular in a cross-section of the slope. In this case, it becomes evident that there is no influence of the geology and any method of circular analysis is applicable; its selection depending only on the degree of accuracy or mathematical premises desired.

Horizontally layered subsoil: In this category, several cases may occur. One of them is when the soil slope presents a "layered" structure of different soils. This may result from different degrees of weathering of residual soils, generally the properties of the soils improving with depth. However, in residual soils derived from rocks with alternating stronger and weaker layers, it is possible to find soils of poor properties

Figure 7.6 Rock slope dipping towards the excavation, Minas Gerais, Brazil

below better soils, constituting a weak zone. The same situations can occur in rock slopes, but with much higher strengths of the materials. In these cases, whenever the weak layer is thick enough, the circular methods could still be used, but only those that allow adopting different parameters for the materials of the individual layers. If the weak layer is thin, it will influence the position of the failure plane; the overall surface may be approximately circular, but in the region of the weak layer it will be planar, coincident with the weak layer. If this weak layer is located at the base of the slope, it is recommended to run additionally a wedge analysis, with the horizontal portion of the failure plane coincident with this layer.

Inclined layered subsoil: In this situation, the worst condition is when there are weak geological features (layers or joints) with direction parallel or almost parallel to the slope, dipping towards the cut (Figure 7.6).

These features represent the weakest plane in the slope mass and, for this reason, the position of the failure plane.

Complex situations: Sometimes, the internal structure is irregular and heterogeneous, not following any of the situations above. It is possible that neither circular nor planar planes are representative of the critical failure surface, which would be irregular.

Rock falls: In some slopes, there may be a possibility of fall of individual or several rock blocks. This situation may happen in jointed rock slopes (toppling failure) or in residual soil slopes with boulders outcropping, which may become unstable and fall or run down the slope, due to progressive erosion of the soil at its base or due to seismic accelerations.

7.4 LANDSLIDE CLASSIFICATION

Table 7.1 presents Cruden and Varnes' (1996) landslide classification according to the kinematics of the movement and material type. It is a new version of Varnes' (1978) similar work.

Figure 7.7 shows the main landslip types according to the kinematics of the slide.

Table 7.1 Landslide classification (*After* Cruden and Varnes, 1996)

Movement	*Material type*	*Soils*	
		Predominantly coarse	*Predominantly fine*
Fall	Rock fall	Debris fall	Earth fall
Topple	Rock topple	Debris topple	Earth topple
Slide	Rock slide	Debris slide	Earth slide
Spread	Rock spread	Debris spread	Earth spread
Flow	Rock flow	Debris flow	Earth flow or mudflow

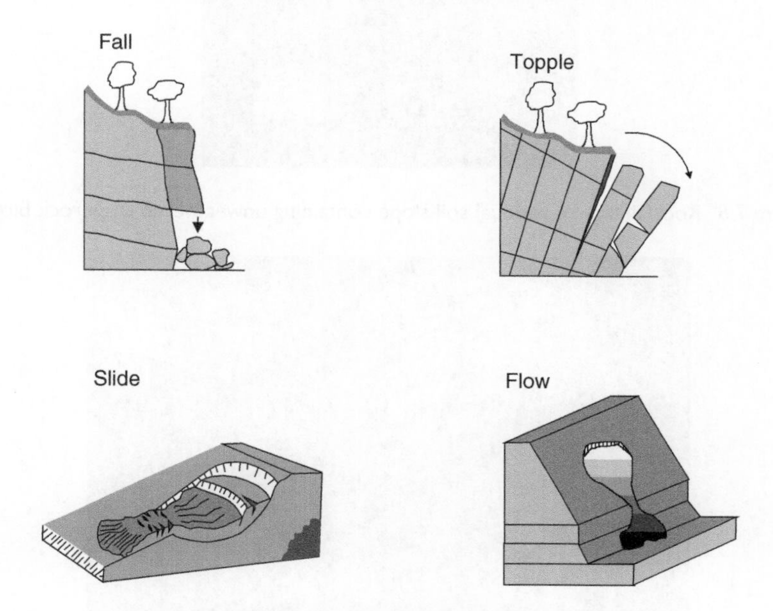

Figure 7.7 Main landslip types (kinematical classification)

Fall

A *fall* occurring in rock slopes involves rock displacement from a slope and is usually associated with very fast rates of movement, of the order of magnitude of metres per second. A fall can also include movements of bouncing and rolling, as it is the case of Figure 7.8, triggered by severe rainfall. Figure 7.9 shows rock fall from jointed rock slope.

Topple

A *topple* of rock blocks or vertical rock slabs occurs in nearly vertical or sub-vertical rock slopes involving a rotation in the direction of the slope. It occurs due to gravity or water thrust in rock joints. Figure 7.10 presents a toppling failure of a slope in Algarve, Portugal, in a vertical jointed rock mass. Figure 7.11 shows a case of toppling failure of a jointed rock slope, in which rock blocks seem stacked on the slope.

Figure 7.8 Rockfall from a residual soil slope containing unweathered large rock blocks

Figure 7.9 Detached rock blocks fall from the jointed rock slope

Slides

Slides are mass movements that present a well-defined failure surface. They are classified as rotational, translational or complex slides, according to the failure surface geometry being respectively circular, polygonal or complex. The slides can also be classified as *shallow* or *deep*, according to the relative depth of the failure surface to the longitudinal length of the landslip.

Figure 7.12 presents geometric features of a slide: crown, main scarp, cracks and toe. The slide is also divided into two main zones: depletion at the top and accumulation at the bottom.

Figure 7.10 Toppling failure, Portugal (courtesy of Dr. R. Pistone)

Figure 7.11 Toppling of rock blocks from jointed rock slope (Carretera Pan Americana, Ecuador)

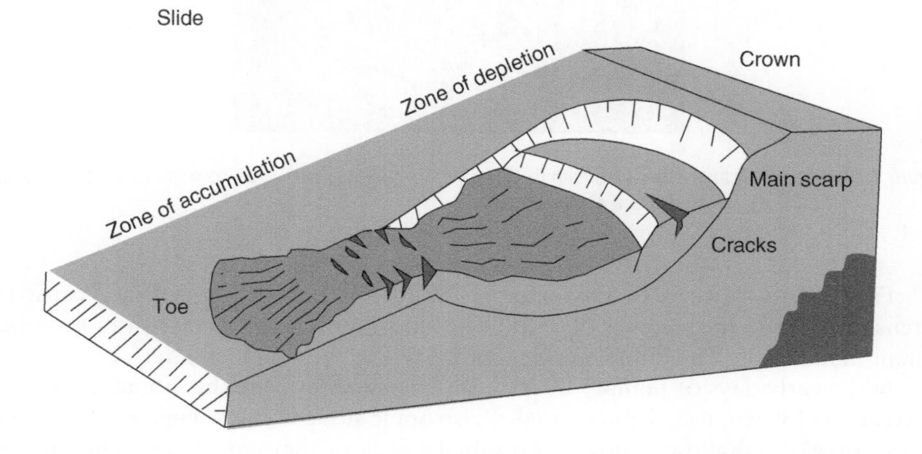

Figure 7.12 Slide features

Figure 7.13 Slide on a cut slope with a well-defined failure surface

Figure 7.14 Shallow translational landslip in Campos do Jordão, Brazil, due to severe rainfall in January 2000 that wiped out a squatter area

Figure 7.13 shows a shallow slide at the BR-040 motorway in Brazil caused by intense rainfall in the summer of 1988. The landslip occurred in saprolites just above a man-made slope. The failure surface took place at the soil-rock interface.

In the early days of January 2000, a summer storm struck the mountain resort of Campos do Jordão, Brazil and caused catastrophic slope failures. Figure 7.14 presents an example of a shallow failure in a residual soil slope that wiped out a church at the top and several houses along its path.

Figure 7.15 Shallow failure at the Cordillera de los Andes, Ecuador

Figure 7.16 Feature of a landslide in Singapore

Figure 7.15 presents an example of failure in the mountains of the Cordillera de los Andes, Ecuador, along the Pan-American Road. It took place during the El Niño meteorological phenomenon, which caused intense rainfall in the region. This landslip seems to have been caused by poor drainage, as there is no surface drainage system at this slope. Above this slide, the photo shows curved surfaces left by previous landslides, demonstrating that the area is very unstable. An example of a landslide in Singapore is featured in Figure 7.16.

Flow

A *flow* is a continuous viscous slide involving soils or rocks (Figure 7.17). It is named *mudflow* if fine soils are predominant, or *debris flow* if it contains a large range of particle diameters.

Figure 7.17 Flow (*After* JSA, 2001)

Figure 7.18 Mudflow, BR-040 Motorway, Brazil

Figure 7.18 shows a mudflow that started on the top of the mountain due to an intense rainfall and reached a road; it was a strip some 200 m long and just a few metres wide, conveying mud and water from the top of the slope.

Figure 7.19 presents the result of a landslip caused by a heavy rainfall of 400 mm in 24 hours that struck Rio de Janeiro in February 1996. The consequence was an exceptional debris flow formed in the catchment area of the Quitite Mountain. Two people were killed and many houses destroyed in this event.

Figure 7.19 Debris flow, Quitite landslide, Rio de Janeiro, 1996

Table 7.2 Landslide velocity scale (*After* Cruden and Varnes, 1996)

Velocity class	*Description*	*Velocity (mm/s)*	*Typical velocity*
7	Extremely rapid	$>5 \times 10^3$	
6	Very rapid	5×10^3	5 m/s
5	Rapid	5×10^1	3 m/min
4	Moderate	5×10^{-1}	1.8 m/h
3	Slow	5×10^{-3}	13 m/month
2	Very slow	5×10^{-5}	1.6 m/year
1	Extremely slow	5×10^{-7}	16 mm/year

Table 7.3 Landslip classification according to soil disturbance

Virgin landslides	Occur in undisturbed soils with mobilised peak soil shear strength
Reactivated landslides	Occur in disturbed soils at a previously occurring landslip surface with mobilised residual soil shear strength

Rate of movement

Cruden and Varnes (1978) presented a classification of landslides according to the rate of movement (Table 7.2). They also give a landslide velocity scale, from one to seven.

Additional classifications

Tables 7.3 and 7.4 present additional landslide classifications according to soil disturbance and drainage conditions.

Table 7.4 Landslip classification according to drainage conditions

Drained (Long term)	Pore pressures due to seepage flow in the soil. No pore pressures during shear or they are fully dissipated
Partially drained (medium term)	Pore pressures generated during shear are partially dissipated
Undrained (short term)	Low permeability soils, pore pressures generated during shear

7.5 LANDSLIDE TRIGGERING MECHANISMS

Guidicini and Nieble (1984) and Varnes (1978) summarised landslide-triggering mechanisms and these are presented in Tables 7.5 and 7.6 respectively. This chapter, however, concentrates on the rainfall-triggered landslides and anthropic action.

The anthropic factors that may cause landslides are:

- Removal of vegetation;
- Spills from water or sewage pipes;
- Incorrect cuts, too steep;
- Very loose fills, not properly compacted;
- Waste deposits, incorrectly placed

Creep

Creep is a slow to extremely slow flow, according to Cruden and Varnes' classification (Table 7.2). They can be shallow or deep, as a function of the relative depth of the failure surface. Insight into this phenomenon will be given by two interesting examples from the authors' experience.

Rainfall

It is well recognised that rainfall is the main triggering factor for slopes in tropical regions and for this reason, the determination of the critical rain intensity is an issue of great interest.

Figure 7.20 presents a summary of the study carried out by Kanji *et al.* (1997) relating critical accumulated precipitation level against time. This work points out that large and catastrophic landslides mostly correspond to high values of precipitation in long periods (several days to weeks). It also shows that debris flows occur mainly under stormy rainfall in short periods of time (minutes or a few hours, at the most), after a preceding period of rains even of small intensity.

On the basis of such graph, it can be seen that the lower boundary of all of the cases registered is a curve that represents the minimum triggering condition. The curve may be expressed by the following equation:

$$P = 22.4t^{0.41} \tag{7.1}$$

where, P is the accumulated rainfall, in millimetres and t is the corresponding period of time, in hours.

Table 7.5 Landslide triggering mechanisms (*After* Guidicini and Nieble, 1984)

Agents					
	Triggering		*Causes*		
Pre-existing	*Before the landslip*	*Triggers*	*Internal*	*External*	*Intermediate*
Geology Morphology Climate Gravity Temperature Vegetation	Rainfall Erosion Freezing, thawing Temperature change Solubility Groundwater change Animal and anthropic activity	Rainfall Snow thawing Erosion Earthquakes Waves Wind Anthropic action	Temperature change Shear strength decrease due to weathering	Geometry change Vibrations	Groundwater level rise Suction reduction Water filling joints Rapid drawdown Piping

Table 7.6 Landslide triggering mechanisms (*After* Varnes, 1978)

Action	*Factors*	*Phenomena*
Load increase	Removal of soil or rock mass	Erosion, landslips, cuts
	Surcharge	Weight of rainfall, ice, snow Natural deposition of soils Weight of vegetation Building of structures, embankments
	Seismic loading	Earthquakes, waves, volcanoes Vibrations, traffic, induced seismic loading
	Horizontal pressure	Water-filled joints, freezing, expansive soils
Strength decrease	Material characteristics	Geotechnical characteristics, stresses
	Changes or variable factors	Weathering, shear strength loss Groundwater rise

The graph of Figure 7.20 shows different zones representing the character or degree of consequence of the landslide triggered by the rain, drawn on the bases of regression analysis.

Of all the rainfall parameters usually used to quantify the effect of rainfall on slope stability or correlate slope failure incidences to rainfall, perhaps rainfall intensity and rainfall amount have received the most attention. In this section, the role of rainfall in inducing slope instability will be highlighted in the context of the boundary-flux variables: rainfall intensity, total rainfall, antecedent rainfall, infiltration and soil properties.

Rainfall intensity and rainfall amount

Brand (1984) in his noteworthy study on landslides in Hong Kong correlated incidence of landslides to rainfall patterns. Simple direct correlation between landslide incidence and rainfall has also been attempted for Brazil (Barata, 1969) and Japan

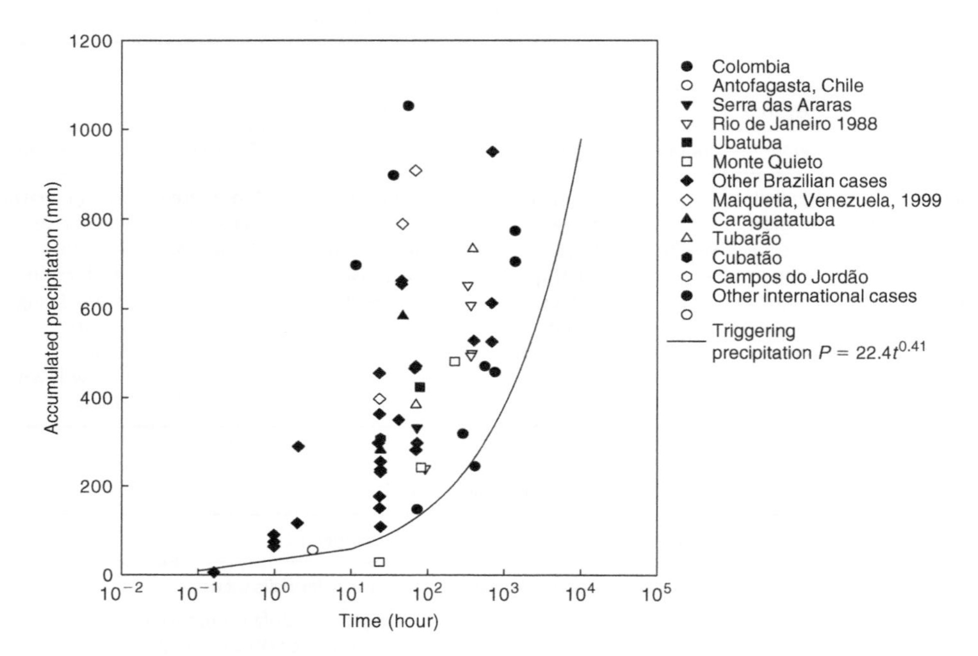

Figure 7.20 Critical precipitation levels that triggers debris flows (*After* Kanji *et al.*, 2000)

(Fukoka, 1980). Sophisticated correlation attempts are also made for Hong Kong (Lumb, 1975; Brand *et al.*, 1984; Au, 1993; Kay and Chen, 1995) and New Zealand (Eyles, 1979; Crozier and Eyles, 1980). From experience in Hong Kong, Brand *et al.* (1984) concluded that the majority of landslides in Hong Kong are induced by short duration rainfall of limited spatial extent but high intensity. The critical rainfall intensity above which landside events would occur was reported to be 70 mm/h. Murry and Olsen (1988) also expressed similar views as they observed that a rainfall intensity of 70 mm/h appeared to have some validity as a criterion for landslides in Papua New Guinea.

Alternatively, from the total rainfall perspective, a significant number of landslides could be expected in Hong Kong if the daily rainfall (24 h rainfall) exceeded about 100 mm (Brand *et al.*, 1984). Au (1993) reported that landslides in Hong Kong typically occurred when the daily rainfall was more than 70 mm and sometimes landslides could occur with daily rainfall as low as 50 mm. However, major landslide incidents are not expected until the daily rainfall exceeds 110 mm. Murray and Olsen (1988) reported that major landslides in Papua New Guinea occurred when daily rainfall was 125 mm.

Although the above discussion seems to indicate some correlations between landslide incidence with critical rainfall intensity or rainfall amount, it is noteworthy that rainfall intensity or rainfall amount alone may not always be indicative of landslide events, as is evident from Figure 7.21. The distribution of total rainfall amount and slope failures in Singapore for the period of December 2006 and January 2007, as shown in Figure 7.21, do not seem to correlate well. Such anomalies are probably the consequence of attempts to correlate landslide incidences with rainfall patterns alone

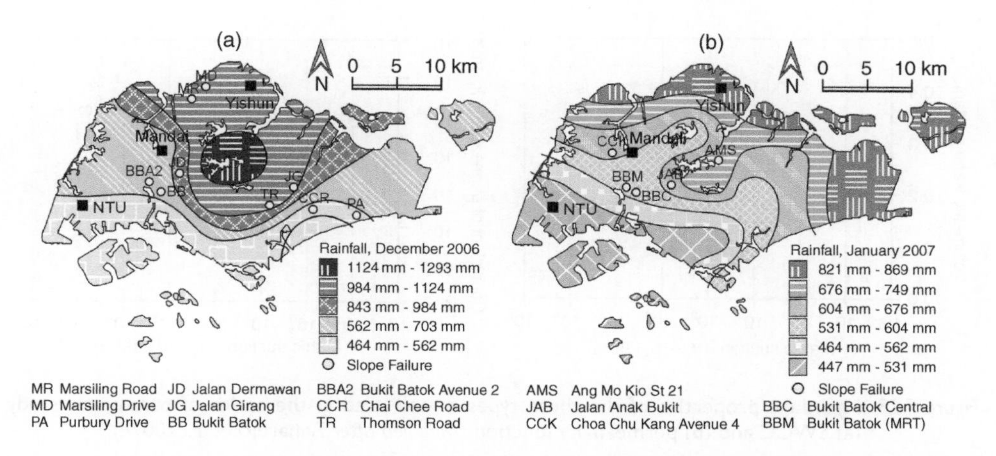

Figure 7.21 Distribution of rainfall and of slope failures in Singapore during the month of (a) December 2006 and (b) January 2007 (*Modified after* Rahardjo *et al.*, 2008b)

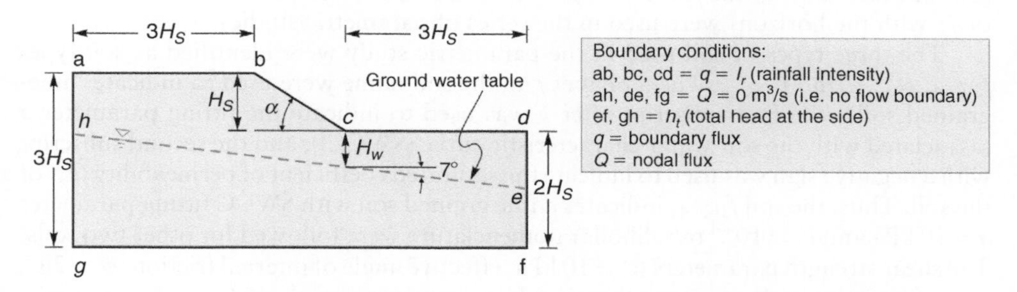

Figure 7.22 Geometry and imposed boundary condition for the homogenous soil slope used in the series of parametric study (*After* Rahardjo *et al.*, 2007c)

rather than evaluating the relative importance of the controlling parameters (rainfall intensity, rainfall amount, antecedent rainfall, infiltration, soil properties, slope geometry) associated with slope failures (Rahardjo *et al.*, 2008a).

A series of parametric studies (Rahardjo *et al.*, 2007c) aimed at identifying the relative importance of the controlling parameters (soil properties, rainfall intensity, initial depth of water table, and slope geometry), in inducing instability of homogeneous soil slopes, indicated that the primary parameters that control the instability of a slope exposed to rainfall are the rainfall intensity and soil properties. The location of initial ground water table and the slope geometry (angle, height) play a secondary role because these parameters control mainly the initial factor of safety of a slope.

The series of parametric studies was performed on a typical geometry of a homogeneous residual soil slope with imposed boundary conditions as shown in Figure 7.22. Four series of parametric studies with a combination of parameters (13 rainfall intensities $I_r = 0.9$, 1.8, 3.6, 5.4, 9, 18, 36, 54, 80, 90, 180, 360, and 900 mm/h each for 24 h duration; 3 soil types namely, $f_{10,-4}$, $f_{50,-5}$, and $f_{100,-6}$; 4 slope angles, $\alpha = 26.6$,

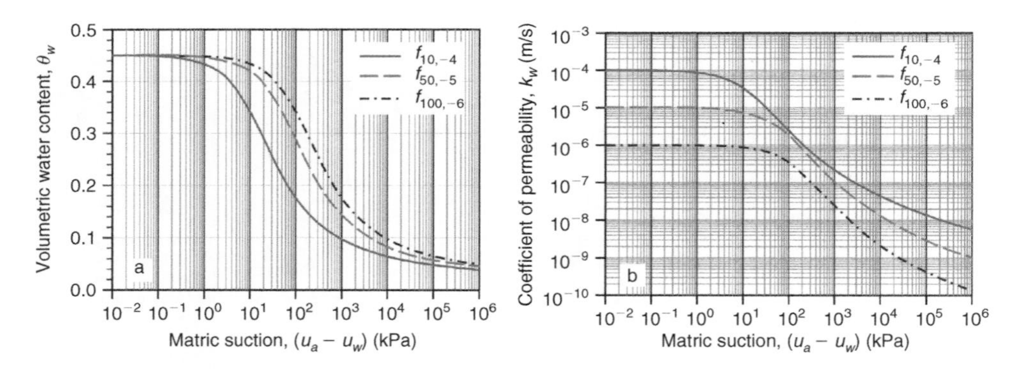

Figure 7.23 Hydraulic properties of the three types of soil used in the series of parametric study (a) SWCC and (b) permeability function (*Modified after* Rahardjo *et al.*, 2007c)

33.7, 45.0, and 63.4°; 4 slope heights, $H_s = 5$, 10, 20, and 40 m; and 5 initial depths of groundwater table at the slope toe $H_w = 2.5$, 5, 7.5, 10, and 15 m with an inclination of 7° with the horizon) were used in the series of parametric studies.

The three types of soil used in the parametric study were identified as soil type-$f_{10,-4}$, $f_{50,-5}$, and $f_{100,-6}$. The alphabet f in the soil name were used to indicate "fine-grained soil", the first subscript after f was used to indicate the fitting parameter a (associated with the soil-water characteristic curve (SWCC)), and the second subscript with a negative sign was used to indicate the saturated coefficient of permeability (k_s) of the soil. Thus, the soil $f_{10,-4}$, indicates a fine grained soil with SWCC fitting parameter $a = 10$ kPa and $k_s = 10^{-4}$ m/s. Similar nomenclature were followed for other two soils. The shear strength parameters ($c' = 10$ kPa, effective angle of internal friction, $\phi' = 26°$, rate of increase in shear strength caused by matric suction, $\phi^b = 26°$, and unit weight of soil, $\gamma = 20$ kN/m^3) and saturated coefficient of permeability (k_s) of soils used in the parametric study were representative of typical residual soils in Singapore. The three types of soils used were also representative of almost all types of soil (in terms of permeability) – from good-drainage soils with k_s of the order of 10^{-4} m/s to poor-drainage soils with k_s of the order of 10^{-6} m/s. The SWCC and permeability function of the three soils used in the parametric studies are shown in Figure 7.23.

The relevant results of the parametric study, to illustrate the role of rainfall intensity on slope stability, are presented herein. The effect of other controlling parameters (slope geometry (angle, height), and location of groundwater table) on slope stability can be found in Rahardjo *et al.* (2007c).

Figure 7.24, obtained from the results of the parametric study, shows the effect of rainfall intensity on the variation of factor of safety. A common pattern on the variation of the minimum factor of safety, $F_{s(\min)}$, with changes in rainfall intensity, I_r, is apparent from Figure 7.24. Regardless of the slope angle or soil type, the higher the rainfall intensity the lower is the minimum factor of safety. This implies that the rainfall intensity primarily controls the minimum factor of safety. A column-wise comparison of the plots in Figure 7.24 shows that under the same rainfall intensity (9 and 80 mm/h) and the same slope angle, the rate of reduction in factor of safety during rainfall is the

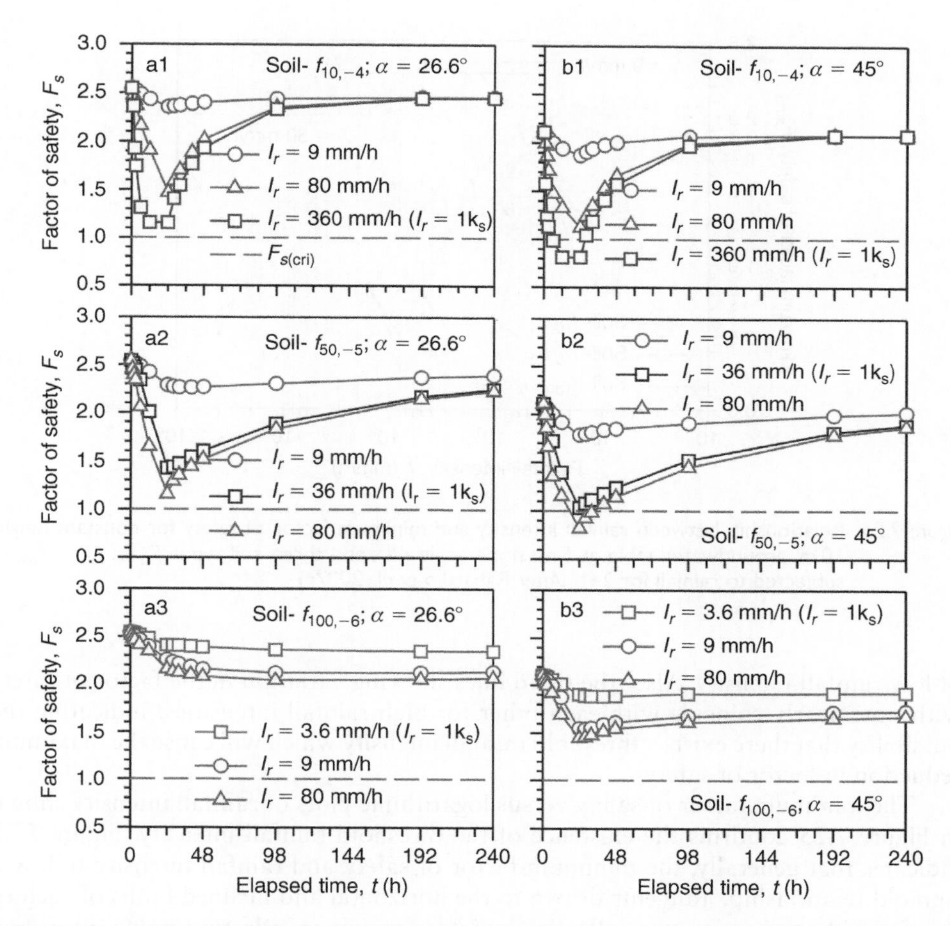

Figure 7.24 Effect of rainfall intensity on the variation of factor of safety with time for a homogenous soil slope of constant height 10 m, groundwater table at 5 m, subjected to rainfall intensity of 9, 80, and $1 \times k_s$ mm/h of respective soil with (a1–a3) slope angle 26.6° and soil types $f_{10,-4}$, $f_{50,-5}$, $f_{100,-6}$; (b1–b3), slope angle 45° and soil types $f_{10,-4}$, $f_{50,-5}$, $f_{100,-6}$ *(Modified after* Rahardjo *et al.*, 2007c)

fastest for soil type $f_{10,-4}$, followed by soil types $f_{50,-5}$, and $f_{100,-6}$. Furthermore, after the cessation of the rainfall (after 24 h), the rate of recovery of factor of safety is the fastest for soil type $f_{10,-4}$, followed by soil types $f_{50,-5}$, and $f_{100,-6}$, regardless of the rainfall intensity. Therefore, the net reduction in the factor of safety, F_s, appears to be a function of both rainfall intensity and soil type.

For a particular homogeneous soil slope, the magnitude and rate of reduction in the factor of safety are directly proportional to the magnitude of rainfall intensity (Figure 7.24). The higher the rainfall intensity, the quicker is the decrease in the factor of safety. For this reason, the slope of the graphs before the rainfall stops (the portion of the graphs at elapsed time $t \leq 24$ h) for high rainfall intensity is steeper than that

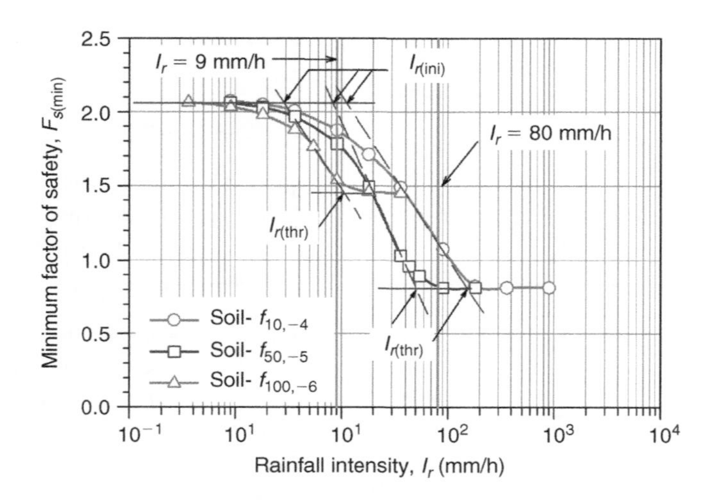

Figure 7.25 Relationship between rainfall intensity and minimum factor of safety for constant height 10 m, groundwater table at 5 m, slope angle 45°, and three soil types $f_{10,-4}$, $f_{50,-5}$, $f_{100,-6}$; subjected to rainfall for 24 h (*After* Rahardjo *et al.*, 2007c)

of low rainfall intensity. Also, the trend lines showing variation in the factor of safety with time nearly coincide with each other for high rainfall intensities; indicating the possibility that there exists a threshold rainfall intensity which will cause the maximum reduction in factor of safety.

The minimum factor of safety versus logarithmic plots of rainfall intensity shown in Figure 7.25 confirms the existence of the threshold rainfall intensity. Figure 7.25 indicates that generally, the minimum factor of safety and rainfall intensity follow a sigmoid relationship. Tangents drawn to the horizontal and inclined limbs of each of the sigmoid curves give two inflection points – an upper inflection point (identified as initiating rainfall intensity, $I_{r(\mathrm{ini})}$) and a lower inflection point (identified as threshold rainfall intensity, $I_{r(\mathrm{thr})}$). The initiating rainfall intensity $I_{r(\mathrm{ini})}$ is characterised as the rainfall intensity at which the minimum factor of safety starts to decrease. The $F_{s(\min)}$ does not start to decrease unless rainfall intensity reaches this $I_{r(\mathrm{ini})}$ magnitude. The threshold rainfall intensity $I_{r(\mathrm{thr})}$ is characterised as the highest rainfall intensity that produces the lowest minimum factor of safety. Further increase in rainfall intensity beyond the $I_{r(\mathrm{thr})}$ magnitude does not produce any further decrease in the $F_{s(\min)}$. Another characteristic from Figure 7.25 of interest is that both the $I_{r(\mathrm{ini})}$ and $I_{r(\mathrm{thr})}$ for the soils used in the study increase in the sequence of $f_{10,-4}$, $f_{50,-5}$, and $f_{100,-6}$ implying that a higher rainfall intensity is needed to destabilise a homogenous soil slope with a high permeability. For homogenous soil slope with a low permeability soil, the threshold rainfall intensity is lower and the reduction in factor of safety is smaller as compared to slopes with a high permeability soil.

From the results of the series of the parametric studies, Rahardjo *et al.* (2007c) highlighted that for a given rainfall duration, a threshold rainfall intensity, $I_{r(\mathrm{thr})}$, exists that produces the lowest minimum factor of safety, $F_{s(\min)}$. The value of $I_{r(\mathrm{thr})}$ increases as the permeability, k_s, of the soil in the slope increases. Rainfall of intensity equal to the saturated coefficient permeability of a soil does not necessarily lead the minimum factor of safety to the lowest value. Generally, for soils with low k_s (of the order of 10^{-6} m/s), a rainfall intensity greater than k_s is required to bring the minimum factor of safety to the lowest value; while for soils with a high k_s, rainfall intensity smaller than k_s is sufficient to bring the $F_{s(\min)}$ to the lowest value.

The results of the parametric studies (Rahardjo *et al.*, 2007c) also indicate the position of water table at the critical condition corresponding to the minimum factor of safety, $F_{s(\min)}$. The results indicate that slopes with a high permeability soil will fail due to mounding of water table; whereas, slopes with a low permeability soil will likely fail due to reduction in matric suction of the unsaturated soil above the water table because of rainwater infiltration from the slope surface.

Antecedent rainfall

In the context of landslide studies, antecedent rainfall is defined as rainfall in the days immediately preceding a landslide event. A landslide-triggering rainfall is the rainfall event (rain that falls at the time the landslide occurs) that is believed to initiate the landslide event. A review of literature suggests that mixed conclusions exist regarding the relative role of antecedent rainfall on slope instability.

Brand (1984) suggested that for landslides in Hong Kong, antecedent rainfall was not a significant factor. The limited influence of antecedent rainfall on landslides in Hong Kong was attributed to high permeability of the local soils (Brand, 1992). Pitts (1985), similarly, concluded that for Singapore the effect of antecedent rainfall is not significant in affecting slope stability. Tan *et al.* (1987) re-examined Pitts' conclusion (Pitts, 1985) and suggested that antecedent rainfall could be significant in affecting the slope stability. Chatterjea (1989) also made a similar conclusion. The case study (Wei *et al.*, 1991) of the Bukit Batok landslide in Singapore revealed that the landslide occurred after a period of heavy rain and there was no rainfall at the time of the landslide. Wei *et al.* (1991) further demonstrated that the rainfall in the preceding months had resulted in raised ground water levels. Morgenstern (1992) attributed such mixed conclusions on the relative role of antecedent rainfall as a result of difference in soil properties from different regions of the world.

The effect of antecedent rainfall on slope stability for Singapore was further examined through case studies of several slope failures in Nanyang Technological University (NTU) campus (Rahardjo *et al.*, 2001) and analyses of long-term field monitoring data on pore-water pressure changes in response to rainfall events (Rahardjo *et al.*, 2008a). Results from both studies demonstrated that antecedent rainfall plays an important role in the slope instability.

A storm on 26 February 1995, resulting in 95 mm rainfall in 2.5 h (average intensity 37.8 mm/h) caused more than 20 shallow landslides in the NTU campus. The nature of some of the failed slopes is shown in Figure 7.26. To identify the link between rainfall event and landslide occurrence, the daily rainfall records from November 1993 through February 1995 (collected by automatic rain gauge stations installed by the School of Civil and Environmental Engineering) were studied. The maximum and minimum monthly rainfalls during the 16 month period was 481.8 mm (in February 1995) and 26.8 mm (in September 1994), respectively. There were four occasions during the

Figure 7.26 Nature and form of a few slope failures (out of 20 slope failures) on 26 February 1995 in the Nanyang Technological University Campus (Photo taken on morning of 27 February 1995)

16 month period in which the daily rainfall was heavy, exceeding 80 mm. These were 94 mm on 15 February 1993, 83.2 mm on 12 November 1994, 86 mm on 4 February 1995 and 94.6 mm on 26 February 1995. The daily rainfall records for November 1993, November 1994, and February 1995 are shown in Figure 7.27.

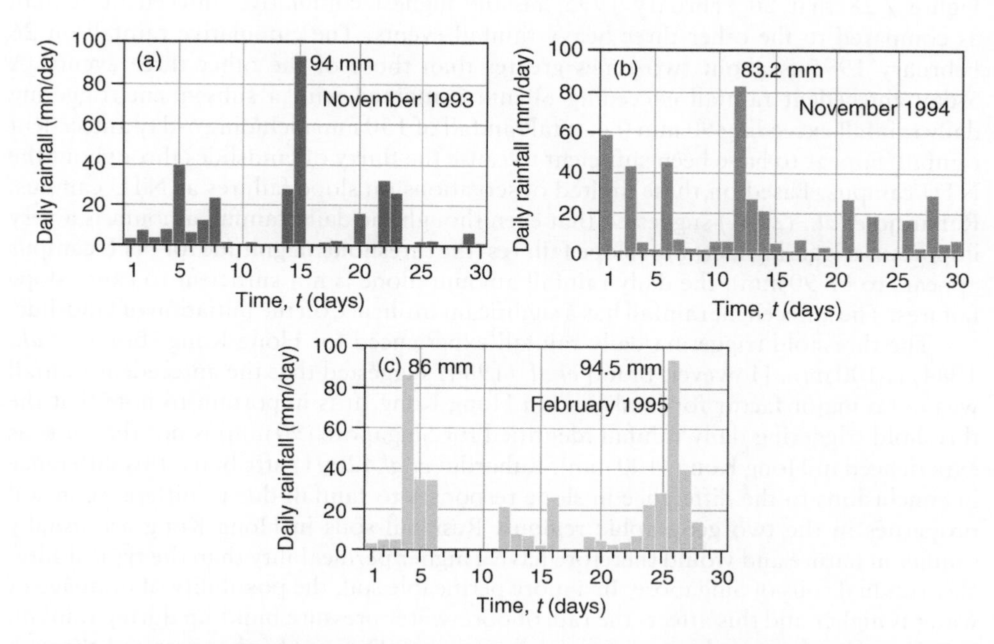

Figure 7.27 Daily rainfall for the months of (a) November 1993; (b) November 1994; (c) February 1995 (*After* Rahardjo *et al.*, 2001)

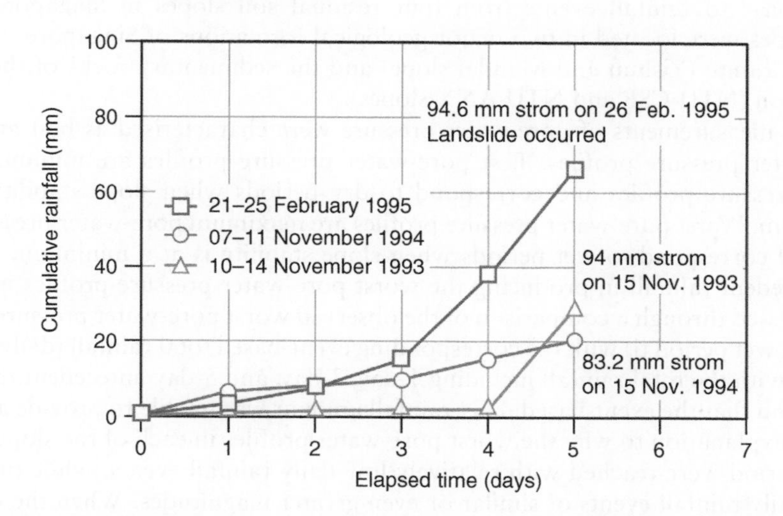

Figure 7.28 Relationship between cumulative antecedent rainfall and the occurrence of landslides (*After* Rahardjo *et al.*, 2001)

When the rainfall data were analysed by taking into account the cumulative antecedent rainfall in the days preceding the heavy rainfall event, the link between rainfall and landslides are evident. Figure 7.28 shows a plot of the cumulative antecedent rainfall in 1 to 5 days preceding the heavy rainfall event. It is clearly evident from

Figure 7.28 that 26 February 1995 has the highest cumulative antecedent rainfall as compared to the other three heavy rainfall events. The cumulative rainfall on 26 February 1995 is about two times greater than those of the other three events. A 5-day antecedent rainfall exceeding 60 mm combined with a subsequent triggering daily rainfall exceeding 90 mm (i.e. total rainfall of 150 mm including 5-day antecedent rainfall) appear to have been sufficient to cause the flurry of landslides throughout the NTU campus. Based on these limited observations on slope failures at NTU campus, Rahardjo *et al.*, (2001) suggested that even though the daily rainfall amount is a very important triggering factor for slope failures (the threshold magnitude for NTU campus appears to be 90 mm), the daily rainfall amount alone is not sufficient to cause slope failures. The antecedent rainfall has a significant influence on the initiation of landslide.

The threshold triggering daily rainfall experienced for Hong Kong (Brand *et al.*, 1984) is 100 mm. However, Brand *et al.* (1984) suggested that the antecedent rainfall was not a major factor for landslides in Hong Kong. It is important to note that the threshold triggering daily rainfall identified for Singapore (90 mm) is not the same as experienced in Hong Kong (100 mm). Rahardjo *et al.* (2001) attributed this difference in conclusions to the difference in slope responses to rainfall due to difference in soil properties in the two geographic regions. Residual soils in Hong Kong are usually sandier in nature and would therefore have a higher permeability than the typical silty-clay residual soils of Singapore. In a more permeable soil, the possibility of drainage of water is higher and this affects the rate of pore-water pressure build-up during rainfall.

The role of antecedent rainfall on slope instability was further assessed through analyses (Rahardjo *et al.*, 2008a) of records of pore-water pressure profile changes in response to rainfall events from four residual soil slopes in Singapore. These four slopes were located in two major geological formations of Singapore, the Bukit Timah Granite (Yishun and Mandai slope) and the sedimentary rocks of the Jurong Formation (NTU-CSE and NTU-ANX slopes).

The measurements of pore-water pressure were characterised as best and worst pore-water pressure profiles. Best pore-water pressure profiles are minimum pore-water pressure profiles and correspond to dry periods when slope stability is at a maximum. Worst pore-water pressure profiles are maximum pore-water pressure profiles and correspond to wet periods when slope stability is at a minimum. The role of antecedent rainfall in producing the worst pore-water pressure profiles in a slope was assessed through a comparison of the observed worst pore-water pressure profiles during a wet period (i) with the corresponding event-based total rainfall (daily rainfall) and (ii) with the total rainfall including 1-day, 3-day, and 5-day antecedent rainfall. It was found that the event-based daily rainfall amount was unable to provide any satisfactory explanation to why the worst pore-water profiles in each of the slopes during a wet period were reached with relatively low daily rainfall events, while there were other daily rainfall events of similar or even greater magnitudes. When the observed highest pore-water pressure profiles were compared with the corresponding total rainfalls including 1-day, 3-day, and 5-day antecedent rainfalls, it was clearly evident that the highest pore-water pressure profiles in the slope were reached when the total rainfall including 5-day antecedent rainfall was a maximum. This clearly shows the role of antecedent rainfall in producing the worst pore-water pressure profiles in a slope.

To quantify antecedent rainfall for landslides in Hong Kong, Lumb (1975) used a 15-day antecedent period. Tan *et al.* (1987) used the same period for Singapore.

Chatterjea (1989) suggested that a 15-days antecedent period is too long and inappropriate for rainfall patterns in Singapore and adopted a 5-days antecedent period. A comparison of antecedent rainfall period from the case study (Rahardjo *et al.*, 2001) and analyses of pore-water pressures and rainfall data from the field measurements (Rahardjo *et al.*, 2008a, b) suggest that the antecedent rainfall period that can influence the stability of a slope could span an interval of 5 days prior to the rainfall event when the slope experiences the worst pore-water pressure conditions or failure. It was also observed through analyses of records of pore-water pressure profile changes that once the highest pore-water pressure profile in a slope was reached, the subsequent rainfall events (before the occurrence of a dry period) do not contribute in further increasing the pore-water pressures. This phenomenon pointed to the existence of a threshold rainfall amount (including 5-day antecedent rainfall) to produce the worst pore-water pressure conditions in a slope.

Studies by Rahardjo *et al.* (2001, 2008a, b) on the role of antecedent rainfall showed that antecedent rainfall, initial pore-water pressures (i.e. indirectly, the antecedent moisture conditions prevailing in the slope) prior to a significant rainfall event as well as the magnitude of the rainfall event play crucial roles in leading a slope to unstable conditions.

Infiltration

Numerous studies (Ching *et al.*, 1984; Brand, 1984, 1992; Tan *et al.*, 1987; Campos *et al.*, 1988; Anderson and Zhu, 1996; Cheng, 1997) have examined the cause of slope failures in tropical regions and concluded that rainwater infiltration is the most prominent triggering factor in the instability of slopes, yet studies aimed at quantifying the infiltration amount in a slope from a rainfall event are limited and scanty. This is primarily due to the fact that infiltration measurements corresponding to rainfall events on slopes are time consuming, tedious and resource intensive and are not readily available from meteorological observation that only provides rainfall intensity or rainfall amount.

In fact, for a given rainfall, the whole rainfall volume delivered by a rainfall may not always contribute to slope instability. It is the infiltration or the amount of water that seeps into the slope during a rainfall event which brings about changes in the soil physical and hydraulic properties and may eventually lead to a slope failure when limiting equilibrium conditions are reached. Field experiments with simulated and natural rainfall events, such as the one shown in Figure 7.29 were performed on several residual soil slopes in Singapore (Rezaur *et al.*, 2003; Rahardjo *et al.*, 2005) for quantifying infiltration, runoff, water content, and pore-water pressure changes in slopes in response to rainfall events.

Results of the experiments with natural and simulated rainfall suggested that a large proportion of the rainfall in Singapore contributes to infiltration. Infiltration and runoff amount as well as the relative increase in pore-water pressures due to a rainfall event are influenced by the antecedent moisture condition in a slope. Assuming interception losses to be negligible, depending on the magnitude of the rainfall, a rainfall event may contribute from 40% to about 100% of its total rainfall as infiltration as illustrated in the relationship between rainfall amount and percent of infiltration for a typical residual soil slope shown in Figure 7.30. Figure 7.30 shows that the

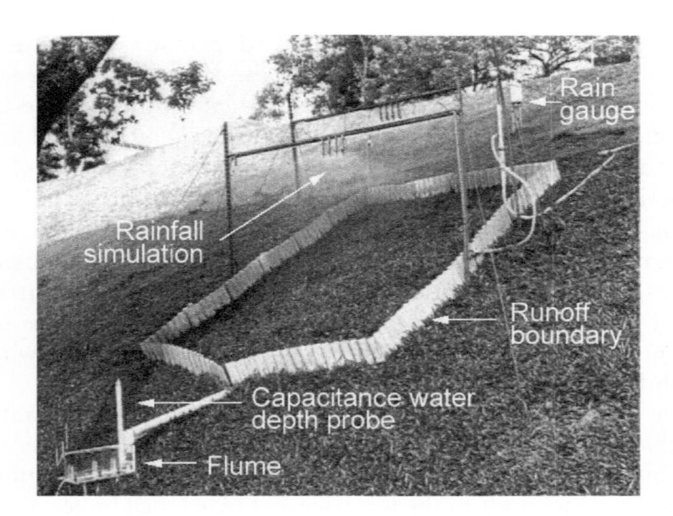

Figure 7.29 Setup for rainfall simulation, runoff, infiltration, water content, and pore-water pressure change measurements on a residual soil slope in Singapore for monitoring slope hydrological responses; tensiometer probes buried in the soil are not visible in the figure; TDR wave guides used to measure water content before and after a rainfall event are also not shown in the figure (*After* Rahardjo *et al.*, 2005)

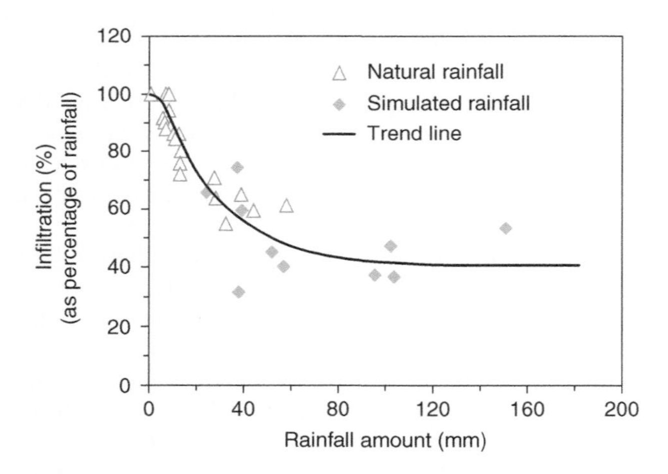

Figure 7.30 Relationship between rainfall amount and percent of infiltration for a typical residual soil slope in the Jurong Formation of Singapore (*After* Rahardjo *et al.*, 2005)

fraction of rainfall contributing to infiltration decreases with an increase in the rainfall amount. Smaller rainfalls may contribute fully to infiltration, while larger rainfalls may contribute more towards runoff generation and relatively less to infiltration. Such information on infiltration amount is useful in seepage analyses that require an estimate of the total infiltration as an input parameter.

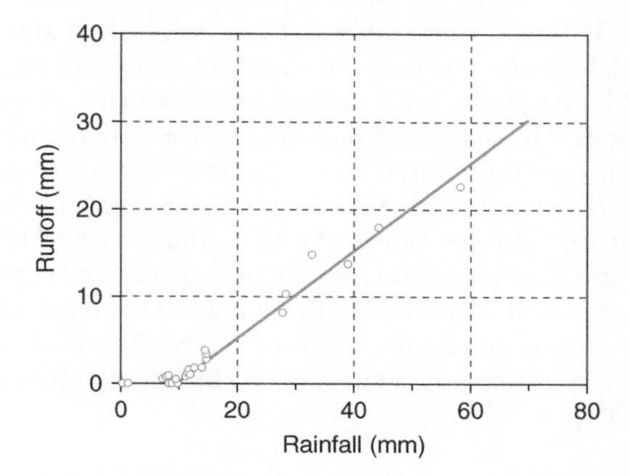

Figure 7.31 Relationship between rainfall and runoff amount for a residual soil slope in the Jurong Formation of Singapore (*After* Rezaur *et al.*, 2003)

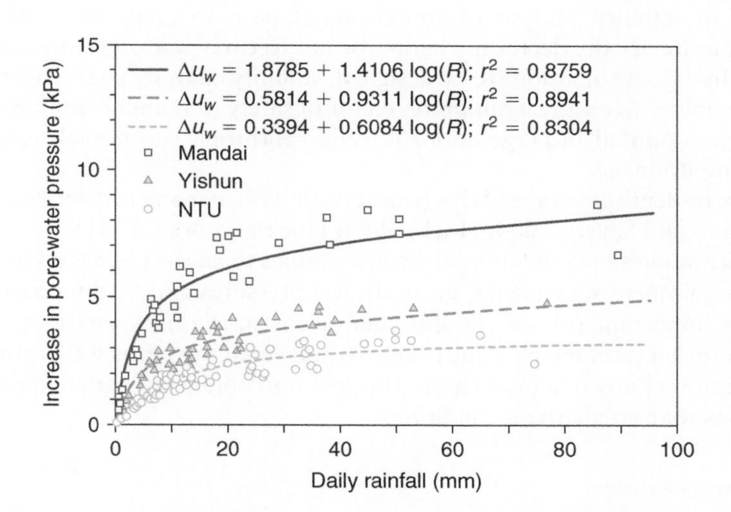

Figure 7.32 Relationship between increase in pore-water pressure at 50 cm depth to daily rainfall amount for typical residual soil slopes in the Bukit Timah Granite (Yishun and Mandai) and Jurong (NTU) Formation of Singapore (*After* Rezaur *et al.*, 2003)

Runoff measurements from slopes in Singapore suggested that rainstorms in excess of 10 mm usually generate runoff, even under dry antecedent moisture conditions. Plots of natural rainfall and corresponding runoff measurements from slopes in Singapore, as shown in Figure 7.31, indicate that a threshold rainfall amount of about 10 mm must be exceeded in order to produce a significant runoff.

Attempts towards correlating the relative increase in pore-water pressure (Δu_w, the difference in pore-water pressure before and after a rainstorm) with rainfall amount resulted in relationships of the form shown in Figure 7.32. The magnitude of the relative increases in pore-water pressures appeared to be highest in the Mandai slope and least in the NTU slope. These differences in Δu_w values among different slopes were found to be related to the soil properties and antecedent moisture conditions of the respective slopes. Mandai slope, being rich in coarse particles, showed higher values of Δu_w than other slopes. This suggested that the relative increase in pore-water pressure is dependent not only on the rainfall amount, but is also sensitive to soil properties and antecedent water content. Although the relative increases in pore-water pressure values are different for different slopes (Yishun, Mandai and NTU), they appear to increase with the rainfall amount and the rate of increase tends to decline at a rainfall amount greater than about 10 mm of daily rainfall. This tendency was found to be related to the near saturation of soils at 50 cm depths. More details on the nature of hydrological responses of slopes in Singapore can be found in Rezaur *et al.* (2003) and Rahardjo *et al.* (2005).

7.6 STABILITY ANALYSES

The scope of stability analysis of an existing slope is to verify its safety condition, which will influence the decision whether or not to carry out preventive or corrective measures. In the case of a slope to be designed, stability analyses enable assessment of a suitable geometry to ensure a minimum factor of safety (FS) under environmental conditions such as rainfall and vegetation, as well as anthropic action such as excavations, loadings and drainage.

A more in-depth coverage of this issue is dealt with in many textbooks and manuals (*e.g.*, Ortigao and Sayao, 2004; Hoek, 1998: Hoek and Bray, 1981).

Stability studies may involve parametric studies to assess FS sensitivity to changes in strength parameters, geometry, groundwater pressures and loading conditions.

A very important role of stability analyses is in the back analyses of landslide failures. In such a case, the FS value is known and equal to one, as well as the geometry prior to failure is known in most cases. The goal is to obtain strength and pore pressure distributions that satisfy these conditions.

Why do landslides occur?

Landslides occur due to factors that induce an increase in loading on a slope which lead to an increase in the shear stresses or reduction in the shear strength. In the first case, it is due to loading; such as a surcharge on the top of a slope, or unloading at the slope base, for example man-made cuts or soil erosion. In the second case, strength reduction takes place due to mineral chemical weathering, disturbance and an increase in pore-water pressure.

Types of stability analyses

There are two ways of carrying out stability analyses. The first is in terms of *total stresses*, corresponding to short-term situations, saturated soils and impeded drainage conditions, such as end-of-construction cases. The second case corresponds to *effective stress analyses* that can be used for long-term stability analyses in which drained conditions prevail, or even short-term cases, when pore pressures can be known with the necessary accuracy.

It is suggested that most natural slopes and also slopes in residual soils should be analysed through the effective stress method, considering the maximum water level that can be reached under severe rainstorms.

Factor of safety definition

There are several ways of defining the FS, each one leading to different FS results. The most usual ways for defining FS are:

a. *Moment equilibrium*: generally used for analysis of rotational landslides, considering a circular slip surface:

$$FS = \frac{M_r}{M_d}, \tag{7.2}$$

where M_r is the sum of resisting moments and M_d is the sum of driving moments, i.e. towards failure.

b. *Force equilibrium*: generally applied to analyse translational or rotational failures considering plane or polygonal slip surfaces:

$$FS = \frac{F_r}{F_d}, \tag{7.3}$$

where F_r is the sum of resisting forces and F_d is the sum of driving forces.

According to these definitions, an unstable slope presents $FS \leq 1$. However, it is possible to have an unstable slope having *FS* greater than one, due to inaccuracy of the analysis methods and uncertainty or scatter in the soil strength parameters.

Establishing an allowable value for the factor of safety (FS_{allow}) will depend on the consequences of failure in terms of property and human losses.

Table 7.7 presents recommendations for allowable *FS* values. The geotechnical designer should consider not only current slope conditions but also future changes, such as the possibility of cuts at the slope toe, deforestation, surcharges and excessive infiltration.

Table 7.7 also applies to temporary slopes and the analyses should take into account loadings during the construction period.

In the case of imminent or pre-existing failures, remediation measures should be a function of the landslip history. In this case, it is fundamental to base the analyses on actual data from site investigation, rainfall and monitoring of pore pressures and displacements. Table 7.8 gives recommended *FS* values in these cases.

Reduction factors, known as partial reduction factors can be used to calculate the shear strength of soils, therefore:

(i) in terms of effective stresses: $\tau = \dfrac{c'}{F_1} + \sigma'_N \dfrac{\tan \phi'}{F_2}$;

(ii) in terms of total stresses: $\tau = \dfrac{s_u}{F_3}$,

Table 7.7 Recommended factors of safety (*After* GEO, 1984)

FS allow		*Risk of human losses*		
		Negligible	*Average*	*High*
Risk of economic losses	Negligible	1.1	1.2	1.4
	Average	1.2	1.3	1.4
	High	1.4	1.4	1.5

i) *FS* for recurrence time of 10 years
ii) For higher risks and soft ground conditions, add 10% increase in FS.

Table 7.8 Recommended factors of safety for rehabilitation of failed slopes (GEO, 1984)

Risk of human losses		
Negligible	*Average*	*High*
$FS > 1.1$	$FS > 1.2$	$FS > 1.3$

Obs.: *FS* for recurrence period of 10 years.

where c' and ϕ' are effective strength Mohr-Coulomb soil parameters, s_u is the undrained shear strength applicable in cases of saturated undrained low permeability clayey soils and F_1, F_2, and F_3 are *reduction factors*, which depend on the accuracy in which they are obtained and range between 1 to 1.5.

Factor of safety calculation

Assessing the factor of safety of natural or engineered slopes, through slope stability analyses, has become a routine engineering practice. While performing slope stability analyses to compute factor of safety, usually the saturated effective shear strength parameters (c' and ϕ') are used. The contribution of the negative pore-water pressure above the ground water table (unsaturated zone) to the shear strength is often ignored by setting its magnitude to zero. The difficulties associated with the measurement of negative pore-water pressure and its incorporation into the slope stability analyses are the primary reasons for this practice. Ignoring the negative pore-water pressure in the slope stability analyses is a reasonable approach when the major portion of the critical slip surface is located below the groundwater table. However, for conditions where the groundwater table is located at a greater depth or where there are concerns about the possibility of a shallow slip surface, ignoring the contribution from negative pore-water

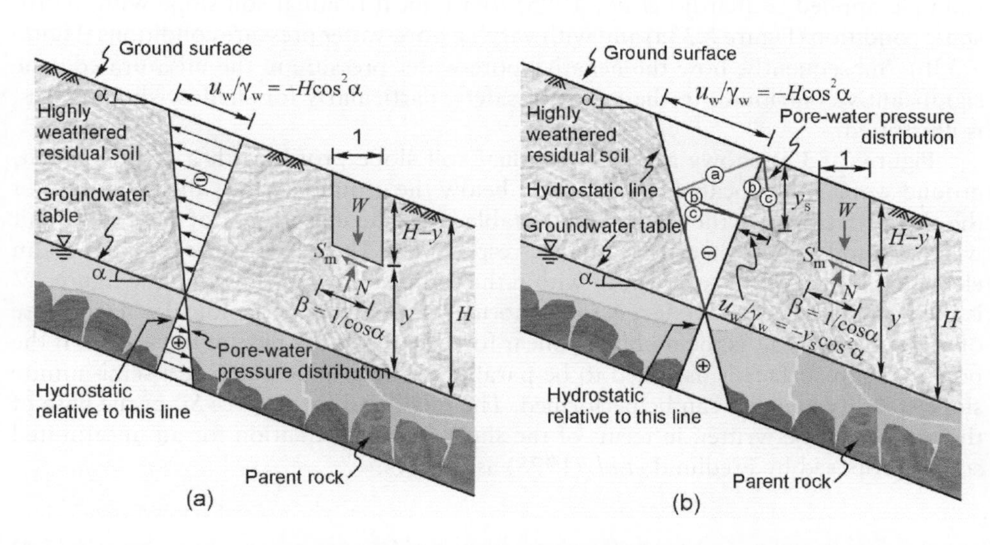

Figure 7.33 Geometry and definition of variables of a residual soil slope for semi-infinite slope stability analysis (a) under hydrostatic pore-water pressure condition and (b) under three possible pore-water pressure conditions above the ground water table due to infiltration (*After* Rahardjo *et al.*, 1995)

pressures to the shear strength of a soil is not a realistic approach. Deep groundwater table conditions often exist in many steep and high residual soil slopes. Recognising the recent advancements in our understanding of the contribution of negative pore-water pressure (matric suction) in increasing the shear strength of soil and wide availability of affordable devices for effective measurement of negative pore-water pressures under harsh field conditions, it is now appropriate to perform slope stability analyses by incorporating negative pore-water pressures.

Rahardjo and Fredlund (1991), Fredlund and Rahardjo (1993), and Rahardjo *et al.* (1995) proposed three approaches for incorporating the negative pore-water pressure into the slope stability analyses, allowing computation of the factor of safety that accounts for the shear strength contributions from the negative pore-water pressures in the unsaturated zone. These approaches are referred as the (i) semi-infinite method, (ii) total cohesion method and (iii) extended shear strength method. These methods are described below.

Semi-infinite method

Rahardjo *et al.* (1995) showed that negative pore-water pressure could be incorporated into the semi-infinite slope stability analyses through the shear strength equation of the soil. With this method, the factor of safety accounting for the negative pore-water pressure can be computed manually without much computational effort. The method of incorporating the shear strength contributions from negative pore-water pressures into semi-infinite slope stability analyses are demonstrated below using an example applied (Rahardjo *et al.*, 1995) to a typical residual soil slope with hydrostatic condition (Figure 7.33a) and with varying pore-water pressure conditions (Figure 7.33b). Subsequently, how the negative pore-water pressure in the unsaturated zone significantly contributes to the factor of safety, particularly for shallow slip surfaces, is illustrated.

Figure 7.33(a) shows a typical residual soil slope profile with a slope angle α, ground water table located at a depth H below the ground surface and the height of the slip surface above the ground water table is y. The unit weight of water and unit weight of soil are denoted by γ_w and γ_s respectively. The various forces acting on an elemental soil wedge of unit width within the soil mass is also shown in the figure. W is the weight of the soil wedge, N is the normal force acting perpendicular to the base of the wedge and S_m is the mobilised shear force acting at the base of the wedge. If the potential slip surface is assumed to be parallel to the ground surface, a semi-infinite slope stability analysis can be performed. The mobilised shear force S_m at the base of the wedge can be written in terms of the shear strength equation for an unsaturated soil as proposed by Fredlund *et al.* (1978) as follows:

$$S_m = \frac{\beta}{F_s}[c' + (\sigma_n - u_a)\tan\phi' + (u_a - u_w)\tan\phi^b] \tag{7.4}$$

where F_s is the factor of safety, defined as the factor by which the shear strength parameters must be reduced in order to bring the soil mass into a state of limiting equilibrium along the assumed slip surface; c' is the effective cohesion intercept; σ_n is the total stress normal to the base of the wedge; u_a and u_w are the pore-air and pore-water pressures, respectively, at the base of the wedge; $(\sigma_n - u_a)$ is the net stress normal to the base of a wedge; $(u_a - u_w)$ is the matric suction; ϕ' is the angle of internal friction associated with the net normal stress state variable; and ϕ^b is the angle indicating the rate of increase in shear strength relative to a change in matric suction $(u_a - u_w)$.

With the shear strength contribution from the negative pore-water pressure being incorporated through Equation (7.4), the factor of safety calculation by the semi-infinite method is extended to two possible pore-water pressure conditions, namely: (i) hydrostatic condition shown in Figure 7.33(a) and (ii) three non-hydrostatic pore-water pressure conditions associated with infiltration into a slope and identified by three possible pore-water pressure profiles; profile-a, profile-b, and profile-c as shown in Figure 7.33(b). Pore-water pressure profile-a represents the situation where infiltration causes the matric suction to decrease to zero at the ground surface and the suction increases with depth until it reaches the hydrostatic line at depth y_s. Pore-water pressure profile-b represents the advancement of a wetting front as suggested by Lumb (1962) and pore-water pressure profile-c represents the situation with a perched water table at a depth y_s.

Assuming that the pore-air pressure is atmospheric (i.e. $u_a = 0$) and the unit weight of the soil γ is constant throughout the soil depth, the factor of safety equation for the hydrostatic condition can be written as:

$$F_s = \left(\frac{\tan\phi'}{\tan\alpha}\right) + \left(\frac{c'}{\gamma(H-y)}\right)\frac{1}{\sin\alpha\cos\alpha} + \left(\frac{y}{H-y}\right)\left(\frac{\gamma_w}{\gamma}\right)\left(\frac{\tan\phi^b}{\tan\alpha}\right) \tag{7.5}$$

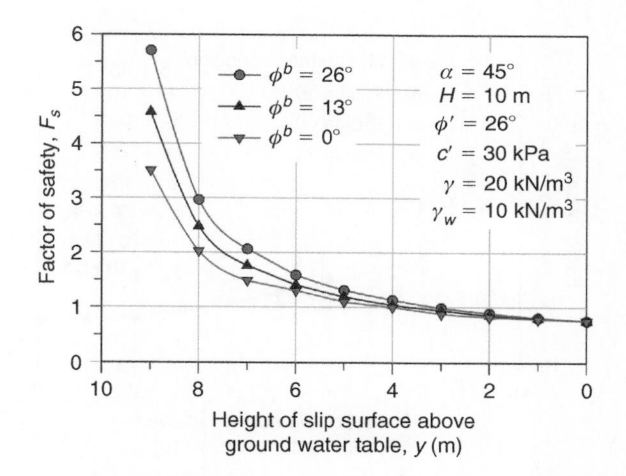

Figure 7.34 Computed factors of safety using semi-infinite method for various ϕ^b values and the hydrostatic condition shown in Figure 7.33(a) (*After* Rahardjo *et al.*, 1995)

Assuming that the depth of water infiltration is y_s (Figure 7.2 (b)) and using the same assumptions as for the hydrostatic conditions (i.e. $u_a = 0$ and γ_w is constant), the factor of safety equation for the three non-hydrostatic pore-water pressure conditions corresponding to the slip surface at a depth of $(H - y) \leq y_s$ can be written as:

For pore-water pressure profile-a:

$$F_s = \left(\frac{\tan\phi'}{\tan\alpha}\right) + \left(\frac{c'}{\gamma(H-y)}\right)\frac{1}{\sin\alpha\cos\alpha} + \left(\frac{H}{y_s} - 1\right)\left(\frac{\gamma_w}{\gamma}\right)\left(\frac{\tan\phi^b}{\tan\alpha}\right) \tag{7.6}$$

For pore-water pressure profile-b:

$$F_s = \left(\frac{\tan\phi'}{\tan\alpha}\right) + \left(\frac{c'}{\gamma(H-y)}\right)\frac{1}{\sin\alpha\cos\alpha} \tag{7.7}$$

For pore-water pressure profile-c:

$$F_s = \left(\frac{\tan\phi'}{\tan\alpha}\right) + \left(\frac{c'}{\gamma(H-y)}\right)\frac{1}{\sin\alpha\cos\alpha} - \left(\frac{\gamma_w}{\gamma}\right)\left(\frac{\tan\phi^b}{\tan\alpha}\right) \tag{7.8}$$

Using Equation (7.5), the factors of safety for the hydrostatic condition shown in Figure 7.33(a) are computed and plotted in Figure 7.34 for different heights of the slip surface (y values) above the ground water table (from 0 to 9 m with increment of 1 m) and three different ϕ^b values of (26, 13 and 0°). The values assumed for other parameters involved in Equation (7.5) are also shown in Figure 7.34. It appears from Figure 7.34 that the factor of safety increases with an increase in the ϕ^b value and the increase in factor of safety becomes more significant as the height of the sliding surface

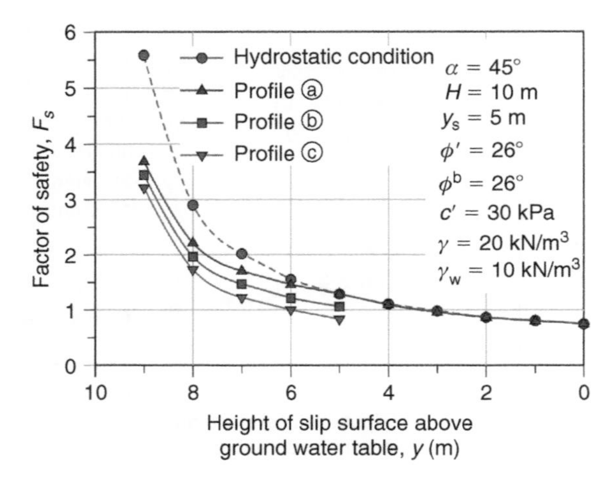

Figure 7.35 Computed factors of safety using semi-infinite method for various pore-water pressure profiles above the ground water table as a result of infiltration into the slope shown in Figure 7.33(b) (*After* Rahardjo *et al.*, 1995)

above the ground water table increases. It is important to note that when $\phi^b = 0$ in Equation (7.5), the contribution of suction towards the slope stability is not accounted for and hence the factor of safety computed using $\phi^b = 0$ is low (Figure 7.34).

Using Equations (7.6 to 7.8), the factor of safety for the non-hydrostatic pore-water pressure profiles shown in Figure 7.33(b) are computed and plotted in Figure 7.35 for 10 different values of y (from 0 to 9 m with an increment of 1 m), a constant $\phi^b = 26°$ and $y_s = 5$ m (assumed). The assumption that the depth of water infiltration, $y_s = 5$ m is based on the justification that further percolation of water to a greater depth may be restricted due to low permeability of the underlying soil. The values assumed for other parameters involved in Equations (7.6 to 7.8) are also shown in Figure 7.35. Figure 7.35 illustrates that for cases with infiltration into slope, the factor of safety depends on the pore-water pressure profiles shown in Figure 7.33(b). For all cases shown in Figure 7.35, the factor of safety decreases rapidly as the slip surface approaches the ground surface (i.e. shallow slides). Theoretically, the slip will occur at a plane where the computed factor of safety is lowest (i.e. within the wetted zone), and the actual location of the slip surface will, however, be governed by the field shear strength profile.

Limit equilibrium method of slices

The total cohesion and extended shear strength methods proposed by Rahardjo and Fredlund (1991) and Fredlund and Rahardjo (1993) require computer coding for the solution of the equations. The incorporation of the matric suction contribution to the shear strength equation and the derivation of factor of safety are based on extension of the conventional limit equilibrium analyses.

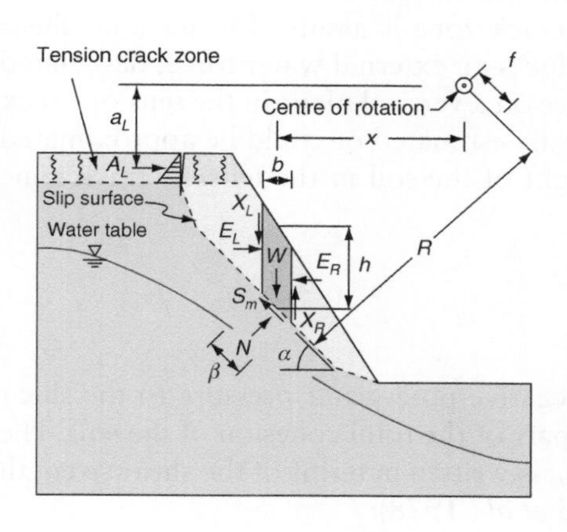

LEGEND

- b = width of the slice
- h = height of the slice or the vertical distance from the centre of the base line of each slice to the uppermost line in geometry (i.e. generally ground surface)
- x = horizontal distance from the centre line of each slice to the centre of rotation or the centre of moments
- f = perpendicular offset of the normal force from the centre of rotation or the centre of moments
- a = perpendicular distance from the resultant external water force to the centre of rotation or to the centre of moments
- β = sloping distance across the base of the slice
- W = total weight of the slice
- N = normal force on the base of the slice
- N = resultant external water forces
- E = horizontal inter slice normal forces (subscripts L and R denote left and right side of the slice, respectively)
- X = vertical inter slice shear force (subscripts L and R denotes left and right side of the slice, respectively)
- R = radius of a circular slip surface or the moment arm associated with the mobilized shear force S_m for any shape of slip surface
- α = angle betwen the tangent to the centre of the base of each slice and the horizon

Figure 7.36 Forces acting on a slice through a sliding mass with a composite slip surface (*After* Rahardjo and Fredlund, 1991)

The limit equilibrium methods of slices for computing a factor of safety is based upon the principle of statics (i.e. static equilibrium of forces and/or moments) and has been commonly used in practice for two main reasons. Firstly, in most cases, where the shear strength properties of soil and pore-water pressure conditions are adequately assessed, the limit equilibrium methods have proved to be a useful tool and provide reliable results (Sevaldson, 1956; Kjaernsli and Simons, 1962; Skempton and Hutchison, 1969; Chowdhury, 1980) in computing the factor of safety of slopes. Secondly, the limit equilibrium method can perform extensive trial and error searches for the critical slip surface with minimum amount of input data and in a short time.

The proposed method (Rahardjo and Fredlund, 1991; Fredlund and Rahardjo, 1993) uses the summation of forces in two directions and the summation of moments about a common point in deriving the factor of safety with the incorporation of negative pore-water pressure. These elements of statics, along with the failure criteria, are not sufficient to make the slope stability problem determinate. Either additional elements of physics or an assumption regarding the direction or magnitude of forces is required to render the problem determinate (Morgenstern and Price, 1965; Spencer, 1967). The proposed method utilises an assumption regarding the direction of the inter-slice forces which has been widely adopted in limit equilibrium methods (Fredlund and Krahn, 1977; Fredlund *et al.*, 1981).

The calculation of the stability of slope approach is made by dividing the soil mass above a slip surface into vertical slices and considering the static forces acting on a single slice within the sliding soil mass (Rahardjo and Fredlund, 1991; Fredlund and Rahardjo, 1993) as shown in Figure 7.36. The forces are considered acting per unit width (perpendicular to paper) of the slope and the variables are defined as illustrated in Figure 7.36. The slope illustrated in Figure 7.36 is typical of a steep slope with a groundwater table located at a greater depth. The slope crest is highly desiccated and there are tension cracks filled with water. The slip surface in the tension crack zone is considered as a vertical line. The tension crack zone is assumed to have no shear strength and water in the tension cracks produces an external water force, designated as A_L, and is computed as the hydrostatic force on a vertical plane in the tension crack zone. The depth of the tension crack is generally estimated or could be approximated analytically (Spencer, 1968, 1973). The weight of the soil in the tension crack zone acts as a surcharge on the crest of the slope.

Total cohesion method

In this method, the contribution from the negative pore-water pressure to the shear strength of unsaturated soil is considered as part of the total cohesion of the soil. The mobilised shear force at the base of a slice, S_m, is written in terms of the shear strength equation for an unsaturated soil as (Fredlund *et al.*, 1978):

$$S_m = \frac{\beta}{F}[c' + (\sigma_n - u_a)\tan\phi' + (u_a - u_w)\tan\phi^b] \quad (7.9)$$

where F is the factor of safety. The factor of safety is equal for all the shear strength parameters and slices involved. Equation (7.9) is the same as Equation (7.4), and the variables have the same meaning as described for Equation (7.4) and therefore are not repeated here.

The visualisation of the mobilised shear force components at the base of a slice are shown in Figure 7.37. The contributions from the total stress and the negative pore-water pressures to shear resistance are separated using the ϕ' and ϕ^b angles and are illustrated in Figure 7.37(b). If the contribution of matric suction is visualised as an increase of the cohesion of the soil, it is then possible to consider the matric suction term, $(u_a - u_w)\tan\phi^b$, in Equation (7.9) as part of the cohesion of the soil, so that; $c = c' + (u_a - u_w)\tan\phi^b$. This is akin to considering the cohesion of the soil c,

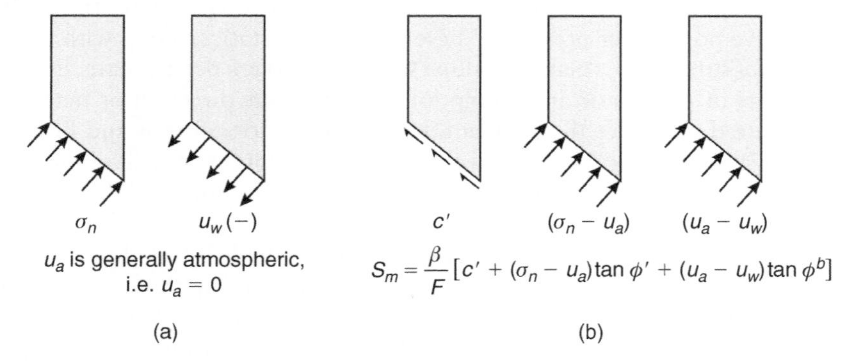

Figure 7.37 Pressure and shear resistance components at the base of a slice; (a) pressure components; (b) contributors to shear strength components (*After* Rahardjo and Fredlund, 1991)

as composed of two components, c′ and $(u_a - u_w)\tan\phi^b$. This consideration allows the mobilised shear force at the base of a slice to be expressed in the conventional form as:

$$S_m = \frac{\beta}{F}[c + (\sigma_n - u_a)\tan\phi'] \tag{7.10}$$

The total cohesion approach has the advantage that the conventional factor of safety equations need not be re-derived and that the shear strength equations retain their conventional form. These advantages therefore provide the added flexibility that a computer program code written to solve saturated soil problems can be utilised to solve unsaturated soil problems. When this is intended, the soils in the unsaturated zone must be subdivided into several discrete layers, each layer with a constant cohesion. The pore-air and pore-water pressures must be set to zero. The total cohesion approach however has the disadvantage that the cohesion is not a continuous function and appropriate cohesion values must be computed manually.

Extended shear strength method

In the extended shear strength approach, the factor of safety equations are derived by taking into account the shear strength contributions from the negative pore-water pressures in the unsaturated zone (Rahardjo and Fredlund, 1991; Fredlund and Rahardjo, 1993). The normal force N at the base of a slice is derived by summing all the forces in the vertical direction. Considering the static equilibrium in the vertical direction for all the forces acting on a slice as shown in Figure 7.36 results in:

$$W - (X_R - X_L) - S_m\sin\alpha - N\cos\alpha = 0 \tag{7.11}$$

Substituting S_m in Equation 7.11 with Equation 7.10 and replacing the term $(\beta\sigma_n)$ with N gives:

$$W - (X_R - X_L) - \left[\frac{c'\beta}{F} + \frac{N\tan\phi'}{F} - \frac{u_a\beta\tan\phi'}{F} + \frac{(u_a - u_w)\beta\tan\phi^b}{F}\right] \times \sin\alpha - N\cos\alpha = 0 \tag{7.12}$$

Rearranging Equation (7.12) for N gives;

$$N = \frac{W - (X_R - X_L) - c'(\beta\sin\alpha/F) + u_a(\beta\sin\alpha/F)(\tan\phi' - \tan\phi^b) + u_w(\beta\sin\alpha/F)\tan\phi^b}{(\cos\alpha + \sin\alpha\tan\phi')/F} \tag{7.13}$$

When considering moment equilibrium conditions, the factor of safety F in Equation (7.13) is equal to the moment equilibrium factor of safety F_m. While dealing with force equilibrium conditions, the factor of safety F is equal to the force equilibrium factor of safety F_f. The pore-air pressure u_a, in most situations, is atmospheric (i.e. $u_a = 0$) and hence Equation (7.13) reduces to:

$$N = \frac{W - (X_R - X_L) - c'(\beta\sin\alpha/F) + u_w(\beta\sin\alpha/F)\tan\phi^b}{(\cos\alpha + \sin\alpha\tan\phi')/F} \tag{7.14}$$

If the base of a slice is located in the saturated soil zone, the term ($\tan\phi^b$) in Equation (7.14) becomes equal to ($\tan\phi'$). Equation (7.14) then reverts to the conventional normal force equation of slope stability analyses for saturated slopes. Computer coding can be written for solving Equation (7.14) with conditions such that ϕ^b angle is used when pore-water pressures are negative or ϕ' angle is used when pore-water pressures are positive. Fredlund *et al.* (1987) suggested that for low matric suction values, ϕ^b, ϕ' can be taken as, $\phi^b = \phi'$ and for high matric suction values ϕ^b, ϕ' can be taken as, $\phi^b < \phi'$. The vertical inter-slice shear forces (X_R and X_L) can be computed by using an assumed inter-slice force function.

The extended shear strength method results in two factor of safety equations; one is derived from the moment equilibrium condition and the other one is derived from the force equilibrium condition.

Factor of safety with respect to moment equilibrium condition

The equilibrium of moment is satisfied by taking moments about an arbitrary point above the central portion of the slip surface and equating the algebraic summation of all moments to zero. For a circular slip surface, the centre of rotation is obviously the centre for moment equilibrium. The centre of moment becomes immaterial when equilibriums of both force and moment are satisfied. When only moment equilibrium is satisfied, the computed factor of safety varies slightly with the point selected for the summation of moments.

Thus taking moments about the centre of rotation for all the forces acting on the composite slip surface shown in Figure 7.36 and equating them to zero for moment equilibrium results in:

$$S_m = \frac{\beta}{F}[c' + (\sigma_n - u_a)\tan\phi' + (u_a - u_w)\tan\phi^b]$$

$$A_L a_L + \sum Wx - \sum Nf - \sum S_m R = 0 \quad (7.15)$$

Substituting for S_m from Equation 7.9 into Equation 7.15, replacing the terms ($\beta\sigma_n$) with N and F with F_m, and rearranging the equation yield:

$$F_m = \frac{\sum\left[c'\beta R + \left\{N - u_w\beta\left(\tan\phi^b/\tan\phi'\right) - u_a\beta\left(1 - \left(\tan\phi^b/\tan\phi'\right)\right)\right\}R\tan\phi'\right]}{A_L a_L + \sum Wx - \sum Nf} \quad (7.16)$$

The F_m notation denotes the factor of safety with respect to moment equilibrium. For condition where the pore-air pressure is atmospheric (i.e., $u_a = 0$), Equation (7.16) reduces to:

$$F_m = \frac{\sum\left[c'\beta R + \left\{N - u_w\beta\left(\tan\phi^b/\tan\phi'\right)\right\}R\tan\phi'\right]}{A_L a_L + \sum Wx - \sum Nf} \quad (7.17)$$

When pore-water pressures are positive, ϕ^b is set equal to ϕ'. For a circular slip surface, the radius R for all slices is constant and the normal force N acts through the centre of rotation (i.e. $f = 0$).

Factor of safety with respect to force equilibrium condition

The factor of safety with respect to force equilibrium conditions is derived by considering all the forces in the horizontal direction acting on the slip surface as shown in Figure 7.36 and setting their algebraic summation equal to zero for equilibrium, which gives:

$$\sum S_m \cos\alpha - \sum N \sin\alpha - A_{\mathrm{L}} = 0 \tag{7.18}$$

The horizontal inter-slice normal forces E_{L} and E_{R} cancel out when summed over the entire sliding mass. Substituting for S_m from Equation 7.9 into Equation 7.18, replacing the terms $(\beta\sigma_n)$ with N and F with F_f, and rearranging the equation yields:

$$F_f = \frac{\sum [c'\beta\cos\alpha + \{N - u_w\beta(\tan\phi^b/\tan\phi') - u_a\beta(1-(\tan\phi^b/\tan\phi'))\}\tan\phi'\cos\alpha]}{A_{\mathrm{L}} + \sum N\sin\alpha} \tag{7.19}$$

The F_f notation denotes the factor of safety with respect to force equilibrium. For conditions where the pore-air pressure is atmospheric (i.e., $u_a = 0$) Equation (7.19) reverts to:

$$F_f = \frac{\sum [c'\beta\cos\alpha + \{N - u_w\beta(\tan\phi^b/\tan\phi')\}\tan\phi'\cos\alpha]}{A_{\mathrm{L}} + \sum N\sin\alpha} \tag{7.20}$$

When pore-water pressures are positive, ϕ^b is set equal to ϕ'. For a circular or a composite slip surface, Equation (7.20) remains the same.

Inter-slice force function

Inter-slice normal forces E_{L} and E_{R} are computed by considering force equilibrium in the horizontal direction for each slice as follows:

$$E_{\mathrm{R}} - E_{\mathrm{L}} + S_{\mathrm{m}}\cos\alpha - N\sin\alpha = 0 \tag{7.21}$$

Substituting for N from Equation (7.14) into Equation (7.21) and simplifying for E_{R} gives:

$$E_{\mathrm{R}} = E_{\mathrm{L}} + [W - (X_{\mathrm{R}} - X_{\mathrm{L}})]\tan\alpha - \frac{S_m}{\cos\alpha} \tag{7.22}$$

The inter-slice normal forces are computed by integrating Equation (7.22) from left to right. The normal force E_{L} on the first slice is equal to the external water force A_{L} or zero when no water is present in the tension crack zone.

The factor of safety equations with respect to moment and force equilibriums (i.e. Equations (7.16) and (7.19), respectively) are non-linear. With the factor of safety being introduced through the normal force equation (Equation (7.13)), the factor of safety either in terms of F_m or F_f appears on both sides of the non-linear equations and therefore, cannot be solved explicitly. An iterative technique is required to solve the non-linear factor of safety equations.

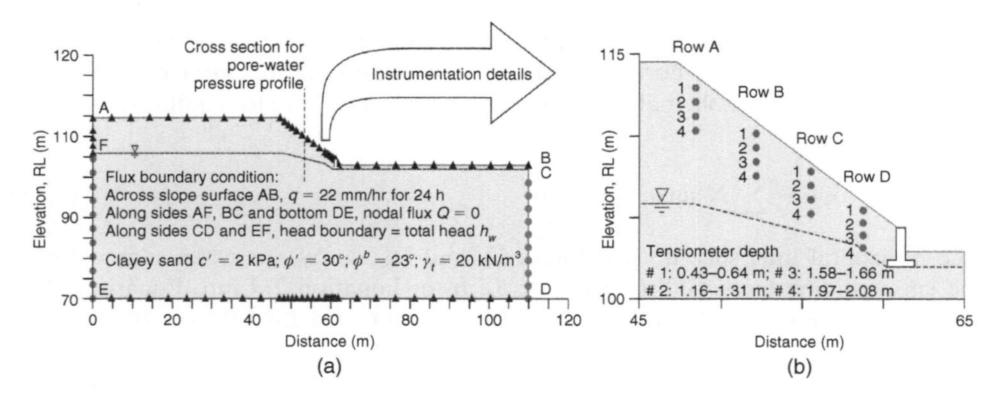

Figure 7.38 Slope parameters and instrumentation details (a) soil profile, properties, and flux boundary conditions used in seepage analyses with the incorporation of unsaturated soil mechanics equations; (b) tensiometer location and depth for direct measurement of matric suction to compare with results from parametric study (*After* Rahardjo *et al.*, 2007a)

Rahardjo and Fredlund (1991) and Fredlund and Rahardjo (1993) methods described above show how the contribution from negative pore-water pressures can readily be incorporated into the conventional slope stability analyses in order to have realistic analyses of stability of residual soil slopes that commonly have a deep ground water table with negative pore-water pressures in the unsaturated zone. Furthermore, the equations and methods introduced above are applicable to both saturated and unsaturated soils and therefore can be used for routine slope stability analyses. The conventional slope stability analyses require only minor modifications when the extended shear strength method is adopted. The total cohesion method requires only subdivision of the soil in a slope into strata in order to account for the matric suction contribution. In addition to the conventional shear strength parameters, both methods require measurements of matric suction and the ϕ^b angle.

Assessment of slope stability during rainfall

Stability of slope during rainfall can be assessed by incorporating matric suction in seepage and slope stability analyses using principles of unsaturated soil mechanics (Fredlund and Rahardjo, 1993). The following case example (Rahardjo *et al.*, 2007a) illustrates the stability assessment of a typical residual soil slope during rainfall. The soil properties, boundary, and rainfall conditions of the residual soil slope are shown in Figure 7.38(a). A second seepage analysis was performed on the same slope for a natural rainfall on 15 June 2007 and the analysis result was then compared with the results from the on-line monitoring of matric suctions in the slope. The details of the on-line monitoring of matric suctions using tensiometers are shown in Figure 7.38(b).

Seepage analyses were conducted to simulate the effects of a rainfall event of 22 mm/h for 24 h on the slope. The analyses were performed with and without incorporating the contributions from negative pore-water pressure in the seepage analyses as shown in Figure 7.39. Comparison of pore-water pressure profiles in the slope during the rainfall event indicates that the rate of rainwater infiltration into the slope is

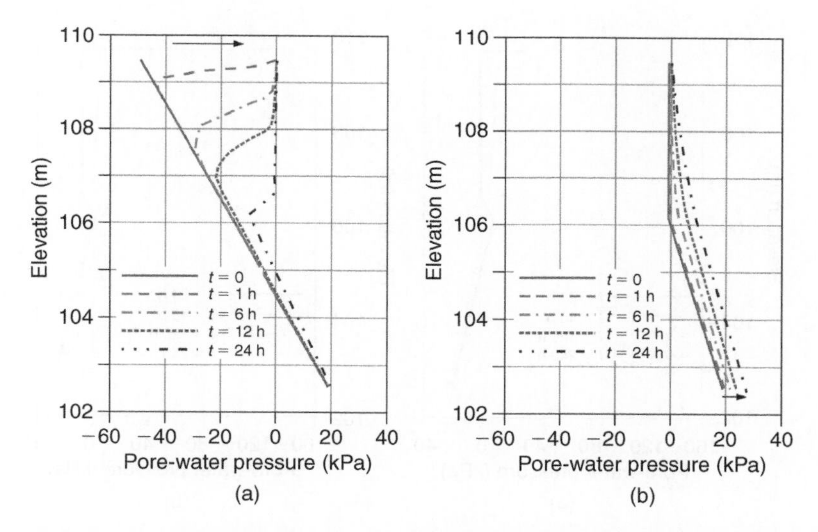

Figure 7.39 Comparison of pore-water pressure profile at the mid section of the slope shown in Figure 7.38(a) during wetting when subjected to a rainfall of 22 mm/h for 24 h. (a) pore-water pressure profiles obtained from seepage analyses considering contribution from negative pore water pressures; (b) pore-water pressure profiles obtained from seepage analyses ignoring contribution from negative pore-water pressures (*Modified after* Rahardjo *et al.*, 2007a)

much affected by the presence or absence of negative pore-water pressure or matric suction in the slope. In other words, assuming the absence of matric suction in the slope may lead to unrealistic assessment of pore-water pressures in the slope during rainfall.

Seepage analyses during the matric suction recovery (drying phase, after the rainfall event) with and without considering the contributions from evaporative flux (Figure 7.40) show that when the contributions from evaporation are taken into account, the rate of matric suction recovery is rapid and significant. This is because the evaporation process is a significant contributor to the matric suction recovery. The evaporation rate used in the numerical analyses was in the range of 5.16–7.53 mm/day, typical of potential evaporation rate in Singapore as calculated by Gasmo (1997).

The changes in factor of safety of the slope under different conditions are shown in Figure 7.41. Figure 7.41 clearly shows that when slope stability analysis was performed by neglecting the contributions from negative pore-water pressures prevailing in the slope, the initial factor of safety was less than 1. This initial factor of safety obtained from the slope stability analysis by ignoring the contributions from negative pore-water pressures does not seem to be realistic, nor represent the actual field condition of the slope where it was stable. This exercise clearly indicates the necessity for considering the contribution from negative pore-water pressure in seepage and slope stability analyses in order to enable realistic modelling of field slope conditions. Figure 7.41 also indicates that the recovery of factor of safety was faster when the effects of both evaporation and negative pore-water pressures are taken into account.

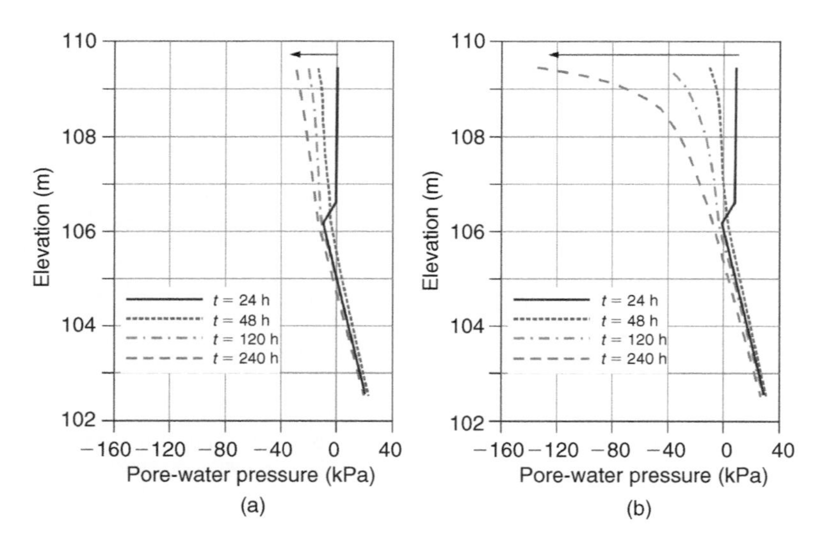

Figure 7.40 Comparison of pore-water pressure profile at mid section of the slope during drying. (a) pore-water pressure profiles obtained from seepage analyses ignoring contribution from evaporation; (b) pore-water pressure profiles obtained from seepage analyses considering contribution from evaporation (*Modified after* Rahardjo *et al.*, 2007a)

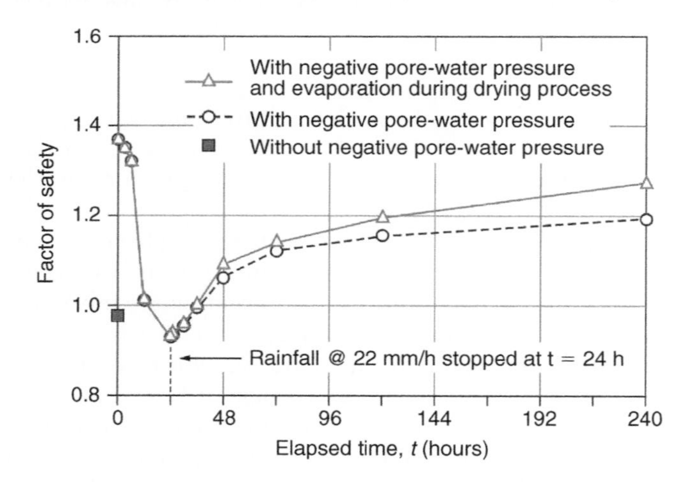

Figure 7.41 Changes in factor of safety of the slope under different conditions (*Modified after* Rahardjo *et al.*, 2007a)

The contribution from evaporation on the recovery of factor of safety becomes more significant about 48 h after the cessation of rainfall.

Figure 7.42 shows a comparison of pore-water pressure contours and pressure head contours obtained from seepage analyses incorporating unsaturated soil mechanics with direct field measurements following a natural rainfall on the slope on 15 June 2007. The results of the seepage analyses with the incorporation of unsaturated soil

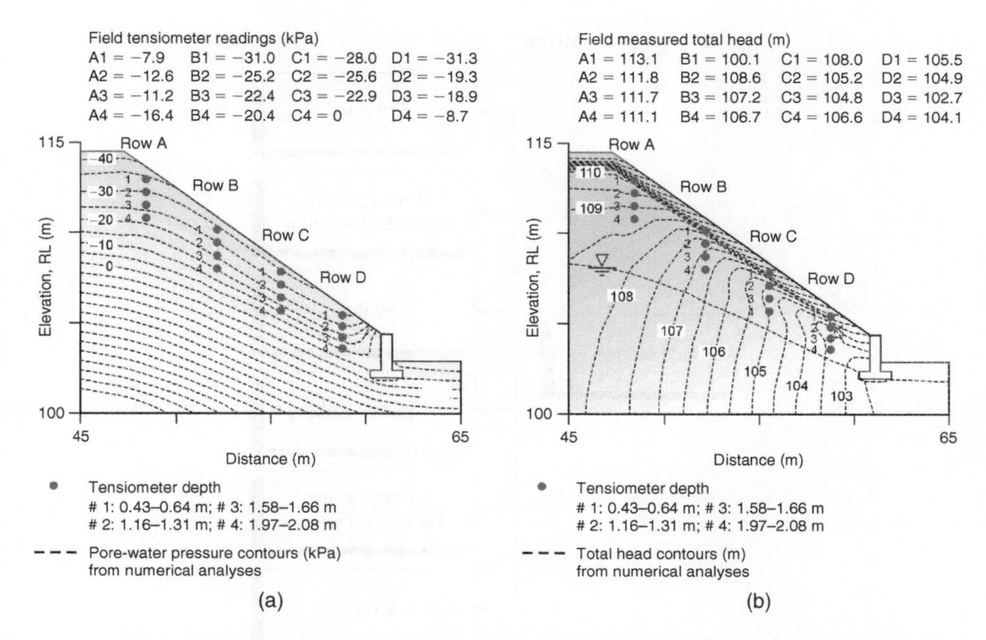

Figure 7.42 **Comparison of results from seepage analyses incorporating unsaturated soil mechanics with direct field measurements; (a) pore-water pressure contours; (b) total head contours (*Modified after* Rahardjo *et al.*, 2007a)**

mechanics (Figure 7.42) appear to agree closely with the field measurements. This example further justifies that when the effects of negative pore-water pressure are incorporated in seepage and slope stability analyses through unsaturated soil mechanics equations, numerical analyses could simulate field processes with reasonably realistic agreement.

7.7 REMEDIAL MEASURES FOR SOIL AND ROCK SLOPES

Ortigao and Sayao (2004) presented an extensive coverage of this issue, and this is briefly summarised here. Figure 7.43 presents several alternative solutions for stabilisation of soil slopes. Drainage and surface protection are always present in stabilisation works and generally they are the most cost-effective. All the others require sound engineering judgement to be applied.

Figure 7.44 presents a flowchart with possible design alternatives.

The following aspects should be taken into account when selecting the stabilisation solutions.

Drainage and surface protection

Generally, this is the most cost-effective solution. In tropical regions, such as Hong Kong and Rio de Janeiro, heavy rains are the most important triggering mechanism for landslips. Therefore, no slope will be safe without a high quality surface and

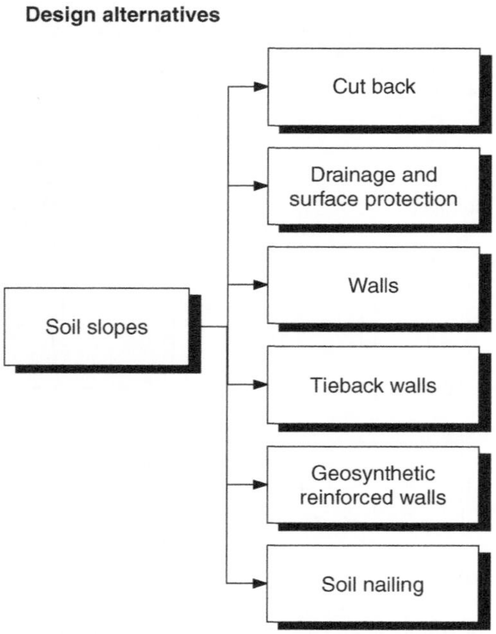

Design alternatives
Soil slopes
Cut back
Drainage and surface protection
Walls
Tieback walls
Geosynthetic reinforced walls
Soil nailing

Figure 7.43 Soil slopes destabilisation alternatives

Soil slopes
Excavations
Fills
Cut back
Soil nailing
Tieback walls
Walls
Geosynthetic reinforced walls
Cut back
Berms or benches
Drainage and surface protection

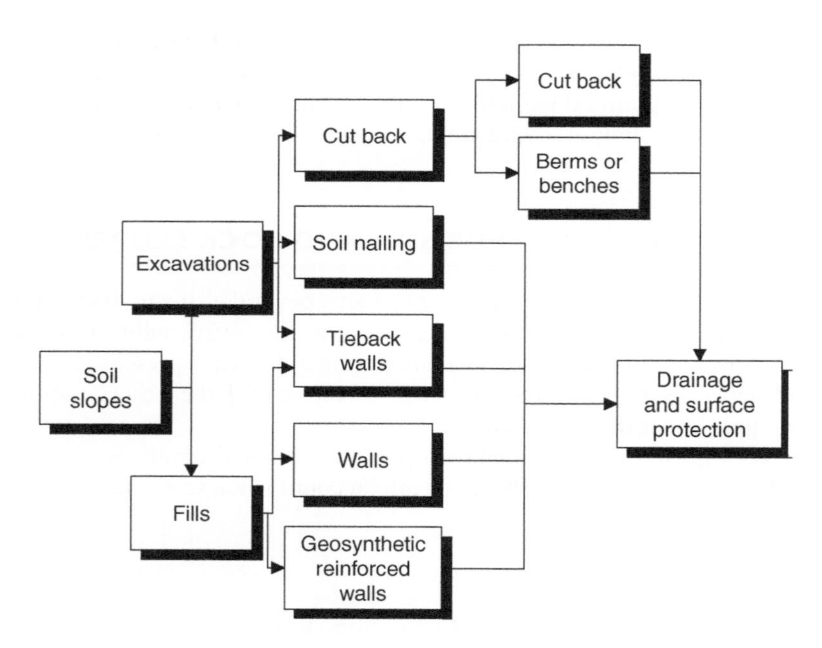

Figure 7.44 Soil slopes: choice of design solution

Figure 7.45 Surface conditions of the instrumented slope (*After* Lim *et al.*, 1996)

internal drainage system. In very large landslides, involving many millions of cubic metres, drainage can be the only cost-effective solution, as in such cases any retaining structure or soil reinforcement method is out of question due to exorbitant costs.

In an attempt to evaluate the effectiveness of slope stabilisation by surface protection, Lim *et al.* (1996) investigated the effect of three different ground surface conditions on variations in matric suction, matric suction profile and total head profile in response to rainfall in a residual soil slope in Singapore. The slope was divided into three sections (i) a canvas-covered grassed surface section; (ii) a grassed surface section and (iii) a bare surface section, and was instrumented to continuously monitor the in-situ matric suction variation at different soil depths in response to rainfall (Figure 7.45). The study indicated that variation in matric suction profile is less significant under the canvas-covered section than under the grass-covered and the bare section of the slope.

Typical daily variations of in-situ matric suction with respect to wetting (rainfall) and drying (evaporation/transpiration) under these three different ground surface conditions are shown in Figure 7.46. Some common characteristics in matric suction change are apparent from Figure 7.46. Irrespective of the ground surface condition, the maximum changes in matric suction occur at a shallow depth (0.5m) near the ground surface and the magnitude of changes in matric suction decreases with an increasing depth. Matric suction at shallow depth near the ground surface is the first to be affected by climatic changes whereas matric suction changes at greater depths show a delayed response.

When ground surface conditions are taken into account, the variations in matric suctions are the most significant in the bare slope section and least significant in the canvas-covered slope section. Even though direct infiltration is prevented in the canvas covered section of the slope, the in-situ matric suction still appears to change during and after rainfall (see Figures 7.46(a2) and b2). This change in matric suction is due to the lateral migration of water from infiltration beyond the edges of the canvas

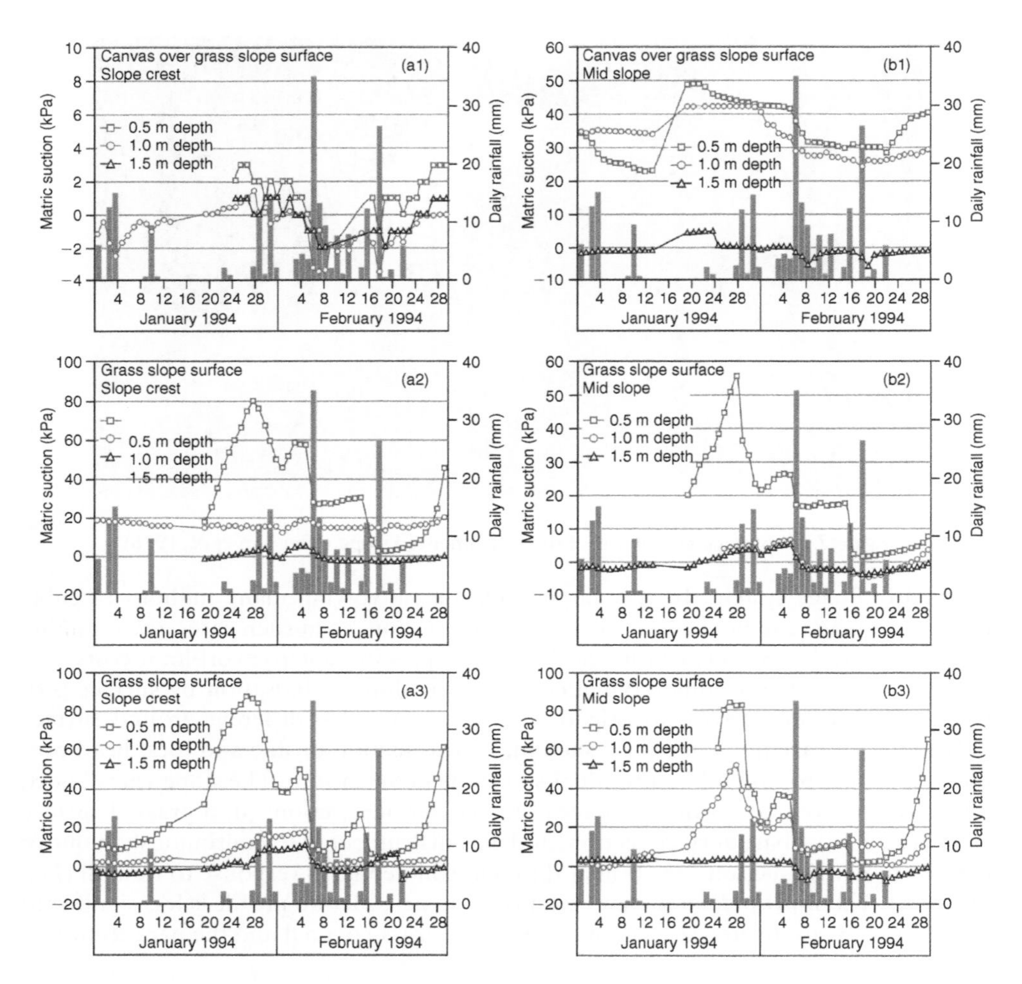

Figure 7.46 Typical daily variations in matric suction with respect to climatic conditions from a residual soil slope in Singapore: (a1–a3) at slope crest with three different ground surface conditions; (b1–b3) mid-slope location with three different ground surface conditions (*Modified after* Lim *et al.*, 1996)

(Figure 7.45). Figure 7.46 also suggests that the amount of decrease in matric suction during a particular rainfall is dependent on the prevailing matric suction prior to the rainfall. Note differences in matric suction axes for Figures 7.46(a1) and 7.46(b3).

Lim *et al.* (1996) also showed that the presence of vegetation in a slope can significantly increase the matric suction and alter the total head profile. Variations in matric suction profile (Figure 7.47(a1–a3)) and total head profile (Figure 7.47(b1–b3)) associated with the wetting events by rainfall (Figure 7.47(a4–b4)) are shown in Figure 7.47. Figure 7.47(a1–a3) shows the progression in matric suction reduction under three surface conditions in response to rainfall events between 27 January and 18 February 1996. A matric suction recovery noticed on February 4 is due to the dry period commencing on 1 February 1996.

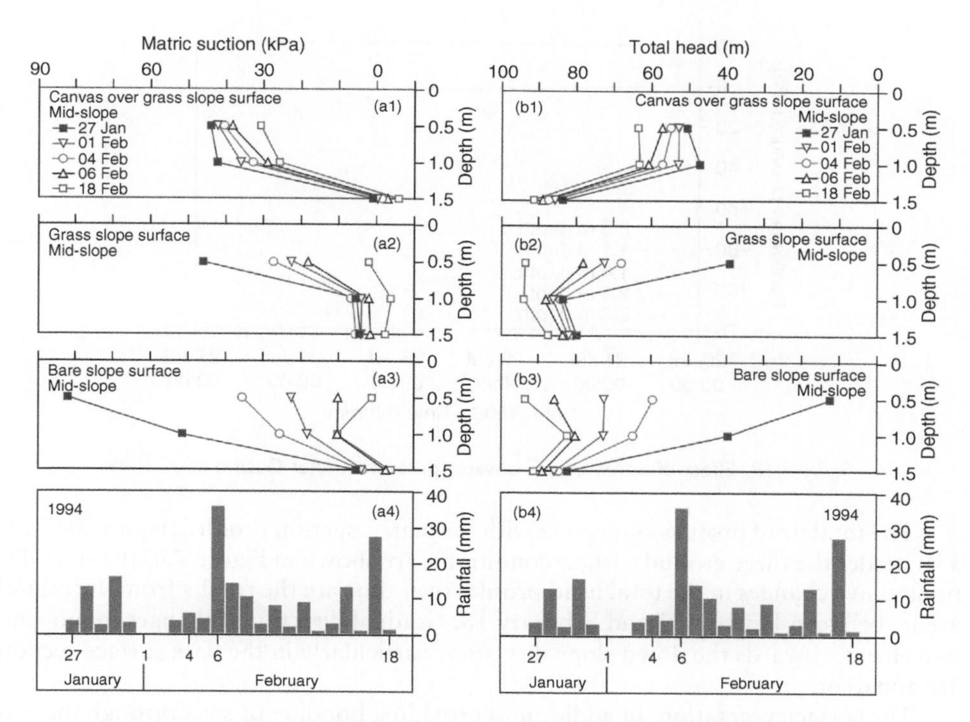

Figure 7.47 Typical changes in matric suction profile and total head profile in response to climatic conditions from a residual soil slope in Singapore. (a1–a3) Matric suction profile at mid-slope location under three different ground surface conditions, (a4) wetting events (rainfall) between 27 January and 18 February 1996; (b1–b3) Total head profile at mid-slope location under three different ground surface conditions, (b4) wetting events (rainfall) between 27 January and 18 February 1996 (Modified after Lim *et al.*, 1996)

The changes in matric suction profiles over time are more significant in the bare surface section of the slope than in the grass-surface section. The matric suction profile from the canvas-covered section shows little variation over time and the variation of matric suction with depth is consistent as a result of negligible infiltration due to the canvas-cover.

The changes in matric suction profile over time in the grass-covered section are significant at shallow depths (near the ground surface, 0.5 m depth) but are insignificant at greater depths. Lim *et al.* (1996) attributed this to the evaporation and transpiration by the grass on the ground surface. The presence of grass on the slope surface accelerates the removal of water by evaporation and transpiration and prevents the wetting front from advancing to greater depths. For the bare section, the removal of water is solely due to surface evaporation; therefore, the advancing wetting front continues to propagate to greater depths after the rainfall. As a result, the reduction in matric suction in the bare slope surface section extends to a depth of 1.5 m or deeper (Figure 7.47(a3)). Therefore, it appears that the surface evaporation alone does not seem to be efficient for matric suction recovery.

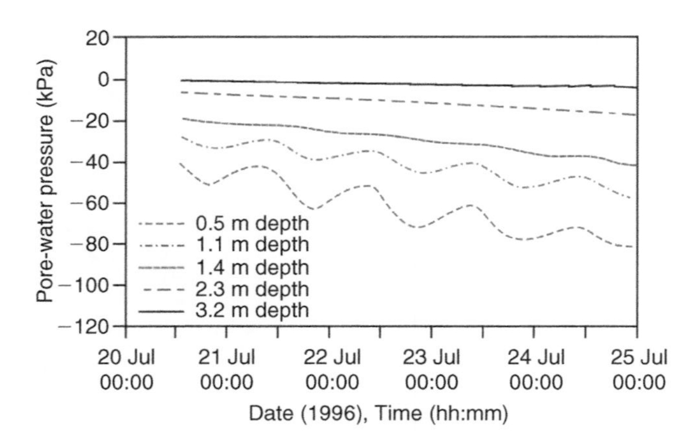

Figure 7.48 Effect of a tree on pore-water pressures (*After* Gasmo *et al.*, 1999)

The total head profiles associated with the matric suction profiles (Figure 7.47(a1–a3)), under the three ground surface conditions, are shown in Figure 7.47(b1–b3). The progressive changes in the total head profiles over time are the results from the rainfall events between January 27 and February 18. Gradual directional changes in moisture movement, towards the down slope direction, particularly in the bare surface section, are apparent.

The surface vegetation, in addition to providing bonding of soils through the root network, also enhances quick matric suction recovery through evapo-transpiration. The role of vegetation on the matric suction recovery in a slope is clearly evident from Figure 7.48 which shows the cyclic pattern of pore-water pressure variation within the root zone of a tree (0–1.4 m depth) located near the toe of a slope as reported by Gasmo *et al.* (1999).

The horizontal drains, defined as holes, drilled into a cut slope or embankment and cased with a perforated metal or slotted plastic liner (Royster, 1980), have commonly been used in tropical regions for stabilizing unsaturated residual soil slopes. The function of a horizontal drainage system is to keep the slope material dry by draining away groundwater. The effectiveness of a horizontal drainage system is a function of many factors which includes the drain location, drain length, drain spacing as well as soil properties and slope geometry (Rahardjo *et al.*, 2003). Typically, effectiveness of horizontal drains is assessed in terms of the increase in factor of safety of a slope with horizontal drains as compared to the factor of safety of the same slope without any horizontal drains.

Parametric studies complemented with field monitoring programs (Rahardjo *et al.*, 2003) on the effectiveness of horizontal drains for slope stabilisation revealed valuable insights on drain location, and identified conditions under which horizontal drains are most effective. The results from the parametric study (Rahardjo *et al.*, 2003) on a slope (with geometry, soil parameters, rainfall input and horizontal drainage configuration shown in Figure 7.49) suggest that horizontal drains are the most efficient both in terms of providing drainage (Figure 7.49(a)) and improving the factor of safety (Figure 7.49(b)) when placed near the bottom of the slope. Figure 7.49 also suggests that the

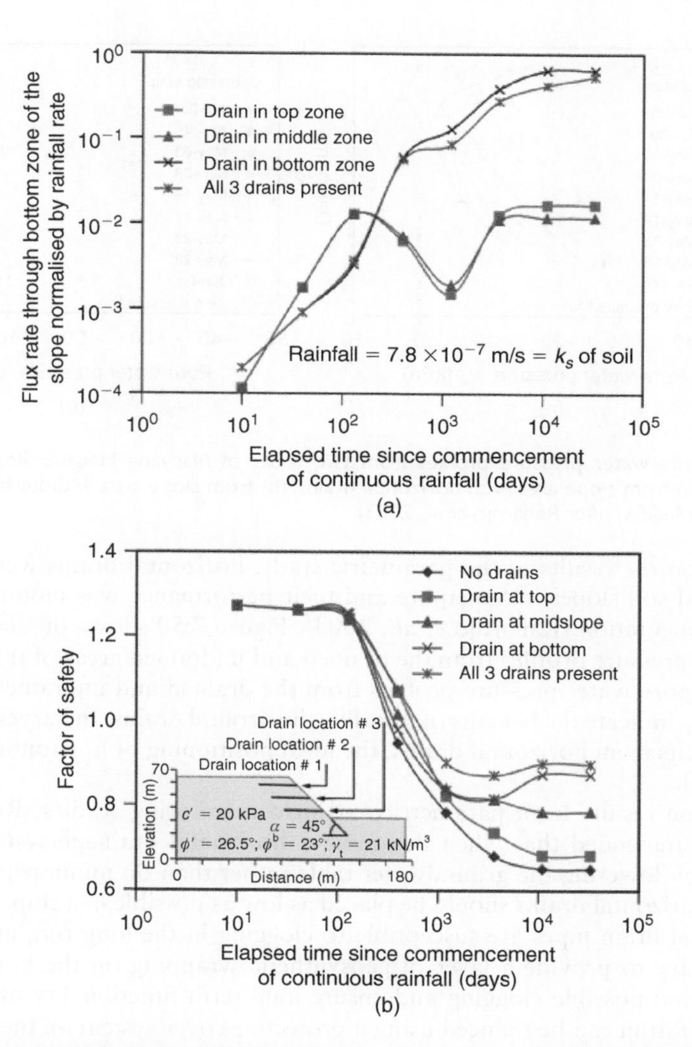

Figure 7.49 **Theoretical effectiveness of horizontal drains in terms of (a) flux through bottom zone of the slope with respect to different drain position; (b) factor of safety with respect to different drain position (*Modified after* Rahardjo *et al.*, 2003)**

least benefit is derived from drains placed on the upper region of the slope and a very small additional benefit in terms of factor of safety (only 3% additional factor of safety) is obtained by combining the bottom drain with two other drains at the top and at the mid-slope. The horizontal drains do not play any significant role in minimizing infiltration but are efficient in lowering the ground water table if placed near the bottom of the slope. The lowering of the ground water table by horizontal drains provides a zone of maintained negative pore-water pressure above the lowered ground water table which contributes to the shear strength of the soil and hence increases the stability as compared to a slope without horizontal drains (Figure 7.49(b)).

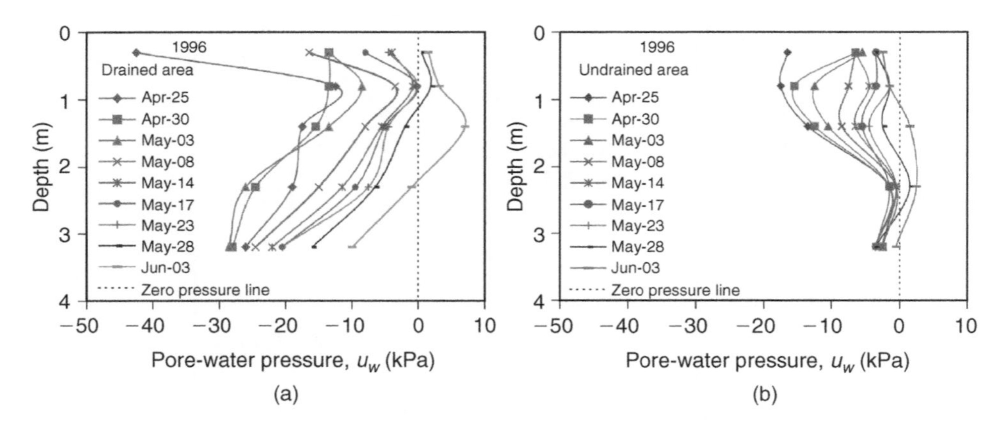

Figure 7.50 Pore-water pressure profiles from two areas of Nanyang Heights Slope in Singapore (a) from slope area with horizontal drains; (b) from slope area without horizontal drains (*Modified after* Rahardjo *et al.*, 2003)

Based on the results of the parametric study, horizontal drains were installed in two residual soil slopes in Singapore and their performance was monitored through field instrumentation (Rahardjo *et al.*, 2003). Figure 7.50 shows the field monitored pore-water pressure profiles from the drained and undrained areas of the residual soil slope. The pore-water pressure profiles from the drained and undrained areas of the slope clearly indicate the benefits of installing horizontal drains. In harvesting the maximum benefits from horizontal drains, the ideal positioning of horizontal drains plays a crucial role.

Based on results from parametric and field monitoring studies, Rahardjo *et al.* (2003) recommended that when designing a horizontal drainage system, emphasis should be on lowering the groundwater table rather than on minimizing the infiltration, and horizontal drains should be placed as low as possible in a slope. Perforations in horizontal drain pipes are susceptible to clogging in the long run, and it is therefore necessary to provide a layer of geosynthetic wrapping on the horizontal drain pipes to avoid possible clogging and ensure long term functionality of the drainage pipes. Infiltration can be reduced using a protective cover system on the slope surface such as "chunam", a thin layer of low permeability cement-lime-soil mixture used in Hong Kong, or surface grouting (shown in Figure 7.51), or a capillary barrier system (Rahardjo *et al.*, 2007(b)) and proper surface drainage on the slope surface.

Studies by Lim *et al.* (1996) on soil cover and Rahardjo *et al.* (2003) on horizontal drains suggest that either surface cover or horizontal drains, when used under appropriate conditions, could provide effective slope stabilisation. In some cases, a combined system of surface cover and horizontal drains as shown in Figure 7.52 can also be used for slope protection.

Capillary barrier

A capillary barrier is a soil cover that consists of two soil layers: a fine-grained soil layer overlying a coarse-grained soil layer that provides a barrier effect to limit the

Figure 7.51 A surface grouted slope in Kyushu, Japan

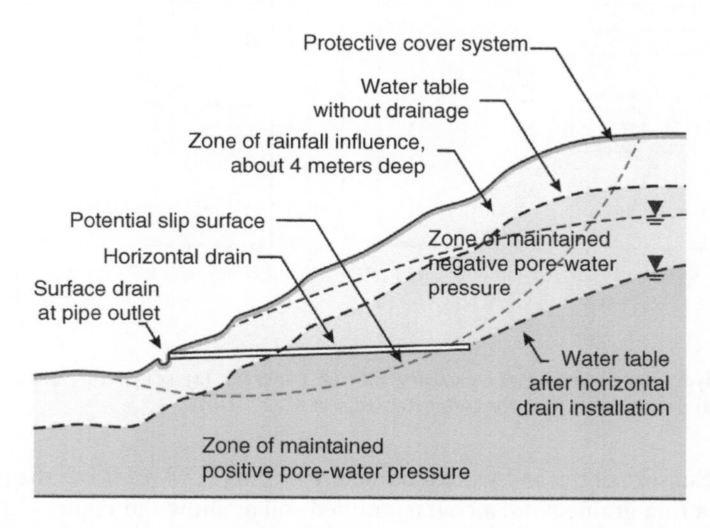

Figure 7.52 Conceptualisation of slope stabilisation using protective surface cover and horizontal drains as a combined slope stabilisation system (*Modified after* Rahardjo *et al.*, 2003)

downward movement of water from the fine-grained soil layer into the coarse-grained soil layer. A basic form of a capillary barrier is shown in Figure 7.53.

The barrier effect, that limits the downward movement of water from the fine-grained soil layer to the coarse-grained soil layer in a capillary barrier system, is induced by the contrast in the hydraulic properties of the two soils. The hydraulic properties of the soils involved are the soil-water characteristic curve (SWCC) and the permeability function. The mechanism of the barrier effect is therefore best understood by

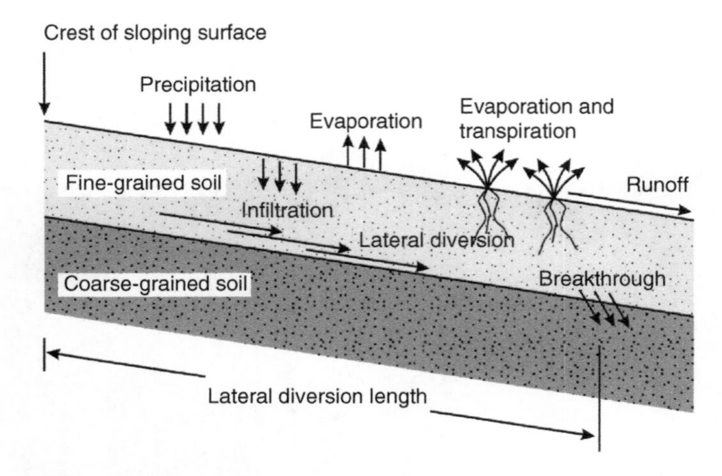

Figure 7.53 Schematic diagram of a capillary barrier and its parameters

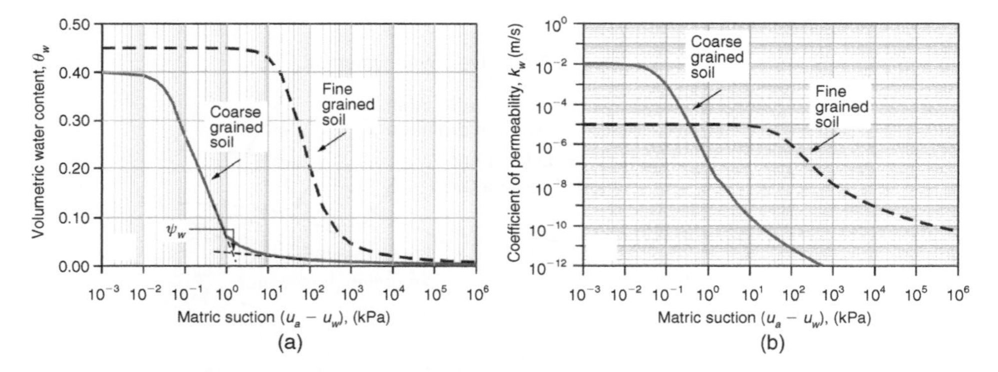

Figure 7.54 Hydraulic properties of capillary barrier material. (a) soil-water characteristic curve; (b) permeability function (*After* Rahardjo *et al.*, 2007(b))

studying the nature of the soil-water characteristic curve (SWCC) and the permeability function of a fine-grained and a coarse-grained soil as shown in Figure 7.54. A SWCC is a relationship that describes the volumetric water (θ_w) content of a soil at various matric suctions (Figure 7.54(a)), whereas a permeability function is a relationship that describes the unsaturated coefficient of permeability of a soil at various matric suctions (Figure 7.54(b)).

The contrast in the unsaturated coefficient of permeability between the fine-grained and coarse-grained soils, at a given matric suction value, which can be seen in Figure 7.54(b), induces the barrier effect in a capillary barrier system. At high matric suction values, the coefficient of permeability of the coarse-grained soil is much lower than that of the fine-grained soil (Figure 7.54(b)). Under these matric suction conditions, the infiltrating water will not flow from the fine-grained soil layer into the underlying coarse-grained soil layer. Instead, the infiltrating water will be retained and diverted laterally through the fine-grained soil layer because the coefficient of permeability of the fine-grained soil is higher than that of the coarse-grained soil layers. If infiltration continues, the volumetric water content and the coefficient of permeability of the fine-grained soil layer will increase. When the infiltrating water reaches the fine-coarse soil interface, the matric suction in the coarse-grained soil layer will decrease. When the matric suction in the coarse-grained soil layer reaches its water-entry value, ψ_w, (Figure 7.54(a)), the coefficient of permeability of the coarse-grained soil layer will increase and at this stage, the infiltrating water will start to flow into the coarse-grained soil layer. If the matric suction in the coarse-grained soil layer continues to decrease further, the coefficient of permeability of the coarse-grained layer will increase rapidly and finally exceed that of the fine-grained soil layer (Figure 7.54(b)). At this condition, the barrier effect from the coarse-grained soil layer will diminish and the infiltration will easily migrate into the coarse-grained soil layer.

From the discussion above on the mechanism of the capillary barrier effect in the context of SWCC and permeability function of the two soils, it appears that the barrier effect in a capillary barrier system can be sustained as long as the matric suction at the fine-coarse soil interface is maintained at a higher value than the water-entry value (ψ_w) of the coarse-grained soil layer (Rahardjo *et al.*, 2007(b)). This requirement can be satisfied by (i) using two soil layers which have marked contrast in their unsaturated hydraulic properties; and (ii) maintaining high matric suction in the fine-grained soil layer through the provision of adequate diversion, drainage and evaporation facilities for rapid recovery in matric suction in the fine-grained soil layer. An appropriate material selection is therefore very important in the design of an effective capillary barrier system.

The effectiveness of a capillary barrier system is usually quantified in terms of its water storage capacity, water storage recovery, diversion capacity, and diversion length. The water storage capacity of a capillary barrier is identified as the maximum soil-pore volume available in the fine-grained soil layer to retain infiltration without allowing percolation into the coarse-grained soil layer. Theoretically, for a unit volume of soil, the maximum water storage capacity is equal to the porosity of the fine-grained soil. Under normal conditions, a capillary barrier seldom stores an infiltration volume equivalent to its maximum storage capacity. Breakthrough occurs before the water storage capacity is reached. This is primarily because when the matric suction in the fine-coarse interface approaches the water-entry value of the coarse-grained soil layer, breakthrough occurs, but the soil in the fine-grained soil layer is still unsaturated (i.e. not all available pore volumes are filled with water).

The release of the infiltration water volume retained in the fine-grained soil layer through evaporation, evapotranspiration or lateral drainage, is called water storage recovery of a capillary barrier system. The water storage recovery of a capillary barrier is essential for its effective performance and longevity. Experimental studies (Krisdani, 2006) have shown that, for tropical regions, surface evaporation alone is not sufficient for the water storage recovery of a capillary barrier. Therefore, fine-grained soils with good drainage characteristics should be used as the top layer to ensure large lateral diversions.

The diversion capacity of a capillary barrier is defined as the maximum lateral flow per unit area of the fine-grained soil layer, above the fine-coarse soil interface, before percolation into the underlying coarse-grained soil layer (Ross, 1990). The diversion length is defined as the distance from the crest of the sloping capillary barrier to the point in the down slope direction where the rate of water percolation into the coarse grained soil layer is equal to the rate of infiltration. It is always desirable to use a combination of fine-grained and coarse-grained soil layers that provide a large diversion capacity and diversion length.

Parametric study of capillary barriers

The capillary barrier has been studied and used as soil cover for landfill and mining wastes under relatively flat slope conditions in arid and semiarid climates (Stormont, 1997; Morris and Stormont, 1998; Yanful *et al.*, 1993; 1999). It has also been proposed (Tami *et al.*, 2004) for slope stabilisation in tropical regions. Recent extensive parametric studies (Tami, 2004; Tami *et al.*, 2004a; 2004b) identified the key parameters for soil selection for a capillary barrier system. One of the parameters identified for selecting appropriate soils for a capillary barrier system is the ψ_w-ratio (Rahardjo *et al.*, 2007b). The ψ_w-ratio is the ratio of the water-entry values between the fine-grained and the coarse-grained soils. The ψ_w-ratio reflects the contrast in the hydraulic properties of the two soils intended for a capillary barrier system. The parametric studies suggested that it is desirable to use a combination of soil layers which gives a large ψ_w-ratio. The larger the ψ_w-ratio, the more significant is the barrier effect produced and the percolation into the coarse grained layer is minimal. Other key parameters that need to be considered during the material selection for a capillary barrier system are the water-entry value for the coarse-grained soil and the saturated coefficient of permeability of the fine-grained soil. It is desirable to use a soil with low water-entry value (preferably <1 kPa) as the coarse-grained soil layer. The reason for this lies in the mechanism of capillary barrier effect (explained in the preceding paragraphs). The saturated coefficient of permeability of the fine-grained soil should not be too low (preferably $>10^{-5}$ m/s) to ensure prompt drainage through lateral diversion of the water retained in the fine-grained soil layer during infiltration.

Physical laboratory scale model study of a capillary barrier

Based on the results of the parametric study, a laboratory scale physical model of a two layered sloping (30°) capillary barrier system (20 cm thick fine-grained soil layer over 20 cm thick coarse-grained soil layer, 0.4 m wide and a horizontal length of about 2.35 m) was constructed (Tami *et al.*, 2004c) in a laboratory infiltration box as shown in Figure 7.55. The infiltration box was equipped with various control and measuring devices as shown in Figure 7.55. Various instrument positions and measurement locations are shown in Figure 7.56. The soils in the capillary barrier system were selected based on the guidelines on material selection. The dimensions of the infiltration box were 0.4 m wide, 2.45 m long, and 2 m in height (Tami *et al.*, 2004b).

A detailed discussion on the results of the laboratory experiment can be found in articles by Tami (2004) and Tami *et al.* (2004a; 2004b). A time series of response of the model capillary barrier to different climatic conditions obtained from laboratory experiments is shown in Figure 7.57. The dynamics of moisture fluxes in the capillary barrier system associated with input-rainfall rate (q_P) and output-lateral diversion rate (q_L), breakthrough rate (q_B), and runoff rate (q_R) are clearly evident from Figure 7.57. There was no runoff generation during any stage of the experiments, therefore, q_R

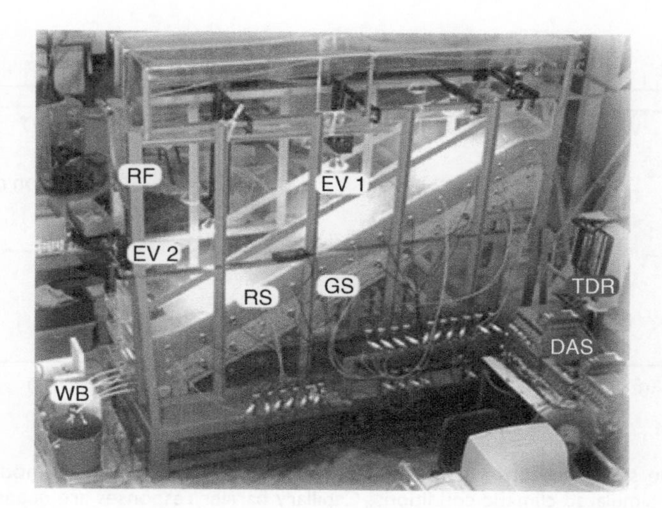

Figure 7.55 A laboratory set-up of a physical model of capillary barrier system for studying the effectiveness of a capillary barrier for slope stabilisation. RS: residual soil layer (fine-grained soil); GS: gravellier sand layer (coarse-grained soil); RF: rainfall simulator; EV1: lamp to enhance drying process; EV2: fan to enhance drying process; WB: weighing balance to measure runoff, lateral diversion, and breakthrough amount; TDR: Array of time domain reflectometry wave guide and tensiometer probe assembly for soil water content and pore-water pressure measurements; DAS: data acquisition system (*After* Tami *et al.*, 2004b)

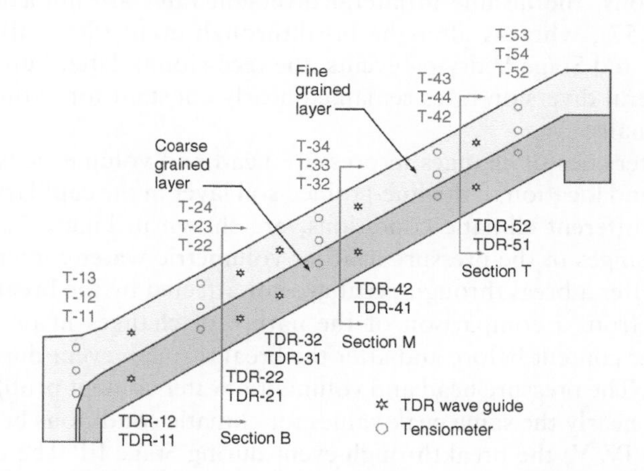

Figure 7.56 Schematic diagram showing locations of the measuring devices in the laboratory scale physical model of the capillary barrier system (*After* Tami *et al.*, 2004b)

do not appear in Figure 7.57. Figure 7.57 suggests that after a breakthrough event, a change in behaviour of the capillary barrier, particularly in the lateral diversion rate (q_L) is expected. This is evident from a comparison of the recession characteristics of lateral diversion rates before and after the breakthrough event during the Stage III wetting process. At the onset of the drying events (D) during Stage I and Stage II

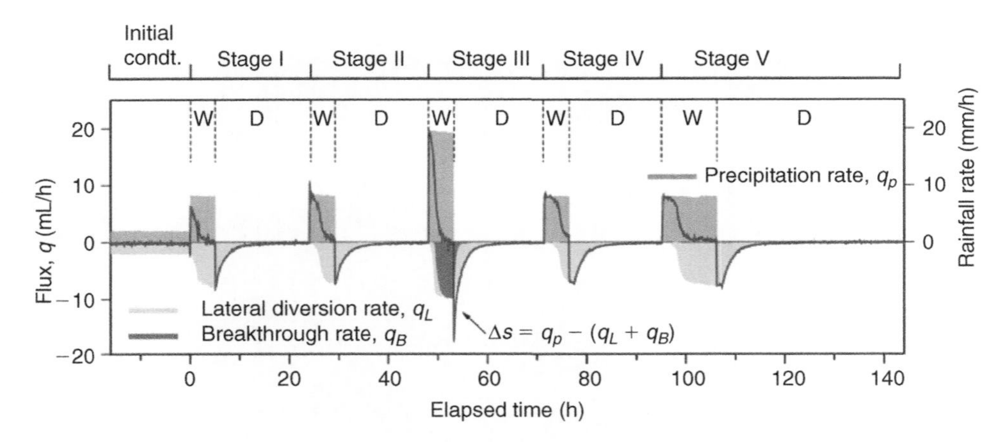

Figure 7.57 Time series response of the laboratory physical capillary barrier model to changes in the simulated climatic conditions. Capillary barrier responses are quantified in terms of lateral diversion rate (q_L), breakthrough rate (q_S), and runoff generation rate (q_R). Simulated climatic conditions are quantified in terms of wetting (W) by precipitation rate (q_P) and drying (D) due to evaporation and drainage after rainfall. Storage rate is quantified as $\Delta s = q_P - (q_L + q_B + q_R)$. Stage I to Stage V represents transient climatic conditions simulated using different rainfall rate and duration. (*After* Tami *et al.*, 2004b)

climatic conditions, the decline in lateral diversion rates are immediate, sharp and rapid (Figure 7.57), whereas after the breakthrough event (Stage III wetting event) during Stage IV and Stage V drying events, the recession in lateral diversion rates are delayed and lateral diversion rates remained nearly constant for about 65 min before decreasing gradually.

The characteristics of changes in pressure head and volumetric water content at various depths and location of the fine-grained soil layer in the capillary barrier model, in response to different climatic conditions, are shown in Figure 7.58. Figure 7.58 suggests that changes in the pressure head or volumetric water content in a capillary barrier system after a breakthrough event are not affected by the breakthrough event. This is evident from a comparison of the nature of changes in pressure head and volumetric water content before and after the breakthrough event during the Stage III wetting process. The pressure head and volumetric water content profiles show similar trend and reach nearly the same peak values for climatic conditions before (Stage I, II) and after (Stage IV, V) the breakthrough event during Stage III. The relatively higher pressure head and volumetric water content profiles observed during Stage III climatic conditions were due to the highest rainfall rate used in this event (20 mm/h for 5 h) that caused breakthrough to occur.

The time series of water balance components and total storage of the laboratory scale capillary barrier model, in response to various climatic conditions, are shown in Figure 7.59. The storage denoted by ΔS (Figure 7.59) was computed as the difference between volume of water received (due to precipitation P) and volume of water released (due to lateral diversion L, breakthrough B, and runoff R) by the capillary barrier system i.e. $\Delta S = P - (L + B + R)$.

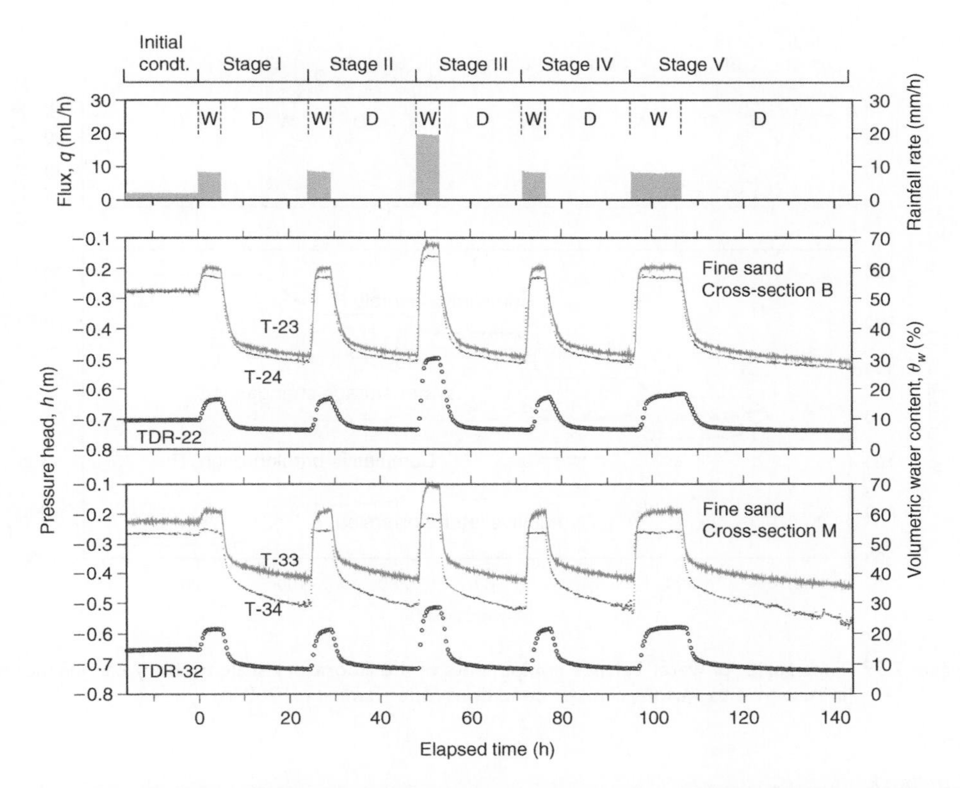

Figure 7.58 Variation in the pressure head and volumetric water content at different depth and location of the fine-grained soil layer of the capillary barrier system in response to variation in climatic conditions (*After* Tami *et al.*, 2004b)

Laboratory experiments (Tami, 2004; Krisdani, 2006) on physical models of a capillary barrier system showed that the performance of a capillary barrier system for slope stabilisation, under high tropical rainfall, is primarily dependent on the storage capacity of the fine-grained soil layer in the capillary barrier system.

Field scale capillary barrier

The application of the concept of capillary barrier for slope stabilisation under tropical climates is relatively new and has not been tested extensively under field conditions. Based on the results from numerical studies and laboratory experiments, a residual soil slope in Singapore was stabilised through the construction of a field-scale capillary barrier system on a slope. The construction details and instrumentation of the slope for monitoring the performance of the field scale capillary barrier system, as a soil cover for slope stabilisation, are shown in Figure 7.60. A satisfactory performance of the capillary barrier has been observed so far.

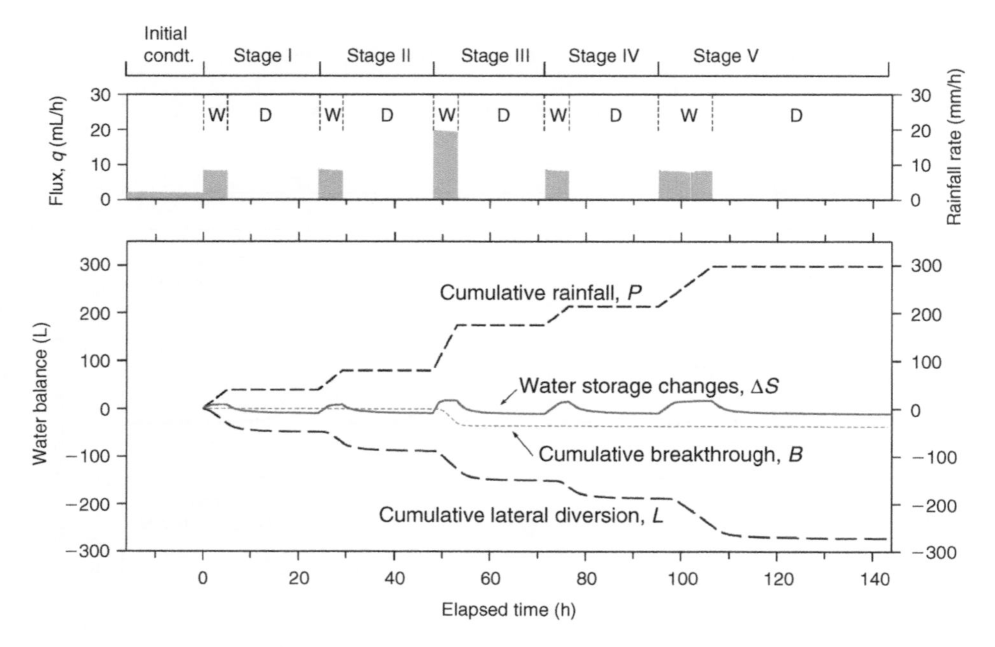

Figure 7.59 Time series of water balance components of the laboratory scale capillary barrier model in response to various climatic conditions (*After* Tami *et al.*, 2004b)

Cut back

A cut back by flattening the slope or creating berms or benches can only be effective in unpopulated areas and where the excavation can be carried out by bulldozers.

Access and means of transportation to job site

If access is difficult, one can use unconventional transportation means. Figure 7.61 presents examples of such difficulty for stabilising slopes in Rio de Janeiro. The closest access road sometimes is too far and the final access has to be by a cable car, mules, helicopters or labour. In such cases, the equipment weight is of major concern, since the cost of transportation is the most important.

Retaining walls and soil nailing

Gravity or concrete walls tend to be economical for small slope heights up to 3 m. Beyond this height, a soil reinforcement solution tends to be much more cost-effective.

Tieback walls, introduced in Brazil back in 1957, became a tradition for stabilising soil slopes in Rio de Janeiro and more than one thousand of these structures exist and show good behaviour, although they are less used in other countries. Pre-stressed soil anchors limit soil displacement, although they require a 30 cm thick wall. The advantage of this system is the flexibility of the construction method which will be *descending*, in case of slopes to be excavated; *ascending*, for fills, or *mixed*.

(a) Placement of top soil on top of the fine-grained layer of the capillary barrier system

(b) Drainage provision

(c) Grass planting on slope surface after placement of the top soil on top of the capillary barrier (boxes on the slope surface are protective covers for tensiometers installed in the slope for measuring the effectiveness of capillary barrier system)

Figure 7.60 Construction of capillary barrier system for stabilising a residual soil slope in Singapore

Figure 7.61 Equipment and materials transportation in difficult access sites: use of helicopter, mules, cable car and workers, GeoRio works in Rio de Janeiro

Soil nailing, in general, presents the minimum cost to stabilise an excavated soil slope. It is easily applicable to vertical and inclined slopes.

Soil reinforcement solutions comprise geosynthetic and other types of reinforcement for walls. They are low-cost solutions and recommended for walls higher than 3 m. The use of non-woven geosynthetics and a segmental concrete block facing leads to a low-cost and effective solution. Alternatively, the use of geosynthetics folding close to the wall face leads to a very flexible wall that can accommodate foundation settlement. In this case, a protective facing is built afterwards to avoid damage to the geosynthetics.

Crib structure (Figure 7.62) is used for slopes susceptible to erosion or crumbling due to weathering. It consists of a network of reinforced concrete beams (crib structure) cast-in-situ on the slope surface. The crib structure is widely used in Japan for slope protection and stabilisation. Additionally, the crib structure can act as a retaining system for soil on the slope surface for slope vegetation.

Figure 7.62 Slope stabilisation by crib structure

Figure 7.63 Rock scarp and talus formation

Slopes with a rock to soil interface

Figure 7.63 presents a slope having a rock–soil interface. A steep rock scarp leads to a talus formation at the toe. The talus is easily identified by the amount of loose rock blocks spread on the top of the soil mass. Rainfall on the rock scarp will find its way into the soil mass at the interface, leading to landslips. Therefore, the most effective way of avoiding landslips in this case, is to build a drainage ditch to collect rainwater before it enters the talus.

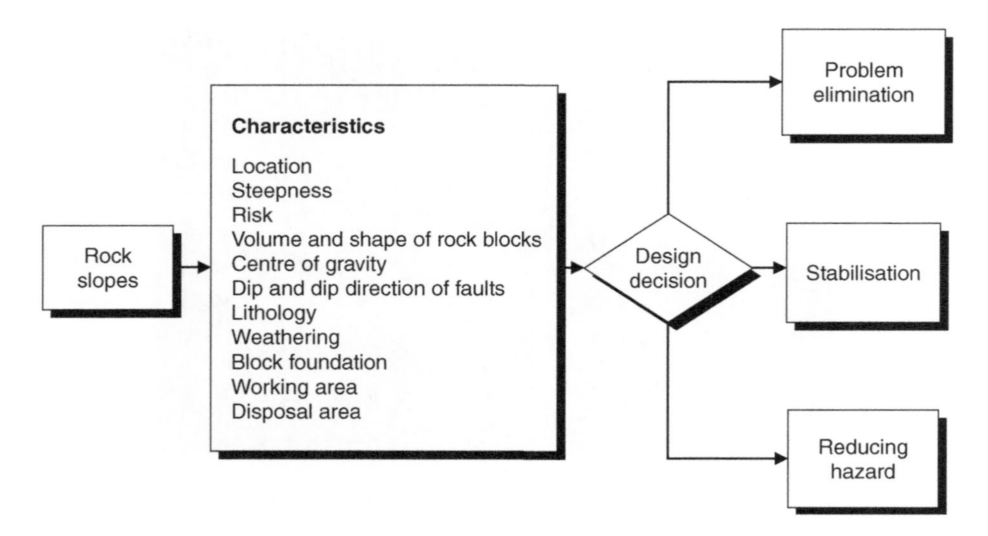

Figure 7.64 Rock slopes: alternative solutions

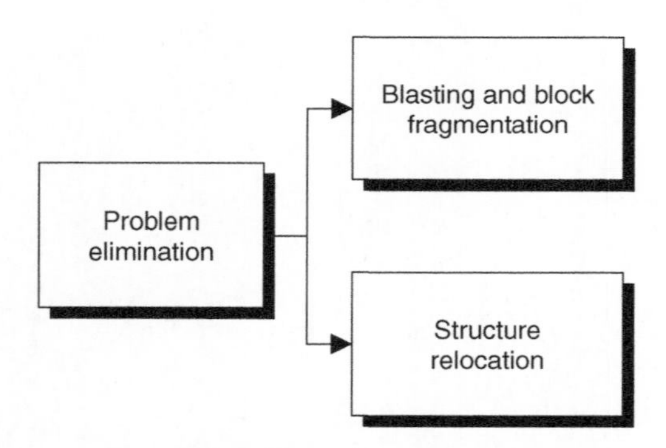

Figure 7.65 Problem elimination

Measures for stabilisation of rock slopes and rockfall

Figure 7.64 presents a range of typical design approaches for rock slope stabilisation. The selection of the stabilisation method depends on slope characteristics, hazard and risk. The design approaches can be summarised into three classes to reduce hazard, namely by *elimination*, by *slope stabilisation* and by *hazard reduction*. Figures 7.65–7.67 summarise the various design approaches.

Hazard reduction methods can be adopted in cases where the number of loose blocks to be individually stabilised is too large.

Summary of stabilisation methods and remedial measures

Table 7.9 summarises the various stabilisation methods.

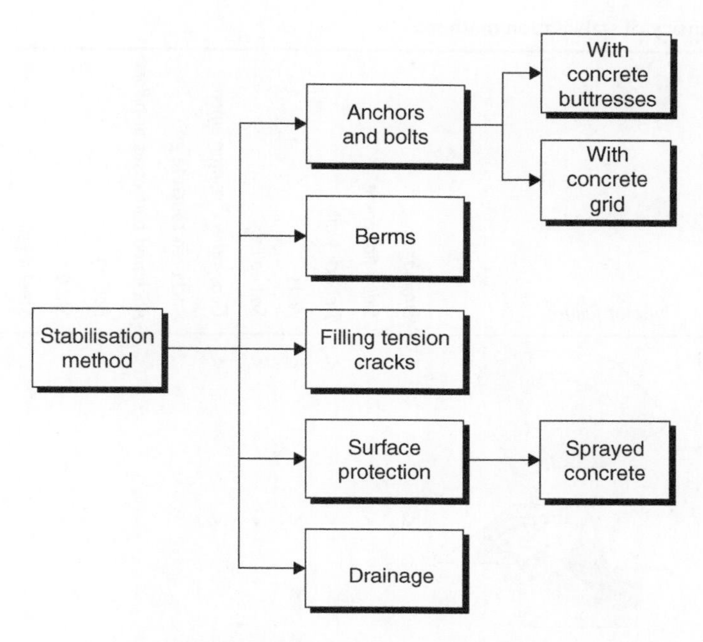

Figure 7.66 Rock slope stabilisation

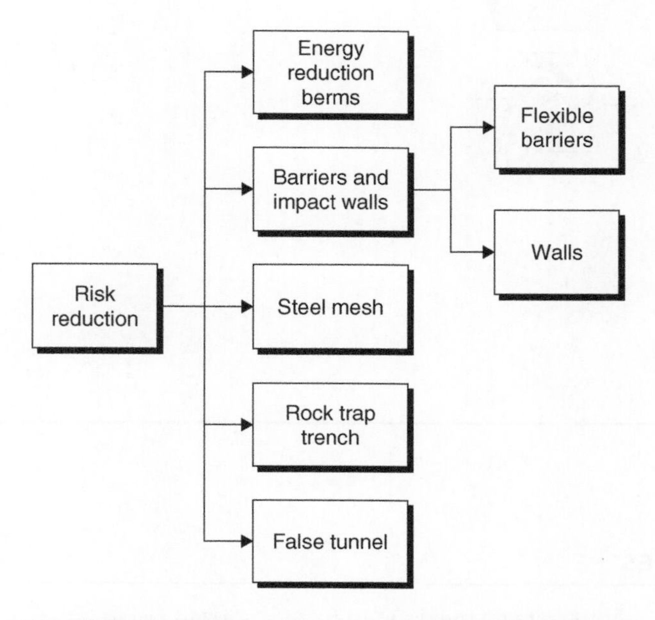

Figure 7.67 Hazard reduction methods, without full stabilisation

Table 7.9 Summary of stabilisation methods

Material	*Type of failure*	*Drainage*	*Slope flattening*	*Tieback walls*	*Walls*	*Soil nailing*	*Geosynthetic reinforcement*	*Anchored concrete grid*	*Anchored buttresses or pillars*	*Bolting*	*Shotcrete*	*Steel mesh*	*Removal of rock blocks*	*Flexible barriers*	*Impact wall*
Soil or severely jointed rock		√	√	√	√	√	√								
		√	√	√	√	√	√								
Rock		√						√	√	√	√				
		√						√	√	√					
									√	√	√	√	√	√	√

REFERENCES

Anderson, S.A. and Zhu, J.H. (1996). Assessing the stability of a tropical residual soil slope. In: Senneset, K. (ed.), *Proc. 7th International Symposium on Landslides*. Balkema, Rotterdam, Netherlands, pp. 1073–1077.

Au, S.W.C. (1993). Rainfall and slope failure in Hong Kong. *Engineering Geology*, 36: 141–147.

Barata, F.E. (1969). Landslides in the tropical region of Rio de Janeiro. *Proc. 7th International Conference on Soil Mechanics and Foundation Engineering*, Mexico City, vol. 2, pp. 507–516.

Brand, E.W. (1984). Landslides in Southeast Asia: a state-of-the-art report. *Proc. 4th International Symposium on Landslides*, vol. 1. Canadian Geotechnical Society, Toronto, pp. 17–59.

Brand, E.W. (1992). Slope instability in tropical areas. In: Bell, D.H., (ed.), *Proc. 6th International Symposium on Landslides*. Balkema, Rotterdam, Netherlands, pp. 2031–2051.

Brand, E.W., Premchitt, J. and Phillipson, H.B. (1984). Relationship between rainfall and landslides in Hong Kong. *Proc. 4th International Symposium on Landslides*, vol. 1. Canadian Geotechnical Society, Toronto, pp. 377–384.

Campos, L.E.P., De Sousa Menezes, M.S. and Fonseca, E.C. (1988). Parameter selection for stability analysis. *Proc. 2nd International Conference on Geomechanics in Tropical Soils*. Publications Committee of 2 ICOTS (ed.), Balkema, Rotterdam, Netherlands, pp. 241–243.

Chatterjea, K. (1989). *Observation on the Fluvial And Slope Processes in Singapore and Their Impact on the Urban Environment*. PhD Thesis, National Univ. of Singapore, Singapore.

Cheng, P.F.K. (1997). Soil infiltration prediction and its significance in slope stability. In: Tan *et al.*, (eds), *Proc. 3rd Asian Young Geotechnical Engineers Conference*, Singapore, pp. 661–669.

Ching, R.K.H., Sweeney, D.J. and Fredlund, D.G. (1984). Increase in factor of safety due to soil suction for two Hong Kong slopes. *Proc. 4th International Symposium on Landslides*, Toronto, pp. 617–623.

Chowdhury, R.N. (1980). A reassessment of limit equilibrium concepts in geotechnique. *Proc. Symposium on Applications in Geotechnical Engineering*. ASCE 1980 Annual Convention, Hollywood, Florida.

Crozier, M.J. and Eyles, R.J. (1980). Assessing the probability of rapid mass movement. *Proc. 3rd Australia New Zealand Conference on Geomechanics*, Wellington, vol. 2, pp. 47–51.

Cruden, D.M. and Varnes, D.J. (1996). Landslide types and processes, chapter 3 in: *Landslides Investigation and Mitigation*, TRB Transportation Research Board, Special Report 247, pp. 36–75.

Eyles, R.J. (1979.) Slip-triggering rainfalls in Wellington City, New Zealand, *New Zealand Journal of Science*, 22: 117–121.

Fredlund, D.G., Morgenstern, N.R. and Wiger, R.A. (1978). The shear strength of unsaturated soils. *Canadian Geotechnical Journal*, 15: 313–321.

Fredlund, D.G. and Krahn, J. (1977). Comparison of slope stability methods analysis. *Canadian Geotechnical Journal*, 14(3): 429–439.

Fredlund, D.G., Krahn, J. and Pufahl, D. (1981). The relationship between limit equilibrium slope stability methods. *Proc. 10th International Conference on Soil Mechanics and Foundation Engineering*, Stockholm, pp. 409–416.

Fredlund, D.G. and Rahardjo, H. (1993). *Soil Mechanics for Unsaturated Soils*. Wiley, New York.

Fredlund, D.G., Rahardjo, H. and Gan, J.K.M. (1978). Non-linearity of strength envelope for unsaturated soils. *Proc. 6th International Conference on Expansive Soils*, New Delhi, India, pp. 49–54.

Fukuoka, M. (1980). Landslides associated with rainfall. *Geotechnical Engineering, Journal of Southeast Asian Geotechnical Society*, 11: 1–29.

Gasmo, J., Hritzuk, K.J., Rahardjo, H. and Leong, E.C. (1999). Instrumentation of an unsaturated residual soil slope. *Geotechnical Testing Journal*, 22(2): 128–137.

Gasmo, J.M. (1997). *Stability of Residual Soil Slopes as Affected by Rainfall*. M.Eng. Thesis. School of Civil and Structural Engineering. Nanyang Technological University, Singapore.

GEO. (1984). *Geotechnical Manual for Slopes*, Geotechnical Engineering Office, Hong Kong, reprinted 2000. [Online] Available from: http://www.cedd.gov.hk/eng/publications.

Guidicini, G. and Nieble, C.M. (1984). *Estabilidade de Taludes Naturais e de Escavação*. Ed. Edgard Blücher Ltda., São Paulo, Brazil.

Hoek, E. (1998). *Rock Engineering – The Application of Modern Techniques to Underground Design*, Course Notes by E Hoek, Brazilian Rock Mechanics Committee, Brazilian Society for Soil Mechanics and Geotechnical Engineering, São Paulo, Brazil.

Hoek, E., Kaiser, P.K. and Bawden, W.F. (1995). *Support of Underground Excavations in Hard Rock*. Balkema, Rotterdam.

Hoek, E. and Bray, J.W. (1981). *Rock Slope Engineering*, 3rd edn, Institution of Mining and Metallurgy, London.

Hutchinson, J.N. (1988). Morphological and geotechnical parameters of landslides in relation to geology and hydrogeology, General Report. *Proc. Int. Symp. Landslides*, Lausanne, Switzerland, vol. 1, pp. 3–35.

Kanji, M.A. (2004). Geology factors in slope stability, chapter 1 in: Ortigao, J.A.R. and Sayao, A.S.F.J. (eds) *Handbook of Slope Stabilisation*, Springer Verlag, Heidelberg. [Online] Available from: www.springeronline.com

Kanji, M.A., Cruz, P.T., Massad, F. and Araujo, H.A. (1997). Basic and common characteristics of debris flows. *Proc. 2nd PSL PanAm. Symp. on Landslides* , Rio de Janeiro, vol. 1, pp. 232–240.

Kanji, M.A., Gramani, M.F., Massad, F., Cruz, P.T. and Araujo, H.A. (2000). Main factors intervening in the risk assessment of debris flows. *Int. Workshop on the Debris Flow Disaster of Dec. 1999 in Venezuela*, Caracas-Venezuela, CD-ROM, p. 10.

Kay, J.N. and Chen, T. (1995). Rainfall landslide relationship for Hong Kong. *Proc. Institution of Civil Engineers, Geotechnical Engineering*, 113(2): 117–118.

Kjaernsli, B. and Simons, N. (1962). Stability investigations of the north bank of the Drammen River. *Géotechnique*, 12: 147–167.

Krisdani, H. (2006). *The Use of Residual Soil and Geosynthetic Material in Capillary Barrier System*. PhD thesis, School of Civil and Environmental Engineering, Nanyang Technological University, Singapore.

Lim, T.T., Rahardjo, H., Chang, M.F. and Fredlund, D.G. (1996). Effect of rainfall on matric suction in a residual soil slope. *Canadian Geotechnical Journal*, 33: 618–628.

Lumb, P. (1962). Effect of rain storms on slope stability. *Proc. Symposium on Hong Kong Soils*. Hong Kong, pp. 73–87.

Lumb, P. (1975). Slope failures in Hong Kong. *Quarterly Journal of Engineering Geology*, 8: 31–65.

Massad, F., Cruz, P.T., Kanji, M.A. and Araújo-Filho, H.A. (2000). Characteristics and volume of sediment transport in debris flows in Serra do Mar, Cubatão, Brazil, *Int. Workshop on the Debris Flow Disaster of Dec. 1999 in Venezuela*, Caracas, JIFI/2000 (CD-ROM).

Morgenstern, N.R. (1992). The evaluation of slope stability-A 25 Year perspective. *Proc. Stability and Performance of Slopes and Embankments*-II. vol. 1, Berkeley, California, pp. 1–26.

Morgenstern, N.R. and Price, V.E. (1965). The analysis of the stability of general slip surfaces. *Géotechnique*, 15: 79–93.

Morris, C.E. and Stormont, J.C. (1998). Evaluation of numerical simulations of capillary barrier field tests. *Geotechnical and Geological Engineering*, 16: 201–213.

Murray, L.M. and Olsen, M.T. (1988). Colluvial slopes – A geotechnical and climatic study. *Proc. 2nd International Conference on Geomechanics in Tropical Soils*, Singapore, vol. 2: pp. 573–579.

Ortigao, J.A.R. and Sayao, A.S.F.J. (2004). *Handbook of Slope Stabilisation*. Springer Verlag, Heidelberg. [Online] Available from: www.springeronline.com.

Pitts, J. (1985). An Investigation of Slope Stability on the NTI Campus, Singapore, *Applied Research Project Report-RPI/83*. Nanyang Technological Institute: Singapore; p. 54.

Rahardjo, H. Krisdani, H., Leong, E.C., Ng, Y.S., Foo, M.D. and Wang, C.L. (2007a). Unsaturated soil mechanics for solving seepage and slope stability problems. *Proc. 12th International Colloquium on Structural and Geotechnical Engineering*. 10–12 Dec. Cairo, Egypt. pp. KNL003/1-38.

Rahardjo, H. and Fredlund, D.G. (1991). Calculation procedure of slope stability analysis involving negative-pore-water pressures. *Proc. International Conference on Slope Stability Engineering Developments and Applications*, 15–18 April. The Institution of Civil Engineers. Isle of Wight, pp. 43–49.

Rahardjo, H., Hritzuk, K.J., Leong, E.C. and Rezaur, R.B. (2003). Effectiveness of horizontal drains for slope stability. *Engineering Geology*. 69: 295–308.

Rahardjo, H., Krisdani, H. and Leong, E.C. (2007b). Application of unsaturated soil mechanics in capillary barrier system. In: Yin, Z., Yuan, J., and Chiu, A.C.F (eds) *Proc. 3rd Asian Conference on Unsaturated Soils*. Nanjing, China, 21-23 April. Science Press, pp. 127–137.

Rahardjo, H., Lee, T.T., Leong, E.C. and Rezaur, R.B. (2005). Response of a residual soil slope to rainfall. *Canadian Geotechnical Journal*, 42: 340–351.

Rahardjo, H., Leong, E.C. and Rezaur, R.B. (2008a). Effect of antecedent rainfall on pore-water pressure distribution characteristics in residual soil slopes under tropical rainfall. *Hydrological Processes*, 22: 506–523.

Rahardjo, H., Li, X.E., Toll, D.G. and Leong, E.C. (2001). The effect of antecedent rainfall on slope stability. *Geotechnical and Geological Engineering*, 19: 371–399.

Rahardjo, H., Lim, T.T., Chang, M.F. and Fredlund, D.G. (1995). Shear-strength characteristics of a residual soil. *Canadian Geotechnical Journal*, 32: 60–77.

Rahardjo, H., Ong, T.H., Rezaur, R.B., and Leong, E.C. (2007c). Factors controlling instability of homogeneous soil slopes under rainfall. *Journal of Geotechnical and Geoenvironmental Engineering*, 133(12): 1532–1543.

Rahardjo, H., Rezaur, R.B. and Leong, E.C. (2008b). Role of antecedent rainfall in slope stability. *Proc. Geotropica*, Kuala Lumpur, Malaysia 26–27 May 2008.

Rezaur, R.B., Rahardjo, H., Leong, E.C. and Lee, T.T. (2003). Hydrologic behaviour of residual soil slopes in Singapore. *Journal of Hydrologic Engineering*, 8(3): 133–144.

Ross, B. (1990). The diversion capacity of capillary barriers. *Water Resource Research*, 26(10): 2625–2629.

Royster, D.L. (1980). Horizontal drains and horizontal drilling: An overview. *Transportation Research Record*, 783: 16–25.

Sayao, A.S.J.F. (2004). Soil slope stability, chapter 5 in: Ortigao J.A.R. & Sayao A.S.F.J. (eds) *Handbook of Slope Stabilisation*, Springer Verlag, Heidelberg. [Online] Available from: www.springeronline.com

Sevaldson, R.A. (1956). The slide at Lodalen, Oct. 6th 1954. *Géotechnique*, 6: 167–182.

Skempton, R.W. and Hutchison, J. (1969). Stability of natural slopes and embankment foundations. State-of-the-Art Report. *Proc. 7th International Conference on Soil Mechanics and Foundation Engineering*. Mexico City. State-of-the-Art vol. pp. 291–340.

Spencer, E. (1967). A method of analysis of the stability of embankments assuming parallel interslice forces. *Géotechnique*, 17: 11–26.

Spencer, E. (1968). Effect of tension on stability of embankments. ASCE *Journal of the Soil Mechanics and Foundation Engineering Division*, 94(SM5): 1159–1173.

Spencer, E. (1973). Thrust line criterion in embankment stability analysis. *Géotechnique*, 23: 85–100.

Stormont, J.C. (1997). Incorporating capillary barriers in surface cover systems. *Proc. Landfill Capping in the Semi-Arid West: Problems, Perspectives, and Solutions*. Environmental Science and Research Foundation, Idaho Falls, Idaho, pp. 39–51.

Tami, D. (2004). *Mechanism of Sloping Capillary Barriers Under High Rainfall Conditions*. PhD Thesis, School of Civil and Environmental Engineering, Nanyang Technological University, Singapore.

Tami, D., Rahardjo, H., Leong, E.C. and Fredlund, D.G. (2004a). Design and laboratory verification of a physical model of sloping capillary barrier. *Canadian Geotechnical Journal*, 41: 814–830.

Tami, D., Rahardjo, H., Leong, E.C. and Fredlund, D.G. (2004b). A physical model of sloping capillary barrier. *Geotechnical Testing Journal*, 27(2): 1–11.

Tan, S.B., Tan, S.L., Lim, T.L. and Yang, K.S. (1987). Landslide problems and their control in Singapore. *Proc. 9th Southeast Asian Geotechnical Conference*, vol. 1. Bangkok, Thailand, pp. 25–36.

Varnes D.J. (1978). Slope movements types and processes, *Special Report 176 Landslides Analysis and Control*, TRB Transportation Research Board, Washington DC.

Wei, J., Heng, Y.S., Chow, W.C. and Chong, M.K. (1991). Landslide at Bukit Batok sports complex. *Proc. 9th Asian Conference on Soil Mechanics and Foundation Engineering* Bangkok, Thailand, vol. 1. Balkema: Rotterdam; pp. 445–448.

Yanful, E.K., Riley, M.D. and Woyshner, M.R. (1993). Construction and monitoring of a composite soil cover on an experimental waste rock pile near Newcastle, New Brunswick, Canada. *Canadian Geotechnical Journal*, 30: 588–599.

Yanful, E.K., Simms, P.H., Rowe, R.K. and Stratford, G. (1999). Monitoring an experimental soil waste cover near London, Ontario, Canada. *Geotechnical and Geological Engineering*, 17: 65–84.

8

Foundations: Shallow and deep foundations, unsaturated conditions, heave and collapse, monitoring and proof testing

A. Viana da Fonseca

Department of Civil Engineering, Faculty of Engineering, University of Porto, Porto, Portugal

S. Buttling

Senior Principal Geotechnical Engineer, GHD, Brisbane, Australia

R.Q. Coutinho

Department of Civil Engineering, Federal University of Pernambuco, Recife, Brazil

8.1 INTRODUCTION

Many aspects of foundation design and construction in tropical soils are the same as those in sedimented soils, about which many text books have already been written. We have therefore tried to emphasise those aspects of foundation engineering on tropical soils which are unconventional, based on our combined experiences in many regions of the world where tropical residual soils exist.

In many parts of the engineering world, limit state design methods are now being used for geotechnical designs. In some places, for example Australia, they have been around for 15 years already. In others, such as Europe, they have only recently been implemented. In the past, when considering shallow foundations on cohesive soils, an adequate factor of safety (probably 3) on ultimate bearing capacity failure was considered enough to also provide a limit on settlement. On cohesionless soils, the ultimate bearing capacities were generally well in excess of what might be required, and allowable settlement would control the design. Hence, the charts produced by Terzaghi which showed the bearing pressure that would lead to 25 mm (1 inch) of settlement for a range of soil densities (SPT N values). Now it is again found that, in many circumstances, the ultimate limit state can be designed for without difficulty, and it is the serviceability limit state which governs. To adequately satisfy this limit state requires a reasonable knowledge of foundation performance, and particularly foundation stiffness. Much of this chapter is devoted to discussing methods of predicting foundation behaviour.

8.2 DIRECT (SHALLOW) FOUNDATIONS

8.2.1 Solutions to foundations on residual soils – factors that affect the concept

Foundations in residual soils might be considered as one more aspect of the broad range of foundation engineering. However, what makes residual soils special is that they contain the characteristics of all the main soil groups (fine or coarse materials, cohesive or granular soils etc.), and they fit in between soils and rock masses in what can be classified as "Intermediate Geotechnical Material" (IGM), see Figure 8.1. As a result, foundation performance can be very variable and designs can often be based on adaptations of designs suited to either soils or rocks. Regional practice and local experience can help to find satisfactory solutions, generally empirical, but they are no substitute for detailed soil investigation of heterogeneities, and monitoring of foundation construction (e.g. by automated pile construction monitoring).

8.2.2 Particular conditions in residual soils

Microstructure, non-linear stiffness, small and large strain anisotropy, weathering and its effects on structure, consolidation characteristics and strain rate dependency

Figure 8.1 Uncertainties due to heterogeneities of residual profiles (*After* Milititsky *et al.*, 2005)

(Schnaid, 2005) are all very important factors in assessing the mechanical characteristics of natural soils, therefore new techniques of measuring soil properties or, better still, new interpretation methods (Viana da Fonseca and Coutinho, 2008) are required. Most of the soils are unsaturated, and need to be dealt with carefully. Since bonded geomaterials, such as residual soils, are highly variable, the interpretation of their mechanical behaviour is complex. Because of the variability, one suitable solution involves cross-correlation of multiple measurements from different tests, but it is even better to have more measurements in one test.

Igneous rocks, like granite, are composed mainly of quartz, feldspar and mica. Quartz is resistant to chemical decomposition, while feldspar and mica are transformed mainly into clay minerals during the weathering process. The effects of temperature, drainage and topography have reduced the rocks in place to residual soils that range from clays to sandy silts and silty sands, grading with depth into saprolite and partially weathered rocks. As weathering proceeds, the reduction in vertical stress as a result of the removal of overburden accelerates the rate of exfoliation (stress release jointing) and the alternate wetting and drying processes in the underlying fresh rock (Viana da Fonseca and Coutinho, 2008). These processes increase the surface area of rock on which weathering can proceed, which leads to deeper weathering profiles (Irfan, 1988; Ng and Leung, 2007b).

Depending on the degree of alteration, some residual soils lose all the features of the parent rock, while others, as illustrated in Figure 8.2, have clear relict structure (Rocha Filho, 1986; Costa Filho *et al.*, 1989; Viana da Fonseca, 2003). Examples of relict structures include evidence of bonding or dissolved bond features, as well as cracks and fissures from the original fractured rock mass (Mayne and Brown, 2003).

Degree of weathering: topographic complexities and characteristics of profiles

Looking at the weathering profile from the bottom upwards, one can find materials grading all the way from fresh rock, through slightly and moderately weathered rock, to soil which retains the characteristics of rock (called young residual or saprolitic soil), to the upper horizon where no remaining rock characteristics can be seen (known as mature residual or lateritic soil). The upper layers may be mixed with transported soil,

Figure 8.2 Parent rock with potentially unstable weak features (Porto, Portugal)

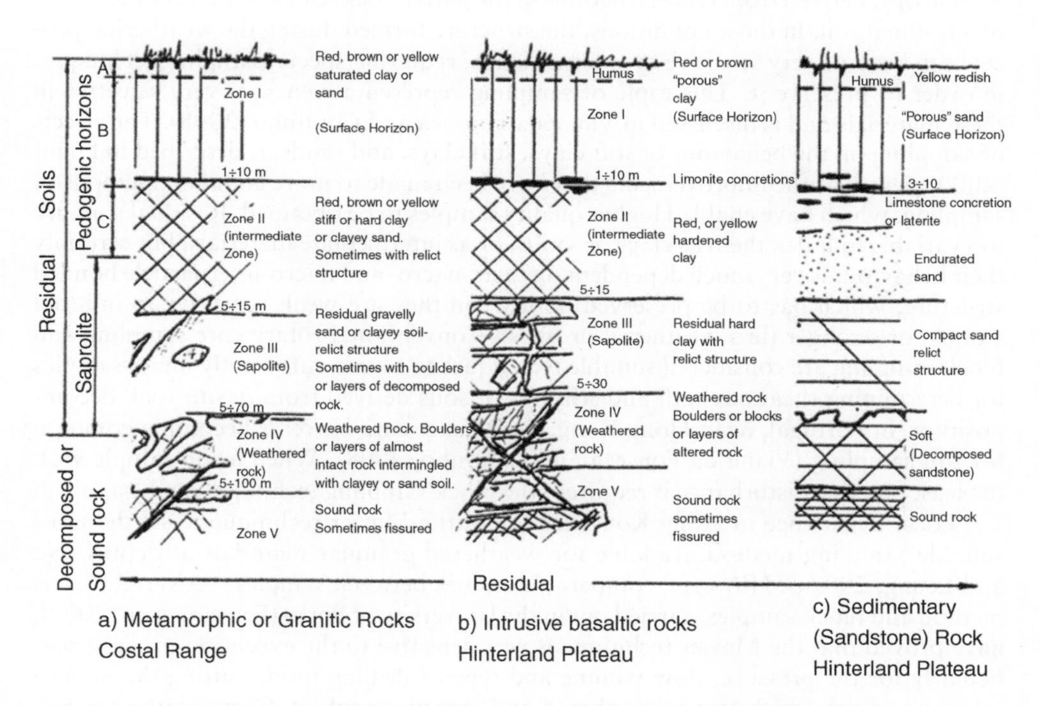

Figure 8.3 Typical profiles of Brazilian residual soils – some zones may be absent (*After* Vargas, 1985)

such as colluvium, which may be difficult to distinguish from the true residual soils. Figure 8.3 presents a proposal for the classification of such profiles. Unfortunately, the sequence is exactly the opposite of that of weathering grades proposed by Dearman (1976), in which Grade I represents Fresh Rock, and Grade VI Residual Soil. Therefore, the term Zone will be used as in the diagram in Figure 8.3.

Zone II (lateritic soil) is usually formed under hot and humid conditions involving high permeability profiles, resulting in a bond structure with high contents of oxides and hydroxides of iron and aluminium (laterisation). Not all soils belonging to this horizon develop enough pedogenetic evolution for laterisation. The Zone III (saprolitic soil) can show a high level of heterogeneity both vertically and laterally as well as a complex structural arrangement which retains the characteristics of the parent rock. The texture and mineralogy of these soils can vary considerably with the degree of weathering and leaching. In a tropical region the weathering profile often shows a very narrow or inexistent Zone V, while in temperate climates this zone can be reasonably thick.

Mitchell and Coutinho (1991), Lacerda and Almeida (1995), and Clayton and Serratrice (1998) each present a general view of many soils which are called unusual soils (Schnaid *et al.*, 2004; Coutinho *et al.*, 2004a), including bonded soils, granitic saprolitic soil and lateritic soils, and unsaturated collapsible soils. Bonding and structure are important components of shear strength in residual soils, since they have a major impact on the cohesive-frictional response (characterised by c' and ϕ'). Anisotropy, derived from relict structures of the parent rock, can also be a characteristic of a residual soil. In those conditions, the structure formed during the weathering process can become very sensitive to external loads, requiring special sampling techniques in order to preserve it. This topic of sampling representativeness is very sensitive in these materials and is discussed in Viana da Fonseca and Coutinho (2008). The effects of sampling on the behaviour of soft clays, stiff clays, and sands, is described in Hight (2000) as well as the improvements that have been made to more common methods of sampling, which have enabled higher quality samples to be obtained. Residual soils are too variable to index them as clays or sands, or as intermediate materials, but certainly their behaviour is very much dependent on their macro- and micro-interparticle bonded structure, which has to be preserved both when they are weak (sensitive to induced strains) or stronger (less weathered profiles). Conventional rotary core sampling and block sampling are considered suitable techniques to obtain sufficiently intact samples for determining shear strength and stiffness of soils derived from *in situ* rock decomposition. In Portugal, as in Hong Kong, the Mazier core-barrel is becoming common for soil sampling (Viana da Fonseca and Coutinho, 2008). When a soil sample with the least possible disturbance is required, the block sampling technique can be applied.

Local experience in Hong Kong has found the Mazier technique to be the most suitable sampling method available for weathered granular materials at depths (Ng and Leung, 2007a). However, comparable studies between samples recovered by this method and block samples, carried out in the University of Porto (Ferreira *et al.*, 2002), have proved that the Mazier technique is very sensitive to the execution process, particularly for the pressure, flow volume and type of drilling fluid, putting the natural structure at risk for the more weathered and granular profiles. Some results are presented in Figure 8.4. In Ng and Leung (2007a), and again in Ng and Leung (2007b), the authors note that the shear-wave velocities of the block specimens were an average 27% higher than those of the Mazier specimens.

Usually, the void ratio and density of a residual soil are not directly related to its stress history, unlike the case of sedimentary clayey soils (Vaughan, 1985; Vaughan *et al.*, 1988; GSEGWP, 1990; Viana da Fonseca, 2003). The presence of some kind of bonding, even weak, usually implies the existence of a peak shear strength envelope, showing a cohesion intercept and a yield stress which marks a discontinuity in stress-strain behaviour. The structure in natural soil has two "facies": the "fabric"

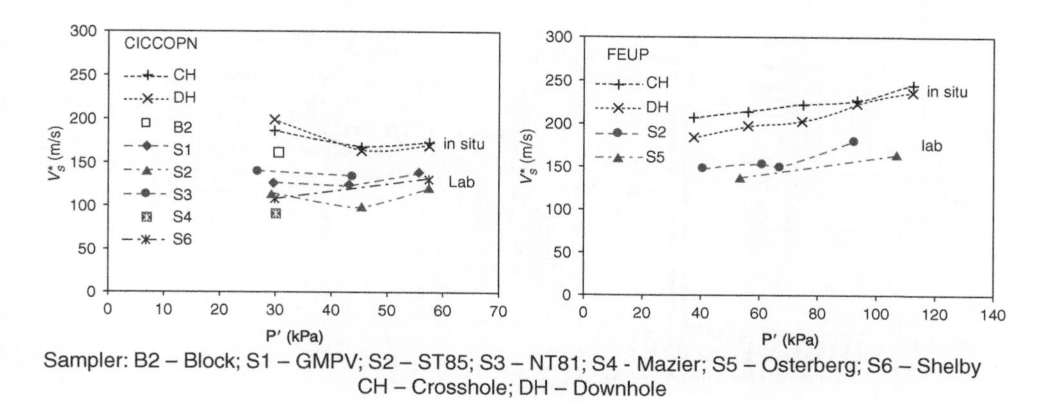

Figure 8.4 Normalised shear wave velocities, for different types of samplers used in residual soil from granite in two experimental sites in Porto (*After* Ferreira *et al.*, 2004)

that represents the spatial arrangement of soil particles and inter-particle contacts, and "bonding" between particles, which can be progressively destroyed during plastic straining (leading to the term "destructuration"). Most, if not all, geomaterials are structured, but the mechanical response of naturally bonded soils is dominated by this effect (Leroueil and Vaughan, 1990). Here the cohesive component due to cementation can dominate soil shear strength, especially in engineering applications involving low stress levels (Schnaid, 2005) or in specific stress-paths where this component is relevant, such as cut slopes.

Trying not to make the wrong choice of foundation type

Residual masses generally exhibit stronger heterogeneity than deposited soils, changing their characteristics gradually both laterally and vertically, especially with regard to their mechanical properties. As a result, it is common to have to adopt different foundation types, such as shallow foundations (footings and mats) and deep foundations (piles), within very limited areas, depending upon the consistency of the overburden soils and the depth to parent rock (Figure 8.5). An accurate mapping of the spatial variability of the mechanical properties, essential for geotechnical design, is very challenging although the situation has improved recently by the use of geophysical methods (Viana da Fonseca *et al.*, 2006). Several *in situ* testing methodologies, such as Standard Penetration Testing (SPT), Cone Penetrometer Testing (CPT), Dilatometer Testing (DMT), Pressuremeter Testing (PMT) and Self Boring Pressuremeter Testing (SBPT), and geophysical survey, surface and borehole seismic tests, electrical resistivity and Ground Penetrating Radar (GPR), have been used to assess the mechanical properties of these particular soils, with varying degrees of success (Figure 8.6).

The water table is in many cases deep in the profile; hence there is a significant layer of unsaturated soil. In this case, the role of matrix suction and its effect on soil behaviour has to be recognised and considered in the interpretation of *in situ* tests. The main difference between saturated soils and unsaturated soils is the existence of

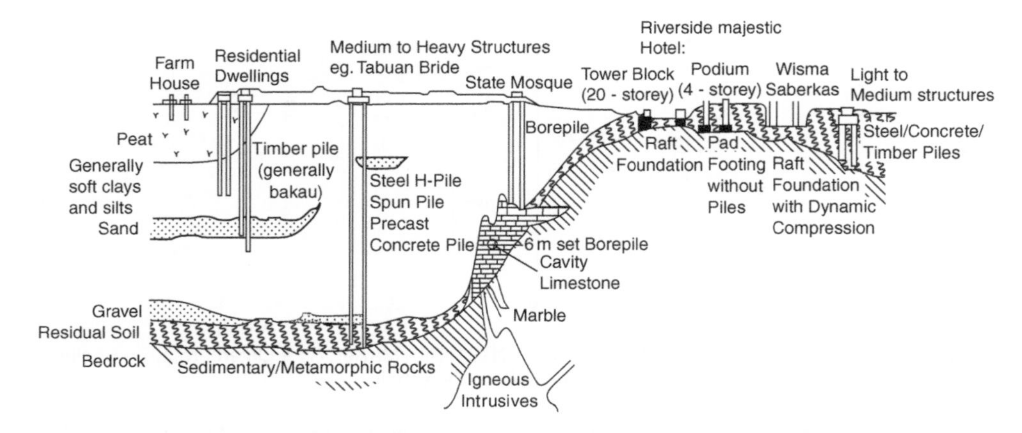

Figure 8.5 Different foundations and relationships to the rockline (*After* Yogeswaran, 1995; Singh *et al.*, 2006)

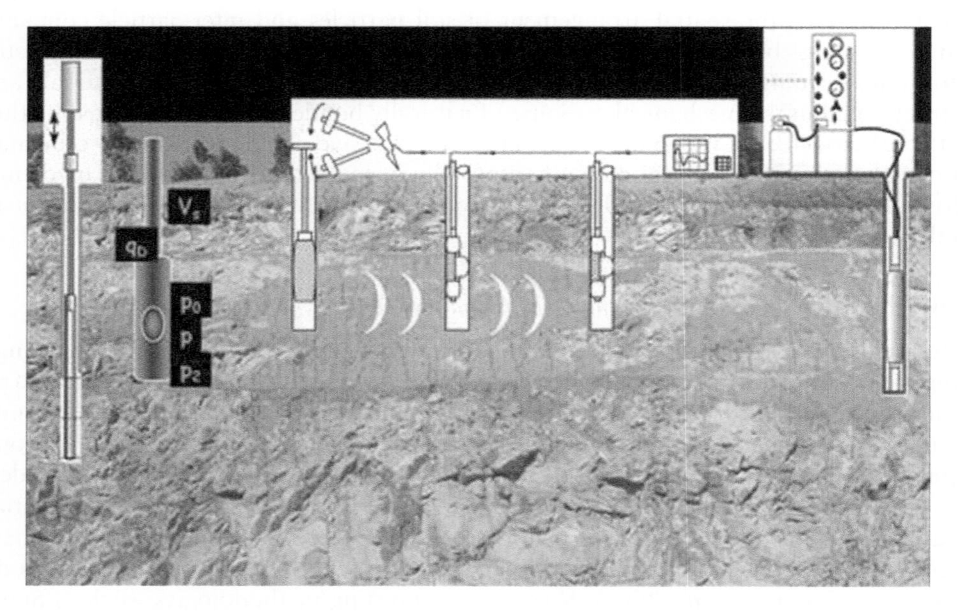

Figure 8.6 Multiple solutions for *in situ* testing of characterisation in complex and highly heterogeneous weathered rocks and/or residual soils (*After* Viana da Fonseca, 2006)

negative porewater pressures, largely known as suction, which tends to increase the effective stresses and hence the strength and stiffness of the soil.

Residual soils derived from a wide variety of parent rocks can also be collapsible, which has serious consequences for foundation behaviour. Collapsible residual soils have a metastable state characterised by a honeycomb structure and partially saturated condition that can develop after a parent rock has been thoroughly decomposed or

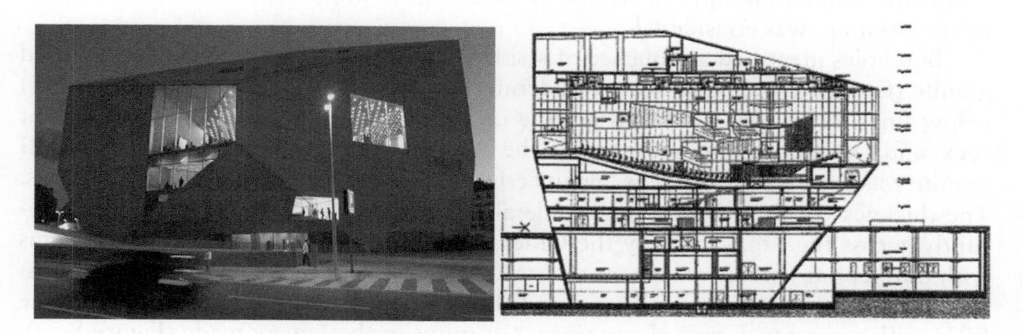

Figure 8.7 View and section through building and car park (*After* Gaba *et al.*, 2004)

while the decomposition is continuing (Vargas, 1971). Commonly, collapsible residual soils form under conditions of heavy concentrated rainfalls in short periods of time, followed by long dry periods, high temperature, high evaporation rates, and flat slopes, so that leaching of material can occur. There are two mechanisms of bonding in the metastable soil structure: soil water suction and cementation by clay or other types of fine particles. Collapsible residual soils usually have low activity and low plasticity. Colluvial deposits (or mature residual soils) can become collapsible in environments where the climate is characterised by alternating wet and dry seasons that cause a continuous process of leaching of the soluble salts and colloidal particles much like residuals soils (Mitchell and Coutinho, 1991).

8.2.3 Main demands for the guarantee of structural limit state conditions

Differential settlements caused by heterogeneity in plan and depth

Foundation displacements can be considered both in terms of displacement of the entire foundation, and differential displacements of parts of the foundation. As well stated in structural codes such as EN1997-1, to ensure the avoidance of a serviceability limit state, assessment of differential settlements and relative rotations shall take account of both the distribution of loads and the possible variability of the ground, unless they are prevented by the stiffness of the structure. Because of the heterogeneity inherent in weathered rock masses and residual soils,as discussed above, where the state of weathering, decomposition and fracturing of the rock may vary considerably in depth and plan, this conditions the design of spread foundations or other mixed solutions.

Foundations for special projects, such as that described by Gaba *et al.* (2004), have to be developed in a way that they ensure strict settlement control. In this case history, the ground investigation, interpretation and foundation design and construction for the Casa da Música do Porto, Portugal, is described. The overall project comprised detailed multi-disciplinary design of the structure, foundations and building services for a 20,000 m^2, high quality concert hall (Figure 8.7).

The ground investigation included boreholes, with SPT and dynamic probing (DPSH and jet grout probing) correlated to be used in the foundation design.

A shallow foundation option, with a methodological improvement of specific zones by jet-grouting, was considered.

Boreholes identified a sequence of residual soil overlying completely decomposed granite (Grade V) over highly and moderately weathered granites (Grade IV and III) below final formation. Grade III granite or better was considered to be competent rock, and a suitable founding material. The boundary between Grade IV and Grade III granite was therefore selected as "engineering rock head" (ERH) for design purposes. The thickness of each weathering grade above this level was found to vary significantly across the site, especially the Grade V granite, presenting potential problems with differential settlements across the structure. Probing was therefore carried out to investigate this variability in more detail. In order to determine the elevation of the ERH with a greater degree of certainty, "jet grout probes" were made (Figure 8.8a). This probing was carried out by drilling through the residual soil and weathered rock using a jet-grouting rig until it met refusal. This technique was chosen because of the ready availability of appropriate plant on site and the relative speed with which the probing could be carried out in comparison with drilling boreholes. This meant that a large number of locations could be probed in a short period of time and the ERH level defined in an expeditious way (Figure 8.8b). The probing was carried out along the lines of load bearing walls and at column locations within the building footprint.

The design of the building used the soil stiffness derived from the penetration tests, and the loading of each zone was modelled using the *Oasys* computer program VDISP (Oasys, 2001). From these analyses, spring stiffness values for each zone were derived and input into a structural model of the building in order to assess the bearing pressures and anticipated settlements. The foundations were represented in the structural analysis as "slab on Winkler soil" finite elements. The design criteria required limits on total and differential settlement. The structural analysis demonstrated that it was not feasible to satisfy the settlement criteria using a shallow foundation scheme, even with ground improvement. A piled solution was therefore adopted, in which end bearing and shaft friction in a socket extending 1 m below ERH provided sufficient geotechnical load capacity for the piles and a stiff loading response, and resulted in the structural capacity of the concrete being the limiting factor in pile design. An allowable working concrete stress of 5 MPa was adopted.

In order to ensure satisfactory construction of the piles, construction controls and acceptance criteria were applied. These included specifying target foundation depths and maximum pile bore penetration rates at the founding level. Integrity testing confirmed that the piles were sound and of good quality construction.

There are several other case histories reported in the literature, where the spatial development of the weathered rock and residual soils is highly irregular and erratic. In the city of Porto, the design and construction of the Metro was based on weathering grades and structural features which were used for the derivation of the design parameters (Babendererde *et al.*, 2004). The highly variable nature of the deeply weathered Porto granite posed significant challenges in the driving of around 7 km of tunnels in largely populated urban areas, involving a large number of underground stations. The change from one weathered zone to another is neither progressive nor transitional (Viana da Fonseca, 2003; Babendererde *et al.*, 2004; Marques *et al.*, 2004), moving abruptly from a fresh granitic mass to a very weathered soil-like mass. The thickness of the weathered parts varies very quickly from several meters to zero. Blocks of sound

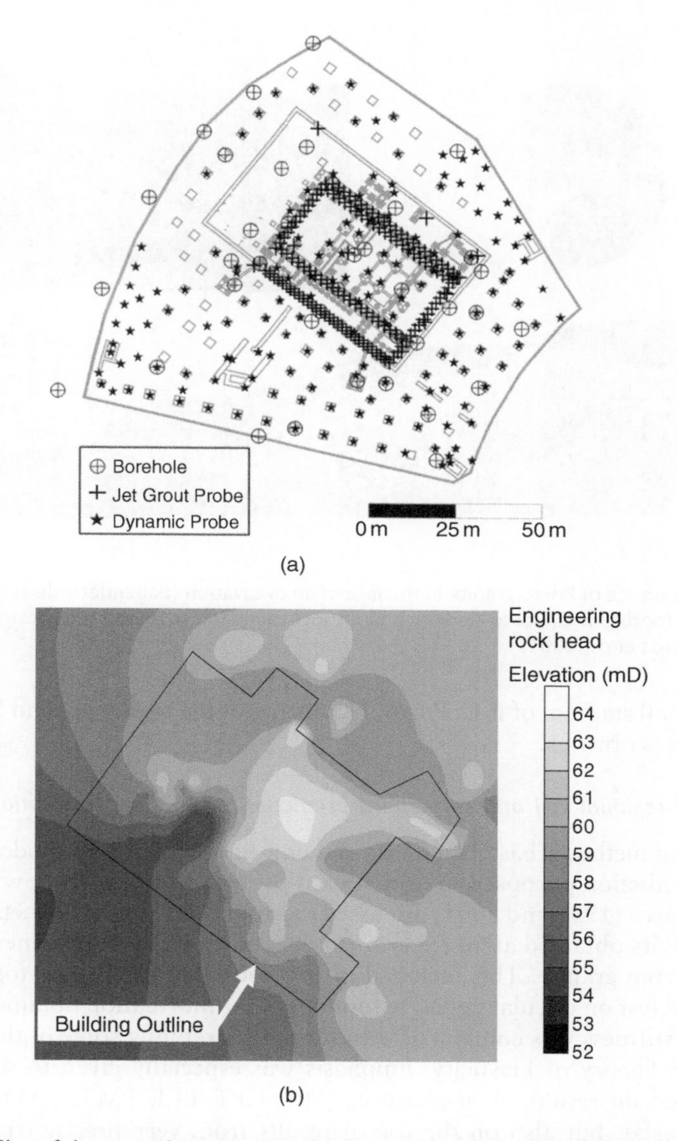

Figure 8.8 (a) Plan of the ground investigation; (b) Contours of engineering rock head within building footprint (*After* Gaba *et al.*, 2004)

rock (boulders or corestones), of various dimensions can "float" inside completely decomposed granite. Weathered material, either transported or *in situ*, also occurs in discontinuities. A particularly striking feature is that, due to the erratic weathering of the granite, weathered zones of considerable size, well beyond the size of typical "boulders", can be found under zones of sound granite. A typical case of this is illustrated in Figure 8.9 showing the appearance of Porto granite in the face of an excavation for

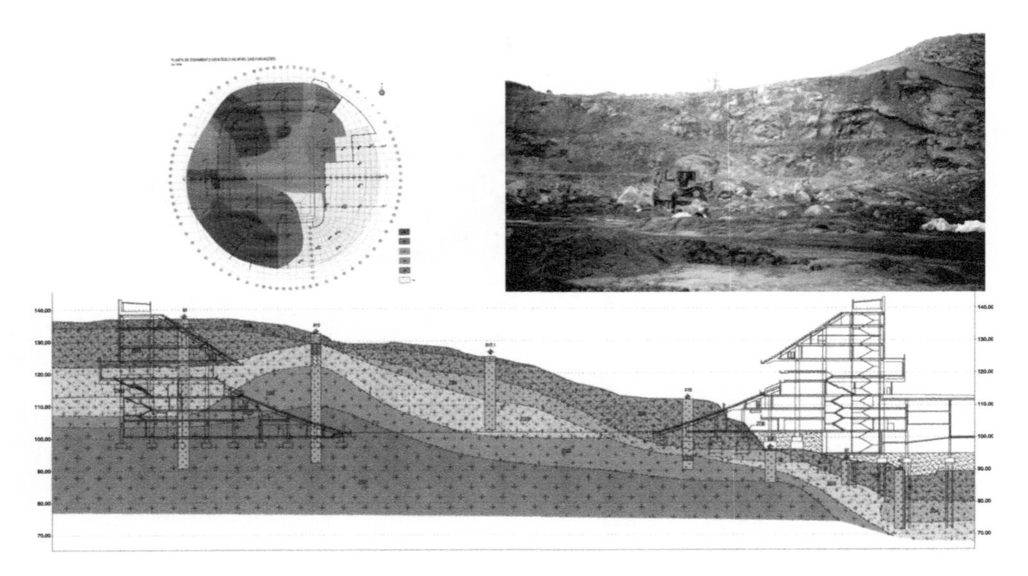

Figure 8.9 Appearance of Porto granite in the face of an excavation (Babendererde *et al.*, 2004) for the new football F. C. Porto stadium and distinct foundations solutions as designed by Campos e Matos *et al.* (2004)

the new football stadium of F. C. Porto. Fracturing of the rock mass and heterogeneity in weathering is obvious.

Load tests on residual soil and settlement prediction on shallow foundation

Semi-empirical methods, based on linear and non-linear behaviour models, mainly for settlement prediction purposes, are often used for the design of shallow foundations. Viana da Fonseca (1996 and 2001) discussed the applicability of such methods, by analyzing the results obtained at an experimental site on a fairly homogeneous saprolitic soil derived from granite. This included *in situ* and laboratory tests, together with a full-scale load test on circular concrete footings. The information obtained in terms of strength and stiffness was combined, with the aim of refining some of the approaches based on the Theory of Elasticity. Emphasis was especially given to semi-empirical methods based on results of *in situ* tests (SPT, CPT, PLT, PMT, DMT and Seismic Cross-Hole tests), but also on the use of results from very precise triaxial tests on high quality samples. Some of the well established methods (Parry, 1978; Burland and Burbidge, 1985; Anagnastopoulos *et al.*, 1991; Schmertmann *et al.*, 1978; Robertson, 1990, 1991; Ghionna *et al.*. 1991; or Wahls and Gupta, 1994 – see below) were tested and some adaptations to parameters and methods were suggested that gave a better fit to the observed behaviour.

Experimental site and analysis of the loading tests

The experimental work was carried out at a site (around 50 m × 30 m) in which a homogeneous saprolitic soil 6 m thick had been identified by a previous site investigation.

Figure 8.10 Overview of the weathered profile revealed in a fresh cutting on the experimental site

Geologically, the parent rock is representative of the granite from Porto region (Viana da Fonseca, 2003). Figure 8.10, taken at the end of the experimental investigation and prior to the construction of a new building for a District Hospital, gives a clear overview of the general saprolitic profile developing with depth.

Apart from the natural spatial variation of the relict structure and fabric of these residual soils, there is evidence of a reasonably homogeneous geotechnical profile, as revealed by results obtained from specimens taken with the SPT sampler and from blocks. A detailed description of the extensive testing programme is given in Viana da Fonseca (2003). Only the results of SPT, CPT and DMT tests and the values of the maximum shear modulus, G_0, obtained from CH tests are shown here in Figure 8.11. It is observed that the CPT cone resistance, q_c, the N_{60} from SPT and p_0 and p_1 from DMT, show a nearly linear increase with depth (or vertical effective stress, σ'_{v0}), whilst G_0 appears to be almost constant.

The loading test of a full scale circular concrete footing, with a diameter of 1.20 m and fully instrumented, was carried out (Figure 8.12a), and is described in detail in Viana da Fonseca (2001). The resulting complete pressure-settlement curve shows a clear increase of the settlement rate with load for pressure values exceeding around 125 kPa (Figure 8.12b).

The time for settlement stabilisation at each load step significantly increased above this pressure, representing a transition from an essentially elastic behaviour, confirmed by the small difference between the inclination of the first loading curve and that of the first unload-reload cycle, to a phase in which the cemented structure of the soil was substantially damaged. Serviceability limit state pressure, applying the criterion proposed by Décourt (1992), is defined by a settlement of 0.75% of the diameter

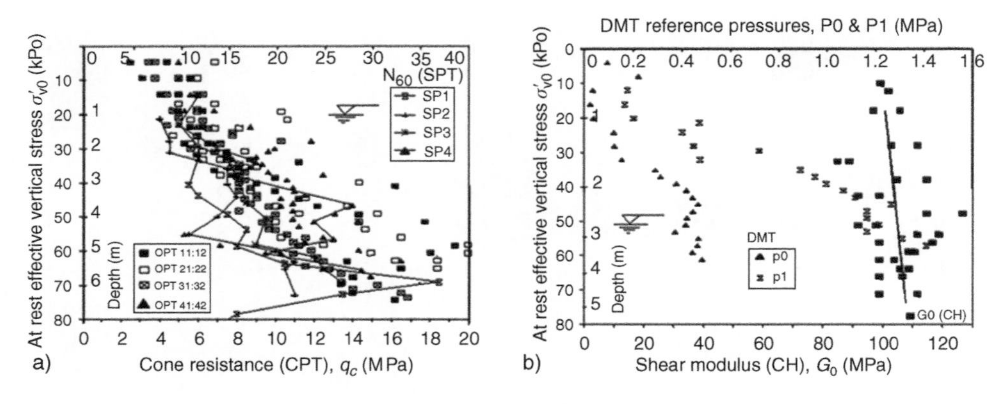

Figure 8.11 *In situ* test results: (a) SPT and CPT; (b) CH and DMT

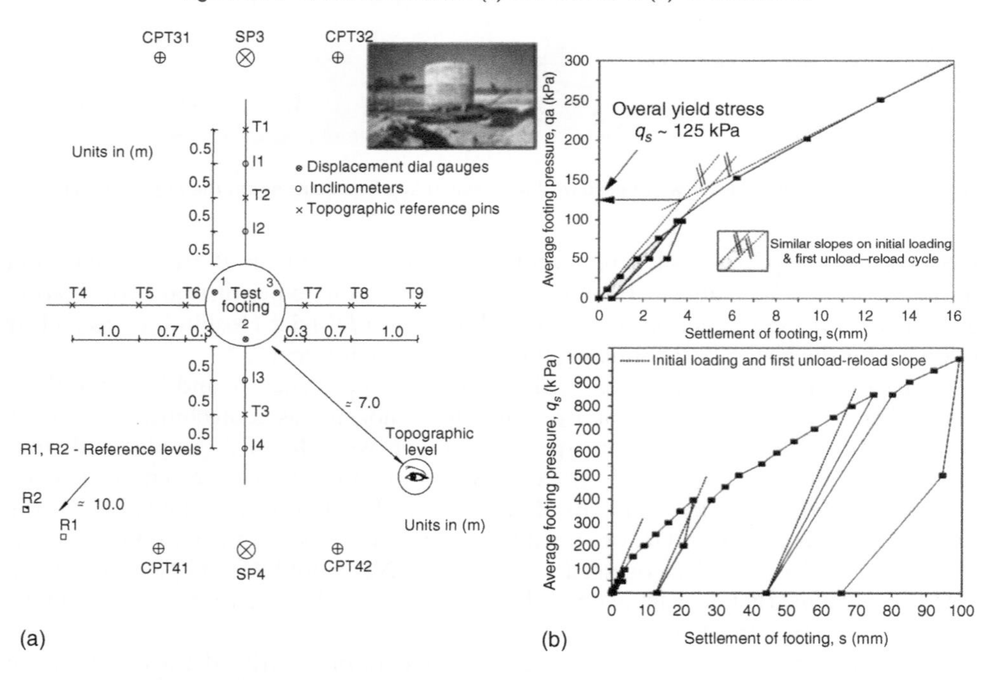

Figure 8.12 (a) Plan of the experimental area; (b) pressure-settlement curve of the footing loading test – general picture and enlargement for $q_s \leq 300$ kPa (*After* Viana da Fonseca, 2001)

of the loading surface, which corresponds to an applied pressure of about 195 kPa. Viana da Fonseca *et al.* (1997) discussed the strain distribution under the centre of the footing obtained from a simplified nonlinear elastic analysis, for applied pressures of 100, 200 and 400 kPa, corresponding to a value near to the serviceability limit state pressure, and to half and twice that value; strain values exceed 10^{-3} only in a restricted zone adjacent to the footing. These results corroborate evidence (Jardine *et al.*, 1986; Burland, 1989; Tatsuoka and Kohata, 1995) of the rather low strain

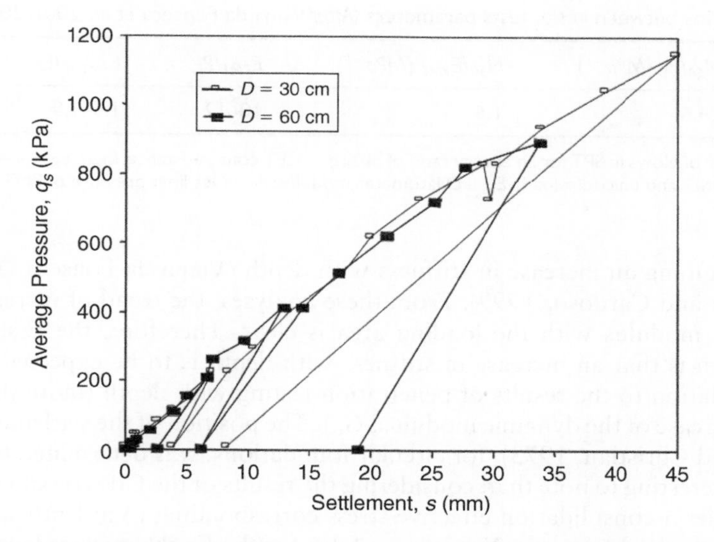

Figure 8.13 Pressure-settlement curve of the plate loading tests ($D = 30$ and 60 cm)

levels involved in a number of soil-structure interaction problems, including shallow foundations, under working conditions. The analysis of vertical displacement of the ground surface around the footing, and the horizontal displacements at the surface and in the ground in its vicinity were also discussed by Viana da Fonseca (2001). Based on the evidence presented, it can be stated that the applicability of elasticity theory to a settlement analysis under service conditions, is reasonable from a practical point of view. In residual soils, the bearing capacity is due to strength factors related to friction and cohesive components of the stressed ground. Reliable results for these can only be determined by two or more *in situ* loading tests, using different sizes and over a homogeneous space.

The execution of more than one loading test to define failure patterns has the advantage of allowing an integrated analysis of different pressure-displacement curves, enabling the importance of both stiffness and depth of influence to be studied. On the same homogeneous soil zone as the footing loading test, two more surfaces were prepared for testing smaller plates of 30 and 60 cm diameter (Figure 8.13). An analysis of punching type failures in these residual soils has been made and from those failure loads (Viana da Fonseca *et al.*, 1997), in the bearing capacity formulation (EN 1997-1), taking account of the water level position, three equations are obtained. These can be optimised to get a range for the two Mohr-Coulomb strength parameters. The derived values were: $c' \cong 7$ kPa and $\phi' \cong 37°$, revealing a fair agreement with the results obtained from extensive laboratory testing of undisturbed samples (Viana da Fonseca, 1998) and from *in situ* testing (Viana da Fonseca *et al.*, 1998).

Deformability characteristics evaluated from loading tests

The common interpretation of the results taken from loading tests on three different sized loaded areas was done by considering Young's modulus constant with depth,

Table 8.1 Ratios between *in situ* tests parameters (*After* Viana da Fonseca *et al.*, 2001, 2003, 2006)

q_c/P_L	N_{60}/P_L (MPa^{-1})	N_{60}/E_{PMT} (MPa^{-1})	E_{PMT}/P_L	E_{PMTUR}/E_{PMT}	E_D/E_{PMT}
4–6	14.6	1.4	10–12	1.4–1.9	$\cong 1.5$

N_{60} – 0 number of blows in SPT for an energy ratio of 60%; q_c – CPT cone resistance; E_{PMT}, E_{PMTUR} – pressuremeter modulus ("elastic" and unload/reload); E_D – Dilatometer modulus; P_L – Net limit pressure of PMT

and by assuming an increase in stiffness with depth (Viana da Fonseca, 1999; Viana da Fonseca and Cardoso, 1999). From these analyses, the trend of increasing values of Young's modulus with the loading area is clear. Therefore, the first conclusion to be drawn is that an increase in stiffness with depth is to be expected and has an obvious relation to the results of penetration testing with depth (more than the very smooth increase of the dynamic modulus, G_0). The position of the settlement centre, z_I (Carrier and Christian, 1973), for circular foundations, was determined to be $z_I \cong B$.

It is interesting to note that, considering the results of the CID triaxial tests (details below) under a consolidation effective stress corresponding to a depth similar to the footing diameter, the tangent Young's modulus for the K_0 shear stress level was found to be $E_{tK0} \cong 8$ MPa. The use of such a modulus in an elastic analysis of the footing loading test would lead to a crude overestimation of the observed settlement. However, if E_{tK0} is multiplied by a factor of "sampling representativeness" (G_0/G_{el}) one obtains a value that would provide a good prediction of the settlement for typical working conditions (the methodology is thoroughly discussed in Viana da Fonseca *et al.*, 1997). This observation suggests that the application of a design methodology that corrects the values of the deformation modulus from triaxial tests by factors referenced to field tests (Cross-Hole tests or similar) may give good results.

Correlations between in situ test parameters

Viana da Fonseca *et al.* (2001, 2003, 2006) made an analysis of two experimental sites on Porto granite saprolitic soils, including derived ratios between PMT and SPT or CPT parameters. Some correlations are included in Table 8.1.

Ratios between distinct values of Young's moduli inferred from the investigations conducted have the obvious interest of fulfilling the needs of geotechnical designers to obtain data from different origins for each specific purpose.

Viana da Fonseca *et al.* (2001, 2003) and Topa Gomes (2009) reported some interesting correlations from data available in local sites: (i) values of Young's moduli determined directly, with no empirical treatment, or without deriving assumptions; (ii) common constant ratios that are assumed to correlate SPT (DP) or CPT parameters with Young's modulus, comparing them with transported soils; (iii) relative values of moduli can be summarised in the way that is expressed in Table 8.2a, while some relations could be pointed out between *in situ* tests, as expressed in Table 8.2b.

It is interesting to note that, for most designs, Sabatini *et al.* (2002) state that the elastic modulus corresponding to 25 percent of failure stress, E_{25}, may be used. In Piedmont residual soils, the use of the dilatometer modulus, E_D, as equal to E_{25} has been shown to provide reasonably accurate predictions of settlement (Mayne and

Table 8.2a Ratios between deformability (Young and Shear) modulus (*After* Viana da Fonseca *et al.*, 1998a, 2001, 2003; Topa Gomes, *2009*)

$E_0(CH)/E_{s1\%}(PLT)$	$E_0(CH)/E_{ur}(PLT)$	$E_0(CH)/E_m$	$G_0(CH)/G_{ur}(SBP)$
≅8–15	≅2–3	≅20–30	≅1.7–3

CH – seismic cross-hole tests; PLT – plate load tests

Table 8.2b Average ratios between Young's modulus and *in situ* "gross" tests (*After* Viana da Fonseca *et al.*, 2003)

$E_0(CH)/N_{60}(SPT)$	$E_0(CH)/q_c(CPT)$	$E_0(CH)/q_d(DPL)$	$E_0(CH)/P_L(PMT)$
≅ 10(MPa)	≅ 30	≅ 50	≅ 8

CH, PLT – ibidem; N_{60}, q_c, q_d, p_L - resistance values

Frost, 1988). However, the specific evaluation of E/E_0 associated with this FS equal to 4 (E_{25}) was found to be 0.34.

The specific application of footing settlement prediction methods based on in situ tests

Viana da Fonseca (2001) adapted some solutions available in the literature that use SPT parameters for settlement evaluation, and the following conclusions were reached.

Methods based on SPT

Terzaghi and Peck (1967)

SPT is really a crude test developed from a method developed by the Raymond Piling Co. in 1912 to obtain samples of the soil at the base of their bored piles. A thick walled 5 cm diameter steel tube was hammered into the ground to obtain the sample, and it was realised that the energy required to cause penetration gave an indication of the strength of the ground. The test was very easy to do because little special apparatus was needed in addition to the heavy well boring equipment already being used on the site. Through the years very many tests were carried out, but research workers were somewhat unhappy about the non-scientific nature of the test, and were doing their best to have it replaced by a penetration test such as the Dutch cone test, when Terzaghi and Peck published their semi-empirical method for estimating settlement in granular materials. As it is recognised today, the predictions are very conservative and, in the present case, the observed settlement would have been predicted under a load of ¼ to ½ of that actually applied. As a result it will not be developed further here.

Parry (1978)

The method of Parry (1978) is based on the expression of the Theory of Elasticity for the calculation of settlements:

$$s = q \cdot B \cdot \frac{1 - \nu^2}{E_s} \cdot I_s \tag{8.1}$$

with the deformation modulus taken as a function of an average value of N_{60} determined in the depth 2B below the footing base ($\overline{N_P}$), where N_{60} is the SPT number of blows, allowing for the method of dropping the driving weight and assuming a 60% efficiency, and the suffix P refers to Parry.

$$E_s = \overline{N_P}/\alpha_P \tag{8.2}$$

In the study presented by Viana da Fonseca (2001), the parameter α_P for the best adjustment ranged from 0.2–0.3, although only for the lower stress levels, becoming strongly non-conservative for stress levels above the "elastic" threshold. Its limitation is mainly due to the fact that it does not consider the non-linearity of stiffness with depth of the influence zone in relation to the dimension of the loaded area, which is not in agreement with reality (Jardine *et al.*, 1986; Burland, 1989; Tatsuoka and Shibuya, 1992). There is a risk therefore, when extrapolating to larger loaded areas (which are usually the case), of overestimating the settlements by calculating them on the basis of an average value of N_{60} over the depth of $2 \cdot$ B, particularly in soils that exhibit increasing stiffness with depth.

Burland and Burbidge (1985)
This proposed method for settlement calculation uses an average value of N_{60}, determined over a depth below the footing base ($\overline{N_{BB}}$), through the following expression:

$$s = \alpha_{BB} \cdot \frac{B^{0.7}}{\overline{N_{BB}}^{1.4}} \cdot q_s \tag{8.3}$$

with α_{BB} varying between 0.93 and 3.09, and 1.71 being the most probable value. In the expression, B denotes footing width, q_s the average contact pressure, and the suffix BB stands for Burland and Burbidge.

When applied to residual soils in Porto, using $\alpha_{BB} = 1.71$, the method was found to be grossly conservative, giving rise, for average service stress conditions, to ratios of 2 to 3 between predicted values and those observed. With the purpose of best adapting the method to suit the experimental results, the values of α_p and α_{BB} were calculated to obtain convergence for the two, following typical values of settlements:

(i) s/B = 0.75%, a level corresponding to a certain "elastic" threshold;
(ii) s = 25 mm, the limit value in accordance with Terzaghi and Peck proposal.

It was then concluded that:

(i) the approach of limiting the settlement to 25 mm produces a reasonable consistency of values in the two cases. This is a consequence of assuming a factor of 0.7 for the minimum size of the loaded area B.
(ii) for the same approach, the method of Parry gives values with slight variations for α_P, resulting in much greater reduction of B (factors of 0.32–0.44) than the maximum proposed by the author (0.30);
(iii) the approach using the "elastic" threshold level ($s/B = 0.75\%$) which is considered more realistic, confirms the good results from the Parry method for loading

over relatively small areas; it being noticed, however, that there is an increasing value of α with increasing foundation size from the plate to the footing, indicating the beginning of a breakdown in the assumption of direct proportionality with B;

(iv) it should be noted that for this approach, the values obtained for α_{BB} were very low (0.63, for the footing, and 0.50 for the plate of 60 cm of diameter) compared with the initial proposals (0.93–3.09).

This last result shows that residual soils studied possess a more pronounced structural stiffness than those analysed by Burland and Burbidge, which did not include residual soils. From the work of Rocha Filho (1986), the application of the Burland and Burbidge proposals to the results of loading tests on shallow foundations and plates with diameters from 0.40 to 1.60 m, carried out on residual soils from gneiss in the university campus of PUC in Rio de Janeiro, resulted in ratios of calculated to observed settlements of between 1.5 and 2.5. The ratios obtained in this study are even larger (2.7–3.4).

Another reason for lack of agreement with the Burland and Burbidge method when applied to residual soils may lie in the fact that the influence zone considered should be smaller than that suggested, due to the higher rate of stiffness increase with depth caused by the simultaneous increase of confining stress as the degree of weathering decreases. It can be considered that the proposal of Burland and Burbidge (1985) will be applicable to larger foundations resting on residual soils (for example $B = 3$–4 m), readopting the average value of coefficient α_{BB}. For the case of $s/B = 0.75$, the following conservative value is suggested:

$$\alpha_{BB}\text{ (saprolitic soil from granite)} = \frac{\alpha_{BB}\text{ (original)}}{2} = \frac{1.71}{2} = 0.85 \tag{8.4}$$

Anagnastopoulos et al. (1991)

The authors processed statistically 150 cases of shallow foundations with several sizes and subjected to different load conditions, mainly on sandy soils (of different origins), and proposed the following expression:

$$s = f\frac{q^{n_q} \cdot B^{n_B}}{N^{n_N}} \tag{8.5}$$

The terms were obtained by multiple regression, with priority for the dependent variables as a function of the relative influence of each one. It should be noted that the expression of Burland and Burbidge (1985) constitutes a particular case of this more general one, with $f = 0.93 - 3.09$; $n_q = 1$; $n_B = 0.7$ and $n_N = 1.4$.

It is interesting to observe that the parameters proposed by Burland and Burbidge are reasonably similar to those indicated by Anagnastopoulos *et al.* (1991)as representative of all sets of studied cases.On the other hand, it is to be noted that there is a high dispersion of the parameters corresponding to the several classes of stiffness, and of the size of the loaded surface. In order to obtain the parameters that give the best agreement with the experimental results, Viana da Fonseca (2001) suggested that the proposals of Burland and Burbidge (1985) be used for the influence depth $z = B^{0.75}$

and for the factor n_N (=1.40). With these factors fixed, multiple regression analyses were carried out assuming the following variants:

(i) the n_q exponent was allowed to be greater than 1, which corresponds to considering the non-linearity of the solution, and the best fit of the curves was found for a bearing pressure $q_s = 400$ kPa;
(ii) a linear solution was assumed ($n_q = 1$) and the convergence for values of the settlement defined for $s/B = 0.75\%$.

Figure 8.14 shows a comparison between the curves obtained for those two hypotheses, illustrating the following: (i) excellent agreement between the theoretical and the experimental results when using the value $n_q = 1.23$. The resulting non linearity causes the value of the constant f to be reduced to 0.18, much lower than the value of 0.60 usually defined in linear elasticity; (ii) the imposition of a linear relationship, as shown by Figure 8.14(b), always implies a subjective approach; the values of the constant α_{BB} obtained for the hypotheses associated with the Burland and Burbidge (1985) proposal, are similar to the one now deduced ($f = 0.60$), depending on the value of B being equally large.

From this, it can be concluded that the sensitivity analysis, by applying exponents to the factors that influence the development of settlements of shallow foundations under service loads, seems to offer a good method for prediction. This, however, needs future confirmation by other experimental studies, in particular those including larger loaded areas. To summarise the results obtained with SPT based methods, it can be stated that:

(i) the depths over which the values of N_{SPT} should averaged, can be accepted in the same terms as stated by Burland and Burbidge (1985);
(ii) the non-linear exponents, in relation to the applied load (n_q) should be assumed to be greater than unity (the deduced resulting value was around 1.23), causing relatively low constants, e.g. for safety, $f = 0.20$;
(iii) although these comparative analyses of the test results indicate n_B values close to 1, smaller values should be adopted (for example, $n_B = 0.7$, as proposed by Burland and Burbidge, 1985) when designing shallow foundations with dimensions generally larger than 2 to 3 m. This method reflects the reducing dependence of settlements on increasing values of B (Bjerrum and Eggestad, 1963).

Methods based on CPT

Schmertmann et al. (1978)

The semi-empirical method of Schmertmann (1970), upgraded in Schmertmann *et al.* (1978), assumes a simplified distribution of the influence factor for the vertical strains under the footing, with these formulations:

$$s = \int_0^{2\cdot B(4\cdot B)} \varepsilon_z \, dz \cong \Delta p \cdot \int_0^{2\cdot B(4\cdot B)} \frac{I_z}{E_z} dz \cong C_1 \cdot C_2 \cdot \sum_{i=1}^{n(2\cdot B;4\cdot B)} \frac{I_{z_i}}{E_{z_i}} \cdot \Delta Z_i \tag{8.6}$$

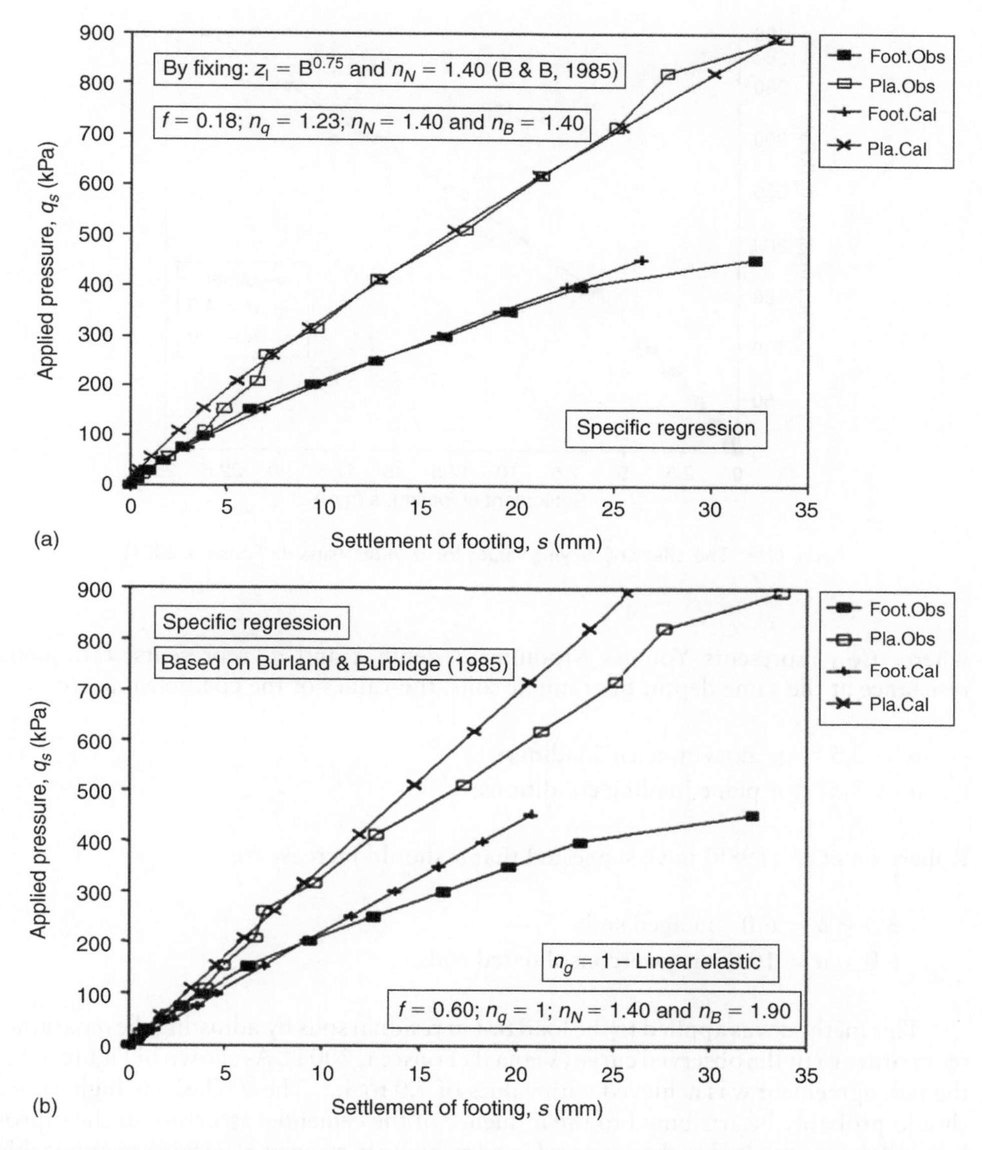

Figure 8.14 **Determination of values for f, n_B and n_q that give best agreement with experimental results (a) for the non-linear relation $n_q < 1$ and (b) the linear relation $n_q = 1$ [n_g fixed at 1.40]**

the values of which are computed on the basis of a deformation (secant) modulus variable with depth, which can be correlated with the CPT cone resistance:

$$E_{S(z)} = \alpha \cdot q_{c,CPT}(Z) \tag{8.7}$$

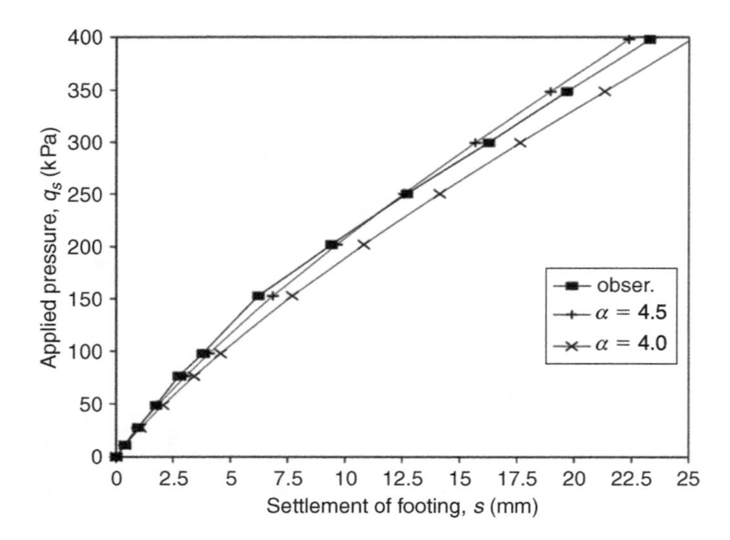

Figure 8.15 The effect of varying values for α (*After* Viana da Fonseca, 2001)

where, $E_{S(z)}$ represents Young's Modulus at depth z and q_c represents static cone resistance at the same depth. In granular soils, the values of the coefficient α are:

$\alpha = 2.5$ for axisymmetric loadings;
$\alpha = 3.5$ for plane loading conditions;

Robertson *et al.* (1988) have suggested that α should increase to:

$3.5 < \alpha < 6.0$ in aged soils;
$6.0 < \alpha < 10.0$ in overconsolidated soils.

This method was applied to the load test in residual soils by adjusting the α parameter in order to fit the observed curve (Viana da Fonseca, 2001). As shown in Figure 8.15, the best agreement was achieved with values of 4.0 to 4.5. These relatively high values should probably be attributed to the influence of the cemented structure of the saprolitic soil, being situated in the range referred to by Robertson *et al.* (1988) as applicable for aged sands.

Robertson method

This method is based on the results of CPT tests carried out under the area to be loaded, and it incorporates factors related to the degree of stress induced by the foundation and the effects of the stress-strain history (including the natural structure of the ground).

Figure 8.16 represents the normalised values of the shear modulus for very small strains, G_0, obtained from cross-hole testing, as a function of the normalised cone resistance, q_{c1}, defined by Robertson (1991).

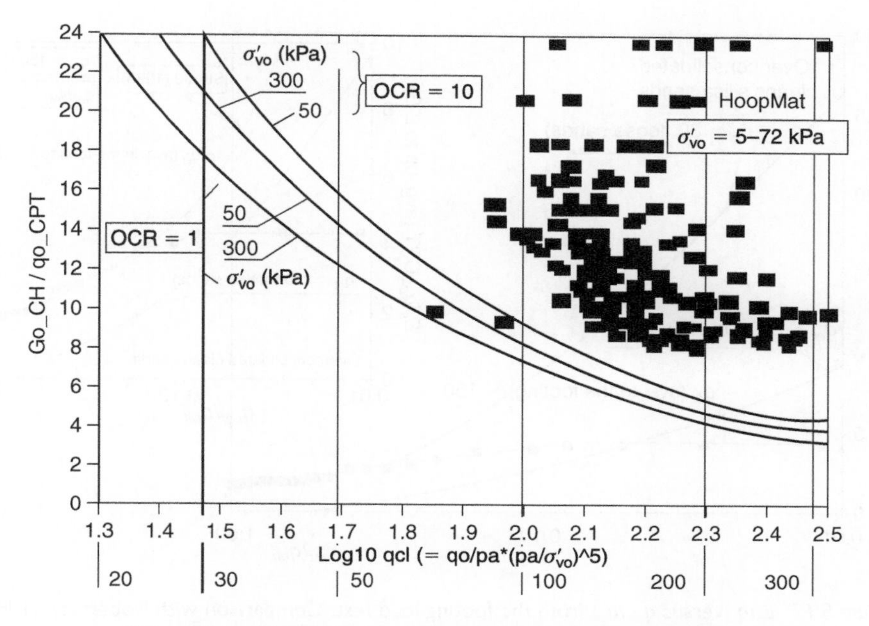

Figure 8.16 G_0/q_c versus q_{c1}; Comparison with Robertson (1991)

From an analytical interpretation of these plotted results, the following is obtained:

$$\left(\frac{G_0/q_c}{q_{c1}}\right)_{porto\ saprolitic\ soil} \cong 1.8 - 2.0\left(\frac{G_0/q_c}{q_{c1}}\right)_{non\text{-}cemented\ sands} \tag{8.8}$$

revealing that the natural cemented structure of these residual soils generally induces higher values of the ratio between the elastic or "pseudo-elastic" stiffness and the strength, than those corresponding to transported soils, either normally or over-consolidated. This tendency is especially noticed at low confinement stress levels, revealing a relative independence of the shear modulus at low strain levels ($<10^{-6}$–10^{-4}) in relation to the at rest stress states (Tatsuoka and Shibuya, 1992). We should also note that the results obtained for the normalised cone resistance, q_{c1}, vary between the values of 100–300, the mode being about 150. This corresponds, in sedimentary soils, to dense sands. Figure 8.17 shows the relationship between E_s/q_c and load level, q_{ser}/q_{ult}, obtained from the results of the footing load test, together with curves indicated by Robertson (1991) for dense sands. Also shown, in the inset, are similar curves proposed by Stroud (1988) for over-consolidated (aged) sands. The curves for dense sand and those obtained from test results shown in the main Figure 8.17 and in the inset are the same, but to a different scale in order to permit a comparison with Stroud.

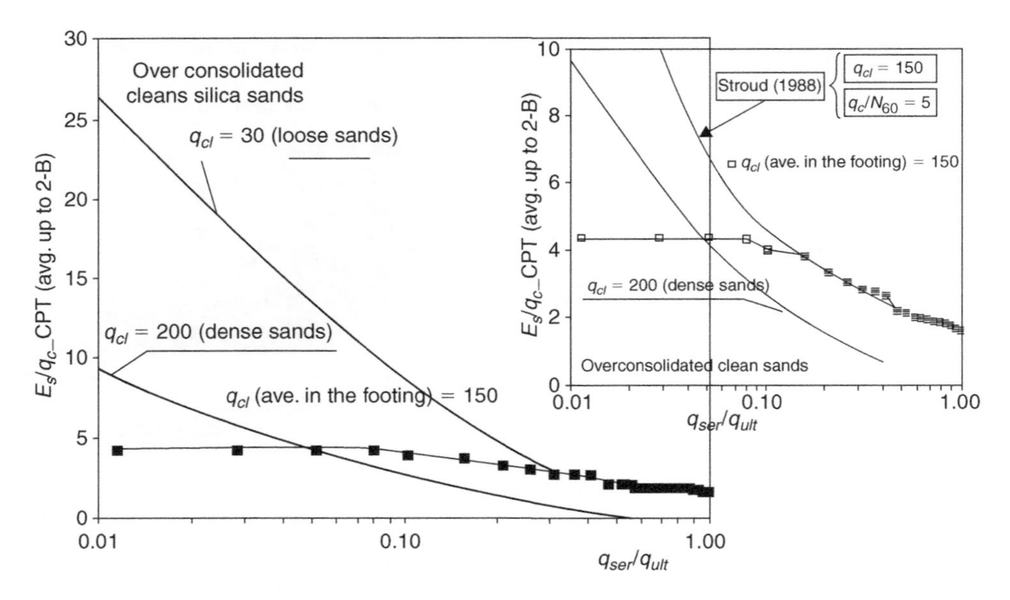

Figure 8.17 E_s/q_c versus q_{ser}/q_{ult} from the footing load test. Comparison with Robertson (1991)

From consideration of the three curves shown by Figure 8.17, it can be concluded that:

(i) for very low load levels, below around 10% of ultimate, the non-linearity of the relationship of E_s/q_c with q_{ser}/q_{ult} is much more accentuated for clean sands, even when over-consolidated (aged), than for the granitic saprolitic soil. This may be the consequence of a larger material stability in the saprolitic soil, due to the cemented structure between particles;

(ii) for higher load levels, there is a good agreement between the experimental curve for the saprolitic soil and the proposal of Stroud (1988) for over-consolidated sands.

It should be noticed however, that the values of the q_{c1} ($\cong$150) indicated by Stroud (1988) for the test results, are typical of the middle range for dense over-consolidated sands, with ageing.

With shallow foundations on saprolitic soils derived from granite, the dependence of the secant deformability modulus on the load level seems to represent, in a consistent way for the highest load levels, the proposal of Stroud, according to Robertson (1991), to use the value $q_c/N_{60} = 5$.

Methods based on PLT

Ghionna et al. (1991)

The Ghionna *et al.* (1991) method considers the dependence of the deformability modulus on the normalised stress-strain levels. It uses a hyperbolic relationship to model

Table 8.3 Hyperbolic K and n parameters for different triaxial testing results

Deformation Modulus	K	n
E_{el} (linear elastic)	35660	0.263
E_{ur} (unload-reload)	19637	0.250
$E_{ti,h}$ (hyperbolic, $q = 70$–95% of q_f)	2749	0.539
$E_{s25\%}$ (secant for $q = 25\%$ of q_f)	1804	0.588
$E_{s50\%}$ (secant for $q = 50\%$ of q_f)	1517	0.504

the behaviour of the soils (Duncan and Chang, 1970) and allows the extrapolation of the results of load tests on foundations of different sizes and shapes for the evaluation of the settlement of larger loaded areas and with different geometries, considering an equivalent homogeneous mean.

In the proposed expression for the evaluation of settlement:

$$s = \frac{1}{K_i} \cdot \frac{q_n \cdot B \cdot I \cdot (1 - \nu^2)}{\overline{\sigma'_{\text{oct}}}^{\,n} \left(q_n \cdot B \cdot I \cdot (1 - \nu^2)\right) / \left(\overline{\sigma'_{\text{oct}}}^{\,1-n} \cdot C_f \cdot H_i\right)} \tag{8.9}$$

the parameters have the usual meaning, but the following ones require special mention:

1 $q_n = q_s - 2/3\sigma'_{v0} = q_s$ because, in our foundation loading tests, σ'_{v0} at the loaded surface, is zero;
2 H_i represents the depth of the load influence zone that, according to the authors, should be considered down to a depth of $2 \cdot B$ from the footing base;
3 n is a suitable hyperbolic exponent;
4 K_i, C_f represent hyperbolic parameters (the first, of stiffness, and the second, of strength) that will be determined from the load tests. There is a larger dispersion of the K_i values, due to the high stiffness sensitivity.

In the present case, the n parameter was determined from the common expression:

$$E = K \cdot (\sigma'_{\text{oct}_0})^n \tag{8.10}$$

produced from the similar analysis of different deformability moduli taken from anisotropically consolidated triaxial compression tests, referred to as CID and CAD tests (Viana da Fonseca, 1996, 2003), on samples taken from the zone of influence of the pilot load tests (see Table 8.3). Values for E_{el} were deduced from the initially linear reload branch of an intermediate unload-reload cycle; those for E_{ur} between vertices of that cycle; E_{ti} obtained from the initial tangent by hyperbolic modeling; and $E_{s25\%}$ and $E_{s50\%}$ from secant values for 25% and 50% of the failure load, respectively.

In the subsequent study, in order to analyse the influence of the parameter n, the following two values were taken:

1 $n_{el} = 0.263$ – corresponding to very low loads, within the elastic threshold;
2 $n_{25\%} = 0.588$ – corresponding to the mobilisation of medium stress levels.

Table 8.4 K_i and C_f values determined from the loading tests ($z_I = B/2$)

	Footing (D 120 cm)		*Plate (D 60 cm)*		*Average values*	
Exponents	K_i	C_f	K_i	C_f	K_i	C_f
$n = 0.263$ (E_{el})	46.3	1.87	45.6	1.95	46.0	1.91
$n = 0.588$ ($E_{s25\%}$)	14.6	1.83	15.6	1.88	14.9	1.86

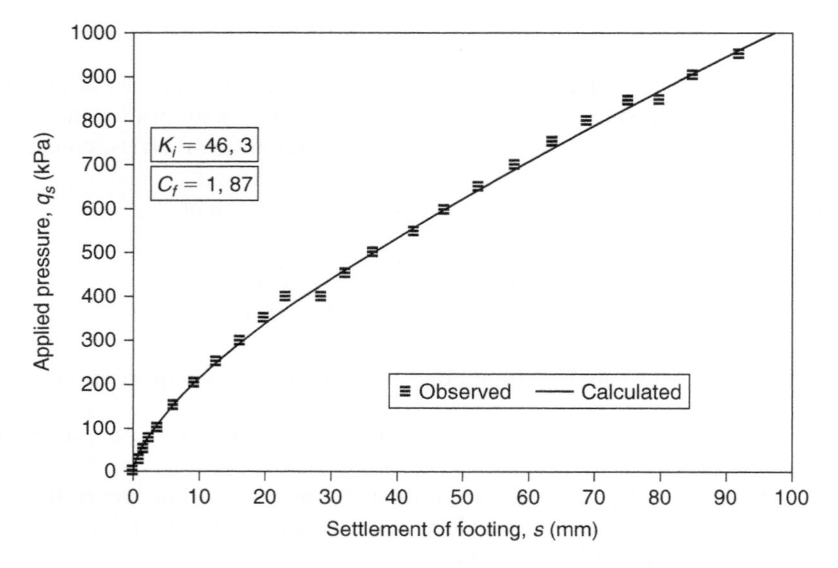

Figure 8.18 Comparison of predictions given by the Ghionna *et al.* (1991) method with observed values from the footing load test ($n = 0.263$)

Based on the results of the loading tests on the footing and the 60 cm diameter plate, the values obtained for K_i and C_f are given in Table 8.4. Depth to the settlement centre was taken as $z_I = B/2$, as assumed in the method of Ghionna *et al.* (1991).

The predicted results, using $n = 0.263$ with the corresponding values for K_j and C_f, are in very good agreement with the observed results from the footing load test, as shown by Figure 8.18.

The general application of this method requires a strict adoption of representative values of the ground in question. It is therefore necessary to use average values of n, K_i and C_f or, alternatively, those critically selected from the available values obtained from the individual analysis of each test. The results of a general analysis of the footing and plate tests, taking into account the average values for n proposed by Ghionna *et al.* (1991) are presented in Viana da Fonseca (2001). It was proved that a low value of n gives the best simulation for the low stress levels. Good agreement is lost for higher stress levels, giving rise to a non-conservative result for the larger footing sizes, which will limit extrapolation for larger size footings. It is clear that this tendency decreases its relevance as, in general, the parameters K_i and C_f refer to the test over the largest

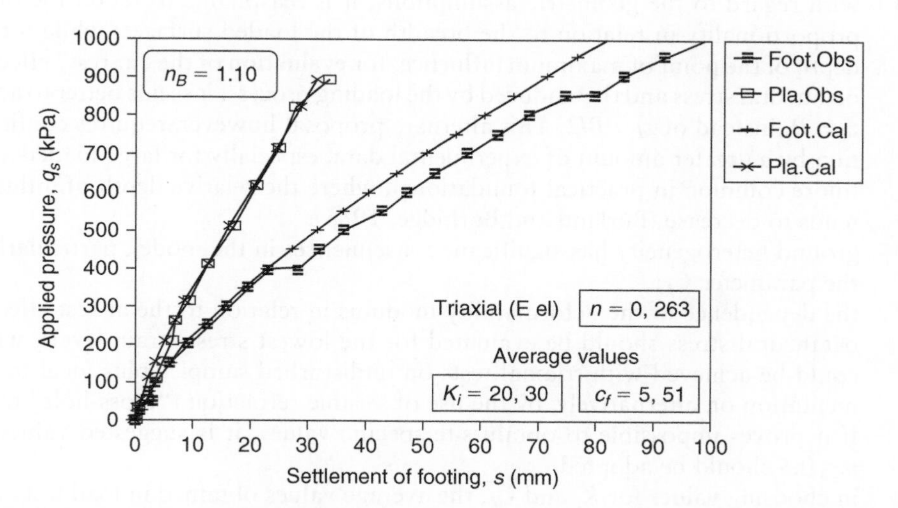

Figure 8.19 Application of Ghionna *et al.* (1991) method by considering a dimension factor, $n_B \neq 1$

size of loaded area. Although it is conservative for small footings, on the whole this approach seems to be acceptable.

To ameliorate the performance of the method, the following modifications to the original method of Ghionna *et al.* (1991) were considered by Viana da Fonseca (2001):

(i) depth of the point of maximum influence: $z_I = B$ instead of $B/2$;
(ii) exponential dependence of the footing breadth by a factor $n_B \neq 1$.

The results obtained from the first modification, taking average values of K_i and C_f, although considerably better, were not satisfactory. The second modification required alteration of the original formulation to produce the following equation:

$$s = \frac{1}{K_i} \cdot \frac{q_n \cdot B^{n_B} \cdot I \cdot (1 - v^2)}{(\sigma'_{oct})^n - (q_n \cdot B^{n_B} \cdot I \cdot (1 - v^2))/((\overline{\sigma'_{oct}})^{1-n} \cdot C_f \cdot H_i)} \tag{8.11}$$

A best fit of the curves was then possible by assuming average parameters of K_i and C_f, making $q_n = q_s$, and introducing n_B as a weighting factor related to the loading area. As seen in Figure 8.19, making $n = 0.263$ and considering the value $n_B = 1.10$ gives a good fit. This tendency for $n_B > 1$ is contrary to what was verified with the other methods under review, because it was considered convenient to maintain the proportionality in relation to B presented by the authors.

To conclude, the applicability of the Ghionna *et al.*'s (1991) method to residual soils can be inferred from:

(i) the model, which integrates results of both *in situ* load tests and laboratory tests, presenting a good approach for foundation settlement prediction; this should be applied exclusively for moderate load levels.

(ii) with regard to the geometric assumptions, it is reasonable to retain the direct proportionality in relation to the breadth of the loaded surfaces, while for the depth of the point of maximum influence, for evaluation of the "at rest" effective octahedral stress and that induced by the loading process, it seems better to adopt $z_I = B$, instead of $z_I = B/2$. This alternate proposal however, requires confirmation by a greater amount of experimental data, especially for large loaded areas (more common in practical foundations), where the relative depth of influence tends to decrease (Burland and Burbidge, 1985).

(iii) ground heterogeneity has significant consequences in the model, particularly in the parameter C_f;

(iv) the dependence of the deformability modulus in relation to the at rest effective octahedral stress should be evaluated for the lowest stress-strain levels, which could be achieved with triaxial tests on undisturbed samples plus local instrumentation or, alternatively, by the use of seismic refraction ("cross-hole" tests). If it proves impossible to obtain site specific values, it is suggested values for $n \leq 0.5$ should be adopted.

(v) in choosing values for K_i and C_f, the average values obtained in load tests with different loading areas should be used, provided that the variation between them is not high. When there is considerable variability of those parameters, the lowest values of K_i and, above all, of C_f should be chosen to ensurea conservative result.

Wahls and Gupta (1994)

The method of Wahls and Gupta (1994) accurately considers the non linear nature of the stress/strain relationships ($\sigma - \varepsilon$). Firstly, as a basic formulation taking account of the penetration testing parameters (N_{SPT} or q_c-CPT) and the resulting correlations with the low strain shear modulus ($G_0 = G_{\max}$) and, secondly, as a law of variation of the secant or tangent shear modulus with the distortion level proposed by other authors (such as Seed and Idriss, 1982). These laws apply only to the materials of the type from which they were developed. Alternatively, the method can be based on the back-analysis of one or more load tests on plates or experimental footings, preferably of different sizes to enable definition of the non-linearity, giving a variation law of Young's modulus with load level (q_{sj}) in relation to failure.

Adapting the Wahls and Gupta method for this last alternative, and assuming a given load step q_{sj}, applied to a foundation on a layer i, of thickness Δh_i and with deformability modulus E_{ij} (i and j translate the dependence in relation to the depth and to a certain load level, respectively), the vertical deformation can be calculated by:

$$\Delta s_{i_j} = \frac{I_{si}}{E_{i_j}} \cdot q_{sj} \tag{8.12}$$

and the corresponding settlement by:

$$s_j = \sum_{i=1}^{n} \Delta s_{ij} \cdot \Delta h_i = q_{sj} \cdot \sum_{i=1}^{n} \frac{I_{si} \Delta h_i}{E_{ij}} \tag{8.13}$$

where n is the number of sub-layers into which the ground is divided within the main settlement influence zone, which should extend to such a depth that the shear stress increment does not exceed the value of initial shear stress, with depths of around $2 \cdot B$ for $L/B \leq 3$, and of $4 \cdot B$, for $L/B > 3$. L and B are the dimensions of rectangular footings.

The greater the number of divisions used, the greater will be the accuracy. I_{si} is the load coefficient for the layer "i", dependent on the size of the loaded area and the value of Poisson's ratio.

The dependence of the deformability modulus on depth can be related to the at-rest octahedral effective stress at the centre of the layer, σ'_{mij}, by means of:

$$E_{ij} = E_{0j} \cdot (\sigma'_{mij})^{\overline{n}} \tag{8.14}$$

with $n = 0.5$, as suggested by the authors.

Dependence in relation to the vertical stress can be expressed by the following relationship:

$$E_{0j} = f\left(\frac{q_{s_j}}{q_{s_{ref}}}\right) \cdot E_{ref} \tag{8.15}$$

or, alternatively, by:

$$E_{0j} = f\left(\frac{(s/B)_j}{(s/B)_{ref}}\right) \cdot E_{ref} \tag{8.16}$$

where $q_{s_{ref}}$, s_{ref} and E_{ref} represent, respectively, the load, the settlement and the deformability modulus, corresponding to certain reference load steps (for example: $s/B = 0.1\%$) and q_{s_j} and s_j the load and the settlement for a generic load level.

Viana da Fonseca (2001) presented a back-analysis of the footing load test and for $s/B = 0.1\%$, a value of $q_{s_{ref}}$ (26 kPa) was obtained and the value of the reference modulus (for $n = 0.5$) deduced from:

$$E_{ref} = \frac{q_{s_{ref}}}{s_{ref}} \cdot \sum_{i=1}^{n} \frac{I_{s_i}}{\sqrt{\sigma'_{m_j}}} \cdot \Delta h_i \tag{8.17}$$

On the other hand, with the pair of values s_j and q_{s_j}, corresponding to each loading step and obtained from the experimental curve, the respective equivalent modulus can be calculated (for an increment from zero to q_{s_j}) from:

$$E_{0_j} = \frac{q_{s_j}}{s_j} \cdot \sum_{i=1}^{n} \frac{I_{s_i}}{\sqrt{\sigma'_{m_i}}} \cdot \Delta h_i \tag{8.18}$$

To define the relationship of non-linear dependence of E_{0j} with the strain level, defined as $(s/B_j)/(s/B_{ref})$, or with shear stress levels, defined by $q_{s_j}/q_{s_{ref}}$, a logarithmic scale was adopted. Two influence depths were considered: $z_I = 2 \cdot B$ and $z_I = 5 \cdot B$.

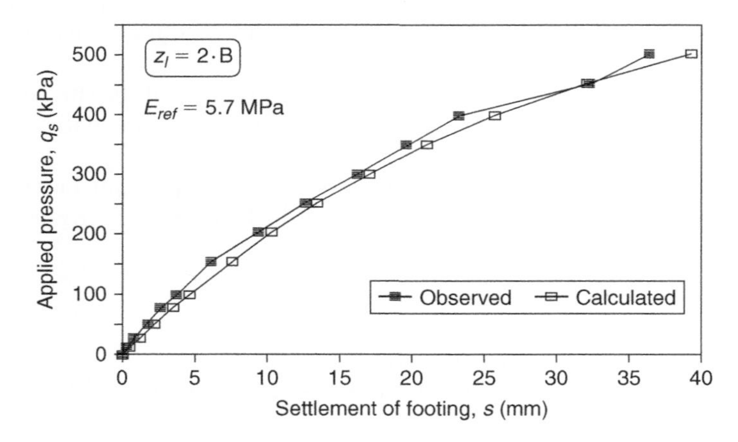

Figure 8.20 Comparison of the experimental curves and those simulated by Wahls and Gupta model for $z_l = 2 \cdot B$

In Figure 8.20, a comparison is made between the experimental and simulated curves, with parameters adjusted to give a good fit. For both influence depths the agreement is excellent, revealing the potentialities of the method to model the non-linearity of the load-settlement behaviour of the experimental footing.

Concerning definition of the reference modulus, there is obvious potential for this formulation in the design of shallow foundations, by means of a specific load test or by pre-loading an experimental footing, using $(s/B)_{ref} = 0.1\%$.

In the work reported before, Viana da Fonseca (2001) proved that the method proposed by Wahls and Gupta (1994) using the results of triaxial tests could be applied to determine the values of Young's modulus from the relevant depth and shear stress level in the soil. These parameters are subsequently used in a simplified nonlinear elastic analysis of the footing load test.

Methods based on DMT

The most widely used methods for the prediction of settlement of shallow foundations based on DMT test results are those due to Schmertmann (1986), and Leonards and Frost (1988). The first is a general method based on the Theory of Elasticity, using weighting factors variable with depth, similar to Schmertmann's CPT method described before, providing a nonlinear pressure-settlement curve, since the strains depend on the ratio between the incremental pressure and the initial effective vertical stress at foundation level.

The method has advantages in relation to the method originally proposed by Marchetti (1980), since the total settlements of common foundations seldom have conditions of lateral confinement (situation represented by the constrained modulus, M_D), so it is more consistent to use a deformation modulus, with analogies to triaxial compression tests. The methods are reasonably adjusted to real situations of isolated footings, in which the stiffness may vary randomly with depth. The groundwater conditions are also integrated in the values of the modulus and the formulation includes the geometric factors of the foundation (shape and embedment in depth).

Table 8.5 Deformation modulus from DMT versustriaxial tests in recent silica sands (*After* Berardi *et al.*, 1991)

OCR	$E_s(0,1\%)/E_{D,DMT} \pm$ *variation*
1	0.99 ± 0.19
1.4–8.8	3.25 ± 0.71

The formulation proposed by Leonards and Frost (1988) is a generalisation of Schmertmann's (1986) proposal, with the following expression:

$$s_j = C_1 \cdot C_2 \cdot q_{effect_j} \left(\sum_0^{H_i} I_{Z_{i_j}} \cdot \Delta Z_i \cdot \left[\frac{R_z(OC)_i}{E_z(OC)_i} + \frac{R_z(NC)_i}{E_z(NC)_i} \right] \right)_j \tag{8.19}$$

where:
C_1: correction for embedment (=1, shallow [surface] foundations);
C_2: correction for time (=1, short term analysis);
q_{effect_j}: effective stress transmitted to the base of foundation;
H_i: depth of influence, similar to the proposal of Schmertmann (1978);
$I_{z_{i_j}}$: influence factor for deformations (Schmertmann, 1978);
Δz_i: depth of sub-layers (20 cm – coinciding with the intervals in DMT);
$R_z(OC)_{ij}$: ratio of stress increment for the overconsolidated portion (OC);
$R_z(NC)_{ij}$: ratio of stress increment for the normally consolidated portion (NC);
being expressed by:

$$R_z(OC) = \left(\frac{\sigma'_p - \sigma'_{v0}}{\sigma'_f - \sigma'_{v0}} \right); \quad R_z(NC) = \left(\frac{\sigma'_f - \sigma'_p}{\sigma'_f - \sigma'_{v0}} \right)$$

where: σ'_f = the vertical effective stress after consideration of final load q_{effect_j}, and: $q_{effect_j} = \sigma'_{v0} + \Delta\sigma'_v$, calculated by the Theory of Elasticity;
$E_{ziD}(OC)$; $E_{ziD}(NC)$: appropriate values of deformation modulus corresponding to the over-consolidated and normally consolidated portions, respectively, for the increment of stress in the layer i for the load q_{effect_j}, and deduced from the correlations between $E_s(n\%)$ and $E_{D,DMT}$.

In Table 8.5, a summary of the correlations obtained in calibration chambers are presented (Berardi *et al.*, 1991).

These formulations have been applied to the case study reported by Viana da Fonseca (2001) taking account of the results of an extensive site characterisation campaign (Viana da Fonseca, 2003), and testing the application of the Leonards and Frost (1988) method. The variation of the Dilatometer modulus with at rest vertical effective stress was expressed by:

$$E_D = 5.54 + 430\sigma'_{v0} \tag{8.20}$$

The resulting pattern of variation is shown in the curves plotted in Figure 8.21.

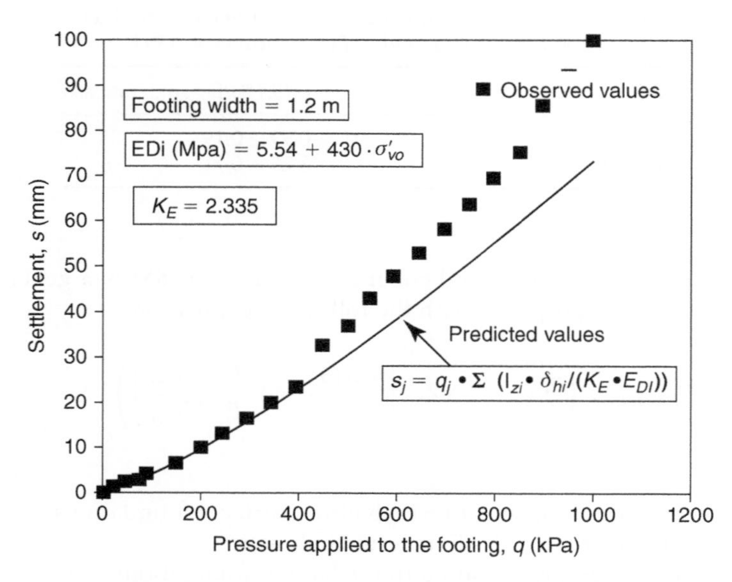

Figure 8.21 Comparison of the measured settlement of the footing (*After* Viana da Fonseca, 2001) with the prediction based on DMT using Leonards and Frost (1988) method

A value of $K_E = 2.34$ was found by fitting a curve through the early pressure-settlement curve, suggesting that:

a) correlations with E_D modulus to be adopted in residual soils, such as these silty saprolites from granite, may be inbetween the proposals due to Berardi *et al.* (1991) for recent sandy soils ($OCR = 1$, $K_E = 0.99 \pm 0.19$) and overconsolidated/aged soils ($OCR = 1.4$–8.8, $K_E = 3.25 \pm 0.75$); these trends are similar to those derived for *CPT* and *SPT* methods presented above; in fact these saprolitic soils may be situated in class 2 of Berardi *et al.* (1991) proposal for the ratio of stiffness to strength (the class 1 for $OCR = 1$ and class 3 for high values of OCR).
b) this value of K_E only applies for moderate values of applied pressure, $q < 35\% q_{ult}$, which is in agreement with the perception of the high level of non-linearity for these young residual soils.

Methods based on pre-bored Pressuremeter (PMT) or Ménard's pressuremeter (MPM)

Many kinds of pressuremeter probes are currently in use (Briaud, 1992; Clarke and Smith, 1993). Their differences are mostly related to the way they are inserted into the ground, such as predrilled hole (PMT), self-bored (SBPT), or pushed-in (CPMT). Since the PMT causes an unavoidable stress relief, and the CPMT causes an unavoidable stress increase, it is obvious that the SBPT is the one that causes the least soil disturbance. Consequently, the SBPT is the only one that allows the measurement of the geostatic total horizontal stress σ_{h0}. It also offers a better interpretation of test results from small to large strain levels. Jamiolkowski and Manassero (1995) summarised the different geotechnical parameters that can be obtained by the three types

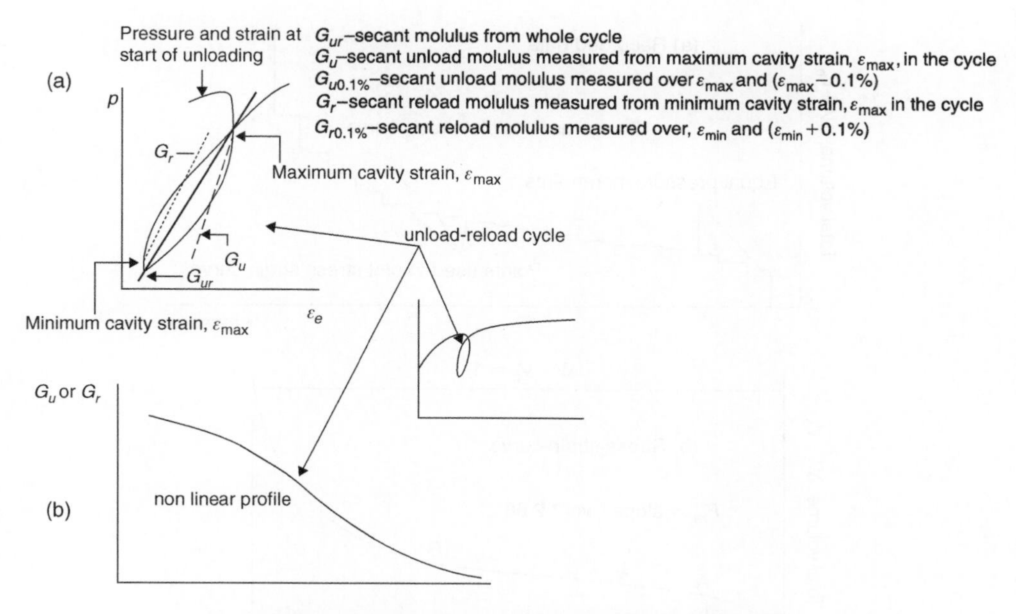

Figure 8.22 Selection of shear moduli (*After* Clarke, 1995)

of pressuremeters. The different moduli that can be obtained by the SBPT are shown in Figure 8.22.

Theoretically, the initial slope of an SBPT pressure/strain curve yields the G_0 value. However, in practice there is still some disturbance (Wroth, 1982) and the modulus must be taken from an unload-reload cycle (G_{ur}). For heavily over-consolidated soils and cemented geomaterials, it could be assumed that $G_{ur} = G_0$ if the strain of one cycle is less than 0.01%.

There are two approaches to the use of G_{ur} in practice:

- To link G_{ur} to G_0 using a determined stress-strain relationship (Bellotti *et al.*, 1994; Ghionna *et al.*, 1991);
- To compare G_{ur} values to the degradation modulus (G/G_0) versus shear strain (γ) curve from the laboratory, taking into account the average values of shear strain and mean plane effective stress associated with the soil around the expanded cavity (Bellotti *et al.*, 1994).

It is not appropriate to obtain G_0 directly from the PMT because of the unavoidable disturbance during predrilling.

By careful testing, with a simple and expeditious methodology, the PMT can be adapted to determine different levels of stiffness and strength in difficult materials such as the highly heterogeneous conditions found in residual profiles.

The routine analysis of PMT tests follows the method originally developed by Ménard (1955). It gives design parameters directly obtained from the pressuremeter

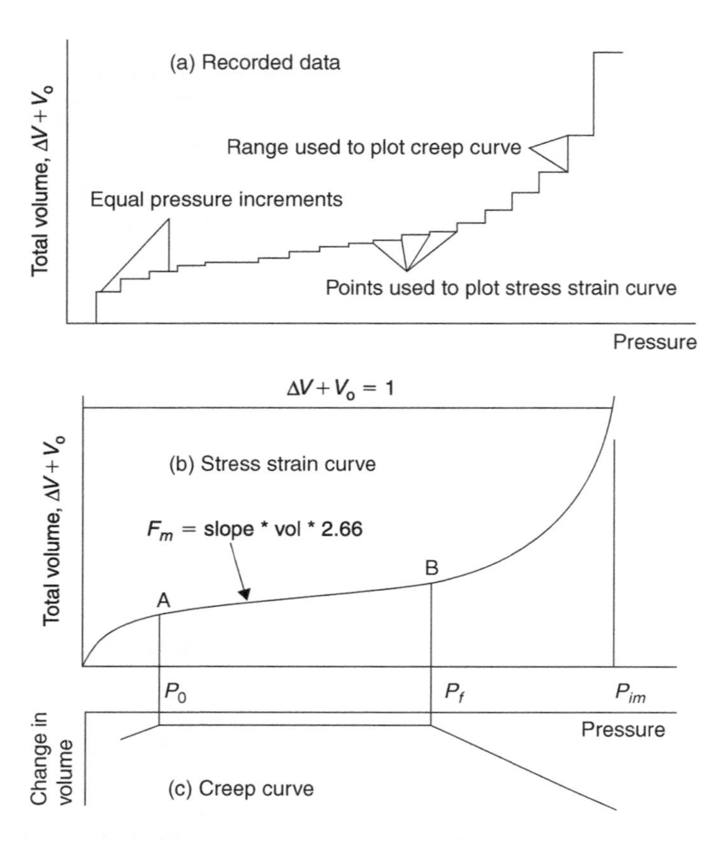

Figure 8.23 Interpretation of PMT according to the ASTM standard (*After* Clarke and Gambin, 1998)

test curve (ASTM, 2004). Figure 8.23 shows the interpretation of the curve and Figure 8.24 shows the procedure to obtain the pressuremeter modulus (E_m), based on the present ASTM (2004) standard.

The interpretation of the results is solely based on the analysis of two curves: the pressuremeter curve (v_i *versus* p_i, recorded at the end of each minute) as shown in Figure 8.24a and the yield curve (the difference between the volumes at 30 sec and 1 min versus pressures) as shown in Figure 8.24b. From these tests, the following parameters are deduced:

- the ***pressure meter modulus*** (E_m):

$$E_m = 2 \cdot (1 + v) \cdot V_m \frac{\Delta p}{\Delta v} \tag{8.21}$$

  where V_M is the volume of the cylindrical cavity in the beginning of the linear behaviour (the pseudo-elastic range), observed between stress and strain, Δp and Δv;
- the ***limit pressure*** (p_L): the pressure necessary to double the initial volume of the original excavated cavity; or the *differential limit pressure* ($p_L^* = p_L - p_0$),

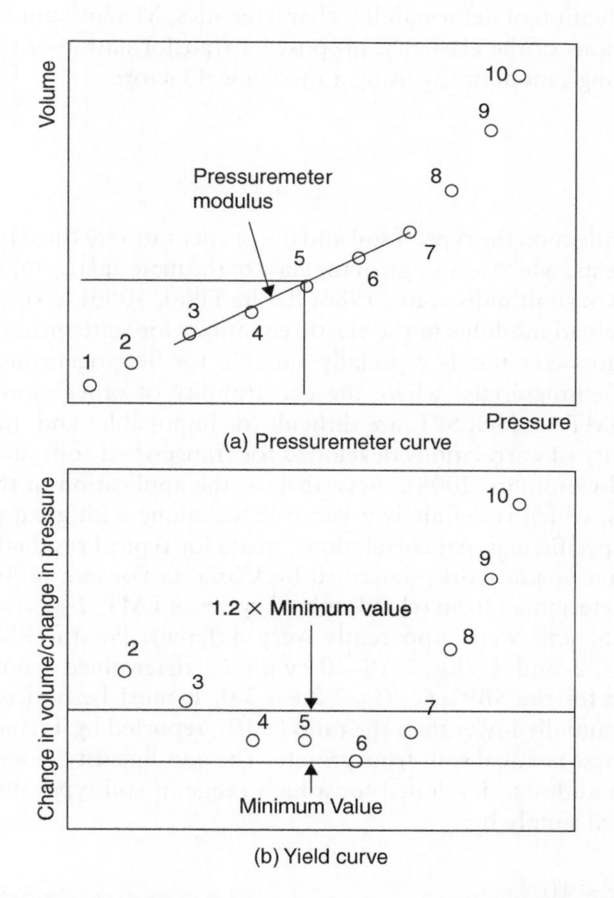

Figure 8.24 Selection of the pressure range to calculate E_m according to the ASTM standard (*After* Clarke and Gambin, 1998)

which is less sensitive to drilling damage or imperfections of the initial shape of cavity;

- a ***yield pressure*** (p_f): the end of the linear range in the curve, corresponding to the value with a clear increase in the change in volume between 30 sec and 1 min.

It must be pointed out that this modulus is related to the average stiffness of the ground associated with a particular strain level. Consequently the use of this value must only be applied in settlement formulae developed by Ménard (Ménard, 1963, 1965), as is done in the French Code for foundation design Fascicule No 62 (Gambin and Frank, 2009). Consequently the PMT modulus must be considered as a test-specific design parameter (Gomes Correia *et al.*, 2004).

For the evaluation of deformability characteristics, Ménard and Rousseau (1962), using the equations of the elasticity, proposed a transformation of the pressuremeter modulus to Young's modulus by using a rheological factor:

$$E = \frac{E_{pm}}{\alpha} \tag{8.22}$$

where α is dependent on the type of soil and the proper ratios defined by the parameters deduced in the tests, which are a sign of the class of the material (E_m/p_L^*). Others authors (Kahle, 1983; Konstantinidis *et al.*, 1986; Rocha Filho, 1986) have suggested the use of the unload-reload modulus in the elastic equations for settlement evaluation.

This pressuremeter test is especially suitable for heterogeneous soils and IGM (Intermediate Geo-materials) where the penetrability of other more common tests, such as CPT, DMT, or even SPT, are difficult (or impossible) and, most importantly, where the validity of correlations developed for transported soils are dubious (Viana da Fonseca and Coutinho, 2008). Nevertheless, the application of this test to design in residual soils, which is definitely a versatile technique with great potential, has to be done using specific regional correlations, made for typical residual materials.

Experimental *in situ* work, described by Viana da Fonseca (2003), showed that the stiffnesses determined from reload-unload cycles of PMT (E_{pmur}) and SBPT tests in saprolitic granite soils were, apparently, very different. For the PMT it was found that $E_{pmur}/E_{pm} \cong 2$ and $E_0/E_{pm} \cong 18$–20, with E_0 determined from seismic survey (G_0-CH), while for the SBPT $G_0/G_{ru} \cong 2.6 - 3.0$. It must be noticed that these last values are substantially lower than the ratio ($\cong$10), reported by Tatsuoka and Shibuya (1992) on Japanese residual soils from granite. The non-linearity model of Akino, cited by the previous authors, developed for a high range of soil types, including residual soils, is expressed simply by:

$$E_{sec} - E_0 \ (\varepsilon \leq 10^{-4}) \tag{8.23}$$

$$E_{sec} = E_0 \cdot \left(\frac{\varepsilon}{10^{-4}}\right)^{-0.55} \ (\varepsilon > 10^{-4}) \tag{8.24}$$

It should be noted however, that pressuremeter data has been used in France (and elsewhere where PMT is common practice) for settlement evaluation, following a specific formulation (known as the "French method") that will be explained below.

EN 1997-2: 2007 (E): Method to calculate the settlements for spread foundations

There is a preference to use results from the PMT (MPM, meaning Ménard Pressuremeter Tests) directly to calculate the settlement, s, of spread foundations using a semi-empirical method developed using influence factors. This is expressed by the following equation:

$$s = (q - \sigma_{v0}) \times \left[\frac{2B_0}{9E_d} \times \left(\frac{\lambda_d B}{B_0}\right)^a + \frac{\alpha \lambda_c B}{9E_c}\right] \tag{8.25}$$

Table 8.6 The shape factors, λ_c, λ_d, for settlement of spread foundations

L/B	*Circle*	*Square*	*2*	*3*	*5*	*20*
λ_d	1	1.12	1.53	1.78	2.14	2.65
λ_c	1	1.1	1.2	1.3	1.4	1.5

Table 8.7 Rheological factor α for settlement of spread foundations

Type of ground	*Description*	E_M/p_{LM}	α
Peat			1.0
Clay	Over-consolidated	<16	1.0
	Normally consolidated	9–16	0.67
	Remoulded	7–9	0.50
Silt	Over-consolidated	>14	0.67
	Normally consolidated	5–14	0.50
Sand		>12	0.50
		5–12	0.33
Sand and gravel		>10	0.33
		6–10	0.25
Rock	Extensively fractured		0.33
	Unaltered		0.50
	Weathered		0.67

where:

B_0 is a reference width of 0.6 m;
B is the width of the foundation (m);
λ_d, λ_c are shape factors given in Table 8.6;
α is a rheological factor given in Table 8.7;
E_c is the weighted value of E_M immediately below the foundation;
E_d is the harmonic mean of E_M in all layers up to $8B$ below the foundation;
σ_{v0} is the total (initial) vertical stress at the level of the foundation base;
q is the design normal pressure applied on the foundation.

In residual soils the values of the rheological factor should be adapted for each situation. Viana da Fonseca (1996) studied thoroughly the application of the elastic formulation:

$$s = \frac{p \cdot B \cdot I}{E_{M/\alpha}} \tag{8.26}$$

By taking representative values for the centre of settlement (as defined by Burland and Burbidge, 1985; or Schmertamn *et al.*, 1970, 1978) and applying them to the footing prototype and large plate load tests, a very clear trend to the rheological coefficient was found, with $\alpha = 0.5$ for service loads, decreasing to $\alpha = 0.33$ for higher loads. These materials have typical values of E_M/p_{LM} in the range of 10–12, agreeing well with the silty materials group in Table 8.7.

It is also interesting to complement the trend of this parameter with the observation that has been made of the applicability of the unload-reload modulus from the PMT (E_{MR}) for the settlement estimation of shallow foundations.

Typical ratios of E_{MR}/E_M of 1.4–2.0 have been found for these soils (Viana da Fonseca and Coutinho, 2008) which, for $\alpha = 0.5$, converts the previous simple elastic formulation to:

$$s = \frac{p \cdot B \cdot I}{E_{MR}} \tag{8.27}$$

The determination of this modulus has been accepted as good practice and defended by the French rules and elsewhere, since it provides a way to solve some of the problems associated with soil disturbance in pre-bores, which affect the "virgin" curve and consequently the E_M.

Conclusion on the methods for prediction of settlement of footings in residual soils

In summary, one can point out the following trends:

- Terzaghi and Peck proposal led to settlements 2 to 4 times higher than observed;
- Parry's (1978) proposal, taking $\alpha = 0.3$, has given reasonable results for the very early load levels (up to 20% of failure, before yield, defined in Viana da Fonseca *et al.*, 1997), but is strongly non-conservative for higher load levels;
- Burland and Burbidge (1985) proposal (average $\alpha = 1.71$) is roughly conservative, with values of predicted settlements 2 to 3 times higher than the observed ones (for loads up to serviceability limits, $s/B = 0.75\%$); a lower value for $\alpha = 0.855$, is in accordance with similar trends in Brazilian residual soils (Rocha Filho, 1986).
- From the CPT based semi-empirical solutions for settlement evaluation, Schmertmann *et al.*'s (1978) method was tested with fine layer discretisation for the most representative PLT ($D = 0.60$ m and 1.20 m). An excellent reproduction of the observed curves was obtained (even in non-linearity terms) when the values of $E/q_c = \alpha$ were modified to 4.0 to 4.5, higher than those proposed by the authors for sandy soils.
- Ménard's rheological factors ($\alpha = E/E_{pm}$) for correction of PMT modulus in order to get the best convergence between observed settlements in PLT tests (at serviceability load levels) and calculated by means of the classical elastic solution taking into account the concept of settlement centre (Viana da Fonseca, 1996), were found to be typical of silty soils ($\alpha = 1/2$), corresponding to the actual grain size distribution of this saprolitic soil. The use of PMT unload-reload modulus happens to give the direct values of the Young's modulus to be taken in the same solutions. On the other hand, the values of SBPT unload-reload moduli reproduce the behaviour of intermediate cycles in PLT tests.
- Finally, a load-settlement analysis of the most significant PLT, similar to CPT interpretation but using DMT Modulus (E_D) was made. The non-linear methods from Leonards and Frost (1988), based on Schmertmann's influence diagrams, and Robertson (1991) were used and the best fit with the experimental results was obtained for a factor of $E/E_D = 2.34$, which is an intermediate value between that for NC and OC sandy deposits (Berardi *et al.*, 1991). The non-linearity of both PLT curves ($D = 0.60$ m and 1.20 m) was also reproduced.

Table 8.8 Secant deformability modulus back-calculated from the footing load test

Load criteria	$s/B = 0.75\%$	$F_s = 10$	$F_s = 5$	$F_s = 2$
E_s (MPa)	17.3	20.7	17.5	11.0

A more detailed analysis of some approaches based on the Theory of Elasticity was developed in another paper (Viana da Fonseca, 2001), comparing semi-empirical methodologies using the results of SPT, CPT, PLT and triaxial tests on high quality samples, with the results from instrumented field tests. Some of the well-established methods (Parry, Burland and Burbidge, Anagnastopoulos *et al.*, Schmertmann *et al.*, Robertson, Ghionna *et al.*, and Wahls and Gupta) were tested and some modifications to the parameters and methods were suggested to give a best fit to the observed behaviour.

- An analysis of the results of the footing test was conducted by reference to the serviceability limit state criteria referred to above (Décourt, 1992). From this load value, which corresponds to a settlement of 0.75% of the loading area diameter, and the loads corresponding to different global safety factors towards bearing capacity failure, it is possible to calculate the secant modulus by elasticity theory (see Table 8.8).
- Accepting that the design modulus E is proportional to q_c ($\alpha = E/q_c$);
- Considering the increase of q_c from CPT with depth (see Figure 11a), a convergence analysis was made, based on an elastic solution, by accepting the settlement centre concept. The procedure was based on the proposal of Burland and Burbidge (1985) for evaluation of the depth of influence as a function of the degree of non-homogeneity, E_0/kD.
- The degree of inhomogeneity, E_0/kD, enabled the determination the position of the settlement centre from Burland and Burbidge's chart. Associating the value of q_c for that depth with the secant deformability modulus back-analysed from the footing load test, the value of α becomes equal to 3 for global safety factor of around 5, and 4 for global safety factor of around 10. The lower value ($\alpha = 3$) is consistent with the serviceability limit criteria, although it can involve significant plastification in the ground. This has been confirmed by numerical elasto-plastic analysis (Viana da Fonseca *et al.*, 1998; Viana da Fonseca and Almeida e Sousa, 2002).
- The method of Schmertmann *et al.* (1978) for settlement evaluation was considered, combining the proposed strain influence factor diagrams with the variation of E over depth. It has to be noted that this approach introduces a non-linearity in stiffness, in spite of being based on a unique equation for E. This formulation was applied to the footing load test results, considering moderate stress levels ($Fs \cong 2.4$, $q_s = 400$ kPa), and revealed excellent agreement between calculated and experimental load-settlement curves for α equal to 4.5, with the non-linear response very well modelled (Figure 8.25).

  The value of α is somewhat higher than that commonly considered for normally consolidated sandy transported soils, due to natural structural factors, associated

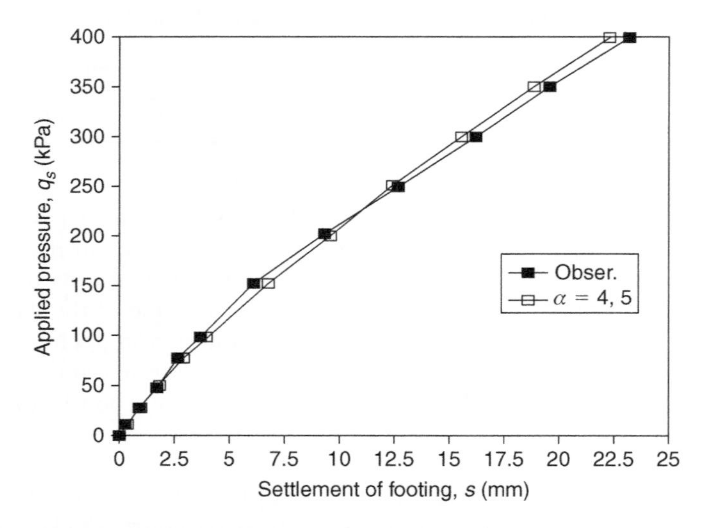

Figure 8.25 Comparison between observed settlement of the footing load test and that calculated using CPT results and adopting Schmertmann *et al.* (1978) coefficients

with the relict interparticle cementation and fabric of residual soils. Robertson *et al.* (1988) had already stated that α values could be as high as 3.5 to 6.0, or 6.0 to 10.0, for aged normally consolidated and highly overconsolidated sandy soils, respectively. The coefficients to be applied to these solutions for settlement evaluation in Brazilian residual soils are in strict accordance with these. Rocha Filho (1986) applied the Robertson *et al.* proposals to the results of loading tests on shallow foundations and plates with diameters from 0.40 to 1.60 m, carried out on residual soils from gneiss in the university campus of PUC, Rio de Janeiro, resulting in ratios of calculated to observed settlements of between 1.5 and 2.5. The ratios obtained in this study are even larger (2.7–3.4).

Note on the strength for ultimate capacity evaluation in residual soils

With regard to strength evaluation, theoretically a peak and a post-peak resistance can be obtained by pressuremeter tests (Gomes Correia *et al.*, 2004). However, because of the influence of disturbance during installation, the peak resistance is usually ignored for PMT. The undrained shear strength can be obtained from the Ménard limit pressure, p_{lm} (Amar *et al.*, 1994):

$$s_u = \frac{(p_{lm} - \sigma_h)}{5.5} \quad \text{for } (p_{lm} - \sigma_h) < 300\,\text{kPa} \tag{8.28}$$

$$s_u = 25 + \frac{(p_{lm} - \sigma_h)}{10} \quad \text{for } (p_{lm} - \sigma_h) > 300\,\text{kPa} \tag{8.29}$$

where p_{lm} is the applied pressure required to double the cavity diameter and σ_h is the estimated *in situ* horizontal stress.

For granular residual soils, Cassan's (1978) assumption that the behaviour of a granular material is a function of the average effective stress $[(\sigma_r' + \sigma_\theta')/2]$ and that the volume change will follow a curve that has two plastic components (one under constant volume and other variables, allowing relationships between the volume and strain to be formulated), together with the plastic criteria, allows the determination of the Mohr-Coulomb strength parameters ϕ' and c':

$$p_L = (1 + \sec\phi') \cdot (p_0 + c' \cdot \cot\phi') \cdot \left[\frac{E_{pm}}{A \cdot (p_0 \cdot \sec\phi' + c' \cdot \cos\phi')}\right]^{(\sec\phi'/1+\sec\phi')} - c' \cdot \cot\phi' \tag{8.30}$$

This method requires more than a single test in a specific horizon, and ideally a multiple regression analysis. In materials where the dilatancy is significant, however, it may have errors of up to 30%.

Method to calculate the bearing resistance of spread foundations

A semi-empirical method to calculate the bearing resistance of spread foundations using the results of an MPM test is as follows:

$$\frac{R}{A'} = \sigma_{v0} + k(p_{LM} - p_0) \tag{8.31}$$

where

R is the resistance of the foundation against normal loads;
A' is the effective base area as defined in EN 1997-1;
σ_{v0} is the total (initial) vertical stress at the level of the foundation base;
p_{LM} is the representative value of the Ménard limit pressures at the base of the spread foundation;
$p_0 = [K_0(\sigma_{v0} - u) + u]$ with K_0 conventionally equal to 0.5, σ_{v0} is the total (initial) vertical stress at the test level and u is the pore pressure at the test level;
k is a bearing resistance factor given in Table 8.9;
B is the width of the foundation;
L is the length of the foundation;
D_e is the equivalent depth of foundation.

Ultimate Limit State defined by excessive deformations – numerical modelling

An FEM numerical analysis of a load test on a carefully instrumented concrete footing, resting on a residual soil from Porto granite, is described in Viana da Fonseca and Almeida e Sousa (2002). The hyperbolic soil model was adopted in order to examine the ideal range of stress-strain level data for parametrical evaluation, when the purpose is to simulate foundation pressure-settlement relations in service conditions. The importance of accurately defining the volume for the overall simulation of such a test, taken up to failure, was emphasised.

A large number of triaxial tests (43) wereperformed with different specimen sizes, consolidation stress conditions and stress-paths. These specimens were obtained by driving thin wall steel tubes (with dimensions equal to those of the specimens to be

Table 8.9 Correlations for deriving the bearing resistance factor, *k*, for spread footings

Soil category	p_{LM} *category*	p_{LM} *(MPa)*	*K*
Clay and silt	A	<0.7	$0.8[1+0.25(0.6+0.4\,B/L)\times D_e/B]$
	B	1.2–2.0	$0.8[1+0.35(0.6+0.4\,B/L)\times D_e/B]$
	C	>2.5	$0.8[1+0.50(0.6+0.4\,B/L)\times D_e/B]$
Sand and gravel	A	<0.5	$[1+0.35(0.6+0.4\,B/L)\times D_e/B]$
	B	1.0–2.0	$[1+0.50(0.6+0.4\,B/L)\times D_e/B]$
	C	>2.5	$[1+0.80(0.6+0.4\,B/L)\times D_e/B]$
Chalk			$1.3[1+0.27(0.6+0.4\,B/L)\times D_e/B]$
Marl and weathered rock			$[1+0.27(0.6+0.4\,B/L)\times D_e/B]$

Note: This example was published in *Fascicule* No 62-V (1993).

used in the triaxial cell) into blocks extracted from a depth of 0.5–1.0 m below the level of the footing base (Viana da Fonseca and Almeida e Sousa, 2002). All the sampling and handling procedures were undertaken with the utmost care in order to preserve, as much as possible, the natural structure of the soil.

The curves resulting from these tests exhibit substantial brittleness for the lowest effective consolidation stress, while for the others the brittle behaviour tends to disappear, probably due to the development of volumetric (collapsible) plastic strains prior to shearing. Thus, for consolidation stresses much higher than those at rest stress state, the stress-strain response tends to be typical of de-structured materials. For the lower consolidation stresses (10 and 20 kPa) peak values of deviator stress are mobilised before the highest value of the dilatancy ratio is reached. This indicates that mechanical behaviour of the saprolitic soil is controlled by cementation between particles rather than by dilatancy phenomena related to particle interlocking. This latter type of behaviour, which is typical of dense granular transported soils, is not compatible with the fabric of saprolitic soil that exhibits a low to medium density ($e = 0.60$–0.85) for an essentially sandy soil.

Figure 8.26a presents stress-strain curves from one of the tests obtained by classical (external LVDT) and local strain measurement systems (Viana da Fonseca, 1996). The former technique leads to rather unrealistic values of stiffness. The curve for the local instrumentation results was assumed as representative and plotted in the modified axis in order to derive the hyperbolic parameters (Figure 8.26b).

As is clear in Figure 8.26b, the Young's modulus deduced from testing results using a hyperbolic trend will depend strongly on the strain range where the model is applied. To best understand this pattern of behaviour, Figure 8.26b may be seen as a good model of what should be expressed as a multiphase solution. The mechanical performance seems to be marked by different trends in three ranges of values that could be associated with three pre-yield zones. Three different initial tangent Young's moduli can then be derived, one for each of these zones, from a classical hyperbolic approach of the stress-stain curve: designated by $E_{ti,0}$, $E_{ti,i}$ and $E_{ti,h}$, these moduli have taken into account, respectively, the lowest shear stress values (where the natural bonding between particles is mostly preserved inside the elastic yield locus), the intermediate states (representative of the metastable condition, where progressive de-structuring

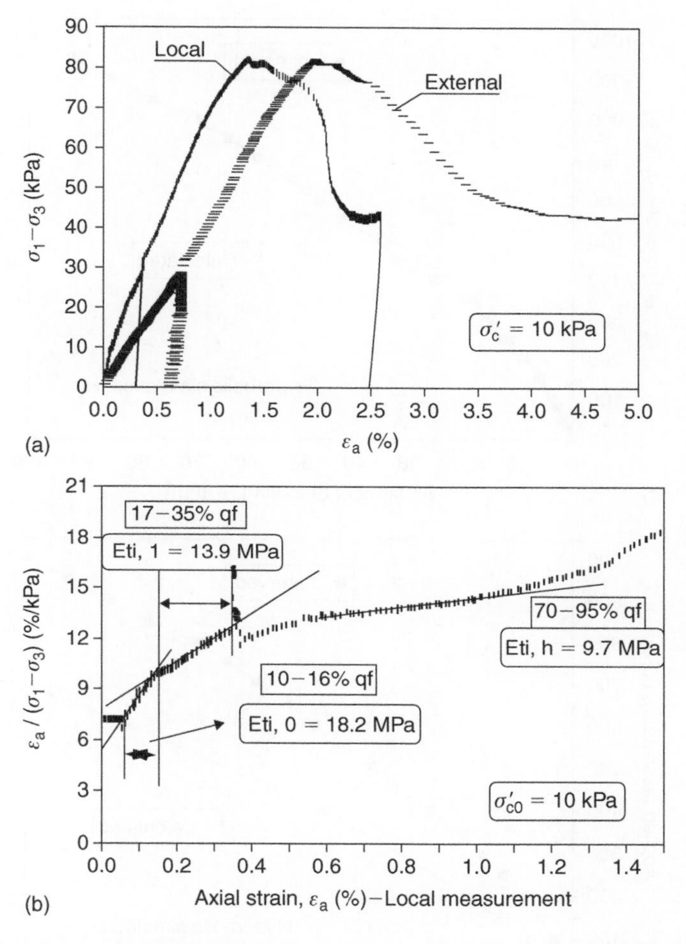

Figure 8.26 CID triaxial test. (a) Stress-strain relations with local and external instrumentation; (b) Hyperbolic tangent Young modulus for different stress ranges: $E_{ti,0}$, $E_{ti,i}$ and $E_{ti,h}$

is observed), and finally the values of shear stress between 70 and 95% of failure and where the behaviour is significantly of a granular type. Figure 8.27 illustrates these different trends and adopted options (Viana da Fonseca and Almeida e Sousa, 2002). The first parameters reflect the lowest stress levels where interparticle structure is more preserved and elasticity seems to prevail. The second value, ranging over the intermediate stress levels, can be associated with the beginning of plastic yielding and progressive breaking down of micro-structure (phenomena that will dominate major zones of the ground below the footing, since an at-rest stress state, characterised by low values of K_0, ~[0.35–0.59] is expected – Viana da Fonseca and Almeida e Sousa, 2001). Finally, the highest stress levels, at the stable classically assumed ranges of 70–95% of failure (Duncan and Chang, 1970) that will be associated with strongly de-structured matrices.

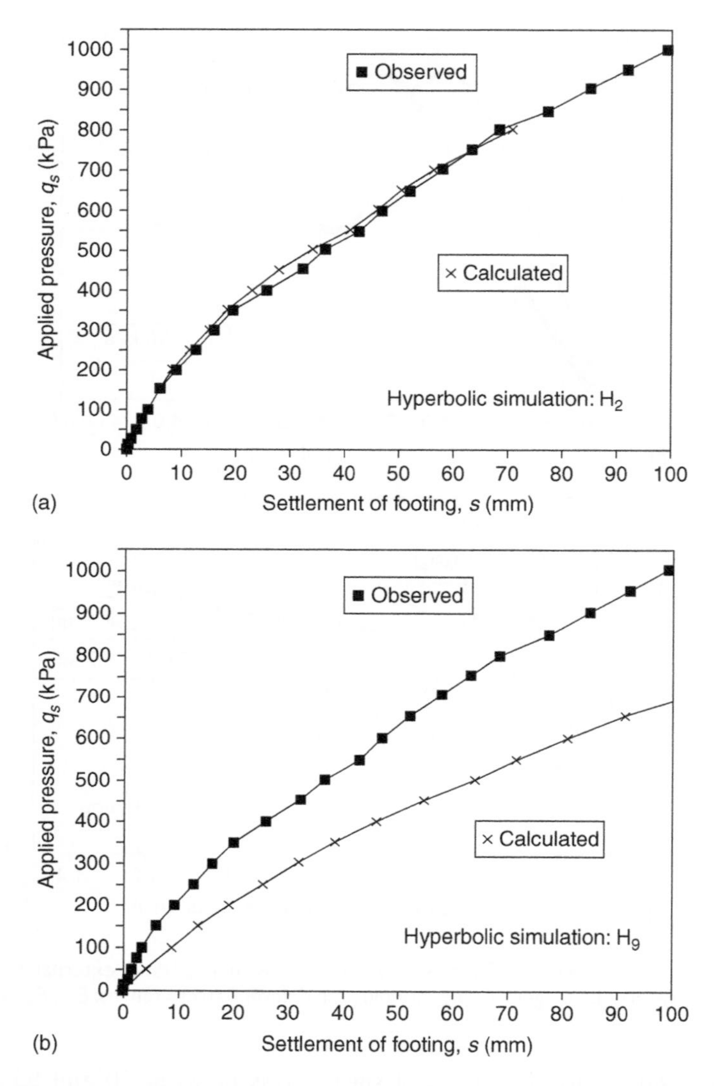

Figure 8.27 Numerical simulation of the footing load test using the hyperbolic model parameters from triaxial tests for the two extreme modelling assumptions: (a) low strain level, and (b) high strain level

The definition of the overall set of hyperbolic parameters for these distinct levels of stress-strain is presented in Viana da Fonseca and Almeida e Sousa (2002), together with a modelling of all significant triaxial tests executed for this study.

From the observation of these curves, it has been found that:

(i) the hyperbolic modelling of the tests with a single set of parameters does not reproduce the overall behaviour of the stress-strain curves;

(ii) modelling the curves from parameters based on $E_{ti,0}$ reproduces with great accuracy the low stress-strain levels but is very conservative on the whole curve;
(iii) on the other hand, the curves modelled from $E_{ti,h}$ values fit very well the later part of the curves, showing important differences when compared to the low to medium levels; the use of intermediate parameters ($E_{ti,i}$) will be a compromise.

Additional studies have been carried out on the importance of the stressed volume on the behaviour of this soil, as it was expected that the ground below the footing would be subjected to increasing average stresses.

The numerical analysis of the concrete footing load test was intended to determine the best parameters for the modelling and prediction of the behaviour of such a foundation. It was presented in the paper by Viana da Fonseca and Almeida e Sousa (2002). Two of the figures presented are combined in Figures 8.27a and b, allowing a comparison between the curves of the applied pressure versus the settlement of the footing obtained experimentally and resulting from the numerical simulation using the hyperbolic model.

There is a clear benefit in showing the sensitivity of the parameters to different levels of stress-strain. A brief summary of the results of the analysis by Viana da Fonseca and Almeida e Sousa is as follows:

(i) only the analyses based on the initial tangent Young's modulus, defined from the low strain ranges of triaxial tests on undisturbed samples ($E_{ti,0}$), represent the footing test reasonably throughout the loading steps;
(ii) the value of $Rf = 0.9$ seems to help the model to fit the data well, particularly in the global definition of the observed curve, while the option for $Rf = 0.7$ shows a poorer fit when considering the deformability parameters for "small" strain ranges, and also in the vicinity of failure;
(iii) analyses that are based on the initial tangent modulus for ranges of "medium" and "large" deformations in the triaxial tests ($E_{ti,i}$ and $E_{ti,i}$) significantly overpredict the observed settlements in the footing load test;
(iv) it is essential to consider the variation of Poisson's ratio along several loading steps, up to values very close to the ultimate load ($q = 950$ kPa), with clear benefit in the definition of the solution of the settlement of the footing base.

It is pertinent to reiterate the reasons why the calculation, based on the parameters deduced from initial tangent modulus for very low stress-strain ranges ($E_{ti,0}$) of triaxial tests, seems to simulate the behaviour of the load test very well. In fact, the parameters deduced from triaxial tests taking higher levels of stress-strain modulus, already include the effect of plastic deformations related to progressive failure of the soil structure, and therefore they represent less satisfactorily the behaviour of the soil in natural conditions. The consequence of this, as was illustrated in Figure 8.9 where modelling parameters were defined from the classic ranges of 70–95% of failure stress, is to significantly overestimate the observed settlements. For these elevated levels of deviatoric stresses the material behaves as a non-structured soil, and the parameters deduced cannot represent foundation behaviour under service conditions.

The influence of different model parameters, in accordance with distinct stress-path zones in the ground, was analysed. This can be relevant in some marginal zones, where stress-paths are rather different from classical compression paths. It was shown that, up to moderate load levels, that is to say, to stress-strain levels not dominated by significant plastic behaviour, this stress-path factor is not important in the simulation of the settlements of the footing base. The influence of the anisotropy of the stiffness characteristics is, however, relevant in the prediction of the settlement of external points to the loaded area, where the stress-paths are significantly different from classic compression.

In tropical residual soils, most shallow foundations will involve dealing with unsaturated soils, and two common problems which flow from this relate to collapsible soils and expansive soils.

8.3 FOUNDATIONS ON UNSATURATED SOILS

According to Fredlund and Rahardjo (1993), matric suction dramatically increases the bearing capacity of the soil, and this is shown in Figure 8.28 which illustrates the effect of various matric suction values on the bearing capacity of shallow foundations. When attempting to arrive at a suitable design value for matric suction, it is useful to construct a plot similar to this figure.

Since shallow footings are normally placed well above the ground water table, if adequate surface and subsurface drainage is provided around the structure, it may be reasonable to assume that negative pore-water pressures will be maintained immediately below a footing. It should also be realised that there may be a fluctuation in the groundwater table as a direct result of building the structure. In some cases, the

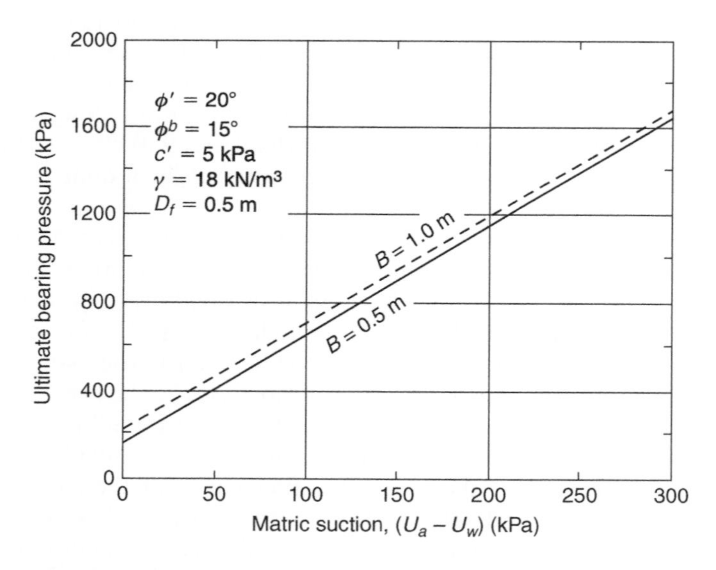

Figure 8.28 Bearing capacity of a strip footing for various matric suction values (*After* Fredlund and Rahardjo, 1993)

groundwater table may be lowered but, more commonly, the water table will rise due to excessive watering of the vegetation surrounding the building.

There are many situations where the groundwater table is far below the ground surface and a hydrostatic profile is not reasonable for design purposes. In these cases, measurements of the *in situ* suction below the footings of existing structures in the vicinity can prove to be of value. These suctions can be relied upon to contribute to the shear strength of the soil, but the decision regarding what value of suction to use in design becomes dependent upon local experience and the microclimate in a particular region.

Design considerations for shallow foundations

The following soil-structure interaction issues should be considered in foundation design. First, soil-structure interaction implies, by definition, that the response of the soil and the structure are interdependent; i.e., the reaction of the soil to the structure affects the performance of the structure, and the reaction of the structure to the soil affects the behaviour of the soil. One of the two following approaches is needed to solve this problem (Houston, 1996).

Method 1: Employ a finite element code which models all components including the structure and the soil in contact with it. However the commercially available codes, being the ones in common use, tend to be either written for structural analysis and hence model structures well (and soils less so), or written for geotechnical analysis and hence model soils well (and structures less so). Ideally, the model for the soil would account for non-linearity in the stress-strain-time constitutive law, including water flow and the strains induced by changes in soil water content. By this method, all responses to load and water would be totally coupled.

Method 2: The response of the soil and structure are decoupled, and then recombined to ensure compatibility via iteration.

Houston (1996) showed an example of this approach with details, and concluded that simplified, and often greatly simplified, versions of this procedure are used in design practice, with corresponding approximations and errors. In most cases, the factors of safety are adequate to accommodate these approximations and errors.

8.3.1 Shallow foundations on collapsible soils

In the case of collapsible unsaturated soils, foundations can behave satisfactorily for some time, and then suddenly suffer significant additional settlement, due to the accidental appearance of a water source that starts to flood through the soil (see Coutinho *et al.*, 2004b).

The amount of volume decrease experienced by a collapsible soil upon wetting under load depends on several factors, including the soil type, the initial water content and dry density, the degree of wetting, and the stress state and boundary conditions.

According to Houston (1996), difficulties associated with estimating collapse settlement include the typical variations in cementation and gradation causing soil properties to change over distances of only a few centimeters both laterally and vertically, and also lack of knowledge of sources of water. The greatest uncertainty in estimating collapse settlement is linked to the uncertainty of the lateral extent and

Table 8.10 Comparison between r_{cpred} from PMT data and r_{cmeas} (After Dourado and Coutinho, 2007)

		Settlement predicted (mm)		
Test	σ_{soaked} *(kPa)*	*Before wetting*	*After wetting*	*Collapse (mm)* ($r_c = r_{P\ soaked} - r_{P\ nat}$)
PC 01	100	0.56	15.06	**14.5**
PC 02	60	0.34	8.84	**8.5**
		Settlement predicted (mm)		
Test	σ_{soaked} *(kPa)*	*Before wetting*	*After wetting*	*Collapse (mm)* ($r_m = r_{m\ soaked} - r_{m\ nat}$)
PC 01	100	1.24	46.24	**45.0**
PC 02	60	0.56	21.06	**20.5**

degree of wetting. The degree of saturation achieved during the conduct of conventional laboratory response to wetting tests is quite high, typically 85 to 95%. Estimated collapse settlements based on full-wetting collapse potential may not be realised in-situ.

The most likely explanation for why the actual field collapse settlement is less than the estimated settlement is that, when a soil is only partially wetted, only a portion of the full collapse potential is realised. In practice, there are many field situations for which only partial wetting occurs. However, not all field wetting conditions result in low degrees of saturation. If the water source is due to a rising groundwater table, or perhaps from long-term, steady state wetting by ponded water, the degree of wetting achieved *in situ* may be adequate to result in essentially full-wetting collapse. In these cases, however, the lateral extent of wetting may still be quite difficult to estimate (Houston, 1996).

Dourado and Coutinho (2007) predicted collapse settlements using PMT results and considering the traditional methodology of Briaud (1992) for footing settlement on sand. The settlements were calculated at both natural water content and fully saturated conditions and the difference in these two conditions was considered as the collapse settlement (rc) of the soil (see Table 8.10).

$$s = \frac{2}{9E_d} \cdot \sigma'_{v0} \cdot B' \cdot \left(\lambda_d \cdot \frac{B}{B'}\right)^{\alpha} + \frac{\alpha}{9E_c} \cdot \sigma'_{v0} \cdot \lambda_c \cdot B \tag{8.32}$$

where

s = footing settlement (final);
E_d = pressuremeter modulus within the zone of influence of the deviatoric tensor;
E_c = pressuremeter modulus within the zone of influence of the spherical tensor;
σ'_{vo} = footing net bearing pressure;
B' = reference width of 0.6 m;
B = width or diameter of the footing ($B > B'$);
α = rheological factor;
λ_d = shape factor for deviatoric term and
λ_c = shape factor for spherical term.

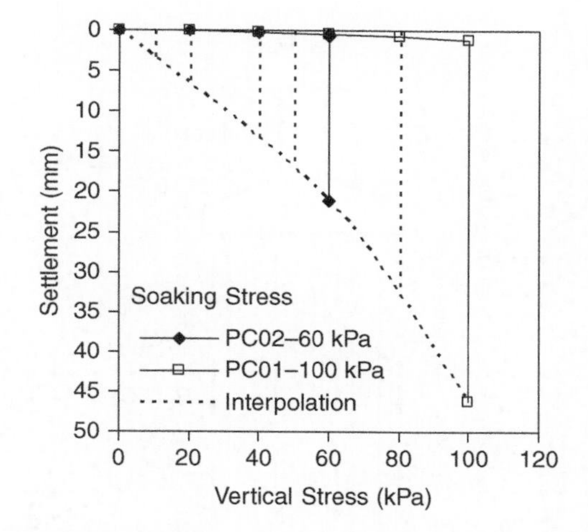

Figure 8.29 Interpolation of the settlement-stress relationship from *in situ* collapse tests (*After* Dourado & Coutinho, 2007)

Table 8.11 Comparison between $r_{c\,int}$ obtained by interpolation and r_c predicted (*After* Dourado & Coutinho, 2007)

Stress (kPa)	$r_{c\,int}$ *(mm)*	$r_{c\,pred}$ *(mm)*	*Factor of increase* $(F_m \cdot r_{c\,pred} = r_{c\,int})$
100	45.0	14.5	3.1
80	31.6	11.4	2.8
60	20.5	8.5	2.4
40	12.5	5.7	2.2
20	6.3	2.8	2.3

Table 8.10 shows that the predicted settlements in the natural condition (0.56 and 0.34 mm) were about 50% of the measured settlements (1.24 and 0.56 mm). It is also seen that the predicted collapse settlements $r_{c pred}$ (14.5 and 8.5 mm) were between about 33 and 50% of the measured settlements $r_{c meas}$ (45 and 20.5 mm).

An increase in settlement is also seen with an increase in soaking stress. Thusthe influence of the soaking stress in the $r_{c pred}$ was evaluated. The data is summarised in Figure 8.29 and Table 8.11.

Dourado and Coutinho (2007) concluded that for any stress, predictions based on PMT data underestimated the r_c measured in the plate load test. The use of a factor of increase (F_m) of 2.5 in predicted collapse settlement $r_{c pred}$ ($F_m \cdot r_{c pred} = r_{c\,int}$) by the PMT, would lead to a better approach to the prediction of r_c for the *in situ* collapse tests. However, the authors cautioned that this factor applied only to the soil studied.

8.3.2 Deep foundations on collapsible soils

Typically, the calculation of load capacity of piles seeks a balance between the applied loads (Q) and the available resistance, made up of Shaft Resistance (R_s) and Toe

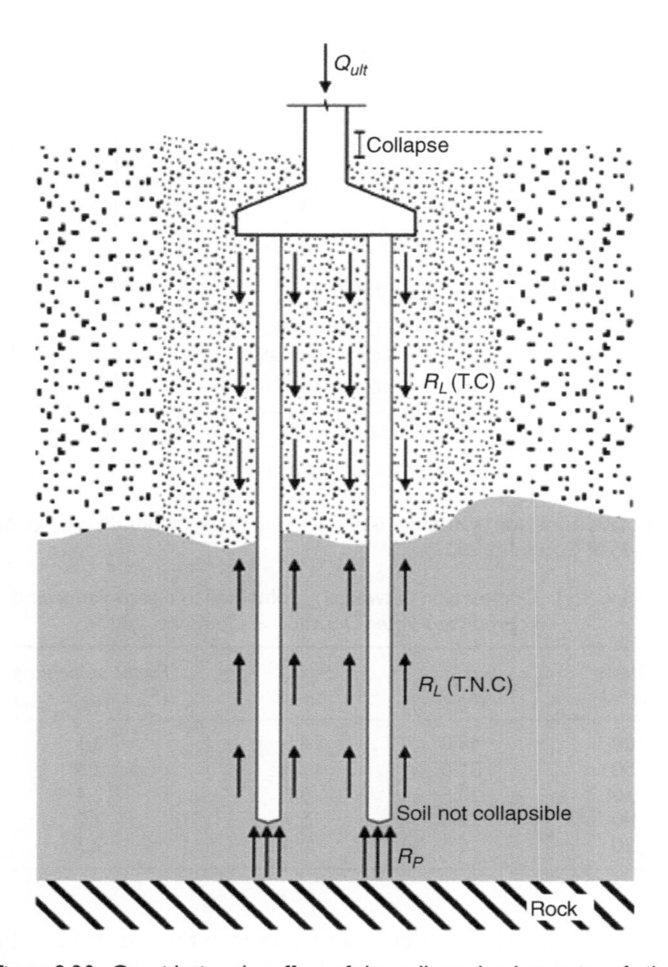

Figure 8.30 Considering the effect of the collapse load capacity of piles

Resistance (R_p) (Figure 8.30). In the presence of collapsible soil, we must consider the additional applied load due to the collapse, in the form of negative shaft friction. This is difficult to determine because of the complexity of the relative movements along the shaft. As a simplified solution, we can calculate the shaft resistance of the collapsible segment and apply this as a distributed load along the shaft. In general we have the following equation:

$$Q_{ult} + R_s(TC) = R_p + R_s(TNC) \tag{8.33}$$

where

Q_{ult} = Maximum applied load acting on the pile;
R_s (TC) = Applied load due to negative shaft friction on the collapsible length;
R_p = Toe resistance;
R_s (TNC) = Shaft resistance of non-collapsible length.

In some standards, such as AS 2159 in Australia, negative shaft friction (NSF) is assumed not to affect the Ultimate Limit State, since the deflections associated with the ULS are generally much greater than those required to fully reverse any NSF. However, NSF can have a significant effect on settlements at working load and hence on the Serviceability Limit State (SLS). Checks on the SLS with full NSF applied are therefore required. The same may well be true with regard to collapse settlements and R_s (TC), but the magnitude of the collapse settlement will need to be checked against the magnitude of the settlement at ULS.

Both shallow and deep foundations in collapsible soil can behave satisfactorily for some time, and then suddenly incur additional settlement of considerable magnitude, due to the accidental appearance of water that starts to flood through the soil.

A geotechnical profile and water content from Eunápolis City was described by Coutinho *et al.* (2010). This site experienced damage during the execution of the deep foundations, and the presence of collapsible soil was considered to be one of the causes. The objective of this study was to identify and characterise the soil and investigate the possibility of influence on the pile foundation.

For this purpose, two static compression pile load tests were performed; one with the soil in natural condition and the other after a flooding process. A system of soaking was used for the flooded load test, through a pit built with soil drains. The pit was subjected to a constant water head, using a water reservoir connected to a car. The total time of soaking was 72 hours, after which the flooded load test was started. Further information is given in Coutinho *et al.* (2010).

Comparing the natural and flooded moisture content profiles, it was seen that flooding was effective to a depth of 10 m, showing that all the length of the pile shaft was flooded. However, due to the higher N_{SPT}, of the order of 50 blows, in the layer at the tip of the piles, the soaking did not reduce the toe resistance.

Analysis of load/settlement curves

Figure 8.31 shows a comparison between the results of the two load tests. Some points of the curve were used to help in the interpretation and comments:

1. At the vertical axial working load of 40 tf (400 kN), there is no significant difference in performance between the piles;
2. Differences in the behaviour of the two piles are observed above an applied load of 51 tf (510 kN);
3. In relation to the settlements, there was a significant increase for the flooded soil; for example, at the load of 64 tf (640 kN), it increased from 10 mm to15 mm;
4. At a load of 72 tf (720 kN), the settlement in the natural soil was 15 mm and in the flooded soil 20 mm. It might have been expected that the difference would be larger, however at this point, the pile toe starts to work for the flooded pile because of the high N_{SPT}.

The blue dotted line shows an extrapolation of the curve for flooded soil without the influence of the strong toe. Considering the pile response in natural soil, the maximum load of 80 tf (800 kN) corresponds to a settlement of 27.1 mm. For this value of settlement to occur, according to the blue dotted line, the applied load would be

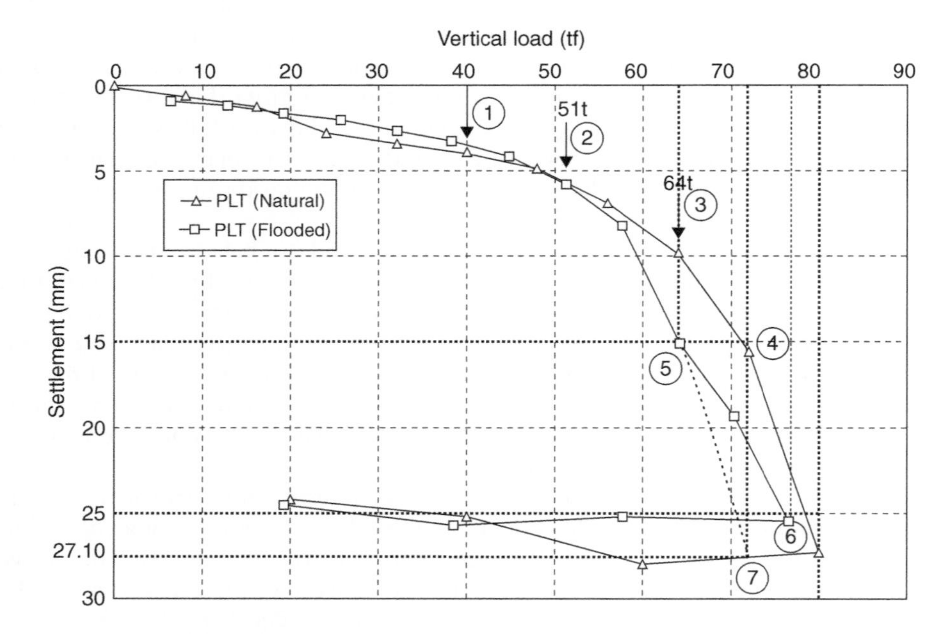

Figure 8.31 Pile load test curves and settlement data for the piles in natural soil and in flooded soil

only 72 tf (720 kN). This represents a reduction in the rupture load of 80 kN, or in the order of 10%.

Load capacity

a) Natural load test (PC – Natural)

Décourt and Quaresma (1978), and Décourt (1996) considered a layered soil profile. The shaft resistance (R_s) was calculated using the average value of N_{SPT} for each layer. For the toe resistance, an undrained shear strength of $s_u = 250$ kPa was used.

In lateritic soils, the shaft resistance of piles can be 2 to 3 times the resistance calculated by conventional prediction methods (Décourt, 2002). The R_s in the layer with iron concretions (between 7.0 and 8.0 m) was calculated separately using a factor of increase of 2.5.

A total shaft resistance (R_s) of 475.4 kN and a tip resistance (R_p) of 227 kN were calculated, making a total ultimate capacity (Q_f) of 702.4 kN (see Table 8.12).

b) Flooded load test (PC – Flooded)

For the flooded soil, a total shaft resistance (R_s) of 291.5 kN and a tip resistance (R_p) of 340 kN were calculated, making a total ultimate capacity (Q_f) of 631.5 kN. The results of both load tests are presented in Table 8.12.

Thus, using the Décourt and Quaresma (1978) method, it can be seen that the predicted shaft resistance reduces from 475.4 to 291.5 kN, or 38%, due to the flooding. The value of the flooded toe resistance (340 kN) was greater than in the natural soil

Table 8.12 Summary of the results

	Décourt and Quaresma (1978) – results (kN)		
Load test	*Shaft resistance* (R_s)	*Toe resistance* (R_p)	*Ultimate capacity* (Q_f)
Natural	475.4	227.0	702.4
Flooded	291.5	340.0	631.5

Table 8.13 Estimates of Q_f obtained and measured

	Rupture Load (Q_r) *(kN)*			
Load test	*Décourt (1996)*	*Van Der Veen (1953)*	*Décourt and Quaresma (1978)*	*Dynamic load test*
Natural	812 RL = 504 kN	809	702.4	770
Flooded	–	836	631.5	–

because of the differences between the soil profiles. However, a reduction in the flooded failure capacity of 10% (702.4 to 631.5 kN) was observed.

The Van Der Veen (1953) method was also used to estimate the ultimate capacity. It gave a good estimate for both curves, with values of $Q_f = 809$ kN (natural) and $Q_f = 836$ kN (flooded).

Van Der Veen method was also applied for the hypothetical case of the corrected flooded curve, the blue dotted line in Figure 8.31. The results showed a Q_f in the order of 700 kN, confirming the conclusion above (reduction of $809 \rightarrow 700 = 13.5\%$).

Table 8.13 presents the ultimate capacities obtained through the three methods utilised in this work (Van Der Veen, 1953; Décourt and Quaresma, 1978; Décourt, 1996). The result obtained by a dynamic load test on the same pile while in a natural condition is also shown.

It can be seen that the values obtained for the failure load were all around 800 kN. The exception was the prediction by the Décourt and Quaresma (1978) method for the flooded pile.

Effect on the ultimate capacity

For the total failure load, the difference is relatively small (10%) because the reduced shaft capacity was replaced by increased toe resistance as a result of the differences between the soil profiles.

The values obtained in the static load tests (Figure 8.31) were of the same order of magnitude. However, if the very resistant layer below of the toe of the flooded pile did not exist, an influence in the order of 10%, after flooding was expected. A similar result is seen in the prediction by the Décourt and Quaresma (1978) method.

Some cases reported in Brazilian literature show a reduction in the results of the total failure load due to the flooding process in the order of 20%. For most well designed foundations, a reduction in ultimate capacity of 10 to 20% would probably not be critical. However, as noted previously, the effect on the Serviceability Limit State could be very important, as suggested by the increase in settlement in the test pile at 640 kN.

8.3.3 Mitigation measures

Several mitigation alternatives are available for dealing with collapse phenomena. Mitigation measures which have been used in the past can generally be fitted into one of the categories below. Often a combination of mitigation measures is used (Houston, 1996):

- Removal of a volume of moisture-sensitive soil;
- Removal and replacement or compaction;
- Avoidance of wetting;
- Chemical stabilisation or grouting;
- Prewetting;
- Controlled wetting;
- Dynamic compaction;
- Pile or pier foundations;
- Differential settlement resistant foundations.

More details of mitigation measures can be seen in Houston (1996).

The following measures for minimising wetting represent good practice. If the wetting is anticipated to occur from surface or near surface infiltration, consider:

1. restricted irrigation watering (e.g. desert landscaping);
2. restricted landscape vegetation adjacent to structures, unless placed in planters;
3. paved surfaces around the structure to the maximum extent practical;
4. use of watertight water and sewer lines in double pipes or troughs, and;
5. replacement or removal and compaction of near surface layers to form a low permeability barrier to water. The barrier to water should be composed of moisture-insensitive soils.

As the probability of wetting is reduced, the risks associated with the collapse phenomena are reduced. In the final analysis, the total costs, including the consequence of estimated collapse settlements, over the lifetime of the structure must be considered and compared for various alternatives.

8.3.4 Recent research and developments for dealing with collapsible soils

The importance of effects of wetting on unsaturated soil response is very well documented. The potential for wetting-induced volume change of all unsaturated soils is becoming widely recognised in the geotechnical profession. Most of the early literature on volume moisture-sensitive soils dealt with naturally-occurring deposits, and widespread recognition of wetting-induced compression and swell of compacted soils is relatively more recent. The importance of coarse aggregate to the swell/compression response of soil upon wetting under load has begun to receive attention. *In situ* methods used in the past, and those presently under development, appear promising for studying the response to wetting of gravelly and other difficult-to-sample unsaturated soils. Lateral movements associated with collapse settlement have been considered in some current and recently completed studies. Lateral strains due to wetting are of particular importance for embankments and slopes, whether made of naturally-occurring or compacted unsaturated soils (Houston, 1996).

Studies of collapsible soils currently in progress and planned for the near future include: (1) the response of collapsible soils to earthquake loading, before and after wetting, (2) strength of collapsible soils after wetting, and (3) constitutive soil model development, considering the role of cementation, including soil suction. Constitutive models incorporating stress state, soil suction, and strain, coupled with unsaturated flow models are desirable. Improved methods for estimating the degree and lateral extent of wetting are needed. Additionally, considerations of risk and integration of risk assessment and total life cycle cost estimates into the mitigation and foundation design selection process are very important to the success of future efforts.

8.3.5 Shallow foundations on expansive soils

Expansive soils, which increase in volume when water is available, but shrink if water is removed, are a continuing source of problems in the design, construction, and maintenance of buildings, buried pipes, roads and airfields, canals, and retaining structures. Wray (1995) reported that a soil is commonly considered to have limited expansive tendencies when its plasticity index (PI) is less than 20. If the plasticity index is greater than 20 but less than 40, the soil is considered to have moderate expansive properties. The soil is considered to be highly expansive if the PI is between 40 and 60. Soils with PI's greater than 60 are considered to be very expansive. Another method often used to classify the expansive potential of soil is the expansion index (EI). The EI is determined from a special laboratory test that is performed in a specified standard manner. A soil with an EI of 50 or less is considered to have low expansion potential. A soil of 91 or greater indicates a soil with high or very high expansion potential.

The properties associated with expansive clays such as heave and swell pressure are dependent on three factors: (i) natural soil properties such as moisture content, dry density, plasticity index and compaction, (ii) environmental conditions including temperature and humidity, and (iii) vertical stresses such as overburden pressure and foundation loading conditions. The following sections explain the influence of some of the above factors on the swell properties of expansive soils.

Many unsaturated soils are also expansive. Environmental conditions such as temperature and humidity influence swell potentials by changing suction in unsaturated expansive soils. The engineering behaviour of unsaturated soils can be interpreted in terms of two key stress state variables, namely: net normal stress and suction (Fredlund and Rahardjo, 1993). The suction changes associated with the movement of water in the liquid and vapour phases are called matric suction and osmotic suction, respectively. The total suction is equal to the sum of matric and osmotic suction. More recently, the focus of research has been directed towards understanding the effects of different types of suctions and their influence on swell characteristics in unsaturated expansive soils.

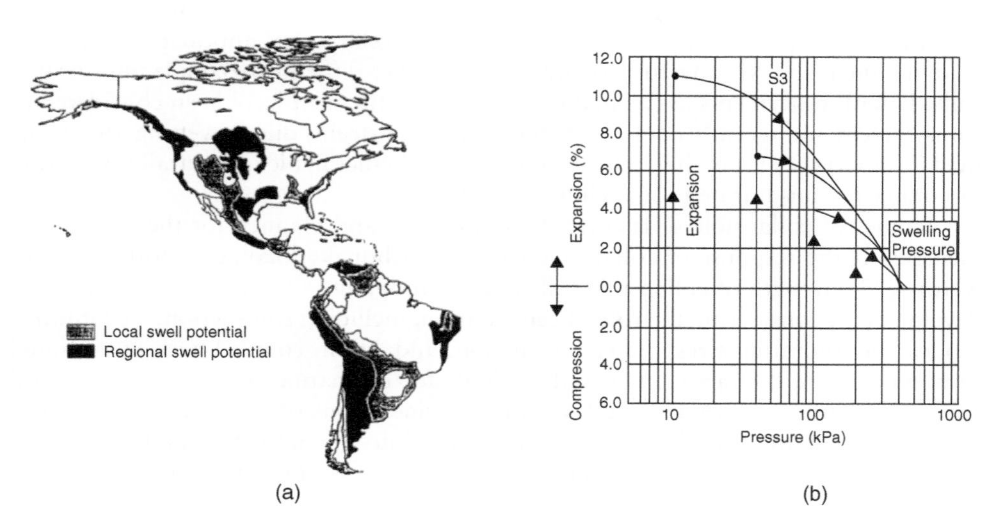

Figure 8.32 (a) Generalised map showing expansive soil distribution in the Americas; (b) Typical results for oedometer tests on samples of an expansive residual soil from Pernambuco, Brazil (*After* Mitchell and Coutinho, 1991)

The final factor that has a major influence on swelling in soils is the effect of overburden pressure on the foundation material at a site. It is well known that the magnitude of swell under confined loading conditions is less than that in unconfined conditions. However, in conventional engineering practice, the majority of laboratory swell tests are conducted either in unconfined conditions with low seating pressures, or in rigid apparatus with unknown lateral confining pressures.

Formation and distribution of expansive soils

Expansive soils are commonly found in arid and semi-arid regions in the world such as Australia, Canada, China, India, Israel, Iran, South Africa, UK and USA. In these areas, the rainfall is moderate, precipitation is seasonal and there are high evaporation rates. Drainage must be sufficiently restricted to permit pore-water salts to remain and become concentrated by evaporation. The topography may be flat, as in bentonite/marine shale deposits, or the grade may be steep, as in volcanic or orogenic settings, where slope stability can be a geotechnical hazard.

Mitchell and Coutinho (1991) included a generalised map showing the distribution of expansive soil in the Americas, compiled on the basis of climate, geology, and engineering experience (Figure 8.32). The distribution in the United States is fairly accurate, since most of the areas are defined from engineering experience. The distributions shown for Canada, Central and South America are less precise, as they rely on climate and geology to a greater degree. Regions of swell (RS) show areas where expansive soils are likely to occur. They do not indicate swelling soils over the entire area. Local areas of swell (LS) represent a more localised potential for expansive behaviour.

Indications of swelling potential may appear from the results of routine tests such as grain size analysis, Atterberg limits, and *in situ* moisture content and dry density.

Table 8.14 Characterisations based on Direct Measurements (from Pupalla *et al.*, 2004)

References	*Test Details*	*Expansion Index*[1]	*Swell Strain, %*	*Characterisation*
Holtz and Gibbs (1956); USBR Method	Oedometer, Zero Lateral Strain; Seating Pressure −7 kPa	—	>30	Very High
			20–30	High
			10–20	Medium
			<10	Low
FHA/HUD classification (1960)[2]	Oedometer, Surcharge of 7 kPa	>130	>12	Expansive
		90–130	7–12	Highly expansive
		50–90	4.7	Moderate expansive
		<50	2.4	Marginal
Ladd and Lambe (1961) (FHA)	Oedometer Setup (Seating Pressures are not known)	—	>6	Very Critical
			4–6	Critical
			2–4	Marginal
			<2	Non-Critical
Chen (1965)	Oedometer, Seating Pressure 48 kPa		>10	Very High
			3–10	High
			1–5	Medium
			<1	Low

[1] Specified by Southern California Local Codes
[2] Federal Housing Administration/Housing and Urban Development

The activity (the ratio of plasticity index to percentage finer than 2 μm) provides a useful measure of expansion potential. Several correlations, based on the above parameters, have been proposed for preliminary estimation of expansive deformations (Snethen, 1984; Chen, 1988). Any expansive soil has a lower and upper limit of moisture content between which swelling and shrinkage can take place. Therefore, if movement is to occur, moisture changes must occur within this critical range.

8.3.6 Characterisation by swell strains

Puppala *et al.* (2004) described two approaches that are commonly used for the characterisation of expansive soils, based on swell strains. The first one is a direct approach using oedometer and/or other test methods to measure volumetric or vertical swell strains in soils. The second is an indirect approach based on soil index parameters or suction based measurements to evaluate the swell strain percentages, and then to address the problematic nature of these soils.

Direct approach

In the direct approach, swell strains are measured using conventional oedometers or other equipment such as confined swell test apparatus and three dimensional free swell test setups. In 3D setups, both lateral and vertical swelling can be measured. The volumetric swell strain information can be calculated from the measurements of lateral and vertical swelling. Table 8.14 presents a few of the characterisation methods available in the literature for classifying expansive soils based on swell strain measurements. Holtz and Gibbs (1956) developed an oedometer test based characterisation method in which a seating pressure of 6.95 kPa was used. Based on this method, a vertical

Table 8.15 Characterisations Based on Suction Measurements (from Pupalla *et al.*, 2004)

Reference	*Soil Suction at Natural Moisture Content, kPa*	*Related Swell Strain (%)*	*Characterisation*
Snethen *et al.* (1977)	>400	>1.5	High
	150–400	0.5–1.5	Marginal
	<150	<0.5	Low

Table 8.16 Characterisations Based on Total Suction Measurements (from Pupalla *et al.*, 2004)

Reference	*Total Suction – Water Content Index (Suction in pf/Water Content in %)*	*Characterisation*
McKeen (1992)	>−6	Very High
	−6 to −10	High
	−10 to −13	Moderate
	−13 to −20	Low
	<−20	Non-expansive

swell strain of 30% or above indicates a very highly problematic soil and a swell strain of less than 10% represents a non-problematic soil.

Chen (1965) applied higher seating pressure of ~48 kPa, for the characterisation of expansive soils. In this approach, soils with swell strains of 10% or above were classified as very highly problematic soils. The soils with swell strains less than 1% were regarded as non-problematic soils. The above two methods can be used in characterising expansive soils for designing both lightly and heavily loaded foundations. The characterisation used for lightly loaded structures can be extended to transportation infrastructure including pavements and runways.

Indirect approach

Several characterisations based on properties from indirect tests are available in the literature. These characterisations are based on soil index parameters and suction measurements. Table 8.15 presents a characterisation based on *in situ* suction using both total and matric suction potentials. This approach requires suction properties to be measured on soil samples using different methods. Instruments such as tensiometers, thermal conductivity sensors or filter paper methods can be used for measuring matric suction. The psychrometer or filter paper method can be used to measure total suction in soils. Table 8.15 can be used to characterise expansive subgrades based on suction potentials. McKeen (1992) developed another procedure based on total suction and water content of soils (Table 8.16) to estimate the swelling potentials.

Fredlund and Rahardjo (1993) provide an approach to calculate swell strain, $\Delta h_i/h_i$ in expansive soils:

$$\frac{\Delta h_i}{h_i} = \frac{C_S}{1 + e_{oi}} \log \frac{P_{fi}}{P_{oi}} \tag{8.34}$$

where

e_{oi} = initial void ratio of the soil layer,

Table 8.17 Characterisations Based on Soil Index Parameters (from Pupalla *et al.*, 2004)

Reference	*Test Type*	*Plasticity Index, PI*	*Characterisation*
Raman (1967)	Atterberg Limits	>32	Very High
		23–32	High
		12–23	Medium
		<12	Low
Chen (1988)	Atterberg Limits	35 and above	Very High
		20–55	High
		10–35	Medium
		0–15	Low

h_i = thickness of the soil layer,
P_{fi} = final stress state in the soil layer,
P_{oi} = initial stress state in the soil layer,
C_S is the swelling index and
Δ_{hi} is swell in the expansive soil layer.

Another indirect characterisation approach is based on soil index parameters, primarily using Atterberg limits or plasticity characteristics. This is the most frequently used method in geotechnical practice since these tests are simple to perform and inexpensive. In a study carried out on San Antonio and Corpus Christi expansive clays, swell characterisation based on Atterberg limit values was used (King *et al.*, 2001). A large scatter was observed, and practitioners should be careful when characterising expansive subgrades using only Atterberg limits. The same study recommends the use of liquidity index and consistency index parameters to aid in the characterisation of expansive soils, and provides a wide range of indices to differentiate between lean, fat and other clays that exhibit swelling behaviour. Table 8.17 presents characterisations based on soil index parameters.

The Potential Vertical Rise (PVR) method also estimates the swell potentials of expansive soils in length units based on Atterberg limits, overburden pressure and loading conditions of various layers. This method has been used in the characterisation of subgrades and design of pavements. The Texas Department of Transportation (TxDOT) is currently involved in the development of new or alternative methods to characterise expansive soils, which may replace the PVR method. Further details on this method can be found in TxDOT testing manual (see TxDOT website).

In addition to the above characterisations, there are a few more empirical relationships available in the literature to predict swell strains, which use soil plasticity index, compaction moisture content and the activity parameter of the soil (Table 8.18). Once the swell strain percentages are determined, they can be used along with Table 8.14 to estimate the severity of the natural expansive subgrade.

8.3.7 Types of foundation that are used in expansive soils

The majority of foundations used on expansive soil sites are one of the following:

- Slab-on-grade
- Slab-on-grade with piles

Table 8.18 Empirical Relations for Percent Swell (S) from soil properties (from Pupalla *et al.*, 2004)

No.	*Description*	*Reference*
1	$S = 0.00216 \times PI^{2.44}$	Seed *et al.* (1962)
2	$S = [(0.00229 \times PI)(1.45 \times C)/w_o] + 6.38$	Nayak *et al.* (1974)
3	$\text{Log } S = 0.9 \times (PI/w_o) - 1.19$	Schneider *et al.* (1974)
4	$S = 83 \times (2 \times A\text{-}1)$	El-sohby (Unknown)
5	$\text{Log } S = 0.08 \times (0.44 \times LL - w_o + 5.5)$	Vijayvergiya *et al.* (1973)

PI = Plasticity Index; C = Clay percent; w_o = Initial water content; A = Activity

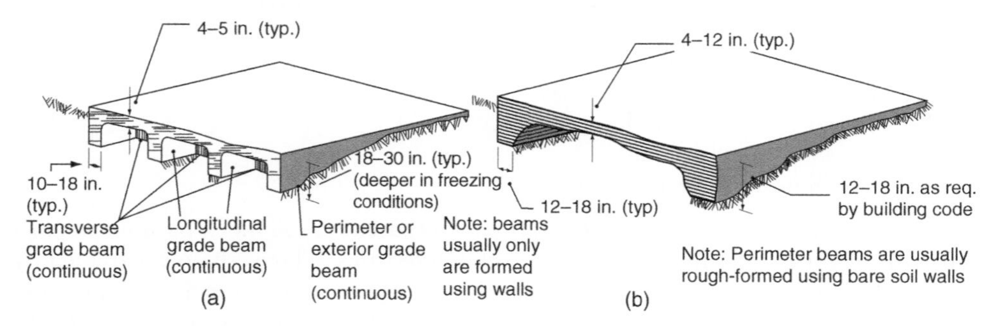

Figure 8.33 Cut-away sketch of a typical stiffened slab-on-grade foundation: (a) thin slab; (b) slab with thickened edges or perimeter beams (*After* Wray, 1995)

- Pile-and-beam
- Basement with wall footings and slab floor
- Basement with wall piles and slab floor
- Structurally suspended floor slab on piles.

Slab-on-grade foundations are usually constructed with either welded wire fabric reinforcement, mild steel bar[1] or deformed bar reinforcement, or with post-tensioning as the reinforcement. Welded wire fabric reinforced slabs are constructed with wire mesh placed in the forms before the concrete is poured over it. Steel bar reinforced slabs are constructed with the rebars fixed and placed inside the concrete forms, and then the concrete is poured and compacted. Post-tensioned slabs are constructed with steel cables covered with a plastic sheathing instead of using steel bars. The cables extend through the sides of the forms and, after the concrete is poured over the cables and has reached its initial strength, the steel cables are tensioned and fixed (see Figure 8.33).

Welded wire fabric-reinforced slabs often do not perform well in expansive soils. Both steel bar and post-tensioned types of slabs can work equally well, but each type has a different technical objective. A structural engineer, working with a geotechnical engineer, should have the principal responsibility for choosing the foundation system.

[1]In some countries mild steel reinforcement is no longer available.

8.3.8 Mitigation and preventive measures

Methods for mitigating the damaging effects of expansive soils include (Chen, 1988):

1. Excavation with or without replacement of the soil by a compacted non-expansive soil;
2. Flooding the in-place soil to achieve swelling prior to construction;
3. Control of compaction water content and density;
4. Mixing the soil with lime or cement before compaction;
5. Using footings or piers that extend below the depth of seasonal moisture change;
6. Using a bearing pressure that is high enough to balance the swell pressure;
7. Use of procedures to minimise changes in the soil water content, including adequate drainage systems and waterproofing of adjacent surface areas;
8. Use of structural mats and slabs that are resistant to differential movements;
9. Use of deep foundations down to non-expansive material;
10. Treatment or replacement of the upper 1–3 m with non-expansive soil (usually granular material) in embankment construction.

There is a "best time" to construct foundations in expansive soils. This is when the soil is neither at its wettest nor at its driest condition, i.e., when the soil is near "equilibrium". However, it is usually impractical to wait for this optimum time to start building. Thus, it is important that a foundation be designed and built so that it will perform adequately under all of the normally expected conditions that the building will experience during its usable lifetime.

Excavation and backfill

If the building is to have a shallow foundation, one solution that has been used successfully is to remove the expansive soil down to a specified depth and replace it with a non-expansive soil. Sometimes the terrain lends itself to constructing an elevated pad of foundation soil beneath the foundation without causing a significant increase in foundation cost. In replacing the expansive soil already existing at the building site, a "non-expansive soil" should be used. A non-expansive soil is usually considered to be a soil that has a plasticity index less than 20 and preferably less than 10. Sandy or coarse silty soils meet this criterion but, because of their relatively high permeability or hydraulic conductivity, if water should ever be introduced into this type of soil, it could travel throughout the soil mass, wet the underlying expansive soil, and cause the heaving that the non-expansive soil was meant to prevent.

If the building has a basement, the backfill soil ideally should be a non-expansive clayey soil, not a sandy soil. Steps should be taken to prevent water from entering the backfill regardless of whether or not the soil is expansive or non-expansive. If the backfill is expansive, then unwanted lateral swelling pressures will be imposed on the basement wall. If the backfill soil is non-expansive and if a considerable amount of water collects in the backfill, the water will impose a hydrostatic pressure against the wall. Water collecting behind the wall can cause damage because basement walls are seldom designed for hydrostatic pressure unless the basement extends below the groundwater table.

"Ponding" the foundation soil before construction

The principle behind the idea of "ponding" is to flood the foundation site and keep it flooded for several weeks so that the water percolates down into the expansive soil and causes it to "pre-heave". Although it is a good idea in principle, it has never been shown to have been successfully applied. If the ponding operation is applied for several weeks, the result will be that the soil near the surface will be very wet. The foundations will then have to bear at a depth below this very wet soil because it is likely to have a low bearing capacity and be highly compressible. In addition, should the ponded soil ever dry out, the foundation will be subjected to shrinkage, which can be just as damaging as the distortion that results from heaving soil. Thus, this method is not recommended.

Soil treated with a stabilisation method

Chemical stabilisation has been used effectively to stabilise highways, airport pavements, and large industrial sites. It has also been used successfully on smaller projects, such as single lot residences and other small buildings, but it is more expensive on a per unit basis than for the larger projects, because of the cost of mobilisation of the necessary plant.

There are two methods commonly employed in chemically treating expansive soils on small lots. One method is to treat the top 12 to 24 inches by mixing lime, cement, or flyash into the soil and then recompacting the mixed soil. This mixing operation requires special equipment which is difficult to operate and manoeuvre on small lots; hand methods of mixing do not do as good a job as the mechanised mixing methods. This method also requires laboratory testing to determine the appropriate amount of chemical to mix with the soil to produce the desired effect. However, these chemicals, when used in the proper proportions, properly mixed, and properly applied to the soil, are known to effectively reduce the amount of shrinkage and heave which the soil might experience if not treated.

The second commonly used method of chemically treating expansive soils is to inject pressurised slurry of water and lime, cement, or flyash to depths of a few metres below the ground. The concept is that the pressure will make the slurry flow through cracks in the expansive soil and effectively seal the soil to the required depth from subsequent penetration of water. The slurry is also expected to interact with the clay particles and reduce the affinity for attracting free water in the same manner as the surface mixing method does. Pressure injected slurry applications have been found to be quite successful in many instances; however, there have also been many instances when the injected slurry seemed to have had no effect on the resulting soil movement.

"Moisture barrier"

A moisture barrier is a structure or material that prevents or retards moisture from moving into or out of the soil. Moisture barriers are used to either prevent moisture from migrating from outside the foundation to a location under the foundation or to prevent moisture from migrating from under the foundation to outside the foundation. Barriers can also be used to prevent roots from trees or bushes from penetrating beneath foundations.

Moisture barriers may be vertical, horizontal, or a combination of horizontal and vertical. The principle behind a vertical moisture barrier is that, if it is attached to a foundation, the distance that water must travel to either get beneath a foundation from the outside or to get out from under the foundation is greatly increased. Because the soil permeability or hydraulic conductivity is so small, it is hoped that the increased distance, which means an increased travel time, will not permit the soil water content to change appreciably from one season to the next and the magnitude of shrinkage or heave correspondingly becomes only a nominal amount.

A similar principle applies to the horizontal moisture barrier, except that only a very wide horizontal barrier will produce the same "time of travel effect" that the vertical moisture barrier produces. The principal advantage of the horizontal moisture barrier is that it effectively moves the edge moisture variation distance out from under the structure to where it is acting under the horizontal moisture barrier, and where there is less concern if the soil heaves and shrinks.

The combination vertical and horizontal barrier might be employed if it is difficult to excavate deep enough to place a deep vertical moisture barrier, or if there are lateral constraints that prevent a full horizontal barrier from being used. One application of the combination barrier is adjacent to a structure but beneath a flower bed or decorative bushes, where the vertical barrier is taken deep enough to allow the plants to grow and then the horizontal barrier is placed beneath the bushes or flower bed which tends to prevent any overwatering from being transported to beneath the foundation.

Other things that can be done to avoid or mitigate any damage

Many things can be done in the design and construction of new buildings that can prove to be beneficial to the long-term acceptable performance of the structure. One of the first things that must be decided is, how much movement, and what effects of that movement, can be tolerated? A stiffer or stronger foundation, particularly with respect to slab foundations, will permit less deflection or distortion in the superstructure of the building which, in turn, will lead to fewer cracks.

Large shrubs, and especially trees, should not be planted close the building. Smaller bushes or flowerbeds adjacent to the house or building should not be watered by "ponding" water in the bed where the bushes or flowers are growing. Trees should be planted so that the drip line of the tree at maturity is still several feet from the edge of the building. If the location of the mature tree's drip line cannot be determined in advance, then a rule of thumb that seems to work well is to plant the tree a distance from the building equal to the mature height of the tree.

Downspouts should discharge at least 1.5 m away from the edge of the structure. Downspouts must also carry water over, and discharge some metres beyond the edge of, any backfilled excavation adjacent to the building, such as for a basement. Porches, steps, sidewalks, patios, and driveways should not be physically connected to the building. These minor structures will move differentially with respect to the house and this can result in damage. Make sure that water cannot pond or pool adjacent to or near the foundation. If a swale was constructed across a property to carry surface runoff water from lots at higher elevations to a storm water sewer or channel, do not alter or change it. Ensure that the gutters and downspouts on the buildings are clean and clear of debris. Make sure that debris or other material has not accumulated in any swales

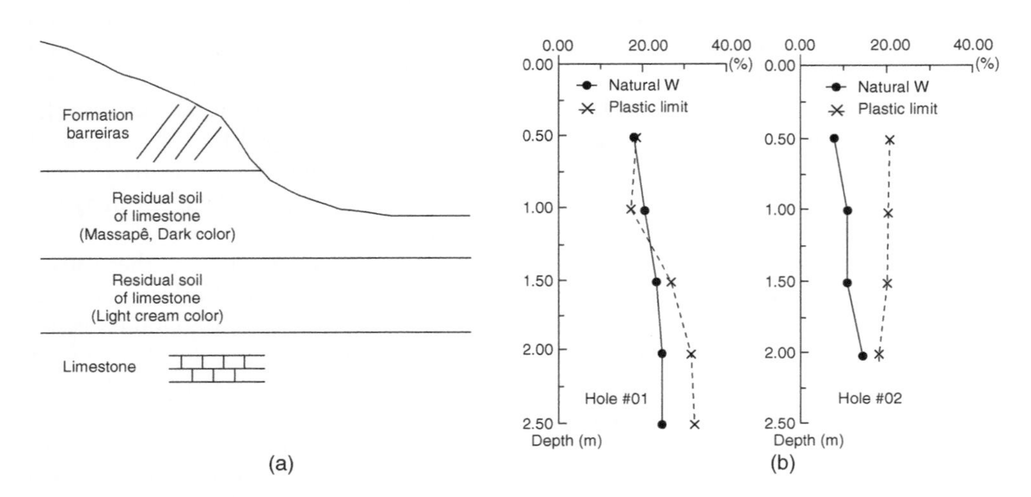

Figure 8.34 (a) Geological section; (b) Estimation of action zone (*After* Gusmão Filho *et al.*, 2002)

that cross the property. Maintaining relatively constant soil water content is a very important task in mitigating or reducing soil shrinkage and heave. This means watering more during hot, dry periods, but it also means continuing to water during cooler ones.

8.3.9 Case histories

Gusmão Filho *et al.* (2002) described two cases which occurred in school buildings, called CAICs (Centers of Integral Education of Children), in the Northeast region of Brazil. The CAICs were similar, with one floor and column loads of approximately 100 kN. They are located in small villages inland to assist the low income population. In this region, seasons produce significant differences in rainfall, so climates vary from super-humid to semi-arid classifications. In this region there are many examples of collapsible and swelling unsaturated soils.

Case history I

This was where the CAIC/Laranjeiras was constructed, in the Sergipe, Brazil. The geological profile is shown in Figure 8.34:

- *Bedrock of limestone*, with some degree of weathering and fracturing;
- *Residual soil of limestone* shows the typical horizons. A transition from horizon C to B is typical of saprolitic soil. The rock appears much altered and can easily be broken by hand. The B horizon has very fine texture and fracturing when it is exposed. Its surface in cuts shows undulation due to the action of water on the clay material. The A horizon is the superficial residual soil, being clayey with contraction fissures up to 1 cm across, dark colour and roots. This indicates a highly expansive material;
- *Formation Barreiras* was deposited over the residual soil and comprises sediments of varied textures. The tops of the hills are made of clay sedimentary deposits; they

have a red colour with many pebbles. A layer of quartz pebbles can be observed at the contact with the residual soil. The terrain shows a much altered residual soil of limestone with cream colour. Several cracks are observed in the ground alongside and parallel to the slopes. On the limestone in the hill, deposits of clay sediments with many quartz pebbles caneasily be seen.

Twelve borings were made in the area of the CAIC buildings. They show an initial layer of clay, locally called "massapê", which is dark in colour, with a consistency increasing with depth and thickness around 2 m. This is underlain by a 3 m thick layer of light yellow medium to stiff silty clay with fine sand, beneath which is the highly decomposed limestone rock.

The terrain is a plateau with higher ground around, so the buildings will be founded in residual limestone soil with a high probability that the soil will be expansive. The investigation programme was planned as follows:

- *Two holes* ($\phi = 10$ cm) to collect samples from depths of 0.5, 1.0, 1.5, 2.0 and 2.5 m;
- *Natural moisture content test* in all samples collected from the two holes (see Figure 8.34b;
- *Wet particle size distribution and index tests* to define the clay fraction and limits W_L, W_p and W_s in the samples collected at depth 0.5 m (dark "massapê"), 1.5 m ("massapê") and 2.5 m (light yellow silty clay);
- *Compaction test, CBR and expansion test* in samples from the Formation Barreiras, in a given section at two different depths.

Special expansion tests in the "massapê" clay were not requested since they take a long time. The results of index tests and field evidence allow the swell potential of the soil to be estimated and suggest solutions for foundations, pavements, buried pipes, etc.

The results for both holes are presented in Table 8.19, and these data allow the swell potential of the soil to be estimated. According to Holtz and Gibbs (1954), the swell potential can be taken as medium to high based on the Ip and high based on the grain size (percentage of fines). Seed *et al.* (1962) considered the Activity, so the soil is medium to a depth of 1 m and high at greater depths.

Predictions of swelling are all subject to errors (O'Neill and Poormoayed, 1980). The method of Vijayvergiya and Ghazzaly (1973) yields percentage swell of a clay sample in a consolidometer, under a 10 kPa surcharge, as a function of moisture content and liquid limit, in the form of a graph, giving the free swell percentage "s1". The method of Seed *et al.* (1962) gives percentage swell of a clay sample, compacted at near optimum moisture content, under 1 kPa surcharge in anoedometer, giving a free swell percentage "s2". The results are shown in Table 8.19.

In the Vijayvergiya and Ghazzaly method (1973), the values in hole #01 are similar, giving an average of 1.8%. However, in hole #02, the values are quite different, from >10% to 3%, with an average around 6%.

In the method of Seed *et al.* (1962), the values for hole #01 range between 2.7% and 6.8%, with an average of about 4%. In hole #02, the two values are more

Table 8.19 Results from holes #01 and #02 – Case history 1 (*after* Gusmão Filho *et al.*, 2002)

Hole	Depth (m)	I_P (%)	% < 1 μm	Activity*	W_{nat} (%)	W_L (%)	Swell (%) s1**	s2**
01	0.0–0.5	20	30	0.83	18.5	39	1.8	2.7
	0.5–1.0	20	35	0.74	21.3	38	1.3	2.9
	1.0–1.5	29	30	1.16	24.2	57	2.2	6.8
	1.5–2.0	24	–	–	26.3	57	2.0	–
	2.0–2.5	23	35	0.77	26.6	57	1.5	4.2
02	0.0–0.5	19	32	0.79	8.1	40	>10	2.4
	0.5–1.0	17	41	0.52	11.3	38	6	–
	1.0–1.5	17	–	–	11.7	38	6	2.1
	1.5–2.0	19	–	–	15.1	38	3	–

*Activity = I_P – [%<2 μm – 10%]
**s1 = free swell percentage (Vijayvergiya and Ghazzaly 1973), $s2 = 3.60 \times 10^{-5} \times D^{2.44} \times C^{3.44}$ (Seed *et al.*, 1962), where $C = \% < 2\,\mu m$ and $D = I_P/[C - 5\%]$.

consistent and the average is around 2.25%. The depth of the active zone can be assumed where the moisture content does not vary with the seasons.

In summary, various methods have been used to determine the swell potential of the material. The results showed a soil having medium to high swell potential.

From the natural moisture content it is estimated that the active layer is about 2.5 m thick, within which the natural moisture content is lower than the plastic limit, explaining why the surface has tile cracks. This layer of soil is susceptible to volume change due to seasonal variation of moisture content. A suitable remedial measure is to excavate the expansive soil and replace it with an inert soil in order to ensure the safety of foundations, pavements, and buried pipes.

Case history 2

The second case was recorded in Nossa Senhora do Socorro/Sergipe where another CAIC structure was built. The terrain has a very uneven topography, with 8 m difference in level between the children's school and the platform below where the rest of buildings are located (Figure 8.35).

The exposed slopes allowed the Formation Barreiras to be identified in the area. The surface soil is a cream colour with 1 m thickness, having a pebble horizon separating it from the lower layer. This is a red lateritic clay and silt matrix. It shows the presence of gravel, included in the laterite, indicating the fluvial origin of the soils.

At the middle of the cut slope, there is a change to a "massapê" clay, which is a type of soil very different from the other soils found in Formation Barreiras.

On the platform, at a depth of 8 m, cracks were observed and some slopes also were fissured indicating their instability. Fifteen borings were made in the area of CAIC buildings. Only two of them were located in the high part of the terrain, including the Formation Barreiras. The other borings were made at platform level. Underlying the Formation Barreiras is a silty clay of variable consistency, around 1–2 m thick, followed by the "massapê" clay which also varies in consistency.

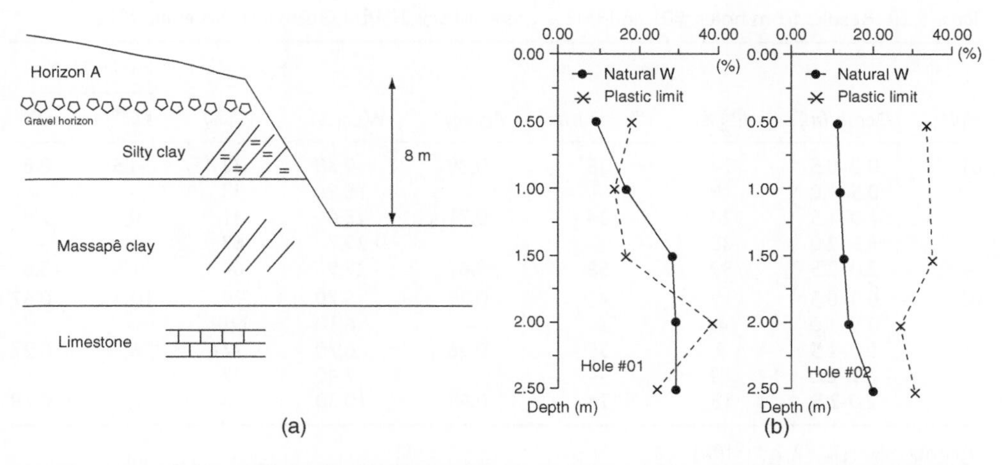

Figure 8.35 Case Number 2: (a) Geological section (b) Estimation of action zone (*After* Gusmão Filho *et al.*, 2002)

The following investigation programme was established in the location of buildings:

- *Two holes* ($\phi' = 10$ cm) to collect samples at depth of 0.5, 1.0, 1.5, 2.0 and 2.5 m;
- *Natural moisture content test* on all samples collected from the two holes;
- *Wet particle size distribution and index tests* to define the clay fraction and limits W_L, W_p and W_s in the samples collected at depth of 0.5, 1.5 and 2.5 m.

As for Case history 1, special expansion tests were not requested since they take a long time. The results of index tests and field evidence allow the swell potential of the soil to be estimated and suggest solutions for foundations, pavements, buried pipes, etc. The results for each hole are presented in Table 8.20, and thesedata allow the swell potential of the soil to be estimated.

According to Holtz and Gibbs (1954), if the I_p of the soil is used as the basis of classification, the swell potential of hole #01 can be taken to be medium to high, and the swell potential of hole #02 is low. Using the activity to classify the swell potential (Seed *et al.*, 1962), in hole #01 it varies from low to high as the depth increases, whereas in hole #02, the swell potential is low at all depths. Figure 8.35b shows the moisture content and plastic limit versus depth, from which the active zone is estimated to reach 2.5 m depth in hole #02.

The swell can be predicted, but not with a great deal of confidence, as was shown before. Table 8.20 shows the free swell percentage according to the method of Vijayvergiya and Ghazzaly (1973), and Seed *el al.* (1962). Both of them show wide scatter in the final results.

Using the Vijayvergiya and Ghazzaly (1973) method, the values for hole #01 vary by a factor of 10, giving an "average" of 3.2%. In hole #02, the numbers are higher, between 6% and 10% with an average of 7.3%.

Table 8.20 Results from holes #01 and #02 – Case history 2 (*After* Gusmão Filho *et al.*, 2002)

Hole	Depth (m)	IP (%)	% < 2 μm	Activity*	W_{nat}(%)	W_L(%)	Swell (%) s1**	Swell (%) s2**
01	0.0–0.5	14	36	0.39	9.30	32	4.5	0.8
	0.5–1.0	19	–	–	16.9	33	–	–
	1.0–1.5	24	34	0.71	28.6	41	0.4	2.9
	1.5–2.0	30	–	–	29.7	69	–	–
	2.0–2.5	39	58	0.67	29.9	64	1.5	15.8
02	0.0–0.5	12	43	0.28	5.70	29	10	0.67
	0.5–1.0	NP	–	–	6.10	NL	–	–
	1.0–1.5	9	35	0.26	6.70	27	6	0.27
	1.5–2.0	13	–	–	7.40	27	–	–
	2.0-2.5	13	24	0.68	10.30	36	6	0.78

*Activity = I_P – [% < 2 μm – 10%].
**s1 = free swell percentage (Vijayvergiya and Ghazzaly 1973), s2 = $3.60 \times 10^{-5} \times D^{2.44} \times C^{3.44}$ (Seed *et al.*, 1962), where C = % < 2 μm and $D = I_P/[C - 5\%]$.

In the method of Seed *et al.* (1962), the values are quite different in hole #01, ranging between 0.8 and 15.8, while in hole #02 they are more consistent. Taking the active zone as 2.5 m deep, the total surface swell is less than 0.5 cm using the Seed *et al.* method in hole #02.

The site for the CAIC building had evidence of being expansive soil. Many methods were used to show the special character of the soil, although the soil was not extreme, having a low swell potential. A suitable remedial measure is to excavate the expansive soil and replace it with an inert soil in order to ensure the safety of foundations, pavements, and buried pipes.

8.4 INDIRECT (DEEP) FOUNDATIONS

8.4.1 General concepts

All types of piles which are used in sedimentary soils and weak rocks can be used in residual soils. The main differences are in the design parameters required to calculate a bearing capacity for the pile. As discussed elsewhere, many of the standard relationships which apply to sedimentary soils do not apply to residual soils because of their history. For example, the conventional relationships between vertical and horizontal stress will not apply where a rock, with high built-in stresses due to tectonic action, weathers to a residual soil. This will in turn affect the shaft friction which can be mobilised on the outside of a pile.

It should also be noted that, in many parts of the world where there are no tropical residual soils, extensive soil investigation, laboratory testing and research over many decades have produced a wealth of knowledge on the fundamental properties of soils. This applies to such soils as the London Clay, Boston Blue Clay, many Scandinavian soils and others. The sampling and testing has been of high quality, and has been supplemented by *in situ* tests, many with sophisticated instrumentation. It will also be shown later that some European countries have built up significant databases using advanced *in situ* testing such as the Cone Penetrometer Test (CPT) or the pressuremeter test (PMT), tied to results of static pile load tests. By comparison, many of the countries where tropical residual soils are prevalent have less well developed geotechnical databases, and this often starts with the basic soil investigation. For example, whereas in stiff clays in Europe it is standard to take undisturbed samples and carry out triaxial testing in the laboratory, in Asia it is more common to rely upon the Standard Penetration Test, often carried out with poor quality equipment. This means that design methods for piles rely in turn on empirical methods and relationships, because fundamental relationships cannot be determined without the parameters on which to base them.

Even this is not as easy as it might be. The Standard Penetration Test itself was developed for the testing of granular soils, and it is normal to stop the test once a soil layer has been proved to be very dense, which is equivalent to a total of 50 blows for a penetration of less than 300 mm. It has since been extended to test cohesive soils, and empirical relationships between blow count N and consistency have been derived and published. It is also widely used in weak rocks, and the work of Stroud (1974, 1988) has established widely accepted correlations. But is there any reason why the same correlations should apply to residual soils? The way in which the soils are produced, by decomposition of a massive rock rather than by deposition and consolidation, would suggest that there may be differences. It should also be noted that the processes at work turning sedimentary soils into rocks are ones which produce an increase in density, while the weathering process will often produce a reduction in density as chemicals are leached out. As a result the standard definitions of dense and very dense will not apply to residual soils, and in some areas it is common to continue the Standard Penetration Test, and to get meaningful results, with blow counts of up to 200. Nevertheless this leads to very heavy wear on the equipment, and certainly the risk that counting may not be too accurate.

Gradually local correlations are being built up based on experience in tropical residual soils, but often this cannot be linked to conventional strength testing because of a lack of comparative data. It is really up to each country to develop their own empirical correlations based on local geology, to publish these within local networks such as conferences and seminars, and to work with neighbours to find common ground in their correlations. Some work was done in the 1980s to establish relationships between shaft friction and pile/soil displacement, often called *t-z* curves, for soils and weak rocks including tropical residual soils in Singapore, and to relate these to ranges of SPT N values. Much of the work in Hong Kong has been reviewed in the two editions of the piling guide, Foundation Design and Construction (GEO 1/2006).

For historical reasons, much of the pile design work in tropical residual soils is based on the concept of undrained shear strength, correlated with shaft friction through a coefficient α, hence these methods are often called the α methods. The SPT N value is then converted empirically to undrained shear strength, and this to shaft friction by multiplying by an appropriate value of α. This concept was initiated by Skempton based on work in London Clay, but has been extended by others and has been shown that α decreases with increasing shear strength. Unfortunately there is no valid reason why the values derived by Skempton for London Clay, and others for sedimented soils, should apply equally well to tropical residual soils, bearing in mind their very different mode of formation and structure.

In fact, the other group of methods, often called the β methods, based on effective stresses, is more likely to be suitable for tropical residual soils, where the weathering process has contributed to the permeability which allows relatively fast drainage of excess pore pressures caused during pile installation.

Displacement piles

All types of displacement piles are used in tropical residual soils, including concrete, cast *in situ* and precast, and both reinforced and prestressed, steel in a variety of shapes including H- or I-sections and pipes, and timber. Steel sections have special design problems, such as how to calculate the end bearing component of an I-section or a pipe, which depends on the balance between the maximum shaft friction available against the surface of the pile and the end bearing of a larger area, such as the whole square for an I-section or the whole circle for a pipe. This determination is further complicated by the differences between static and dynamic behaviours, such that a pipe pile may be plugged statically at a certain length, where the internal shaft friction is enough to overcome the end-bearing over the whole area, and yet the plug continues to rise up under driving because the dynamic action reduces the shaft friction.

Driven piles are often noted to be affected by "set up" or "soil freeze", in which the capacity in both shaft friction and end bearing is reduced by the excess pore pressures mobilised during driving. As these dissipate, the capacities are noted to increase. At the site of the Bishan Depot for the Singapore MRT, a large number of square section precast concrete piles had been driven into the weakly cemented sand called the Old Alluvium. When dynamic pile testing was attempted a few days later, most of the pile heads broke before any measurable pile movement could be achieved, because of the set up which had gripped the outside of the pile.

At another site in the Gulf of Thailand, steel pipe piles were being driven with a hydraulic hammer from flying leads for a mooring dolphin at a fuel unloading facility, but the piles, which were being continuously monitored by dynamic pile testing, failed to reach the required capacity. Driving was stopped when the instruments, near the top of the pile, were about to become submerged in sea water. The piles were already 45 m long, to pass through sea water and mud before reaching firm founding strata, and were extended by welding on another 12 m length over the following day. By the time driving started again, some 36 hours from when it had stopped, it was not possible to drive the pile any further and the dynamic pile testing showed that the pile had more than reached the required capacity.

Dynamic pile testing has given us a unique opportunity to investigate this phenomenon practically in the field, since we can now make good estimates of pile capacity at any instant of time, from the end of initial driving till as long as we want to wait and carry out a restrike. This is an invaluable tool in finding the optimum length and driving conditions for any pile, since the pile design is always for the long term value and we do not need to be concerned if the capacity at the end of initial driving is lower. However the problem will always occur, both for land piles and in marine applications, that it becomes extremely difficult to get the driving plant, be it a crane or a barge, back onto the previously driven pile. It therefore becomes necessary to carry out local calibrations, on as many piles as possible and with varying lengths of delay, to find the capacity at end of initial drive which will lead to the required final capacity after

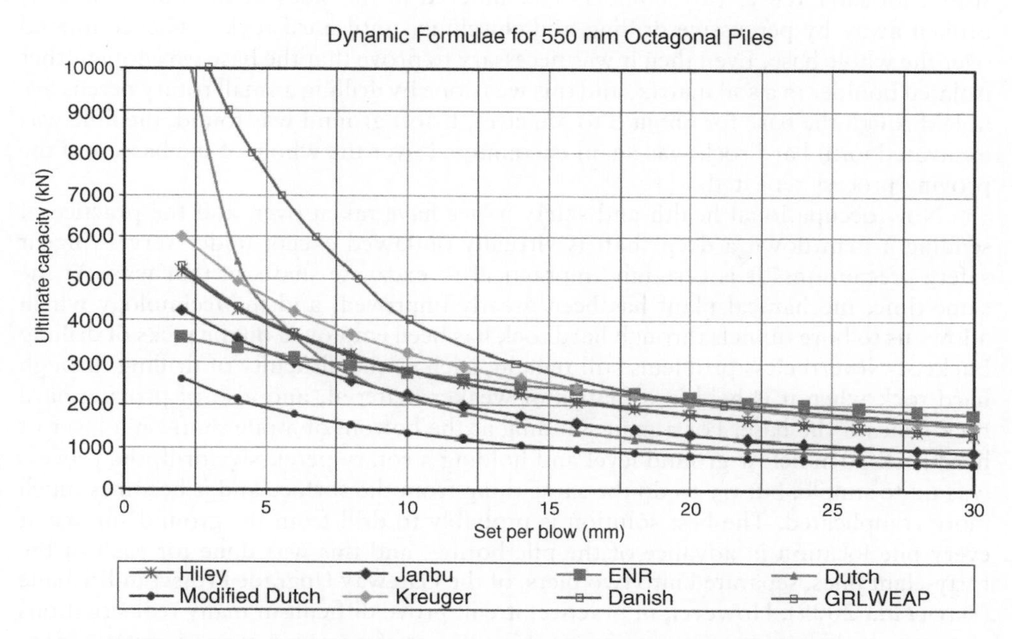

Figure 8.36 Relationship between set per blow and ultimate pile capacity by dynamic formulae

a period of time. This sort of data can very usefully be built up locally, and shared through local publications, conferences and seminars.

It has been noted that, of the many dynamic formulae available, there is a very strong tendency throughout South East Asia, and further a field, to favour the Hiley formula over all the others. Figure 8.36 shows the relationship between set per blow and ultimate pile capacity for a selection of 7 well known dynamic formulae. It is to be expected that each formula has, for at least one set of conditions related to hammer type, pile type and soil type, given acceptable results, since all are essentially empirical. It is equally to be expected that no single formula will work with equal accuracy for all types of hammer and all types of pile material in all types of soil. It is recommended that dynamic pile testing is used to give a more reliable indication of capacity, and that a suitable calibration is then obtained for a dynamic formula which gives the best agreement.

Replacement piles

One of the major problems with replacement piles in tropical residual soils, in this case often bored piles, has been the presence of corestones or boulders within the weathered rock profile. This led historically to the use of techniques such as the hand dug caisson, in which a man excavated the soil and weathered rock about 1 metre vertically at a time, from the top downwards. The spoil would be shovelled into a small skip, hauled to the surface and tipped by his partner, often his wife. As he went, he would build a support with *in situ* concrete, formed behind a tapered shutter. Next morning, as he continued the excavation, the formwork would drop away from the taper and was stored for later reuse. Any boulders encountered in the sides of the hole would be broken away by percussive drilling and chiselling, until hard rock was encountered over the whole base. Even then it was necessary to prove that the base was not another isolated boulder in a soil matrix, and this was done by drilling a small rotary percussive hole through the base for about 3 to 5 metres. If soft ground was found, the hole was excavated until hard rock was again encountered over the whole of the base, and the proving process repeated.

Now occupational health and safety issues have taken over, and the practice of sending a man down a deep shaft is virtually outlawed except under very stringent safety precautions. It is certainly impractical to excavate shafts in that way. At the same time, mechanical plant has been greatly improved, and the technology which allows us to bore tunnels through hard rock has been employed on the picks of drilling buckets. Nevertheless problems still remain, such as the difficulty of drilling through hard rock when it is held in a matrix of weaker material, and also of proving hard rock beneath the base. For a man standing at the bottom of a pile shaft on a layer of hard rock 30 m below ground level and holding a rotary percussive drill, the process was quite simple, but try to do the same thing from the surface and it becomes much more complicated. The best solution is probably to drill from the ground surface at every pile location in advance of the pile boring, and this was done for each of the forty-eight piles, separated into two piers, of the Gateway Upgrade Project in Brisbane (Day *et al.*, 2009). However, in practice, it will prove difficult in many real situations to persuade the project owner to invest this amount of money in ground investigation, and it will generally involve a two stage process which absorbs crucial amounts of time. It is necessary to make a preliminary investigation which will allow the best type and size of pile to be determined, and to make an estimate of the depth based on available information. This will then provide the locations for the piles, and the estimated depth can be used to decide the level at which the boreholes can be terminated. Getting the required amounts of time and money into a project budget is a major challenge.

8.4.2 Pile design

At this time, with the increased importance of the Serviceability Limit State, understanding the performance of a pile under load is probably more important than ever. According to Baguelin *et al.* (1978), the *in situ* testing equipment which helps the most to visualise the way soil behaves around a pile is probably the Ménard pressuremeter (MPM). Although this statement was made primarily in respect of sedimented soils, it is even more applicable when we deal with more complex materials such as naturally structured soils, like weathered rocks and residual soils, which are sensitive to penetration processes (for SPT or CPT) or sampling methodologies. The possibility of reaching most ground horizons by pre-boring in the case of the PMT (or MPM) prior to the insertion of a testing probe and executing the expansion test, allows very reliable and valuable information to be obtained, expressed as a pressure-displacement curve, which is very much better than any "simple" correlation with SPT or CPT.

There are criticisms of this test, bearing in mind that it may be a complex and time consuming process, for some materials. When dealing with extreme geomaterials, such as soft clayey soils or loose sands where penetration tests are possible, on the one hand, and rocks where rock coring and UCS tests for classification purposes are readily carried out, the PMT may be of less benefit. However, in IGM (Interface Geomaterials) or weathering profiles, the criticisms are unfair, or even erroneous. The feasibility of testing at all in such a heterogeneous environment, and with materials behaving under particular levels of cementation, and micro- and macro-structures and fabric, makes this test very valuable.

Piles in residual soils: Effects of installation process

The design of piles in residual soil is usually based on empirical correlations, mostly using the results of the standard penetration test, regardless of the construction method of the bored pile (Chen and Hiew, 2006). The use of "rational" (classical) methods or empirical solutions for residual soils, adapted from those for sedimented sandy or clayey soils, may overestimate the shaft friction, particularly with wet construction processes, and make the evaluation of base resistance very unreliable, dependent particularly on the cleaning of the pile base before concreting. The empirical correlations of the unit shaft friction (f_s) and base resistance (q_b) with SPT-N values, are commonly used loosely, regardless of the construction method, as follows:

$$f_s = K_s \cdot N \tag{8.35}$$

$$q_b = K_b \cdot N \tag{8.36}$$

K_s and K_b are the empirical factors for shaft friction and base resistance, respectively, in kPa. The K_s values generally vary from 2.0 to 3.0, but need to be treated with extreme caution. They are genuinely empirical, based on field tests, and therefore strictly relate only to similar piles, especially in terms of soil type and construction method. Although it may happen that similar K_s values will apply to residual soils in Singapore, Malaysia and Hong Kong, for example, there is no guarantee that this is the case. Local tests will always be required in confirmation but, if that confirmation is achieved, then the wider database of results may be able to be used. It also needs to be noted that some authors report the results of tests, where the K_s values refer to what was measured, whereas others refer to recommended design values, where more conservative values are used.

Lei and Ng (2007) describe the practice in Hong Kong, where excavated rectangular barrettes and large diameter bored piles have been commonly adopted, in the last two decades, as the foundations for tall buildings and heavy infrastructure projects. There is generally the aim of embedding the tip in sound rocks, because current design procedures assume a heavy reliance on end bearing for the Ultimate Limit State. When this is not possible, for technical or economic reasons, piles are designed relying on shaft resistance in the overlying deep-seated thick saprolites.

For private building works in Hong Kong, shaft resistance in excess of 10 kPa is not normally permitted by the local regulations, without performing a load test on site. While some studies into the behaviour of large diameter bored piles in saprolitic soils are reported in Ng *et al.* (2001a & 2001b) and Lei and Ng (2007), there is a re-evaluation of the behaviour in terms of fully mobilised or substantially mobilised shaft resistance in these local saprolites for rectangular barrettes under bentonite and circular bored piles, comparing it with other soils elsewhere. In these studies, instead of relating the shaft friction to the SPT-N value, by K_b, it is related to the effective vertical stress by β. The conclusions are as follows:

a. for the barrettes and bored piles in saprolites in Hong Kong, a moderately conservative local displacement for the mobilisation of the ultimate shaft resistance is found to be approximately 20 mm, which is larger than that found elsewhere;
b. the average K_s and β values for barrettes are 1.28 and 0.27, and for bored piles 1.2 and 0.30, respectively;
c. compared with barrettes and bored piles in Old Alluvium, residual soils and weathered granites in Singapore, and bored piles in residual soils in Malaysia, the K_s values for barrettes and bored piles in saprolites in Hong Kong are generally low. The range of β values for barrettes and bored piles in Hong Kong is comparable only to the β values recommended for the design of bored piles in loose sand elsewhere;
d. compared with non-grouted barrettes and bored piles, shaft-grouted barrettes and bored piles show a relatively stiffer unit shaft resistance response to local displacement and a higher ultimate unit shaft resistance.

From a loading test on a barrette in a residual soil named Guadalupe Tuff in the Philippines, a relatively low value of β of about 0.2 was back-calculated by Fellenius *et al.* (1999). For the design of bored piles in loose sand, Davies and Chan (1981) recommend β values of 0.15–0.3, and for the design of cast-in-place piles in loose sand, while others recommend β values of 0.2–0.4.

There are several cases reported in the bibliography that indicate a very clear dependence on construction methods, for the results of pile load tests when founded in residual soils (Chen and Hiew, 2006; Fellenius *et al.*, 2007; Viana da Fonseca *et al.*, 2007). The installation of piles by pushing or driving pre-cast units, by using bored cast-in-place methods with temporary steel casing or with bentonite or polymer as the stabilising fluid, or by using Continuous Flight Auger (CFA) methods, will lead to very different soil-pile interaction behaviours.

A series of published papers based on work in South East Asia started from testing associated with the Singapore MRT in the early 1980s. These included Buttling (1986), Buttling and Robinson (1987), Buttling and Lam (1988), Toh *et al.* (1989), Buttling (1990), Chang and Broms (1991), Phienwej *et al.* (1994), and Chen and Hiew (2006). Toh *et al.* (1989) reported K_s values for the Kenny Hill Formation in Malaysia of about 2.5–2.7 for SPT-N values less than 120 based on nine fully instrumented tested piles. A study by Chang and Broms (1991) on Singapore residual soil, suggested a design value of $K_s = 2$ for SPT-N values of less than 150. Phienwej *et al.* (1994) carried out a further study based on 14 fully instrumented bored piles constructed using both dry and wet methods, and reported a measured value of $K_s = 2.3$ for SPT-N values below 120. Tan *et al.* (1998) studied 13 bored piles constructed using both dry and wet methods, also suggested adopting $K_s = 2$ for design purposes, while limiting the maximum unit shaft friction to 150 kPa.

Note that the Chang and Broms suggestion refers to SPT-N values of 150, while that of Tan *et al.* (1998) implies SPT-N values of 75. Both are well above the standard limit of 50 blows and need to be treated with caution. Whereas in Europe, it is normal practice to stop the SPT when the blow count reaches 50, in other places, notably

Hong Kong, it has been the practice to carry on to blow counts as high as 200. One of the reasons for this is that the SPT was developed for testing sedimented granular soils, where a blow count of 50 shows a very dense soil and further driving will achieve very little. In a tropically weathered profile, it may be argued that the way in which the soil behaves is very different, and that useful information with regard to soil strength is gained up to higher blow counts. However, great care is needed with the higher numbers, as there is a great difference between a measured SPT-N value of 175, and one which has been extrapolated from 50 blows over a penetration of less than 300 mm. This has become common practice in some areas, but is highly questionable. Unless the 50 blows were achieving a similar penetration for each blow, then there is clearly no justification for extrapolation. If, for example, blow counts are recorded every 75 mm, and they show 19, 21, 10 for 30 mm, then some sort of extrapolation may be justified. However, if they are recorded every 150 mm, and show 6, 44 for 40 mm, they paint a very different picture. To turn the latter numbers into 50 blows for 190 mm, and therefore 79 blows for 300 mm, and use this to calculate a shaft friction could be very misleading.

It is also to be noted that there is widespread concern about the effect of support fluids, especially bentonite, on the shaft friction achieved. Bentonite is a slippery fluid if you place your hand in it, and we know that it forms a filter cake on the sides of a hole in permeable soils, where the water within the fluid flows out into the soil and leaves the bentonite behind, forming a very low permeability skin which enhances stability. Both of these factors make designers worry that shaft friction will be reduced. The major question which remains is "reduced with respect to what?" Since bentonite is normally used where other methods, such as boring in the dry, are not applicable, it is very difficult to obtain two numbers which can reasonably be compared. Equally, it is not of much help to know that a higher shaft friction could be used by boring in the dry, if boring in the dry is not practicable. This is also strength of empirical relationships, because any effects of using bentonite are built into the factors produced by back analysis of appropriate test results.

This is reinforced by Touma and Reese (1972), who performed load tests on drilled shafts constructed in sands with the use of slurry. They inferred from the results that there was no clear reduction of the shaft friction due to use of slurry in construction.

Fleming and Sliwinski (1977) studied 21 pile load tests in clays, 9 pile tests in sands, and three pile tests in chalk. In clays, the piles were constructed using both dry and wet methods, and in sands they were constructed using the wet method, and using temporary casing. Where used, the temporary casings were driven without pre-excavation with slurry. Fleming and Sliwinski (1977) concluded that in sands, a thin membrane, or filter cake, is formed at the soil/pile interface. The test results indicated that shaft friction at high displacements could be reduced about 10 to 30 percent compared with that developed using a temporary casing; however, they could not determine if that was because of the bentonite slurry, or because of some other factors, such as the larger diameter hole left behind when the temporary casing is removed.

CIRIA Report 77 (Fearenside and Cooke 1978) studied seven piles formed in London clay specifically to check the effects of bentonite on pile capacity; three were constructed using bentonite and four under dry conditions. There was no evidence that the use of bentonite adversely affected skin friction. In fact, piles constructed with slurry had higher capacities than those with the same diameter and length constructed

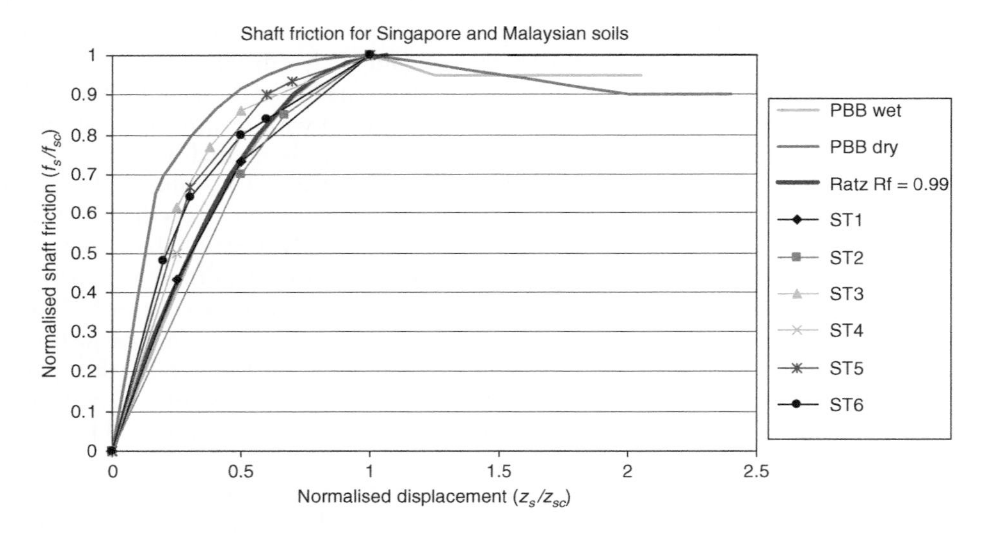

Figure 8.37 Variation of shaft friction with displacement

dry. However, it was suggested that this was due to the difference in sidewall roughness in shafts constructed with slurry using a drilling bucket, compared with shafts drilled dry using an auger.

As noted above, Phienwej *et al.* (1994) reported on 14 instrumented test piles formed in the Kenny Hill Formation in Kuala Lumpur. They summarised the data onto a plot of normalised shaft friction against normalised displacement, and this is reproduced as Figure 8.37. Although there is some scatter, it is small considering that the results from several layers in at least 10 piles on six different sites are included. It also appears to show a significant difference between the piles constructed wet and dry, with the dry piles showing a much stiffer response.

Also shown in Figure 8.37 is the shaft friction/displacement relationship, often referred to as a *t-z* curve, used in the program RATZ, which performs the same function as the load transfer relationships referred to by several of the authors listed above (Buttling, 1990; Chang and Broms, 1991; Phienwej *et al.*, 1994). It can just be made out that the shaft friction used by RATZ, with an Rf value of 0.99, agrees very well with the field data for wet piles, marked as PBB wet. In addition, a number of other curves have been plotted, based on instrumented pile tests on piles formed under bentonite in Singapore residual soils, taken from load transfer curves in Buttling (1987). These are seen to lie between the two extremes of the dry and wet curves, with a tendency to be nearer the wet curve. These curves are all normalised, such that actual shaft friction values can only be determined if the peak value of shaft friction is known, and this is where the importance of the K_s value lies. It is very interesting to note that, while Phienwej *et al.* (1994) show that the performance of the shaft is much stiffer for piles constructed in the dry, the data to support the relationship between peak shaft friction (f_{sc}) and SPT-N value appears to be independent of the process.

Table 8.21 K_s values for bored piles

Depth range (m)	*Empirical factor K_s*
Piles bored under bentonite	
12–18	0.87
18–21	0.45
21–24	1.18
24–26	0.69
Piles bored in the dry	
12–20	1.18
20–27.6	3.56
27.6–35.4	0.49

Chen and Hiew (2006) compared the results of static pile load tests on two fully instrumented bored piles in residual soil, also of the Kenny Hill Formation, using different methods for temporary support. The residual soil was mostly loose to medium dense silty sand, with SPT-N varying from 4 to 13. A very hard silt layer with SPT-N of more than 50 was found at 12 m below the ground surface.

The first was a test pile to failure, 1000 mm in diameter and 27 m long. It was started with a temporary casing 6 m long and, after the soil inside the casing was removed by auger, the bore was filled with bentonite slurry. Drilling was completed with a bucket, and the loose materials at the base were removed using a mechanical cleaning bucket. Concrete was placed using the tremie method.

The second was a working pile, 1200 mm in diameter and 36 m long. Because of a deep cut off, a temporary outer casing was installed to debond the upper part of the pile from the soil. The soil inside this casing was removed, and a concrete was formed inside this casing about 1 m thick. A smaller diameter casing was then installed and the boring work continued to the design depth in the dry. Concrete was placed using the tremie method after the base had been cleaned.

The results of the static load tests were analysed and compared, and it was found that the shaft friction was higher for the bored pile constructed using the dry method. The K_s values for different depth ranges in the two piles are presented in Table 8.21.

Chen and Hiew concluded that the K_s values for the dry pile may suggest that a value greater than 2 could be used, noting that this pile was not tested to failure so peak shaft frictions were not mobilised. However, it is hard to justify the value of 3, which they recommend for general use, on the basis of only this evidence. They also conclude that a lower value should be used for piles bored under bentonite, and suggest a K_s value of 2. While this is in keeping with the proposals of other authors referred to, as noted above, it cannot be justified on the basis of their test results. Noting that the piles were only 5 m apart, and that the larger diameter pile was bored in the dry, it is not at all clear why bentonite was used in the smaller test pile loaded to failure. Its relevance to the general design of piles bored under bentonite in residual soil is therefore questioned.

As pile diameters become larger, the relative importance of the end bearing capacity also increases. This is a point of major concern to designers for a number of reasons:

1 The contribution of the base resistance to the ultimate geotechnical capacity of the pile can be very significant;

2 This peak base resistance will usually occur at a significant displacement, which is often assumed to be about 10% of the pile diameter;
3 This displacement is often an order of magnitude greater than the displacement required to reach peak shaft friction, and this leads to a potential anomaly when simply adding peak shaft friction capacity to peak end bearing capacity and dividing by a simple factor, since the two peak values very rarely coexist;
4 The end bearing resistance which is available in practice is highly subject to quality of workmanship;
5 One of the common factors, which has been widely written about, is the cleanliness of the pile base at the time of placing the concrete via a tremie pipe;
6 Another factor, which has received much less attention, is the potential for loosening of material at the base of the pile by stress relief.

There is significant practical experience to show that, if the base of a pile is not properly cleaned prior to concreting, then the resulting layer of debris, usually comprising silt, sand and perhaps lumps of clay, in a loose state, will lead to sudden unacceptable pile head settlement once the shaft friction has been fully mobilised. The means of preventing this are not complicated, but require attention to detail, and an understanding of the processes involved.

Where bentonite is used as the drilling fluid, this has two main purposes. One is to maintain stability of the bore in cohesionless soil layers by applying an excess internal fluid pressure to the inside of the filter cake. The other is to hold soil particles in suspension through the combined effects of fluid density and viscosity. With the particles in suspension, they can be removed from the hole by exchanging the bentonite column. Thus the larger particles will fall under gravity and be removed by a mechanical cleaning bucket, while the smaller particles are distributed vertically throughout the bentonite column, as a result of the removal and reinsertion of the drilling tool. The viscosity of the bentonite means that it will take an unacceptably long time for particles of sand and silt near the surface to settle to the bottom to be removed mechanically. The correct process is therefore to remove the heavier, dirty, contaminated bentonite from the base of the pile, clean it in a desanding plant and hydrocyclone, and replace it with cleaned bentonite at the surface. The removal can be achieved by use of a submerged pump, but the air lift is generally more popular. Even this simple piece of plant is not widely understood in practice by the field operatives required to use it, and there can be a tendency to use too much air. The principle is based on simple fluid mechanics, which requires that, if a steel tube (the air lift pipe) contains a fluid with air bubbles, and the fluid outside the tube (in the pile shaft) does not contain air bubbles, then the difference in density between the two fluids will cause the lighter fluid to rise in the tube and be replaced at the base by the heavier fluid. For this reason very small air flows will work, and allow the base to be cleaned without applying high velocities or suctions. On the other hand, since the clean drilling fluid is lighter than the contaminated fluid, it cannot be displaced from above, and all the attempts to pump clean bentonite down from the surface, even through trumpet-shaped end pieces, will be to no avail. The table lamps containing different coloured oils, with the warmer, lighter one rising through the heavier one, are a good demonstration of what will happen when this is tried in the field.

Alternatively, increasing use is being made these days of polymer-based drilling fluids. These are made from long chain polymers, which can have the ability to behave quite unlike normal fluids. It is clear that polymers which, like bentonite, had their origins in the oilfield drilling industry, also, like bentonite, help to improve the stability of pile bores, especially in cohesionless soils below the water table. What is not quite so clear is exactly how they do this, because it is certainly not through the development of a filter cake and the creation of an internal excess fluid pressure. It seems likely that there is some way in which the long chains of the polymer are forced out into the interstitial voids of the cohesionless soil, and thereby create a form of soil reinforcement. If this is the case, it also explains why they perform well in circular holes, but not so well when it comes to rectangular holes such as diaphragm walls, where bentonite still seems to be the preferred fluid. It is also clear that we tend to use the same tests to check on the qualities of a drilling fluid, even though the way in which it works is completely different.

For example, as stated previously, we rely on the viscosity, and the shear strength, of bentonite to support soil particles, and we generally measure the "apparent viscosity" based on the time it takes for a known volume of bentonite to pass through a small hole, as in a Marsh Cone. However the polymer-based drilling fluids do not have an increased viscosity, indeed particles are not held in suspension and settle out faster than in water, yet the long chains still slow the progress of the fluid through the small hole of the Marsh Cone, giving another "apparent viscosity", though for a different reason and in a different range from that expected of bentonite.

Where polymers are used, air lifts or submerged pumps are generally not required, since the particles will settle to the base very quickly. Instead a mechanical cleaning bucket can be positioned at the base of the hole to trap debris as it settles, and rotated gently to pick up the last few lumps before being slowly raised to the surface. There is not the tendency of the whole column of drilling fluid to become contaminated with debris, as there is with bentonite, so the cleaning process can be much quicker. Other benefits include a much smaller requirement for plant and storage, compared with bentonite silos, recirculating tanks, and desanding equipment, and also an easier disposal problem at the end of the project, since most polymer drilling fluids are biodegradable.

Even with the best cleaning techniques, it is not possible to counter the effects of stress relief. If the soil at the base of the pile is a very dense cohesionless material, at a depth of, say, 50 m, it is likely to be under a vertical effective stress of 400 to 500 kPa. If it is overconsolidated, and that applies to residual soils just as much as to sedimented soils, then it could be a higher value. By the time a hole has been bored and filled with drilling fluid with a density of about 1.03 g/cc, the vertical effective stress is only about 10 to 15 kPa, and this can cause the soil to swell, and to take in drilling fluid, whether it be bentonite or polymer, as it does so. This has been noted particularly in sedimented soils, such as piles in London taken down to the Thanet Sands, and piles in Bangkok where a series of pile tests on many sites for the Second Stage Expressway System, undertaken in about 1990, showed base stiffnesses in the range of 25 to 40 kPa/mm, for bases in very dense sands. It can reasonably be assumed that the same process will occur in residual soils, where there is limited cohesive material present to hold the material together. Since, if it occurs, this is a localised effect at the base, it may be corrected by base grouting after pile installation. On the other hand, with the use of limit state design, it may be unnecessary. As long as the ultimate geotechnical capacity is still present, even if at a large displacement, the serviceability limit state is still satisfied, which will probably be dependent on the shaft friction and then the piles will still work satisfactorily. It must also be remembered that in practice, where piles operate in groups of significant size, the overall behaviour is the result of the application of stress to a much larger zone than the base of the piles, and performance at working stress may not be affected by base stiffness.

Having said all that, the base resistance for a bored pile is difficult to estimate, because the resistance is very dependent on workmanship. Unlike the small variation in the K_s values, the K_b values vary significantly. The K_b values as reported by Toh *et al.* (1989), based on two test piles that were taken to failure, are between 27 and 60. They also suggested that, unless there is confidence in the cleaning of the base, bored pile design should consider shaft friction only. Chang and Broms (1991) had suggested K_b of 30–45 for the design of bored piles. However, Tan *et al.* (1998) found that the K_b values can be as low as 7–10, possibly as a result of poor cleaning. Chen and Hiew recommended a K_b value of 10 for preliminary design. They suggested a higher K_b value of 30 could be adopted if cleaning of the pile base before concreting could be carried out effectively.

A series of axial load tests on drilled shafts in a Piedmont residual soil are reported by Brown (2002). Several construction techniques were used to install the shafts in an attempt to identify the significance of installation technique on performance under axial load. These included:

(i) slurry shafts constructed using a truck-mounted drilling rig with a conventional soil auger. A casing was used in the upper 1 to 2 m and extended approximately 1 m above ground to facilitate testing. The bottom of each shaft was cleaned successively using a clean-out bucket and an air-lift. The slurry varied from high grade commercial bentonite to a polymer in dry pellet form;
(ii) four of the shafts were constructed using temporary casing advanced ahead of the excavation; and,
(iii) one CFA pile, constructed by screwing the hollow stem augers into the ground and filling the void with concrete during withdrawal.

Excavation and visual inspection of the concrete/soil interface provided additional insight into the effects of installation technique on side shear resistance. The following conclusions were drawn (Brown, 2002), and might be applied to soil conditions similar to this site:

- In these Piedmont fine-grained silty soils, the shafts installed using bentonite slurry had a reduced capacity compared to other installation techniques. This effect appears to be largely related to the presence of a thin film of bentonite left at the concrete/soil interface as a result of filter cake formation during drilling. This observation seems consistent with several other limited studies in granular soils, but the surprising finding at this site was that the effect seemed to be pronounced even for shafts with very limited exposure times. This effect may not extend to soils of lower hydraulic conductivity, which may reduce the tendency for filter cake formation.

Table 8.22 Measured and Computed Unit Load Transfer Values

Method	*Unit side shear (kPa)*	*Unit end bearing (kPa)*
Measured average (excluding bentonite)	55	850
FHWA 1999 guidelines, cohesive soils	51	830
FHWA 1999 guidelines, cohesionless soils	82	800

- The polymer slurry materials appeared to promote an excellent bond between the concrete and soil. There was a distinct tendency for shafts constructed using these materials to exhibit strain softening behaviour, although the mechanism for this effect is unclear. The strain softening may be related to dilatant behaviour of the relatively undisturbed Piedmont residual soil in the near field around the shaft during shear.
- The use of casing advanced ahead of the shaft excavation resulted in axial shaft capacity which was comparable to that of the polymer shafts, but without the strain softening tendency. The rotation of the casing during installation was observed to remould and distort the soil in the near field around the shaft and the casing produced a smooth concrete/soil interface. However, the twisting of the casing during extraction and the cutting teeth on the bottom of the casing left a rough macrotexture on the sidewall. The surface texture left by the cutting teeth appeared to be beneficial; a smooth steel casing which might be extracted by a vibrohammer could produce less desirable side shear capacity, particularly in cohesive soils (Camp *et al.*, 2002).
- The occurrence of soil inclusions with cross-sectional area of up to 20% of the shaft cross-section had no effect on the short-term performance under axial load. It should be noted, however, that the average compressive stress was only around 4.3 MPa, and so structural failure of the shaft column was not an issue.

Brown (2002) presents comparisons with recommendations used for routine design of drilled shafts in residual soil. Since Piedmont silty soils are intermediate between clay and sand, in practice design may include both a total stress and effective stress analysis.

For piles or drilled shafts in cohesive soil, the current FHWA guidelines suggest a unit shaft friction of $f_{\max} = \alpha \cdot s_u$, where $\alpha = 0.55$ and $s_u =$ average undrained shear strength, $= 92$ kPa in this case. Computed unit end bearing for cohesive soils is taken as $9 \cdot s_u$ kPa. For cohesionless soil, the current FHWA guidelines suggest a unit shaft friction of $f_{\max} = \beta \cdot \sigma'_v$, where $\beta = 1.5 - 0.245[z]^{1/2}$, $z =$ depth in metres to middle of cohesive layer, and $\sigma'_v =$ effective vertical stress at depth z. Computed unit end bearing for cohesionless soils is taken as $57.5 \cdot N_{SPT}$ kPa, where N_{SPT} is the standard penetration test resistance within two diameters below the base (around 14 in this case). These guidelines are based on the computed unit end bearing mobilised at a displacement of approximately 5% of the base diameter. The computed values of axial unit shaft friction and end bearing using the Federal Highway Administration (FHWA) guidelines (O'Neill and Reese, 1999) are provided in Table 8.22.

Figure 8.38 Installation of the bored piles with temporary casing

ISC'2 Pile Prediction Event in residual soil in FEUP, Porto

Scope and steps of the programme

In the Fall of 2003, the Faculty of Engineering of the University of Porto (FEUP) and the High Technical Institute of the Technical University of Lisbon (ISTUTL) invited the international geotechnical community to participate in a Class A prediction event on pile capacity and pile response to an applied loading sequence for Bored, CFA and Driven Piles. An extensive site investigation had been carried out, in which several *in situ* testing techniques were used. Undisturbed samples were recovered and an extensive laboratory-testing programme was carried out.

Two each of three different kinds of piles were installed. 600 mm diameter bored piles (E0 and E9) installed using a temporary casing, 600 mm diameter CFA piles (T1 and T2), and 350 mm square, driven, precast concrete piles (C1 and C2). Of these, 3 piles were tested dynamically (E0, T2 and C2), and 3 were loaded axially up to failure (piles E9, T1 and C1).

Piles E0 and E9 were constructed in August 2003 by first using a rotary drilling rig to install a temporary casing that was cleaned out using a 500 mm cleaning bucket (Figure 8.38). The external diameter of the cutting teeth at the bottom of the temporary casing was 620 mm. Concrete, with slump 180 mm, was placed by tremie. The casing was withdrawn on completion of the concreting and "overconsumption" was below 10%.

Piles T1 and T2 were constructed in August 2003 using a rotary drilling rig and a 600 mm continuous flight auger (see Figure 8.39). Concrete grout was ejected with a pressure of 6000 kPa at the beginning of the grout line and the flow rate was steady at 700 l/min. Concrete slump was 190 mm and "overconsumption" was less than 6%.

The C-piles, were driven on September 17, 2003 with a 40 kN drop hammer (Figure 8.40).

The piles were exhumed after testing, and Figure 8.41 shows the outside appearance of each pile type.

Figure 8.39 Installation of CFA pile

Figure 8.40 Installation of precast concrete driven pile

The predictions

For the Pile Prediction Event, the participants were provided with information on pile geometry, soil profile, equipment and high strain dynamic test results. They were challenged to predict the performance of the piles under static load by submitting a table of load vs. settlement at the pile head, and also:

(i) parameters and models used;
(ii) calculation methodology;
(iii) pile base resistance and shaft resistance, separately if applied;

Figure 8.41 Some of the differences in the three types of piles after extraction

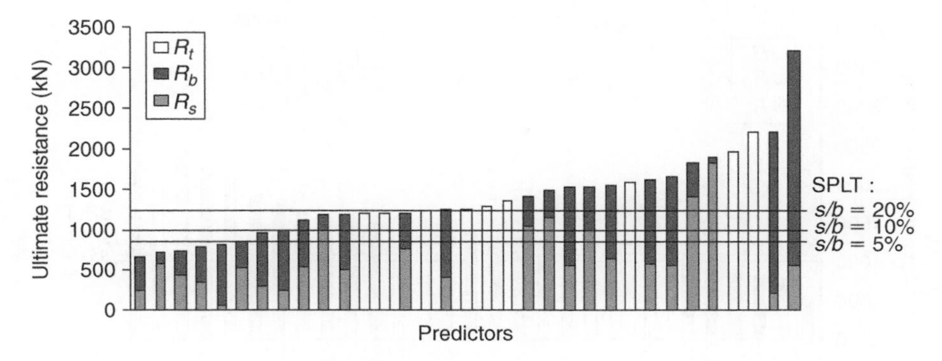

Figure 8.42 Ultimate resistance: predictions for bored pile E9

(iv) ultimate compressive resistance and indication of criteria used to determine this;
(v) allowable bearing capacity, and factor of safety, if applied, used to determine this;
(vi) explanation of the methods used to make the predictions.

In December 2003, a total of 33 persons from 17 countries submitted predictions, and static loading tests on piles E9, T1 and C1 were then performed in January 2004. A summary of the predictions and the test results is given in Viana da Fonseca and Santos (2008).

A compilation of the predictions for the ultimate compressive resistance is represented in Figures 8.42, 8.43 and 8.44 for piles E9, T1 and C1, respectively. The pile base resistance (R_b) and shaft resistance (R_s) are indicated separately when applied; otherwise the total resistance (R_t) is shown. The predictors applied different calculation methods, such as analytical or empirical methods, results of dynamic load tests or a combination of both. It is also important to note that different criteria or calculation approaches were used to define the ultimate compressive resistance.

The predictions presented in the figures are very scattered demonstrating that the accurate estimation of pile axial capacity is still a very difficult task in residual soils. For each pile type one prediction was optimistic, and the means and standard deviations of the remainder are shown in Table 8.23.

For the non-displacement piles (bored piles and CFA piles), most predictions have overestimated the ultimate capacity, probably because the settlements induced in the static tests have not mobilised the failure mechanism assumed in theoretical formulations for the evaluation of base resistance. On the other hand, the predictions for the displacement piles (precast driven piles) were conservative, probably because there was an underestimation of the strength gains due to pile installation effects.

Predictions based on analytical methods give a large scatter, since there is a great risk with such methods in using the input data with no judgement. It is generally preferable to use directly the results of *in situ* tests for prediction of ultimate resistance of piles. Synthesis of that data for the ISC'2 event site is presented in Viana da Fonseca and Santos (2008).

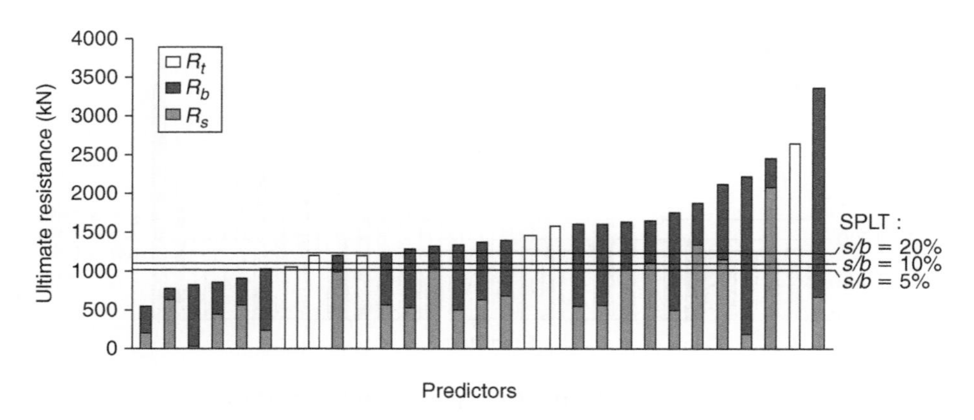

Figure 8.43 Ultimate resistance: predictions for CFA pile T1

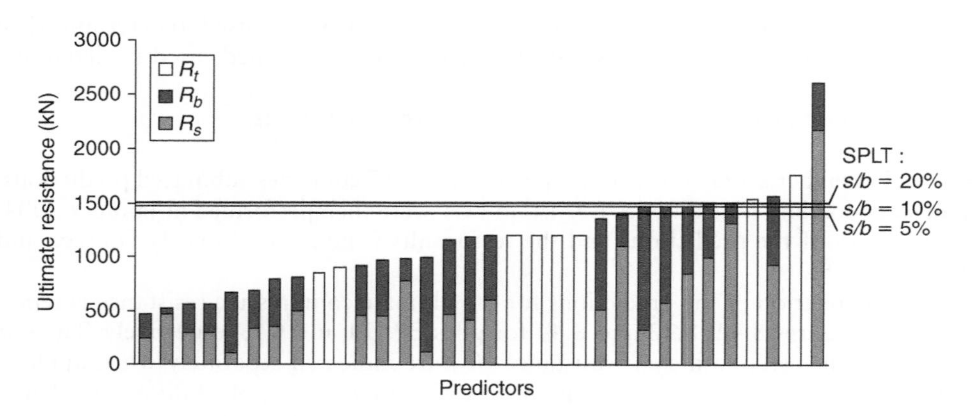

Figure 8.44 Ultimate resistance: predictions for driven pile C1

The SPT-based predictions performed fairly well for the non-displacement piles, being in fairly good agreement with the load test results for the bored pile (E9), while generally over-predicting the ultimate capacity of the CFA pile (T1). As for CPT based predictions, these show good agreement with the load test results for the bored pile (E9), with some scatter, but only reasonable agreement for the CFA pile (T1), for which there were fewer predictions. For the driven pile C1, the predicted ultimate capacity values, of which there were a large number, were very low when compared with the load test results, with a few exceptions (Viana da Fonseca and Santos, 2008).

The results of the PMT-based predictions and their interdependence with the pile construction process in residual soil are interesting. The results generally agree very well with the Q_{SPLT} ($s/b = 20\%$), being far better than the SPT and CPT based predictions, especially for the non-displacement bored pile (E9 see Figure 8.45a) and CFA pile (T1 see Figure 8.45b). For the displacement pile (C1 see Figure 8.45c), the agreement is good for ½ of the predictions.

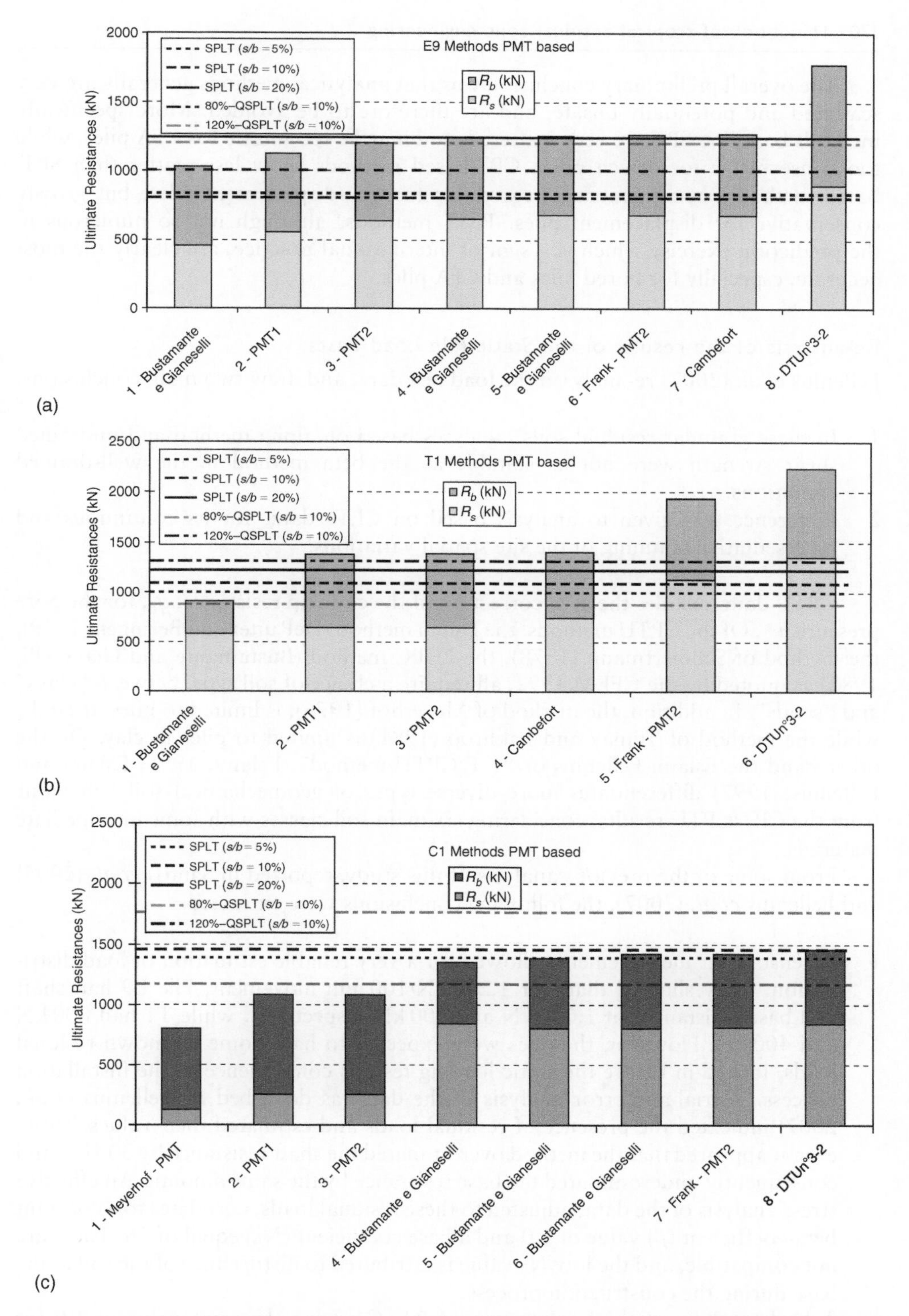

Figure 8.45 Predictions from PMT results of ultimate resistance of: (a) bored pile, E9; (b) CFA pile, T1; (c) precast driven pile, C1

The overall preliminary conclusion was that analytical methods generally are very scattered and potentially unsafe, and are therefore to be avoided. More specifically methods based on SPT are very scattered, and may be too unsafe for CFA piles, while too conservative for driven piles. CPT-based methods show less scatter than SPT-based methods, being generally reasonable for non-displacement piles, but grossly conservative for displacement piles. PMT methods, although not so numerous in the prediction exercise which is a sign of international practice, are clearly the most accurate, especially for bored piles and CFA piles.

Re-analysis of the results of the Static Pile Load Tests

Fellenius *et al.* (2007) re-analysed the load test data, and drew two main conclusions:

1 In these granular residual soils, analyses based on alpha method and undrained shear strength were not as suitable as the beta method in the well-drained conditions.
2 Preference was given to analysis based on CPTu data, for its continuous and representative scanning of the site spatial variations.

CPTU data includes the area correction of the cone tip resistance, q_c, for the pore pressure, u_2. Of the CPTU methods, the Dutch method (DeRuiter and Beringen, 1979), the method of Schmertmann (1978), the LCPC method (Bustamante and Gianeselli, 1982) as quoted by the CFEM 1992, all require a choice of soil type, between "clays" and "sands". In addition, the method of Meyerhof (1976), is limited to piles in sands, while the method of Tumay and Fakhroo (1981) is limited to piles in clay. On the other hand the Eslami-Fellenius or "E-F CPTU method" (Eslami, 1996; Eslami and Fellenius, 1997) differentiates more diverse types of geomechanical soil behaviour from the CPT/CPTU results, considering six main soil classes with some intermediate materials.

From some of the previous analysis of this study, reported in Santos *et al.* (2005) and Fellenius *et al.* (2007), the following conclusions were extracted:

- Extensometer measurements allowed for a very reliable estimation of load distribution. They showed that, for 1,200 kN/100 mm movement, Pile E9 had shaft and base resistances of 1,000 kN and 200 kN respectively, while T1 had 800 kN and 400 kN. However, the piles were expected to have some unknown residual loads, locked-in before the static loading test, as consequence of the installation process. A trial-and-error analysis of the data (as described by Fellenius *et al.*, 2007) indicated the presence of residual loads and estimated their values. However, it appeared that the method overestimated the shaft resistance by 300 kN and consequently underestimated the base resistance by the same amount. An effective stress analysis of the data, adjusted to these residual loads, correlates to a constant beta-coefficient (β) value of 1.0 and a base coefficient (N_b) equal of 16. These are not compatible, and the low N_b value is attributed to disturbance of the soil at the base during the construction process;
- A back-analysis of the loading test on Pile C1 using the same value of 1.0 for β as that derived from the load distribution of Piles E9 and T1, indicates total

shaft and base resistances of 520 kN and 980 kN, respectively. This base resistance corresponds to a N_b of 70, which is compatible with the β of 1.0.

A new analysis was carried out by Viana da Fonseca *et al.* (2007), with the purpose of having a more fundamental evaluation of the residual loads and, consequently, the load distribution between ultimate shaft resistance and tip resistance for 100 mm top settlement. This was possible by applying a mathematical model that allowed a back analysis of the top load-settlement curve.

A model to predict single pile performance under vertical loading was proposed by Massad (1995), which included many aspects of load transfer phenomena, considered previously by Baguelin and Venon (1971), such as pile compressibility and progressive failure. In addition, it takes into account the presence of residual stresses due to driving or subsequent cycling loadings. The solutions are analytical, in closed form, and were derived using load transfer functions based on Cambefort's Laws, accounting for the current knowledge of the shaft and tip displacements needed to mobilise the full resistances. They may be applied to bored, jacked or driven piles subjected to a preliminary monotonic loading and/or subsequent loading-unloading cycles. The soil is supposed to be homogeneous with depth, along the entire pile shaft. A thorough description of the model is given in Viana da Fonseca *et al.* (2007).

For the bored pile E9 and the CFA pile T1, a comparison was possible between the results of these analyses and the measurements from the extensometers installed within the piles. With these results, it was possible to evaluate the load distribution, for a reference top settlement of 100 mm, into an ultimate shaft resistance and base resistance, as shown in Table 8.23. These values are in close agreement with the ones from the previous analysis (Fellenius *et al.*, 2007), and the differences are due to the estimation of the residual loads. With this new numerical model, residual load values derived for the bored (E9) and CFA (T1) piles were around 150 kN, whereas Fellenius *et al.* (2007) had estimated them at 300 kN. The value of 150 kN for the base residual load was consistent with the measured value. In addition, for the driven pile (C1), the model defined an upper bound value for residual load of 500 kN. If the model is assumed to be valid, the ultimate unit shaft resistance for bored (E9) and CFA (T1) piles will be about 60 kPa. For the driven (C1) pile, this value may be taken as a lower limit.

Returning to the proposed relationships between SPT N value, shaft friction and end bearing in equations (33) and (34) results, in this case, in a K_s value of around 3.75, while K_b is around 70.

The K_s value of 3.75 is higher than the ranges previously reported (Table 8.24).

However, some caution is necessary. The higher values are measured values in only two tests, while some of the tabulated values are measured and some are proposed values to be used in design. The latter should always include a degree of conservatism.

Table 8.23 Load Distribution for 100 mm pile head settlement

Pile	*Shaft load (kN)*	*Base load (kN)*	*Total load (kN)*
E9 (Bored)	696	481	1177
T1 (CFA)	703	499	1202
C1 (Precast)	511 to 1021	1004 to 494	1515

Table 8.24 Values of K_s by different researchers

Author	K_s
Toh *et al.* (1989)	2.5–2.7
Phienwej *et al.* (1994)	2.3
Tan *et al.* (1998)	2
Chang and Broms (1991)	2
Chen and Hiew (2006) dry process	3
Chen and Hiew (2006) wet process	2

Table 8.25 Values of K_b by different researchers

Author	K_b
Toh *et al.* (1989)	27–60
Chang and Broms (1991)	30–45
Chen and Hiew (2006) preliminary	10
Chen and Hiew (2006) with good cleaning	30

The measured values for a dry process bored pile and a CFA pile are therefore compatible with the proposed design value.

The values of K_b are also much higher than those previously suggested (Table 8.25). This would imply that the bored pile tested in the ISC'2 programme was subject to effective base cleaning, as would be expected for a dry process pile, and that the CFA pile was installed carefully without loosening of soil at the base.

Design of axially loaded piles using the LCPC method

The advantage of using pressuremeter tests when dealing with residual soils, but also the possibility of having piling design rules for a wide variety of construction methods, makes the LCPC method very interesting and quite universal. The method was developed by the Laboratoire Central des Ponts et Chaussées, the French Highway Agency, and is also known as the French Design Rules for Foundations, Fascicule 62-V. As described by Bustamante *et al.* (2009), this method is based on the results of 30 years of pile loading tests on prototype piles constructed by more than 20 different techniques, mostly in Europe, the soil having been previously subjected to a site investigation using the Ménard pressuremeter (MPM). The study looked at the analysis of 550 load tests, of which 350 piles were instrumented to record the shaft friction of each soil layer and the base pressure; so as to compare these with the MPM design rules initiated by Louis Ménard in the 1960s, based on both the theory of cavity expansion in soils and his own experiments. Recently, additional experimental data, including comprehensive site investigations with PMT, CPT and SPT, have been gathered about full-scale pile load tests on deep foundations with instrumentation to:

(i) include the most recent installation/construction techniques used in common practice; and
(ii) refine the values of the limiting shaft friction q_s and the pile tip bearing factor K_p.

The principles

The principles of the PMT method used for designing foundations for civil engineering works in France are given by Gambin and Frank (2009). During most pile load tests, the end of the test occurs when the pile head starts to sink rapidly. The load related to this particular threshold is called the limit load Q_L. Conventionally, Q_L is defined as being the load for which the head settlement s_L is given by:

$$s_L \geq \frac{D}{10} + \Delta e \tag{8.37}$$

where D is the diameter of the pile, and Δe is the pile elastic shortening. The bearing capacity of a pile is expressed by:

$$Q = AK_p[(p_{LM} - p_0)_e] + P\sum(q_{si} \cdot z_i) \tag{8.38}$$

where

A = pile tip area
K_p = tip bearing factor
$[(p_{LM} - p_o)_e]$ = net equivalent Ménard limit pressure under the pile tip
P = pile perimeter
z_i = thickness of soil layer "i"
q_{si} = uniform shaft friction of soil layer "i".

The parameters K_p and q_s, which are essential in this equation, are measured on prototype piles load tested to failure with strain gauges.

The universality of this method has enabled it to be cited as a "common method to calculate the ultimate compressive resistance, Q, of piles" in the European standard for geotechnical design, EN 1997-1. In the future French standard to complement Eurocode 7, the factors are being readjusted to take into account the additional full scale tests available. The adaptation of these rules to the action and material factors of Eurocode 7 requires, in particular, having a precise knowledge of the scatter of the ratio of predicted bearing capacity to the measured bearing capacity.

The types of piles and the construction methods included in the design rules for axially loaded piles recommended by the French code covers all types of bored cast-in-place and driven piles, of any cross section (circular, polygonal, barrette, etc.); steel sheet piles and reinforced concrete diaphragm walls, when used as bearing elements (i.e. carrying vertical loads), are considered to belong to "piling".

Applicability of the method

The range of deep foundations to which this method may be applied is particularly varied including: pre-manufactured or cast-in-place; driven, jacked, jetted, screwed, bored or excavated; supplemented by grouting; piles with cylindrical section, but also H-section steel piles and sheet piles, as well as bored bearing elements (barrettes) of any shape; constitutive materials of concrete, grout, mortar, steel, or combinations thereof (timber piles are very rare in Europe).

The classification system for the design methods, presented by Bustamante and Frank (1999), is as follows:

Preformed piles:	*driven precast concrete* *driven steel (caisson, H, tube)* *prestressed tube* *concreted driven steel (H, tube)*
Rotary bored or excavated piles:	*bored without casing* *bored with casing* *flush bored (including barrettes)* *continuous flight auger* *cast-in-place screwed*
Micropiles:	*micropile types I, II, III and IV*
Piers	
Jacked piles:	*concrete jacked* *steel jacked*
Driven cast-in-place piles:	*internal drop hammer driving* *with a sacrificial shoe*

Basis of the method

There are some definitions in this method for design of axially loaded piles:

for compression piles Q_c : creep load
and Q_u : limit load
for tension piles Q_{tc} : tension creep load
and Q_{tu} : tension limit load.

The creep load Q_c may be used to characterise the axial behaviour of piles. ISSMFE (1985) defines the creep load as "a critical experimental load beyond which the rate of settlement under constant load takes place with a notably increased increment". In case of a static load test performed using the maintained load procedure, it is easily determined by a simple graphic construction (for instance AFNORNFP94-150).

The limit load is defined as the load at which the displacement of the head of the pile is 10% of the width or diameter of the pile. When deriving Q_c, Q_u, Q_{tc} or Q_{tu} from static load test(s), a reduction is applied to the measured value(s), related to the uncertainty of the result.

1 According to the Frenchstandard Fascicule 62-V, if only one single pile load test is performed:

$$Q = \frac{Q_m}{1.2} \tag{8.39}$$

where Q_m is the measured value.

If several load tests are performed:

$$Q_k = Q_{\min}\left(\frac{Q_{\min}}{Q_{\max}}\right)^{\xi'} \tag{8.40}$$

Table 8.26 Values of factor ξ' (Fascicule 62-V, 1993)

Number of load tests	2	3	4	5
ξ'	0.55	0.20	0.07	0.00

Table 8.27 Correlation factors ξ to derive characteristic values from static pile loadtests (*n* – number of tested piles)

ξ *for n*	*1*	2	3	4	≥ 5
ξ_1	1.40	1.30	1.20	1.10	1.00
ξ_2	1.40	1.20	1.05	1.00	1.00

where Q_{min} and Q_{max} are the minimum and maximum measured values, and ξ' is given in Table 8.26.

2 EN 1997-1 adopts similar concepts by stating that, when deriving the characteristic ultimate compressive resistance $R_{c;k}$ from measured values $R_{c;m}$, from pile load tests and from comparable experience, including one or more site specific pile load tests, an allowance shall be made for the variability of the ground and the variability of the effect of pile installation, by applying correlation factors as follows:

$$R_{c;k} = \min\left\{\frac{(R_{c;m})_{\text{mean}}}{\xi_1}; \frac{(R_{c;m})_{\text{min}}}{\xi_2}\right\} \tag{8.41}$$

where ξ_1 and ξ_2 are correlation factors that depend on the number of tests. The values of ξ_1 and ξ_2 are presented in Table 8.27.

On the limit states

Bustamante and Frank (1999), and Frank (2008) define Design Loads, Q_d, by the following:

ULS (ultimate limit state)

Under fundamental combinations: $Q_{tu}/1.4 \leq Q_d \leq Q_u/1.4$
Under accidental combinations: $Q_{tu}/1.3^{(1)} \leq Q_d \leq Q_u/1.2$
where

$Q_u = Q_{pu} + Q_{su}$ is the ultimate limit load in compression (end bearing + skin friction),

and

$Q_{tu} = Q_{su}$ is the ultimate limit load in tension (skin friction alone).

SLS (serviceability limit state)

Under rare combinations: $Q_{tc}/1.4^{(2)} \leq Q_d \leq Q_c/1.1$
Under quasi permanent combinations: $0^{(3)} \leq Q_d \leq Q_c/\ 1.4$

---

[1]For micropiles, $-Q_{tu}/1.2$; [2]For micropiles, $-Q_{tc}/1.1$; [3] For micropiles, $-Q_{tc}/1.4$

where Q_{cc} or Q_{tc} are the corresponding (compression and tension) creep loads:

$Q_{cc} = 0.5Q_{pu} + 0.7Q_{su}$ and $Q_{tc} = 0.7Q_{su}$ for non-displacement piles
$Q_{cc} = 0.7Q_{pu} + 0.7Q_{su} = 0.7Q_u$ and $Q_{tc} = 0.7Q_{su}$ for displacement piles

It should be noted that the way of treating serviceability limit states in the French standard is different from Eurocode 7. The former is based on the creep load, easily determined from bearing capacity approaches, while the latter relies essentially on settlement estimates.

The evaluation of Q_{pu} and Q_{su} is made by deriving the limiting values for the base pressure (q_u) and the shaft friction (q_{si}) by the formulae:

$$Q_{pu} = q_u \cdot A_p \tag{8.42}$$

$$Q_{su} = \sum_{0}^{i} q_{si} \cdot A_{si} \tag{8.43}$$

where A_p is the area of the base, A_{si} the shaft surface area in layer i. The values of q_u and q_{si} are derived by specific correlations to the *in situ* tests as described below. For tension piles, the limiting shaft friction is assumed to be the same as for compression piles.

PMT (MPM) method

The last update reported by Bustamante *et al.* (2009) included a total of 561 tests, of which 276 tests (or 49%) were taken to their limit load. For the remainder, the load was extrapolated up to this value by one of the analytical methods. Finally, 13% of the piles were subjected to tensile tests.

The Base Bearing Factor K_p
The K_p value can be chosen from Table 8.28.

Table 8.28 Direct design with PMT data in weathered rock

Group Code	K_p
1	1.6
2	2.0
3	2.3
4	2.4*
5	1.1*
6	1.4*
7	1.1*
8	1.5*

*If a higher K_p value is to be used, it must be proved by a load test.

Table 8.29 Description and characteristics of 138 analysed piles

Group Code	*Type No.*	*Piles* [2] *Qty*	*B*[3] *(mm)*	*D*[4] *(m)*	*Pile Description*
1	1	8	500–2,000	11.5–23	Pile or barrette bored in the dr
	2	64	270–1,800	6–78	Pile and barrette bored with slurry
	3	2	270–1,200	20–56	Bored and cased pile (permanent casing)
	4	28	420–1,100	5.5–29	Bored and cased piler (recoverable casing)
	5[1]	4	520–880	19–27	Dry bored piles/or slurr bored piles with grooved sockets/or piers (3 Types)
2	5[1]	50	410–980	4.5–30	Bored pile with a single or double-rotation CFA (2 types)
3	7	48	310–710	5–19.5	Screwed cast-in-place
	8	1	650	13.5	Screwed pile with casing
4	9[1]	30	280–520	6.5–72.5	Pre-cast or pre-stressed concrete driven pile (2 types)
	10	15	250–600	8.9–20	Coated driven pile (concrete, mortar, grout)
	11	19	330–610	4–29.5	Driven cast-in place pile
	12	27	170–810	4.5–45	Driven steel pile, closed tip
5	13	27	190–1,22	8–70	Driven steel pile, open end
6	14	23	260–600	6–64	Driven H pile
	15	4	260–430	9–15.5	Driven grouted[5] or [6] H Pile
7	16	15	—	3.5–2.5	Driven sheet pile
1	17		80–140	4–12	Micropile type I
	18	8	120–810	8.5–37	Micropile type II
8	19	23	100–1,220	8.5–67	SGP[5] micropile (type III) or SGP pile
	20	20	130–660	7–39	MRP[6] micropile (type IV) or MRP pile

The Ultimate Shaft Friction q_s

According to Bustamante *et al.* (2009), the parameter q_s is chosen as follows:

- select the pile type from Table 8.29 and get the applicable Q_i from Table 8.30
- use Figure 8.46 to obtain, on the selected Q_i curve, the q_s for the Ménard limit pressure (p_{LM}) measured at the appropriate depth.

Contrary to the set of discontinuous straight lines which are shown on the similar graph in the French codes, Fascicule 62-V and AFNOR DTU-13.2, it should be noticed that in Figure 8.46, the q_s lines are continuous, which avoids any ambiguity when choosing this parameter.

Application of CPT method to the results of ISC'2-PPE static pile load tests

For CPT values, the correlations with the unit shaft and base resistances may be expressed, following the same pattern as the French/LCPC method (Bustamante and Frank, 1999), as the following:

$$q_s = \min\left[\frac{q_c}{\beta}; q_{s\max}\right] \tag{8.44}$$

Table 8.30 Direct design with PMT Data in weathered rock

Pile Type No.	Q_i
1	Q6*
2	Q6*
3	Q1*
4	Q4*
5	Q6
6	Q5*
7	Q4*
8	Q2*
17	Q6*
18	Q6*
19	Q9*
20	Q10*

** Use of a higher value must be proved by a load test.

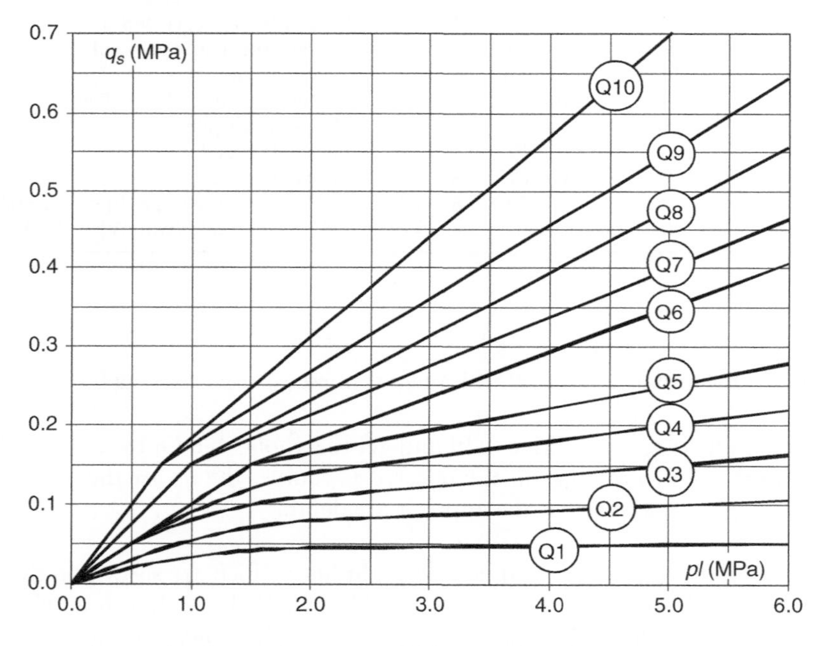

Figure 8.46 Direct design using PMT Data. Chart for shaft friction q_s

$$q_b = K_c \cdot q_{ce} \tag{8.45}$$

where q_c is the average CPT static cone resistance in each assumed layer and q_{ce} is the cone resistance around the base – measured between half a diameter above the base and one and a half diameters below); K_c is the base bearing factor, which is a function of the type of soils, and the pile type and method of execution (values are available for non-displacement piles and displacement piles); β is a correlation factor that depends

Table 8.31 Unit shaft friction from CPT data (*After* Bustamante and Frank, 1999)

Type of piles	*Soils* *Range of* q_c *(MPa)*	*Clays and Silts* <3	*3–6*		*>6*		*Sands and Gravels* <5	*8–15*	*>20*
A: Bored without casing	β	—	—	75			200	200	200
(dry method)	$q_{s\,max}$(kPa)	15	40	80	40	80			120
B: Bored with temporary	β	–	100	100	—	100	250	250	300
casing	$q_{s\,max}$(kPa)	15	40	60	40	80	—	40	120
C: Driven precast	β	—	75		—		150	150	150
concrete	$q_{s\,max}$(kPa)	15	80		80		—	—	120

on the pile type and the soil; $q_{s\max}$ is the limiting value of shaft friction, which again depends on the type of soil and the method of pile construction.

In the specific case of the non-displacement piles tested in the residual soil of Porto granite:

$$\beta \cong 55; K_c \cong 0.35$$

For the displacement piles, where the shaft friction is reduced by $\beta \cong 120$, then $K_c \cong 0.75$.

These may be compared to some reference values (Bustamante and Frank, 1999), which would suggest for sand or silty sand, the values for base resistance would be $K_c = 0.15$, for non-displacement piles, and $K_c = 0.50$, for displacement piles. For shaft resistance the patterns for β and $q_{s\max}$ proposed in the LCPC method are given in Table 8.31 (considering the three classes closer to the three types of piles used in the experimental site).

These appear to show, considering that typical values of CPT are in the mid-range, that for the bored and CFA piles, the values of the shaft coefficient ($\beta \cong 55$) are more in agreement with the behaviour of silty soils, while for the driven piles, the values of the shaft coefficient ($\beta \cong 120$) are much more in agreement with the behaviour of sandy soils.

It must be noted that all the reported values were obtained for one static pile load test. Therefore, even considering the extensive *in situ* tests, the values of the design coefficients must be reduced, as required by French standards as well as in EN 1997-1.

Driving formulae and wave equation analysis

Since the early 20th century, many different dynamic formulae have been proposed as a simple means of determining pile capacity from measurements made on site, of hammer weight, drop height, final set (in mm per blow) and temporary compression. There are plenty of formulae (Poulos and Davis, 1980), including the Hiley, Janbu, Danish, Dutch, Crandall and modified Engineering News Record formulae. Not surprisingly, considering the range of soil types, of hammer types, and of pile types, none has proved to be sufficiently reliable in all situations. They are commonly used for harbour, water-front, canal and off-shore constructions for which driven steel profiles, i.e. pipes, H-sections and sheet piles, are employed. They are also very widely used in conjunction

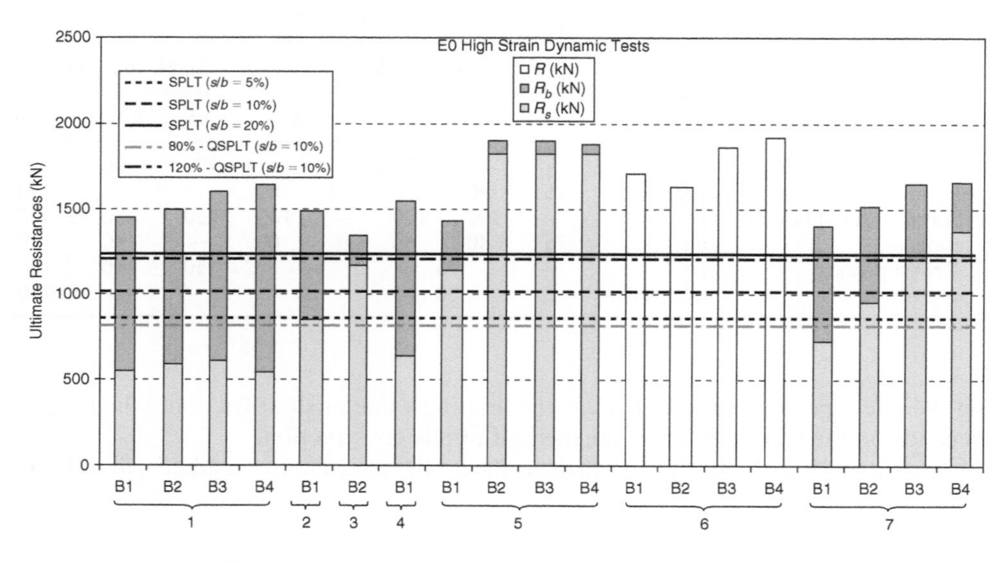

Figure 8.47 Predictions from HSDT to ultimate resistance of pile E0 (bored)

with other forms of testing, since it is generally not feasible to carry out a static or dynamic load test on every pile. Presently, there is an increasing trend to use dynamic load testing, in which the interpretation is generally made using methods based on wave propagation theory, such as the CASE method, CAPWAP analysis, TNO and Simbat. Whereas CASE, CAPWAP and TNO analyses are mainly used for displacement piles, Simbat has been developed for bored piles. A recommended procedure is to verify a static pile design by carrying out dynamic load tests on a suitable number of piles, depending on the size of the project. The French Code and EN 1997-1 give guidance on how many piles should be tested, by varying the correlation coefficient with the number of tests. The Australian Standard AS 2159:2009, introduces a geotechnical resistance reduction factor, within the range of 0.4 to 0.9, and the larger numbers are only possible with reasonable amounts of testing. While carrying out the dynamic testing, the chosen dynamic formula can be calibrated, and the values of set and temporary compression used to check the capacity of all other piles at time of driving through a dynamic formula. There are criticisms of dynamic pile testing and its correlation with static load testing because of the very short time during which the force is applied. Attempts have been made to overcome this with Statnamic testing, in which the force, while still dynamic, is applied over a longer time by the explosive expansion of gases which raise a heavy weight against the resistance of the pile. It is becoming popular, but is still expensive to set up and operate and is limited in availability.

State of the art of stress wave testing

From the ISC'2 event, Viana da Fonseca and Santos (2008) reported and analysed the contribution of participators that used High Strain Dynamic Tests (HSDT) for the prediction of the bored, CFA and precast driven piles. Figure 8.47 presents the 7

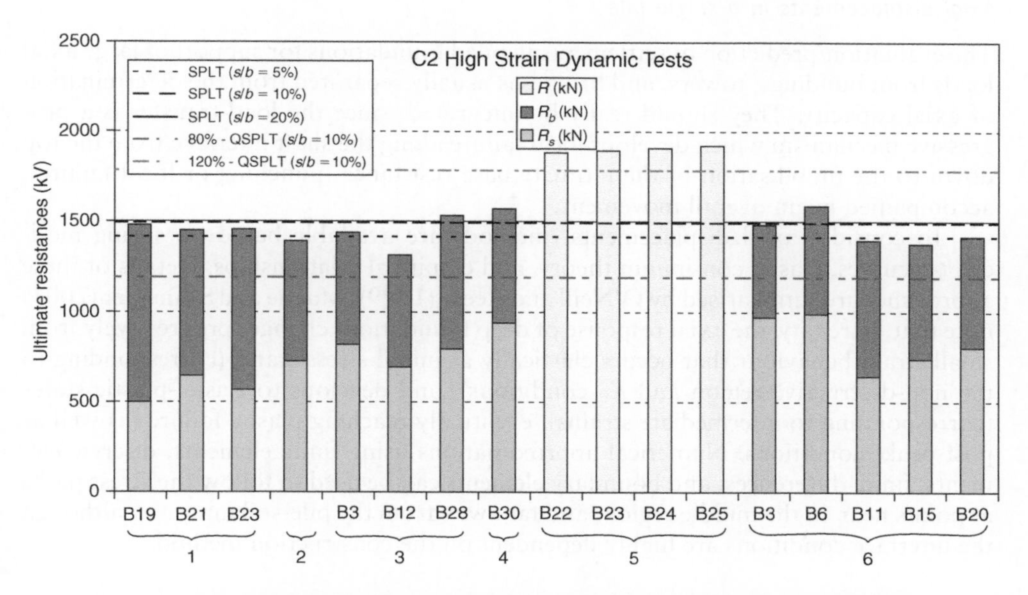

Figure 8.48 Comparison between HSDT on pile C2 and SPLT on pile C1 (driven)

predictions for the bored pile, E0, compared with the ultimate resistance obtained in the Static Pile Load Test (SPLT) on E9. There were four blows, and four predictors used all four, while three used only B1. It can seen that all the predicted values were higher than those obtained in the SPLT, at a settlement = 10% of diameter, by between 34 and 91%. It can also be seen that the predicted values are within quite a narrow range, about ±15%.

A similar analysis is presented for CFA piles. The HSDT predicted values were, generally, higher than those obtained from SPLT for a settlement = 10% of diameter. In this case, the predicted values cover a wider range, about ±33%, and were between 62 and 189% of the SPLT value.

Finally, the report presented the results of the application of HSDT to precast driven piles. Most of the results are in the range of 88–128% Q_{SPLT} (Figure 8.48). The exception is one predictor who proposed the highest capacity for all three types of pile. The values for the remaining predictors are similar, within a range of about ±15%. Clearly, the HSDT is best at predicting the capacity of driven piles, for which it was developed. This may be because the dynamic load test (with consecutive dynamic blows with increasing energy) can modify the subsequent pile resistance compared to the static "virgin resistance" of the pile, the "virgin resistance" being defined as the pile resistance obtained during its first loading. This is particularly true for the bored and CFA piles, in which the mobilisation of base resistance is highly affected by the load history and, as a consequence, the HDST with increasing energy can lead to unsafe predictions of the static "virgin resistance". The difference between the values of HSDT and SPLT can probably be attributed to differences of soil characteristics between piles C2 and C1.

Axial displacements in a single pile

The evaluation/prediction of settlements in piled foundations for support of large axial loads from buildings, towers, and bridges is usually separated from the determination of axial capacity. They should really be integrated, since the load transfer is a progressive mechanism which develops local failure along the shaft interface from the top down to the mobilisation of the ultimate base resistance (punching or local failure), accompanied by an overall movement.

To consider axial displacements, methods are available based on spring models, *t-z* curves, elastic continuum theory, and empirical relationships. Details of these approaches are summarised by O'Neill and Reese (1999). Mayne and Schneider (2001) note that, in reality, the axial response of deep foundations changes progressively from small strain behaviour that occurs elastically at initial stress states (corresponding to the non-destructive region and K_0 conditions), and develops to elastic-plastic states (corresponding to intermediate strains), eventually reaching plastic failure (as well as post-peak) conditions. Numerical approximations using finite elements, discrete elements, finite differences, and boundary elements can be used to follow the stress paths at points near, in the middle region, and far away from the pile-soil interface, although the interface conditions are highly dependent on the construction method.

Geomechanics characteristics of geomaterials

The practice of foundation analysis demands that even simplified analytical methods include key features of geomaterial behaviour. These features include the well established idea that soils and weathered rocks exhibit highly non-linear behaviour under load, with linear elastic behaviour observed only at very small strains in the vicinity of the operational G_{max} (G_0). This small-strain stiffness, which can be determined from shear wave velocity measurements ($G_0 = \rho \cdot V_s^2$, ρ being the mass density of the soil) applies to the initial static monotonic loading, as well as the dynamic loading of geomaterials (Burland, 1989; Tatsuoka and Shibuya, 1992; Lo Presti *et al.*, 1993). The equivalent elastic modulus is found from: $E_0 = 2 \cdot G_0(1+\nu)$, where ν is the value of Poisson's ratio of geomaterials at small strains. Mayne and Schneider (2001) illustrate this pattern of behaviour and tentatively suggest the relative position of the most common *in situ* tests in terms of the stress-strain curves (see Figure 8.49).

This highly nonlinear stress-strain-strength-time response of soils depends upon loading direction, anisotropy, rate effects, stress level, strain history, time effects, and other factors. So, as Mayne and Schneider (2001) commented, it is difficult to recommend a single test, or even a suite of tests, that directly obtains the relevant E_s for all possible types of analyses in every soil type. They admit that, in certain geomaterials, it has been possible to develop calibrated correlations between specific tests (e.g. PMT, DMT), since they generate the stress-strain-strength curve, generally at an intermediate level of strain, allowing a fair correlation with full-scale structures, including foundations and embankments, or with reference values from laboratory tests.

A more comprehensive analysis is proposed by some (Mayne *et al.*, 1999b; Mayne and Schneider, 2001, amongst others), which takes advantage of the significance the small-strain modulus (G_0) attains as a reference starting point, and makes it possible to use a generalised approach where the initial modulus (E_0, related to G_0) is degraded to an appropriate stress level for the desired FOS.

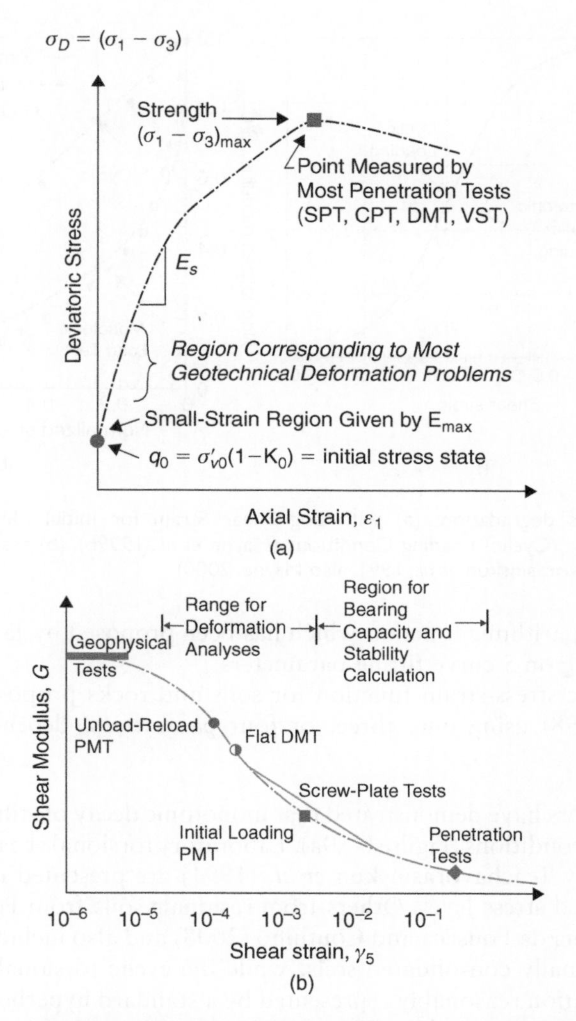

Figure 8.49 (a) Stresses and stiffnesses of soils at small and large strains, (b) Variation of shear modulus with strain level and relevance to *in situ* tests (*After* Mayne and Schneider, 2001)

There are several expressions that have been proposed to represent modulus degradation. Mayne (2000) presents a good overview of some of these proposals:

- the simple hyperbola (Kondner, 1963) requiring only two parameters: shear modulus, G_{max}, and maximum shear stress, or shear strength, τ_{max}; this fails to adequately model the complex behaviour of natural soils with other non-linearities;
- modified hyperbolic laws (Duncan and Chang, 1970; Hardin and Drnevich, 1972; Prevost and Keane, 1990; Fahey and Carter, 1993; Mayne, 1994; Fahey and Carter, 1993), which increase the number of required parameters to either 3 or 4;

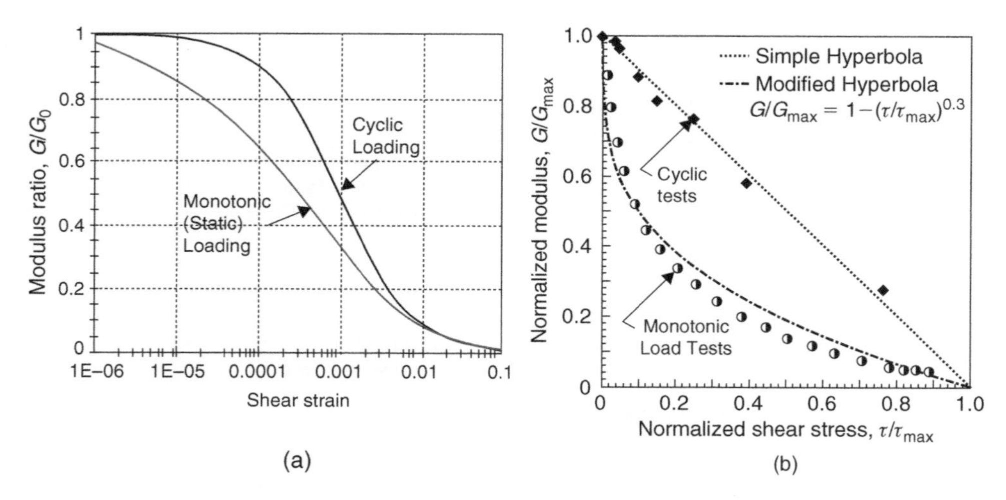

Figure 8.50 Modulus degradation: (a) with Log Shear Strain for Initial Monotonic (Static) and Dynamic (Cyclic) Loading Conditions (Mayne *et al.*, 1999b); (b) results for Toyoura sand (Teachavorasinskun *et al.*, 1991; also Mayne, 2000)

- a periodic logarithmic function which has been proposed by Jardine *et al.* (1986, 1991), relying on 5 curve-fitting parameters;
- a logarithmic stress-strain function for soils and rocks proposed by Puzrin and Burland (1998) using one, three, or four parameters, depending on available information.

Several authors have demonstrated that monotonic decay of stiffness is faster than in cyclic loading conditions (Figure 8.50a). Laboratory torsional shear tests on Toyoura sand, reported by Teachavorasinskun *et al.* (1991) are presented in Figure 8.50b in terms of mobilised stress level. Others from residuals soils from Porto Granite were presented by Viana da Fonseca and Coutinho (2008) and also included in Figure 8.51.

Even in normally consolidated soils, while the cyclic torsional shear tests show modulus degradation reasonably represented by a standard hyperbola, the monotonic tests lose stiffness more quickly (Figure 8.52). For monotonic torsional shearing of normally-consolidated sands, the modified hyperbola proposed by Fahey and Carter (1993) takes one of the two forms:

$$\frac{G}{G_0} = 1 - f\left(\frac{\tau}{\tau_{max}}\right)^g \quad \text{or} \quad \frac{E}{E_0} = 1 - f\left(\frac{q}{q_{ult}}\right)^g$$

where f and g are fitting parameters and q/q_{ult} is the reciprocal of the safety factor. Values of $f = 1$ and $g = 0.3$ appear reasonable first-order estimates for unstructured and uncemented geomaterials (Mayne *et al.*, 1999a).

Evaluating axial displacements

The axial load-displacement behaviour of single piles can be predicted by elastic continuum theory through the solutions developed by Poulos and Davis (1980), by

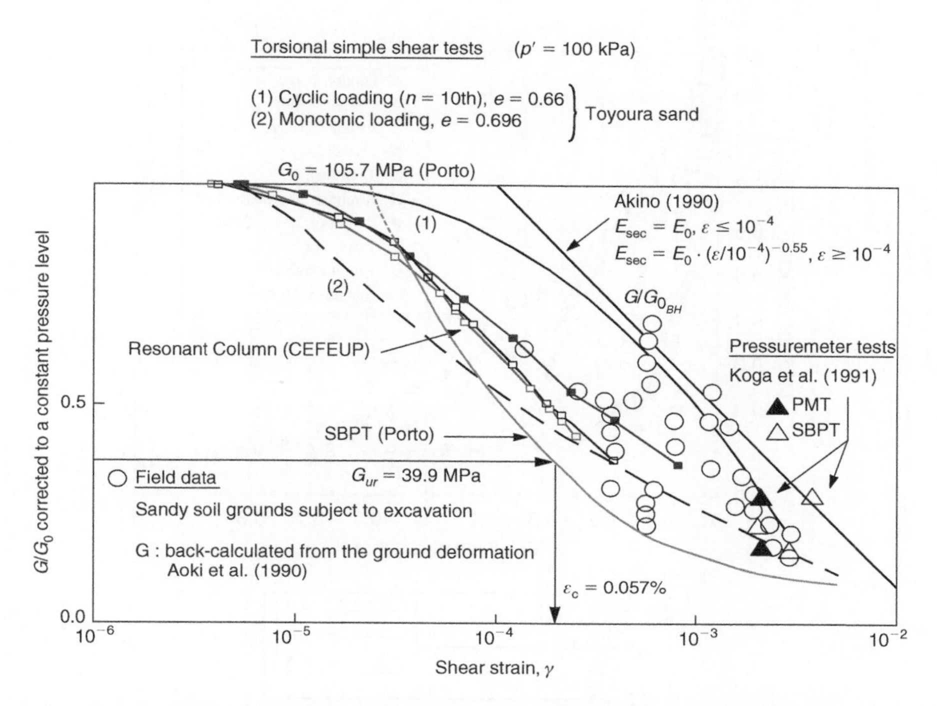

Figure 8.51 Comparison of secant shear modulus as a function of shear strain level: results of resonant column results in residual soils from granite of Porto with data from other testing conditions in sandy soils (*After* Aoki *et al.*, 1990; Tatsuoka and Shibuya, 1992)

finite elements (Poulos, 1989), and by approximate closed-form analytical solutions (Randolph and Wroth, 1978, 1979; Fleming *et al.*, 1985). Continuum theory characterises the soil stiffness by two elastic parameters: an equivalent elastic soil modulus (E_s) and Poisson's ratio (ν_s). Four generalised cases are considered: (1) a homogeneous case where E_s is constant with depth; (2) a Gibson-type condition where E_s is linearly-increasing with depth; (3) friction or floating-type piles; and (4) end-bearing type piles resting on a stiffer stratum.

The vertical displacement of a pile foundation subjected to axial compression can be expressed (Poulos, 1987, 1989) as:

$$s = \frac{P_t \cdot I_\rho}{(E_{sL} \cdot d)} \tag{8.46}$$

where P_t is the applied axial load at the top of the shaft, E_{sL} stands for the soil modulus along the sides at full depth ($z = L$), d = foundation diameter, and I_ρ = displacement influence factor. The factor I_ρ depends on the pile slenderness ratio (L/d), pile material, soil homogeneity, and relative soil-pile stiffness, and is given in chart solutions, tables,

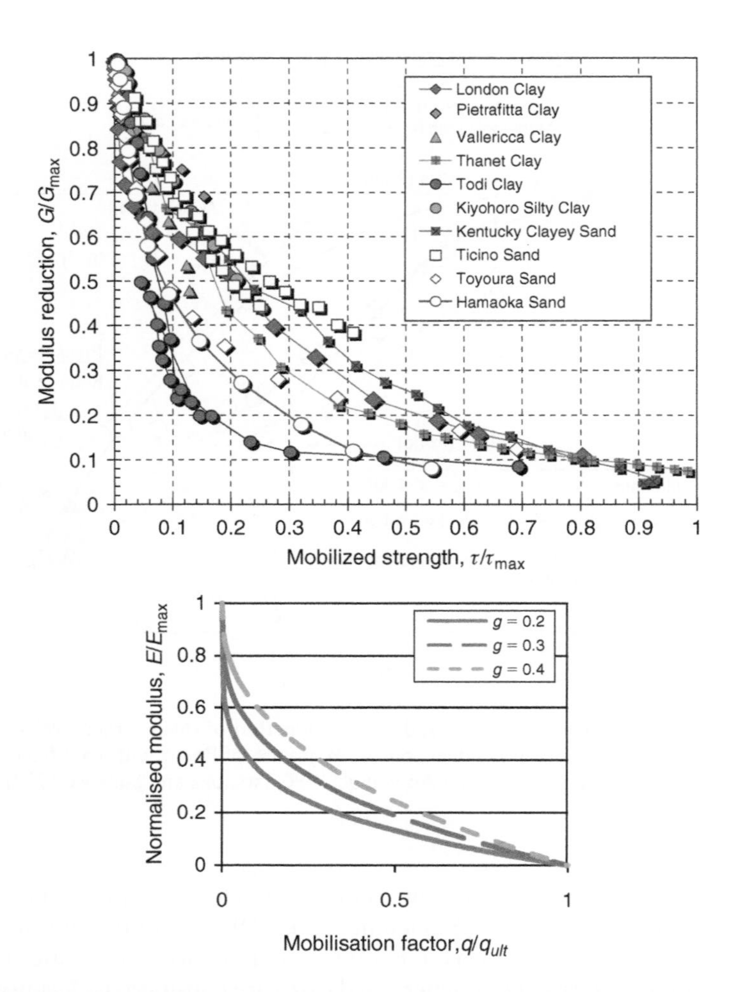

Figure 8.52 Modulus degradation in triaxial tests on uncemented and unstructured geomaterials and modified hyperbolae with $g = 0.2, 0.3,$ and 0.4 (*After* Mayne *et al.*, 1999b)

or approximate closed-form. The last is given by the solutions of Randolph and Wroth (1978, 1979) and Poulos (1987):

$$I_\rho = 4(1+v_s)\cdot\frac{[1+1/\pi\lambda\cdot 8/(1-v_s)\cdot\eta_1/\xi\cdot\tanh(\mu_L)/\mu_L\cdot L/d]}{[4\eta/(1-v_s)\xi+(4\pi\cdot\rho\cdot\tanh(\mu_L)\cdot L)/\zeta\cdot\mu_L\cdot d]} \tag{8.47}$$

where $\eta = d_b/d =$ eta factor ($d_b =$ diameter of base, so that $\eta = 1$ for straight shafts).
$\xi = E_{sL}/E_b =$ xi factor ($\xi = 1$ for floating pile; $\xi < 1$ for end-bearing).
$\rho^* = E_{sm}/E_{sL} =$ rho factor ($\rho^* = 1$ for uniform soil; $\rho^* = 0.5$ for simple Gibson soil).
$\lambda = 2\cdot(1+v_s)\cdot E_p/E_{sL} =$ lambda factor.

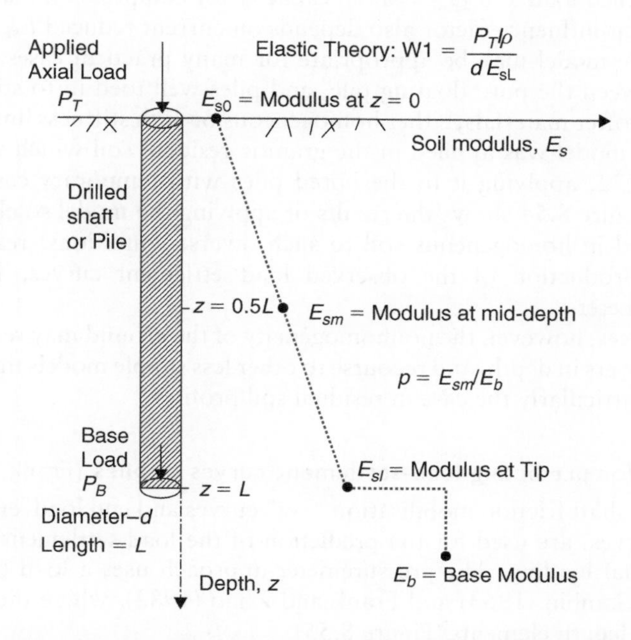

Figure 8.53 Terms used in elastic continuum model (*After* apud Mayne, 2000).

$\zeta = \ln\{[0.25 + (2.5\rho^*(1 - \nu_s) - 0.25) \cdot \xi]\ (2 \cdot L/d)\} =$ zeta factor.
$\mu_L = 2(2/(\zeta \cdot \lambda))^{0.5} \cdot (L/d) =$ mu factor.
$E_p =$ pile modulus (concrete plus reinforcing steel).
$E_{sL} =$ soil modulus value along pile shaft at level of base.
$E_{sm} =$ soil modulus value at mid-depth of pile shaft.
$E_b =$ soil modulus below foundation base (Note: $E_b = E_{sL}$ for floating pile).
$\nu_s =$ Poisson's ratio of soil.

Figure 8.53 shows the generalised stiffness profile for these cases, with corresponding definitions of moduli input for the analysis.

Elastic continuum methods also provide an evaluation of axial load transfer distribution. The fraction of load transferred to the pile base (P_b) is given by Mayne and Schneider (1999), based on the work of Fleming *et al.* (1985):

$$\frac{P_b}{P_t} = \frac{4/(1-\nu_s) \cdot \eta/\xi \cdot 1/\cosh(\mu_L)}{4/(1-\nu_s) \cdot \eta/\xi + 4\pi\rho/\zeta \cdot \tanh(\mu_L)/\mu_L \cdot L/d} \tag{8.48}$$

Additionally, a realistic non-linear load-settlement model of a pile can be determined by inserting a modified hyperbolic model, as introduced above, in the following expression:

$$s = \frac{Q \cdot I_\rho}{d \cdot E_{\max} \cdot [1 - f(Q/Q_{ult})]^g} \tag{8.49}$$

where Q = applied load and Q_{ult} = axial capacity for compression loading. Note that the displacement influence factor also depends on current reduced E_{sL}.

This simple model may be appropriate for many practical cases of foundations which are between the pure floating pile, and piles well toed in to stiff bases with a shaft through softer materials, either homogeneous or with stiffness linearly increasing in depth. This model was applied in the granitic residual soil which was thoroughly studied for ISC'2, applying it to the bored piles with temporary casing, CFA, and driven piles. Figure 8.54 shows the results of applying the model solely considering a shaft embedded in homogeneous soil to such diverse conditions, resulting in fairly reasonable reproduction of the observed load/settlement curves, but using very different parameters.

In other cases, however, the nonhomogeneity of the ground may require modelling of successive layers in depth, and recourse to other less simple models may be necessary. This may be particularly the case in residual soil profiles.

PMT method for predicting load-settlement curves of piles (Frank, 2008)

In this method shaft friction mobilisation "τ-z" curves and end load–end displacement curve (q-s_p curves) are used for the prediction of the load-settlement curve of single piles under axial loading. The pressuremeter approach uses a load transfer method developed by Gambin (1963) and Frank and Zhao (1982), where the pile is cut in a series of equal length elements (Figure 8.55).

According to Frank and Zhao (1982) the settlement s_p of the pile base is given by:

$$s_p = \left(\frac{B}{k_p}\right) q_p \tag{8.50}$$

where B is the diameter of the pile and q_p is the pile base pressure, while the skin friction q_{si} mobilised during the settlement of each pile shaft element i is then obtained as a function of z_{si} which is the local displacement of the ith shaft element against the adjacent soil layer (Figure 8.55). k_t and k_q are functions of the pressuremeter modulus E_M and the diameter B of the pile, with models proposed by Frank and Zhao and shown in Figure 8.56:

$$k_t = 2.0E_M/B \quad \text{and} \quad k_q = 11.0E_M/B \quad \text{for fine soils}$$
$$k_t = 0.8E_M/B \quad \text{and} \quad k_q = 4.8E_M/B \quad \text{for granular soils}$$

Examples of the use of this PMT method for predicting load-settlement curves of piles are given in Frank (1984) or Bustamante and Gianeselli (1993), and one is reproduced in Figure 8.57.

Lateral pile loading

Piles are often subject to lateral loading, and should be designed considering the lateral interaction with the ground. The methods using the subgrade reaction modulus (or p-y reaction curves, where p = reaction pressure, and y = horizontal displacement) are well known for the design of piles under lateral loads. These methods, which

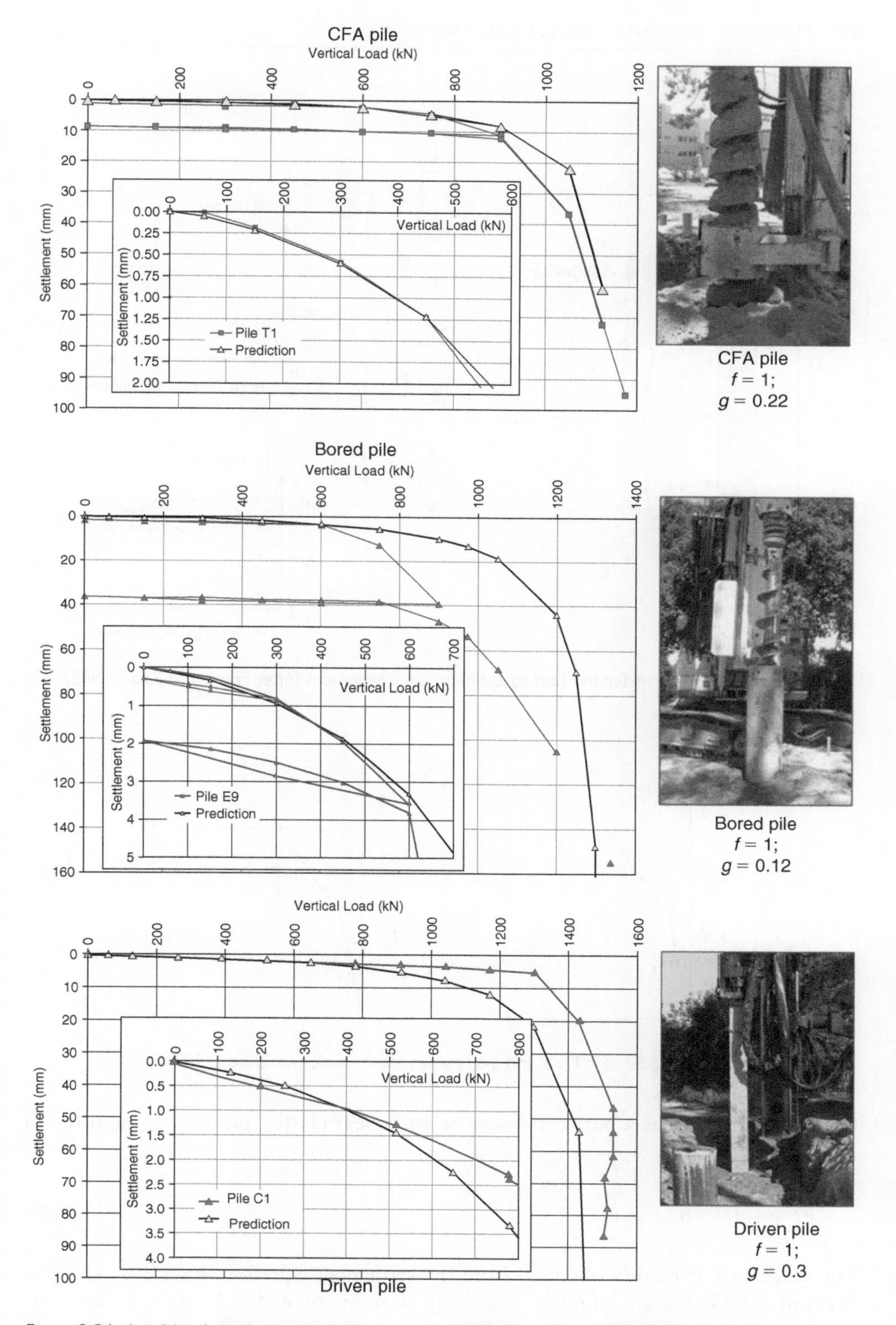

Figure 8.54 Axial load-displacement behaviour of single piles calculated by elastic continuum theory with non-linear stiffness integration (*After* Mayne and Schneider, 1999).

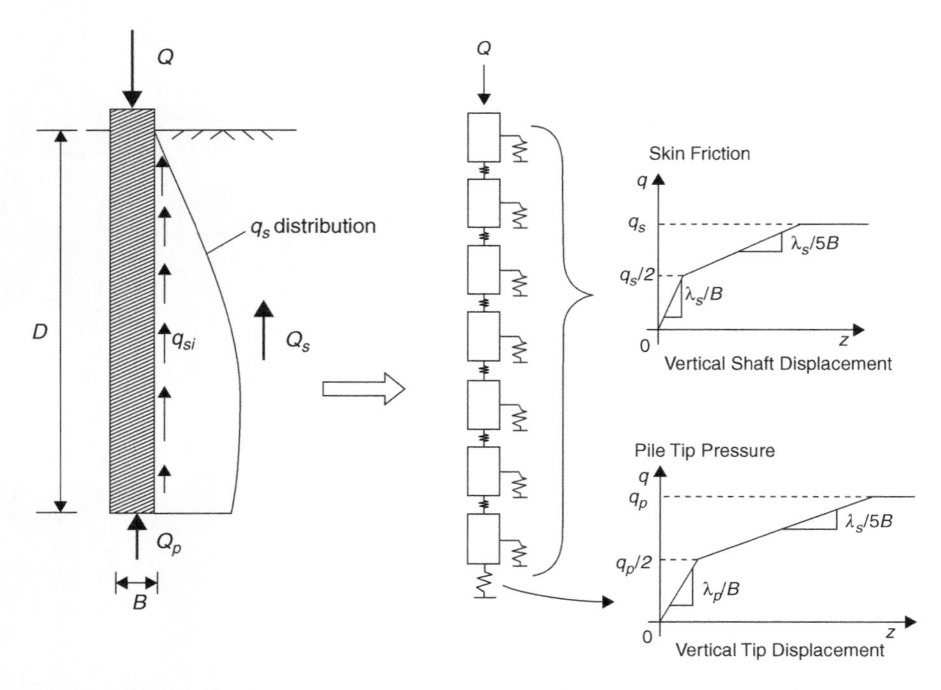

Figure 8.55 Load transfer method to estimate pile settlement (*After* Frank and Zhao, 1982)

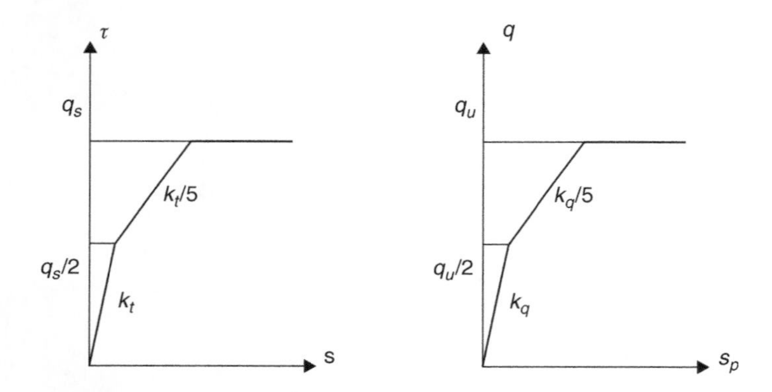

Figure 8.56 PMT τ-s and q-s_p curves after Frank and Zhao (1982)

consider the pile to be a beam on linear or non-linear elastic spring, are supported by the Winkler theory:

$$EI\frac{d^4y}{dz^4} + k \cdot B \cdot y = 0 \tag{8.51}$$

The value of k is readily obtained from the settlement equation $w = f(E_M)$ given by Ménard (1963) for an infinitely long strip footing, of width B, since $k \cdot B = p/w$.

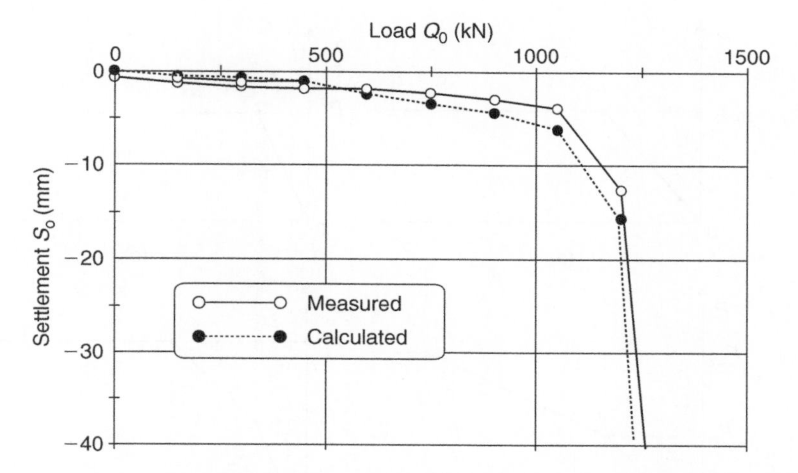

Figure 8.57 Comparison of measured and calculated load-settlement relationship for the Koekelare pile (*After* Bustamante and Gianeselli, 1993)

In this condition, the ground reaction is similar to that observed under a PMT borehole expansion (Gambin and Frank, 2009). When E_M values are averaged, for B larger than 0.6 m, below the critical depth:

$$k \cdot B = E_s = E_M \cdot \left\{ \frac{18}{[4 \cdot (2.65 \cdot B/B_0)^{\alpha} \cdot B_0/B + 3\alpha]} \right\} \tag{8.52}$$

where α is the Ménard rheological factor ($1/4 < \alpha < 2/3$) and B_o is a reference diameter equal to 0.6 m. According to Gambin and Frank (2009), this was checked as early as 1962 using PMT data on various laterally loaded prototype piles. The present design rules (Frank, 1999), using generalised p-y curves, include the decrease of k as y increases. In Figure 8.58, four different loading conditions are shown to be dependent on the creep (or yield) pressure p_c (which can be estimated as: $p_c = p_L/2$).

Gambin and Frank (2009) present a small modification to Equation (51) where a soil applies a horizontal thrust to a pile (Figure 8.59). The equation becomes:

$$EI \cdot \frac{d^4y}{dz^4} + k \cdot B \cdot [y(z) - g(z)] \tag{8.53}$$

where $y(z)$ is replaced by $[y(z) - g(z)]$, $g(z)$ being the free horizontal displacement of the soil without pile.

A description of the data interpretation and analysis of the behaviour of laterally loaded piles in ISC'2 experimental site using PMT and DMT based methods has been given by Tuna (2006) and Tuna *et al.* (2008).

The bored piles E0 and E9 had 6 m net embedment depth, and reaction piles E1 to E8, all 22 m long, were embedded in bedrock. They are described in detail in Tuna *et al.* (2008), as well as the CFA piles (T1 and T2), about 6 m depth and 60 cm diameter,

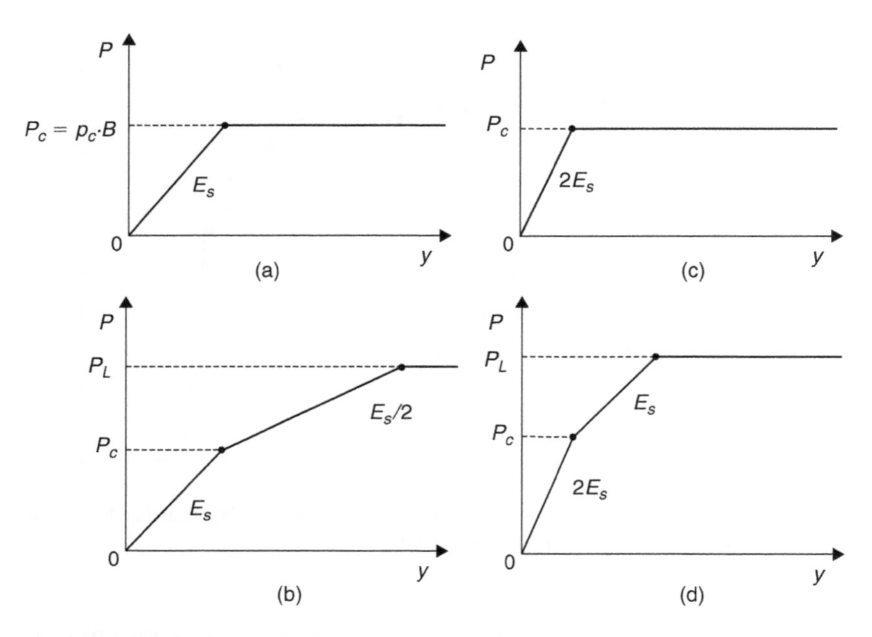

Figure 8.58 Soil reaction against lateral displacement for actions at head level (*After* Frank, 1999), for: (a) permanent actions at pile head; (b) soil lateral thrust; (c) short time actions at pile head; (d) unexpected instant actions at pile head

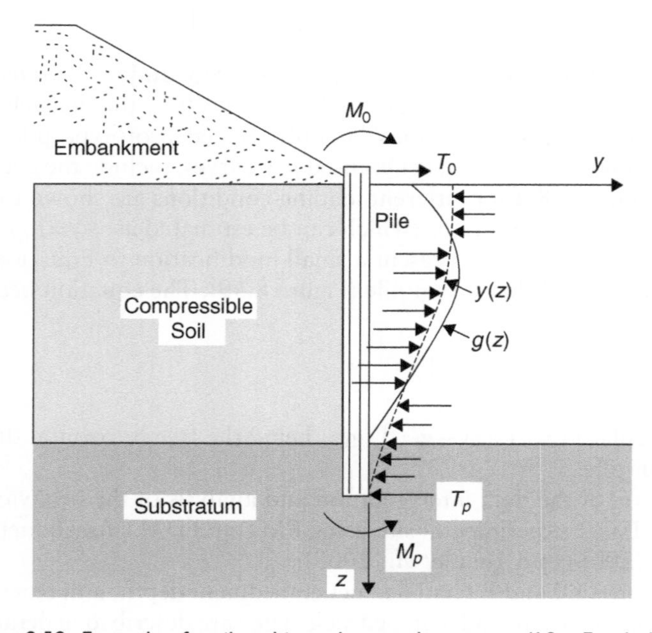

Figure 8.59 Example of a pile subjected to earth pressure (*After* Frank, 1999)

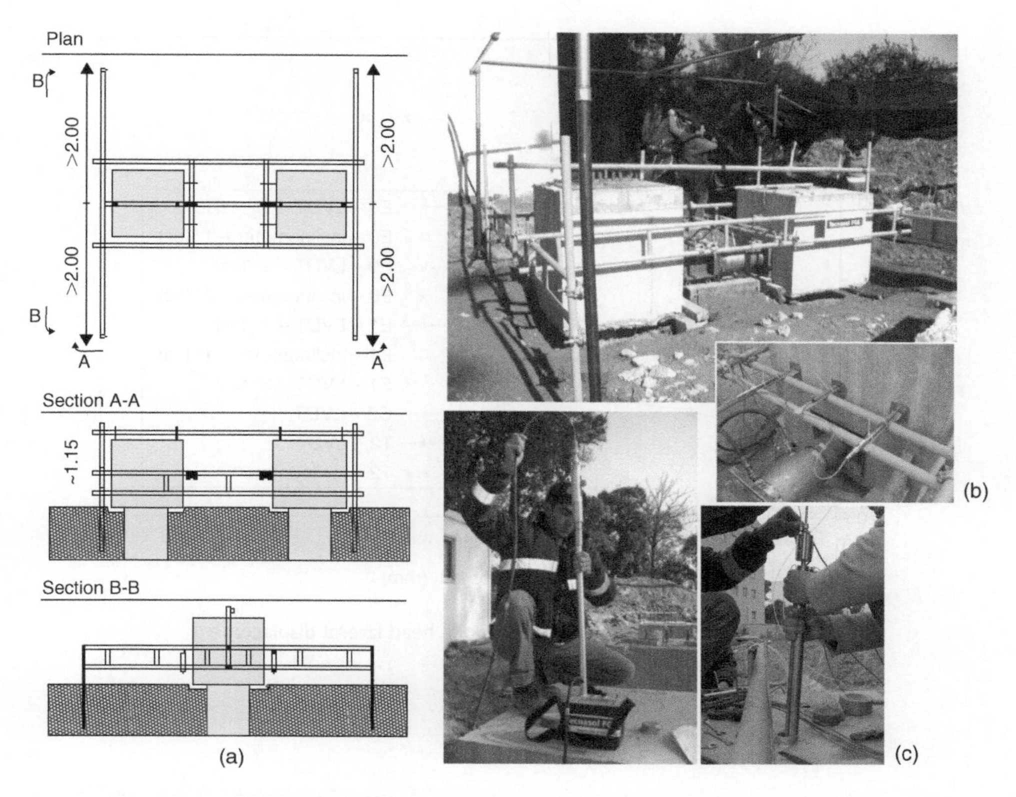

Figure 8.60 (a) Assemblage of the test with reference beam; (b) structure; (c) instrumentation: inclinometer, retrievable extensometer and LVDT mounted horizontally.

and the driven piles (350 mm square). The piles were laterally loaded by pushing each pair apart (E0-E1, E7-C2 and E8-T2) using a hydraulic jack. Measurement of displacements (vertical and horizontal) and rotations of pile heads were made using LVDTs. Detailed inclinometer measurements were made in piles E0, E1 and T2, to define the development of these displacements and rotations through the full length of the piles. In order to determine the bending moment in pile E0, strains in the concrete were measured using retrievable extensometers (Figure 8.60).

Pile head lateral displacements, as measured using the LVDT and also from inclinometer integration, are presented in Figure 8.61.

Tuna *et al.* (2008) used the *p-y* curve method based on PMT (MPM) parameters suggested by Ménard *et al.* (1969) (see Figure 8.62) through the following equations, where B is the pile diameter, B_0 is the diameter of the reference pile (taken equal to 0.6 m) and α is Ménard's rheological factor, varying between 0.25 and 1.

$$\text{For } B > B_0: \quad K = \frac{18E_M}{4\,(2.65(B/B_0))^{a}\,B_0/B + 3\alpha} \qquad (8.54)$$

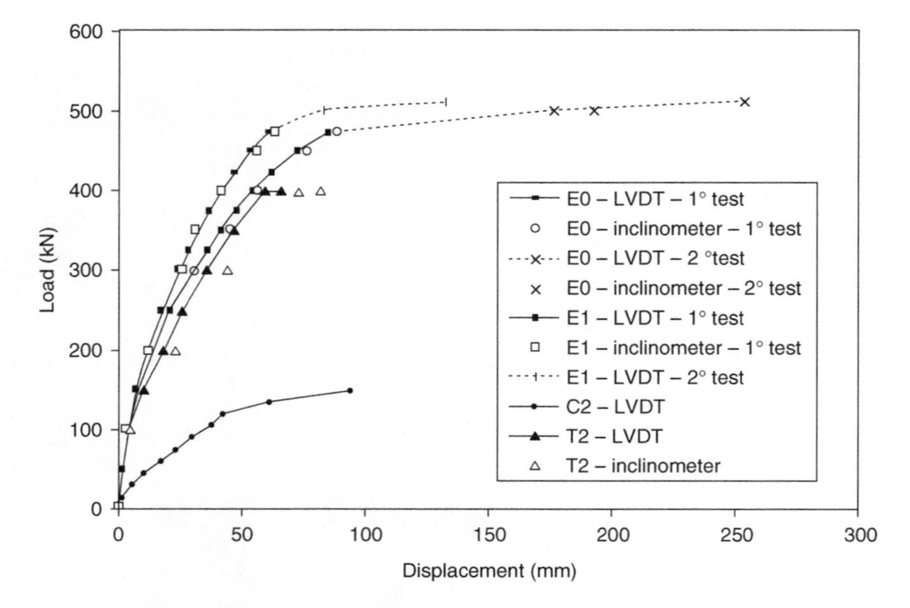

Figure 8.61 Load versus observed pile head lateral displacement

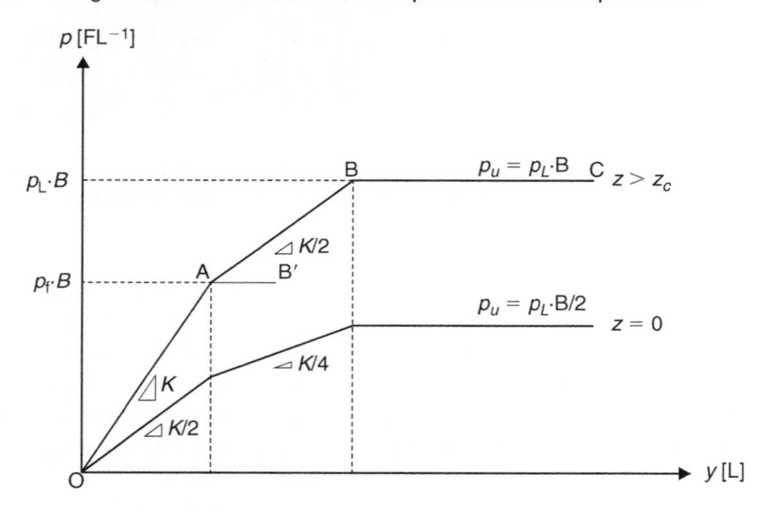

Figure 8.62 P-y curve by Ménard *et al.* (1969)

$$\text{For } B < B_0: \quad K = \frac{18E_M}{4(2.65)^{\alpha} + 3\alpha} \tag{8.55}$$

The reaction curve is modified for depth values, z, less than a critical depth z_c, by multiplying the pressure by a coefficient equal to $0.5(1 + z/z_c)$. For cohesive soils, z_c is about $2 \cdot B$ and for granular soils about $4 \cdot B$.

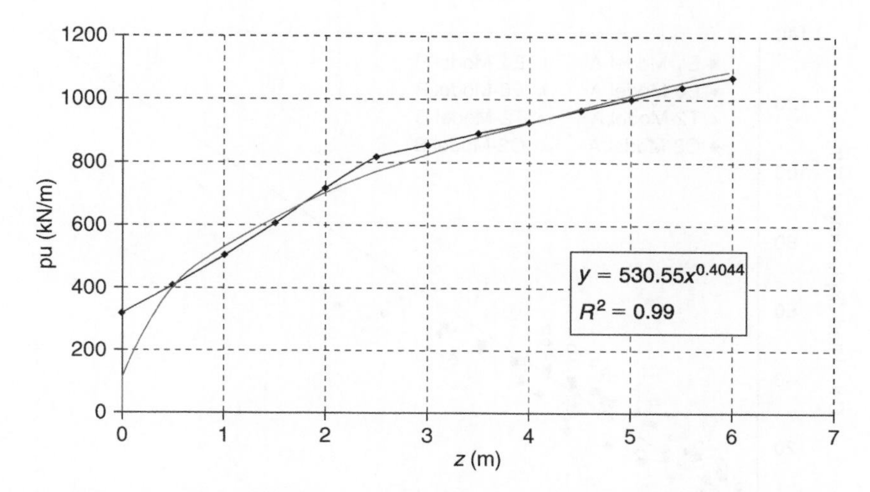

Figure 8.63 Pile E0 – linear relationship of p_u with depth, with power law and trend line

The development of limit pressure (p_L), creep pressure (p_F) and pressuremeter modulus (E_M), with vertical effective stress (depth), is defined by the following equations (a value of 0.33 was adopted for α):

$$p_L\,(\text{kPa}) = 6.4\sigma'_{v0}\,(\text{kPa}) + 1017.2 \tag{8.56}$$

$$p_F\,(\text{kPa}) = 2.4\sigma'_{v0}\,(\text{kPa}) + 365.3 \tag{8.57}$$

$$E_M\,(\text{kPa}) = 92.5\sigma'_{v0} + 12949 \tag{8.58}$$

Additionally, a study was made in order to adjust the soil strength to the points A and B (see Figure 8.62) by power laws. The main purpose was to decrease the high values of p_u, obtained for depths less than 0.5 m. For example, for the bored pile E0, p_u was approximated by a power law as shown in Figure 8.63.

Displacements of pile heads were calculated by the application of the original method proposed by Ménard *et al.* (model A) and by applying the power law approximation (model B). Figure 8.64 shows the calculated and observed displacements.

For the driven pile, the displacements by both methods are generally underestimated. For the bored and CFA piles, model A overestimates the calculated displacements for small loading levels, but underestimates then for higher loading levels. Model B overestimate the displacements for the bored piles, while it predicts fairly well for CFA piles.

In conclusion, this work has indicated that the PMT method, originally proposed by Ménard *et al.* (1969), generally underestimates horizontal displacements in residual soils, mostly at high loading levels, but this can be corrected by decreasing soil strength for shallow depths. Both the Ménard proposal and the one designated as "modified", overestimated displacements for very small loading levels, as a consequence of the excessively low value considered for the reaction modulus (K). Even so, it can be seen

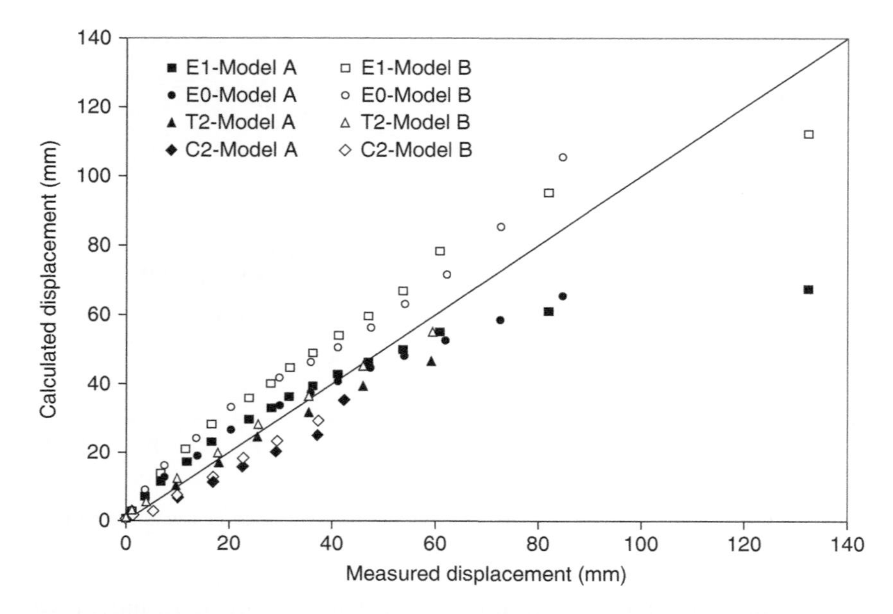

Figure 8.64 Measured and calculated displacements using Ménard *et al.* (1969) method

that this method leads to a reasonable simulation of the behaviour of bored (E0, E1) and CFA (T2) piles. For driven piles (C2) this method is unreliable, because the power law considered for the variation of p_u with depth, is inadequate to model the behaviour of this type of pile.

Quality control and monitoring

Piles normally provide an extremely important function, in supporting a structure such as a bridge or a building. They are also, generally speaking, out of sight once formed so that it is impossible to carry out a detailed inspection. It is therefore, commonly accepted in piling engineering, both by public authorities and by the private sector, that piles have to be subjected to very specific and demanding quality control processes. Two types of tests are in common use: (i) integrity tests (destructive or non-destructive); and (ii) load tests (static or dynamic). These can be done separately or in conjunction, depending on the type of piles, the ground conditions, the importance and the size of the project, the anticipated construction defects and workmanship of the piling contractor.

Integrity testing

The most commonly used methods of non-destructive integrity testing are the following:

a) low strain integrity tests (sometimes referred to as seismic tests);
b) sonic-logging (sometimes referred to as sonic coring);

c) vibration tests (referred to as impedance or transient dynamic response);
d) high strain integrity tests (PDA);
e) parallel sonic method (MSP).

Among these methods, the first four are the most frequently used, the last (MSP) being recommended in special cases when actual length of pile embedment is in dispute.

The most popular is the low strain integrity test, in which a small hammer, which can be fitted with an accelerometer, is used to strike the top of the prepared pile. Another accelerometer is used to record the passage of the stress wave as it travels down the pile and reflects back to the surface. This record can be used to make an assessment of the impedance of the pile at any section, and the impedance is a function of the stiffness and the area, both of which are affected by pile integrity. The test is very simply and quickly carried out, making about ten blows on each pile head which can be processed later, and more than 100 piles can be tested in one day if access is suitable. It is therefore very economical. With good data and experienced interpretation, it is possible to detect defects such as bulging, cracking, and low modulus concrete, possibly caused by honeycombing in bored piles, but it is hard to get signals from depth. Aspect ratios of 20 to 1 are normally good, and up to 30 to 1 may be acceptable. A commonly used apparatus is the Pile Integrity Tester (PIT) produced by Pile Dynamics Inc., but other similar devices are available. In special cases, this method can be augmented by down-hole tests as illustrated in Figure 8.65 (CIRIA, 1997) in which a signal is picked up at various levels though a piezo-electric receiver lowered into an adjacent hole.

In major projects (highway and bridges, railways, harbours, etc.) sonic coring is preferred, but this has two main disadvantages. The method requires access tubes, made from plastic or, more usually, steel to be cast into the pile or diaphragm wall, by fixing them to the reinforcing cage prior to concreting. This is an inconvenience during construction, and therefore a significant expense. It is also much less practical for CFA piling. As a result, it is normal to preselect piles for integrity testing, which takes away the random aspect which is an essential feature of quality control. Based on acoustic principles, one measures the propagation time of sonic transmission between two piezo-electric probes, which are simultaneously raised from the base of the pile to the top. The travel time of the signal is a function of velocity, and hence concrete modulus, and distance; so, if the two tubes being used are parallel, the arrival time of the first signal should be at a constant time interval. It is easy to detect depths at which signals are delayed. In spite of the cost it is a reliable method which gives results which are independent of depth, although it is very sensitive and has been found to detect lack of bond between sound concrete and plastic access tubes.

The impedance method was more popular in the seventies and seems to be hardly used now, having given way to the low strain integrity test. High strain integrity tests, or dynamic methods, are normally used as load testing. Some other methods are still applied to some particular problems and as supplementary tests: (a) echo method (usually limited to short piles, not exceeding 20 m); and, (b) gamma ray method (for detecting defects in bored piles and diaphragm walls and controlling proposed remedial works).

All the integrity tests described are non-destructive, and are indirect, requiring subjective interpretation. If possible defects are located it may be advisable to confirm them with direct integrity testing, by coring over the whole or part of pile length. This is

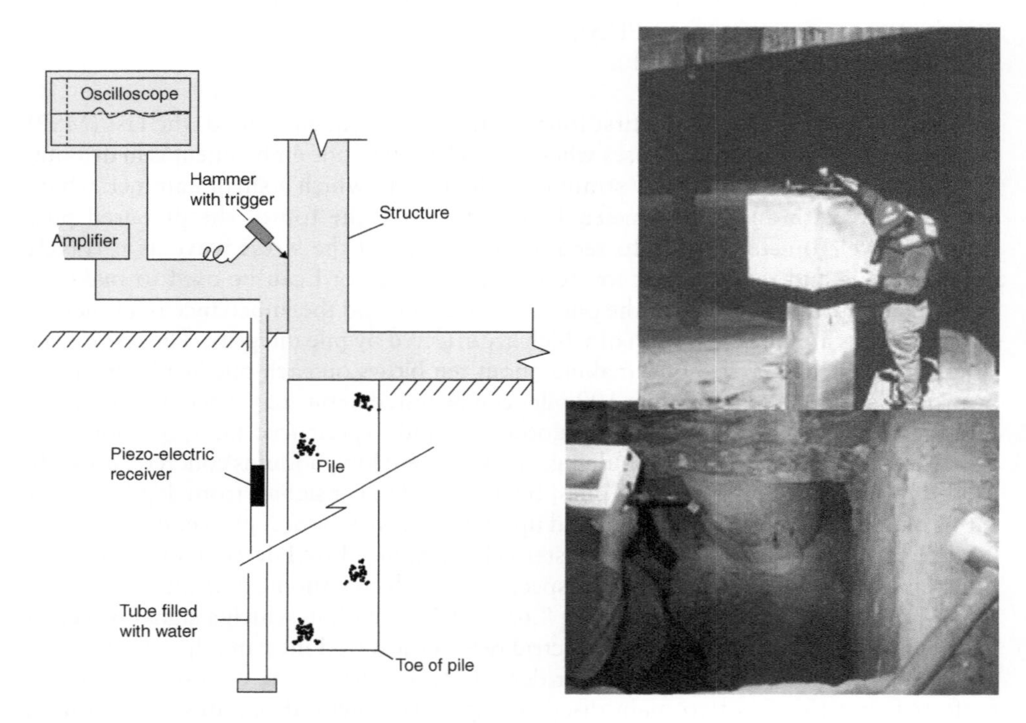

Figure 8.65 Down-hole low strain integrity tests (scheme CIRIA, 1997; pictures GEOMEC, 2004)

a costly method, but is often been prescribed in Europe for bored piles of large diameter and heavy loads (Bustamante and Frank, 1999). It allows the evaluation of concrete quality (homogeneity or segregation and discontinuity) and may show contact of the soil/concrete in the base, if the base material is of a similar strength and consistency to concrete. Coring of the pile shaft is usually imposed when non-destructive methods are not effective or the importance of the defect detected is high. In special cases a down-hole television camera may be used to allow the sides and the bottom of the core hole to be inspected.

Pile load tests

For major or complex projects, it is commonly accepted that, in addition to integrity tests, full scale (static) load tests have to be carried out to verify workmanship and, more importantly, design. Current structural codes, such as recent European Code for Geotechnical Design (CEN, 2004), are very clear in the indication that the design shall be based on one of the following approaches:

- *the results of static load tests, which have been demonstrated, by means of calculations or otherwise, to be consistent with other relevant experience;*
- *empirical or analytical calculation methods whose validity has been demonstrated by static load tests in comparable situations;*

- *the results of dynamic load tests whose validity has been demonstrated by static load tests in comparable situations;*
- *the observed performance of a comparable pile foundation, provided that this approach is supported by the results of site investigation and ground testing.*

The recent Australian Standard AS 2159-2009 "Piling – Design and installation" states in the Preface that the major changes from the previous version include:

a) *Requirement for some testing to be "normative";*
b) *Inclusion of new types of test including rapid pile testing.*

Further, when determining an ultimate geotechnical capacity from a calculated capacity using a "geotechnical resistance reduction factor", this value is restricted to 0.4 if no testing, even proof load testing, is carried out. However, it can be up to 0.9 if testing is used and all other conditions are favourable.

In Singapore, following a number of significant construction problems in the past, it has become compulsory to carry out static load testing unless very low prescribed strength values are used. As a result, about 500 to 550 static load tests are carried out each year in the island state alone, with capacities up to 6,000 tonnes.

Static load tests may be carried out on trial piles, installed for test purposes only, before the design is finalised, or on working piles which are part of the foundation. If on trial piles, they are usually taken to failure, to record the ultimate geotechnical capacity of the pile, but if on working piles they are normally restricted to less than 1.5 × working load in order to verify performance. In terms of the EN 1997-1, these tests are to be carried out in the following situations: (i) when using a type of pile or installation method for which there is no comparable experience; (ii) when the piles have not been tested under comparable soil and loading conditions; or (iii) when the piles will be subject to loading for which theory and experience do not provide sufficient confidence in the design.

There are two main methods for carrying out pile tests: the Maintained Load (ML) and the Constant Rate of Penetration (CRP). In the first, loads are applied in increments, and maintained for a certain period of time. This method is generally preferred and widely used. Some recommendations that can be referenced include ISSMFE (1985), AFNOR NFP94-150, ASTM D1143-81and De Cock *et al.* (2003), and problems can arise with the definition of when a new load increment is to be applied. It is also conventional to hold the load for an extended period at some increments, such as the working load or some higher multiple of it, and this then makes it difficult to compare the settlement at different load increments. For piles and soil conditions where this is important such as in stiff clays, the CRP test is more effective, though it is more complex to carry out. A special gauge is used to measure settlement, and this is linked to a clock which drives a second pointer according to time. The load is then applied, and constantly increased and monitored, such that the actual settlement follows the clock pointer at a "constant rate of penetration". This continues until no significant increase in load is required to achieve further penetration.

The pile load test procedures, particularly with respect to the number of loading steps, the duration of these steps and the application of load cycles, should be such that conclusions can be drawn about the deformation behaviour, creep and rebound of a piled foundation from the measurements on the pile. For trial piles, the loading should be such that conclusions can also be drawn about the ultimate failure load.

Another variation which is gaining in popularity in certain parts of the world is the Bi-Directional test, better known as the Osterberg Cell or O-Cell test after its inventor, the late Jorge Osterberg. In this test, one or mare hydraulic jacks, or layers of jacks, is cast into a bored pile, along with a full array of instrumentation including extensometers and strain gauges. No surface reaction is needed since, during the test, one of the layers of jacks is expanded pushing part of the pile up against the reaction of the rest of the pile going down. In the most informative test, a lower layer of jacks near the base is expanded first, measuring the resistance of the base and some small amount of the shaft against the middle and upper shaft portions. Once ultimate resistance has been determined, the middle of the shaft is pushed downwards against the reaction of the upper section. Finally, the lower section is joined with the middle section, by closing the valves on the lower jacks, to provide reaction to raise the upper section. By suitable combining these results, the ultimate capacity of the whole pile can be estimated. This method is particularly suited to applications such as in rivers for major bridges, where conventional reaction systems are prohibitively expensive (Randolph, 2003).

High Strain Integrity Tests or Dynamic Load Tests may be used to estimate the compressive resistance of a pile. These were developed from the impact hammers and stress wave theory applied to driven piles, but has been widely extended to cover bored and CFA piles by the use of special drop hammers. The two leading exponents are probably Pile Dynamic Inc. from USA and TNO from the Netherlands. They can give very useful results, because the post-processing of the signal can model the behaviour of a layered soil, and allow reasonable estimates of shaft friction, end bearing, and even load/settlement behaviour. For best results, an adequate site investigation is needed and the method should be calibrated against Static Load Tests, see ASTMD4945.

The results of full scale static load tests with instrumentation by means of extensometer system and strain gauges in a large number of piles have been collected in Europe over more than 30 years. This has allowed the creation of a database with the measured values of unit shaft friction and base resistance values which could then be used in design charts such as those developed by LCPC in France, for the PMT and CPT methods referred above (Frank, 2008; Bustamante and Frank, 2009).

REFERENCES

Amar, S., Baguelin, F., Canépa, Y. & Frank, R. (1994). Experimental study of the settlement of shallow foundations. In: *Vertical and Horizontal Deformation of Foundation and Embankments*, ASCE Geotechnical Special Publication No. 40. ASCE, New York, 2, 1602–1610.

Anagnastopoulos, A.G., Papadopoulos, B.P. & Kawadas, M.J. (1991). Direct estimation of settlements on sand, based on SPT results. In: *Proceedings of the 10th European Conference on Soil Mechanics & Foundation Engineering*, Firenze, Italy. Balkema, Rotterdam. 1, 293–296.

Aoki, M., Shibata, Y. & Maruoka, M. (1990). Estimation of ground deformation during construction period (part 1), *Proceedings of the Annual Conference of AIJ*. B, 1649–1650 (In Japanese).

Babenderde, S., Hoek, E., Marinos, P. & Cardoso, A.S. (2004). Characterization of Granite and the Underground Construction in Metro do Porto, Portugal. In: Viana da Fonseca, A. & Mayne, P.W. (eds) *Geotechnical & Geophysical Site Characterization*. Rotterdam, Millpress. 1, 39–47.

Baguelin, F., Jézéquel, J.F. & Shields, D.H. (1978). *The Pressuremeter and Foundation Engineering*. Clausthal, Germany, Trans Tech Publications.

Baguelin, F. & Venon, V.P. (1971). The influence of piles' compressibility on the resistant stress mobilization.In: '*Le Comportement des Sols Avant la Roture*'. Special number of the Bulletin Liaison de Laboratoire des Ponts et Chaussés. Paris, LPC.

Balakrishnan, E.G., Balasubramaniam, A.S. & Phienwej, N. (1999). Load deformation analysis of bored piles in residual weathered formation. *ASCE Journal of Geotechnical and Geoenvironmental Engineering*, 125(GT2), 122–131.

Bellotti, R., Fretti, C., Jamiolkowski, M. & Tanizawa, F. (1994). Flat dilatometer tests in Toyoura sand. In: *Proceedings of the 13th International Conference on Soil Mechanics & Foundation Engineering, New Delhi*. Rotterdam, Balkema. 4, 1779–1782.

Berardi, R., Jamiolkowski, M. & Lancellotta, R. (1991). Settlement on shallow foundations in sands.Selection of stiffness on the basis of penetration resistance. *Geotechnical Special Publication* 27. ASCE, New York. 1, 185–200.

Bjerrum, L. & Eggestad, A. (1963). Interpretation of loading test on sand. In: *Proceedings of the 1st European Conference on Soil Mechanics & Foundation Engineering*, Wiesbaden. 1, 199–203.

Briaud, J.L. (1992). *The Pressuremeter.* Rotterdam, Trans Tech Publications.

Brown, D.A. (2002). Effect of Construction on Axial Capacity of Drilled Foundations in Piedmont Soils, *ASCE Journal of Geotechnical and Geoenvironmental Engineering*, 128 (GT12), 967–973.

Burland, J.B. (1989). Small is beautiful: The stiffness of soils at small strains. *Canadian Geotechnical Journal*, 26 (4), 499–516.

Burland, J.B. & Burbidge, M.C. (1985). Settlement of foundations on sand and gravel. *Proceedings of the Institution of Civil Engineers*. 78, 1325–1381.

Bustamante, M. & Doix, B. (1985). Design method for ground anchors and grouted micropiles. *Bulletin Liaison Laboratoire Central des Ponts et Chaussées*, 140, 75–92 (in French).

Bustamante, M. & Frank, R. (1999). Current French design practice for axially loaded piles. *Ground Engineering*, 32 (3), 38–44.

Bustamante, M. & Gianeselli, L. (1982). Pile bearing capacity predictions by means of static penetrometer CPT. In: *Proceedings of the 2nd European Symposium on Penetration Testing, ESOPT II, 24–27 May 1982, Amsterdam*. Rotterdam, A. A. Balkema. 2, 493–500.

Bustamante, M., Gambin, M. & Gianeselli, M. (2009). Pile design at failure using the Ménard pressuremeter: an up-date. In: Iskander M., Debra F., Laefer D.F. & Hussein M.H. (eds), *Contemporary Topics in In situ Testing, Analysis, and Reliability of Foundations, Proceedings of the International Foundation Congress and Equipment Expo'09, IFCEE'09, 15–19 March 2009, Orlando, Florida*. ASCE Geotechnical Special Publication No. 186.

Buttling, S. (1986). Instrumentation and testing of bored piles. In: *4th International Geotechnical Seminar: Field Instrumentation and In situ Measurements*. Singapore, Nanyang Technological Institute.

Buttling, S. (1990). The calibration and use of a finite difference model of single axially loaded piles. In: *International Symposium on Developments in Laboratory and Field Tests in Geotechnical Engineering Practice*, South East Asian Geotechnical Society, Bangkok.

Buttling, S. & Robinson, S.A. (1987). Bored piles – design and testing. In: *Singapore Mass Rapid Transit Conference*, Singapore.

Buttling, S. & Lam, T.S.K. (1988). Behaviour of some rock-socketed piles. In: *5th Australia-New Zealand Conference on Geomechanics*, Australian Geomechanics Society, Sydney.

Camp, W.M., Brown, D.A. & Mayne, P.W. (2002). Construction methods effects on drilled shaft axial performance. In: O'Neill, M.W. & Townsend, F.C. (eds), *Deep foundations 2002*, Geotechnical Special Publication No. 116, ASCE, Reston, Va., 193–208.

Campos e Matos, A., Pimenta, P. & Marques, H. (2004). Structural conception of the Dragon Stadium, Porto. In: Figueiras, Faria, Samento, Pipa, Henriques & Cachim (eds) *Proceedings of the Portuguese Congress of Strctural Concrete*. FEUP (in Portuguese).

Carrier, W.D. & Christian, J.T. (1973). Rigid circular plate resting on a non-homogeneous elastic half-space. *Géotechnique*, 23 (1), 67–84.

Cassan, M. (1978). *Les Essais In situ en Mécanique des Sols. Vol. 1: Realization et Interpretation*. Editions Eyrolles, Paris.

Chang, M.F. & Broms, B.B. (1991). Design of bored piles in residual soils based on field-performance data. *Canadian Geotechnical Journal*, 28 (2), 200–209.

Chen, F.H. (1965). The use of piers to preventing the uplifting of lightly structured founded on expansive soils. In: *Proceedings of the Engineering Effects of Moisture Changes in Soils International Research on Engineering Conference on Expansive Clay Soils, College Station*, 108–119.

Chen, F.H (1988). *Foundations on Expansive Soils*. 2nd edition, Elsevier Science Publications, New York.

Chen, C.S. & Hiew, L.C. (2006). Performance of bored piles with different construction methods. *Geotechnical Engineering*. 159 (3), 227–232.

Clarke, B.G. (1995). *The Pressuremeter in Geotechnical Design*. Glasgow, Blackie Academic and Professional.

Clarke, B.G. & Gambin, M.P. (1998). Pressuremeter testing in onshore ground investigations: A report by the ISSMGE Committee TC 16. In: Robertson, P.K. & Mayne P.W. (eds) *Geotechnical Site Characterisation (ISC'98)* Rotterdam, Balkema. 1429–1468.

Clarke, B.G. & Smith, A. (1993). Self-boring pressuremeter tests in weak rocks. In: Cripps *et al.* (eds) *The Engineering Geology of Weak Rocks*. Rotterdam, A.A. Balkema.

Clayton, C.R.I. & Serratrice, J.F. (1997), General report session 2: The mechanical properties of hard soils and soft rocks. In: Anagnastopoulos *et al.* (eds) *Geotechnical Engineering of Hard Soils and Soft Rocks*. Rotterdam, Balkema. 3, pp. 1839–1877.

Costa Filho, L.M., Döbereiner, L., De Campos, T.M.P. & Vargas, E. Jr. (1989). Fabric and engineering properties of saprolites and laterites. *Session 6 – Invited lecture. Proceedings of the 12th International Conference on Soil Mechanics & Foundation Engineering. Rio de Janeiro*. 4, 2463–2476.

Coutinho, R.Q., Castro, B.P.F. & Dourado, K.C.A. (2010). Identification, characterization and analysis of deep foundation in collapsible soil – Regional prison of Eunápolis, Bahia, Brazil. In: *Fifth International Conference on Unsaturated Soils, Barcelona*, 1161–1167.

Coutinho, R.Q., Dourado, K.C.A. & Souza Neto, J.B. (2004a). Evaluation of effective cohesion intercept on residual soils by DMT and CPT. In: *Proceedings of ISC-2 on Geotechnical and Geophysical Site Characterization, September 2004, Porto, Portugal.*

Coutinho, R.Q., Souza Neto, J.B. & Dourado, K.C.A. (2004b). General report: Characterization of non-textbook geomaterials. In: *Proceedings of ISC-2 on Geotechnical and Geophysical Site Characterization, Porto, Portugal*. 2, 1233–1257.

Davies, R.V. & Chan, A.K.C. (1981). Pile design in Hong Kong. *Hong Kong Engineer*. 9 (3), 21–28.

Day, R., Johnston, I. & Yang, D.Q. (2009). Design of foundations to Second Gateway Bridge – Brisbane. *Proceedings of the 7th Austroads Bridge Conference: Bridges Linking Communities*.

De Cock, F., Legrand, C. & Huybrechts, N. (2003). Axial Static Piles Load Test (ASPLT) in compression or in tension – Recommendations from International Society for Soil Mechanics & Geotechnical Engineering Subcommittee ERTC3-Piles. In: Vanicek *et al.* (eds) *Proceedings of the 13th European Conference on Soil Mechanics & Geotechnical Engineering, Prague*. 3, 717–741.

De Ruiter, J. & Beringen, F.L. (1979). Pile foundation for large North Sea structures. *Marine Geotechnology*, 3 (3), 267–314.

Dearman, W.R. (1976) Weathering classification in the characterisation of rock: A revision. *Bulletin International Association of Engineering Geologists*, 13, 373–381.

Décourt, L. (1992). SPT in non classical material. In: Lima, Nieto, Viotti, Buena (eds.) *Applicability of Classical Soil Mechanics Principles in Structured Soils, Proceedings of the US/Brazil Geotechnical Workshop, November 1992, Belo Horizonte, Brazil*. Mato Grosso, Univ. Fed. Viçosa. pp. 67–100.

Décourt, L. (1996). Foundation failures assessed on basis of the concept of stiffness (in Portuguese) *SEFE III* São Paulo. 1, 215–224.

Décourt, L. (2002). Capacidade de carga de estacas executadas no campo experimental de engenharia geotécnica da U.E. de Londrina. Algumas Ponderações (in Portuguese). *Proceedings of the 12th Congresso Brasileiro de Mecânica dos Solos e Engenharia Geoténcica, São Paulo/SP, Brasil*, 3, 1545–1555.

Décourt, L. & Quaresma, A.R. (1978). Bearing capacity of piles from SPT values, *5th Panamerican Congress of Soil Mechanics and Foundations Engineering, Buenos Aires*. 1, 45–54.

Dourado, K.C.A. & Coutinho, R.Q. (2007). Identification, classification and evaluation of soil collapsibility by Ménard pressuremeter. *Proceedings of the 13th Panamerican Conference on Soil Mechanics & Foundation Engineering. Isla de Margarita, Venezuela*. 724–730.

Duncan, J.M. & Chang, C.Y. (1970). Nonlinear analysis of stress and strains in soils. *ASCE Journal of the Soil Mechanics & Foundations Division*. 96 (SM5), 1629–1653.

Eslami, A. (1996). *Bearing Capacity of Piles from Cone Penetrometer Test Data*. Ph.D. Thesis, University of Ottawa, Department of Civil Engineering.

Eslami, A. & Fellenius, B.H. (1997). Pile capacity by direct CPT and CPTu methods applied to 102 case histories. *Canadian Geotechnical Journal*. 34 (6), 886–904.

Fahey M. & Carter J.P. (1993). A finite element study of the pressuremeter test in sand using a non-linear elastic plastic model. *Canadian Geotechnical Journal*, 30 (2), 348–362.

Fearenside, O.R. & Cooke, R.W. (1978). *The skin friction of bored piles formed in clay under bentonite, CIRIA Report No 77*. London, Construction Industry Research and Information Association.

Fellenius, B.H., Altaee, A., Kulesza, R. & Hayes, J. (1999). O-cell testing and FE analysis of 28-m-deep barrette in Manila, Philippines. *ASCE Journal of Geotechnical and Geoenvironmental Engineering*, 125 (GT7), 566–575.

Fellenius, B.H., Santos, J.A. & Viana da Fonseca, A. (2007). Analysis of piles in a residual soil – The ISC'2 prediction. *Canadian Geotechnical Journal*. 44, 201–220.

Ferreira, C., Mendonça, A.A. & Viana da Fonseca, A. (2004). Assessment of sampling quality in experimental sites on residual soils from Porto granite. *Proceedings of the 9th Portuguese Conference on Geotechnics, Aveiro, Portugal*. Lisbon, SPG. 1, 27–38.

Fleming, W.G.K., Weltman, A.J., Randolph, M.F. & Elson, W.K. (1985). *Piling Engineering*, New York, Wiley & Sons.

Fleming, W.G.K. & Sliwinski, Z.J. (1977).*The use and influence of bentonite in bored piles*. London, Construction Indsutry Research and Information Association (CIRIA).

Frank, R. (1984). *Theoretical Studies of Deep Foundations and Selfboring in situ Tests in LPC and Practical Results (1972-1983)* Research Report LPC No. 128, Laboratoire Central des Ponts et Chaussées, Paris (in French).

Frank, R. (1999). *Design of Shallow and Deep Foundations*. Paris, Presses de l'Ecole Nationale des Ponts et Chaussées et Techniques de l'Ingénieur, ENPC (in French).

Frank, R. (2008). *Design of foundations in France with the use of Ménard pessuremeter tests.* Invited lecture, Jornada el ensayo presiométrico en el proyecto geotécnico, 24 June 2008, Madrid, CEDEX-UPC, pp. 1–16.

Frank, R. & Zhao, S.R. (1982). Estimation par les paramètres pressiométriques de l'enfoncement sous charge axiale de pieux forés dans des sols fins. *Bulletin Liaison Laboratoire Central des Ponts et Chaussées.* 119, 17–24 (in French).

Fredlund, D.G. & Rahardjo, H. (1993). *Soil Mechanics for Unsaturated Soils.* New York, John Wiley & Sons Inc.

Gaba, A., Pickles, A. & Oliveira, R., (2004). Casa da Música do Porto: site characterisation. In: Viana da Fonseca, A. & Mayne, P.W. (eds) *Geotechnical and Geophysical Site Characterization. ISC'2.* Rotterdam, Millpress. 2, 1089–1096.

Gambin, M. (1963).The Ménard Pressuremeter and the Design of Foundations. *Actes Journées des Fondations*, Paris, Laboratoire Central des Ponts et Chaussées. (in French).

Gambin, M. & Frank, R. (2009). Direct design rules for piles using Ménard pressuremeter test. Contemporary Topics in *In situ* Testing, Analysis, and Reliability of Foundations. In: Iskander M., Debra F. Laefer D. F. & Hussein M.H. (eds.), *Proceedings of the International Foundation Congress and Equipment Expo'09 (IFCEE'09), 15–19 March 2009, Orlando, Florida.* ASCE Geotechnical Special Publication No. 186. Reston, Va, pp. 111–118.

GEOMEC (2004). *P.I.T., Pile Integrity Testing; L.S.T., Low Strain Dynamic Pile Integrity Testing.*Notes on FUNDEC Short Course, IST. Technical University of Lisbon.

GSEGWP (1990). Report on tropical residual soils – Geological Society Engineering Group Working Party. *The Quarterly Journal of Engineering Geology*, 23 (1), 1–101.

Ghionna, V.N., Manassero, M., & Peisino, V. (1991). Settlement of large shallow foundations on partially cemented gravely sand deposit using PLT data. *Proceedings of the 10th European Conference on Soil Mechanics & Foundation Engineering, Firenze.* Rotterdam, Balkema. 1, 1417–1422.

Gomes Correia, A., Viana da Fonseca, A. & Gambin, M. (2004). Routine and advanced analysis of mechanical *in situ* tests. Keynote lecture in Viana da Fonseca, A. & Mayne, P.W. (eds) *Geotechnical and Geophysical Site Characterization, ISC'2.* Rotterdam, Millpress. 1, 75–95.

Gustmão Filho, J.A., Gusmão, A.D. & Veloso, D.A. (2002). Case Records Involving Foundation in Swelling Unsaturated Soils in Brazil. In: Jucá, de Campos & Marinho (eds) *Unsaturated Soils, Lisse, Swets & Zeitlinger.* 877–882.

Hardin, B.O. & Drnevich, V.P. (1972). Shear modulus and damping in soils. *ASCE Journal of the Soil Mechanics & Foundations Division.* 98 (SM7), 667–692.

Hight, D.W. (2000). Sampling methods: evaluation of disturbance and new practical techniques for high quality sampling in soils. Keynote lecture in *7th National Congress of the Portuguese Society of Geotechnics.* Lisbon, SPG.

Holtz, W.G. & Gibbs, H.J. (1953). Engineering properties of expansive clays. *Reprinted in Award Winning ASCE Papers in Geotechnical Engineering, 1977, ASCE, New York,* 1950–1959.

Holtz, W.G. & Gibbs, H.J. (1956). Engineering properties of expansive clays. *Transactions of the American Society of Civil Engineers*, 121 (1), 641–663.

Houston, S.L. (1996). Foundations and pavements on unsaturated soils – Part 1: Collapsible soils. *Proceedings of the 1st International Conference on Unsaturated Soils – Unsat' 95, Paris – France*, 1421–1437.

Irfan, T.Y. (1988). Fabric variability and index testing of a granitic saprolite. *Proceedings 2nd International Conference on Geomechanics in Tropical Soils, Singapore.* Rotterdam, A.A. Balkema. 1, 25–35.

Jamiolkowski, M., Lancellotta, R., LoPresti, D.C.F. & Pallara, O. (1994). Stiffness of Toyoura sand at small and intermediate strain, *Proceedings of the 13th International Conference on Soil Mechanics & Foundation Engineering, New Delhi.* 3, 169–173.

Jardine, R.J., Potts, D.M., Fourie, A.B. & Burland, J.B. (1986).Studies of the influence of non-linear stress-strain characteristics in soil-structure interaction. *Géotechnique*, 36 (3), 377–396.

Jardine, R.J., Potts, D.M., St. John, H.D. & Hight, D.W. (1991). Some practical applications of a non-linear ground model. *Proceedings of the 10th European Conference on Soil Mechanics & Foundation Engineering, Firenze*, Rotterdam, Balkema. 1, 223–228.

Kahle, J.G. (1983). Predicting Settlement in Piedmont Residual Soil with the Pressuremeter Test. *Transportation Research Board Meeting*. Washington.

King, P., Stuzyk, K., Lara, O. & Steiben, G. (2001). Under slab in situ moisture content. *Proceedings, ASCE Texas Chapter Meeting*, San Antonio, Texas, 99–108.

Kondner, R. (1963). Hyperbolic stress-strain response: cohesive soils. *ASCE Journal of the Soil Mechanics & Foundations Division*. SM1, 115–143.

Konstantinidis, B., Schneider, J.P. & Van Reissen, G. (1986). *Structural Settlement of Shallow Foundations on Cohesionless Soils: Design and Performance.* Geotechnical Special Publication No 5. New York, ASCE.

Lacerda, W.A. & Almeida, M.S.S. (1995). Engineering properties of regional soils: residual soils and soft clays. *Lecture at the 10th Panamerican Conference on Soil Mechanics and Foundation Engineering, International Society for Soil Mechanics & Geotechnical Engineering, México*, 4.

Ladd, C.C. & Lambe, T.E. (1961). The identification and behavior of compacted expansive clays. *Proceedings of the 5th International Conference on Soil Mechanics and Foundation Engineering. Paris*, 1, 201–205.

Lei, G.H. & Ng, C.W.W. (2007). Rectangular barrettes and circular bored piles in saprolites. *ICE Journal of Geotechnical Engineering*. 160 (4), 237–242.

Leonards, G.A. & Frost, J.D. (1988). Settlement of shallow foundations on granular soils. *ASCE Journal of the Geotechnical Engineering Division*. 114 (GT7), 791–809.

Leroueil, S. & Vaughan, P.R. (1990). The general and congruent effects of structure in natural clays and weak rocks. *Géotechnique*, 40, 467–488.

Lo Presti, D.C.F., Pallara, O., Lancellotta, R., Armandi, M. & Maniscalco, R. (1993). Monotonic and cyclic loading behavior of two sands at small strains, *ASTM Geotechnical Testing Journal* 16 (4), 409–424.

Marchetti, S. (1980). *In situ* tests by flat dilatometer. *ASCE Journal of the Geotechnical Engineering Division*. 106 (GT3), 299–321.

Marques, E.A.G., Viana da Fonseca, A., Carvalho, P. & Gaspar, A. (2004). Example of erratic distribution of weathering patterns of Porto granite masses and its implication on site investigation and ground modelling. In: Viana da Fonseca, A. & Mayne, P.W. (eds) *Proceedings of ISC-2 on Geotechnical and Geophysical Site Characterization, Porto, Portugal*. Rotterdam, Millpress. 2, 1293–1299.

Massad, F. (1995). Pile analysis taking into account soil rigidity and residual stresses. *Proceedings 10th Pan-American Congress on Soil Mechanics and Foundation Engineering, November 1995, Guadalahara, México*. 2, 1199–1210.

Mayne, P.W. (1994). CPT-based prediction of footing response. *Measured & Predicted Behavior of Five Spread Footings on Sand (GSP 41)* Reston, Va, ASCE. 214–218.

Mayne, P.W. (2000). Geotechnical site characterization by seismic piezocone tests, *Proceedings of the 4th International Geotechnical Conference*, Soil Mechanics & Foundations Research Laboratory, Cairo University, Giza, January 24–27, 2000. 91–120.

Mayne, P.W. & Brown, D.A. (2003). Site characterization of Piedmont residuum of North America. *Characterization and Engineering Properties of Natural Soils*. Lisse, Swets & Zeitlinger. 2, pp. 1323–1339.

Mayne, P.W. & Frost, D.D. (1988). Dilatometer experience in Washington, D.C. *Transportation Research Record 1169*, Washington, D.C., National Academy Press. 122 (10), 813–821.

Mayne, P.W., Martin, G.K. & Schneider, J.A. (1999a). Flat dilatometer modulus applied to drilled shafts in the Piedmont residuum. *Behavioral Characteristics of Residual Soils, Geotechnical Special Publication* No. 92, ASCE, Reston, Va, pp. 101–112.

Mayne, P.W. & Schneider, J.A. (2001). Evaluating axial drilled shaft response by seismic cone modulus. In: Brandon, T.L. (ed.) *Foundations & Ground Improvement, Geotechnical Special Publication No 113*, GeoInstitute – ASCE, Reston, Va, pp. 655–669.

Mayne, P.W., Schneider, J.A. & Martin, G.K. (1999b). Small- and large-strain soil properties from seismic flat plate dilatometer tests. *Pre-Failure Deformation Characteristics of Geomaterials, Torino*. Rotterdam, Balkema. 1, 419–426.

McKeen, G.R. (1992). A model for predicting expansive soil behavior. *Proceedings of 7th International Conference on Expansive Soils, Dallas, Texas*, 1–6.

Ménard, L. (1955). *Pressuremeter*. French patentof invention. No 1.117.983 (in French).

Ménard, L. (1963). Estimation of load bearing capacity of foundations from pressuremeter test results. *Sols Soils* 5, 9–32 (in French).

Ménard, L. (1965). Rules for the estimation of load bearing capacity and settlements of foundations from pressuremeter test results. *Proceedings of the 6th International Conference on Soil Mechanics & Foundation Engineering*, Montreal. 295–299 (in French).

Ménard, L., Bourdon, G. & Gambin, M. (1969). Méthode générale de calcul d'un rideau ou d'un pieu sollicité horizontalement en fonction des résultats pressiomètriques. *Sols Soils No* 22/23 (6), 16–29.

Ménard, L. & Rousseau, J. (1962). L'évaluation des tassements. Tendances Nouvelles. *Sols Soils 1 (1)*, 13–30. Paris (in French).

Meyerhof, G.G. (1976). Bearing capacity and settlement of pile foundations. The 11th Terzaghi Lecture, November 5 1975, *ASCE Journal of Geotechnical Engineering*, 102 (GT3), 195–228.

Milititsky, J., Consoli, N. & Schnaid, F. (2005). *Pathologiesof Foundations*. Edições Oficina de Textos, S. Paulo, Brasil (in Portuguese).

Mitchell, J.K. & Coutinho, R.Q. (1991). Occurrence, geotechnical properties, and special problems of some soils of America. *Proceedings of the 9th Pan-American Conference on Soil Mechanics and Foundation Engineering, Viña del Mar, Chile*, 4, 1651–1741.

Ng, C.W.W. & Leung, E.H.Y. (2007a). Small-strain stiffness of granitic and volcanic saprolites in Hong Kong. In: Tan, Phoon, Hight & Leroueil (eds.) *Characterization and Engineering Properties of Natural Soils*. London, Taylor & Francis. 4, 2507–2538.

Ng, C.W.W. & Leung, E.H.Y. (2007b). Determination of Shear-Wave Velocities and Shear Moduli of Completely Decomposed Tuff. *ASCE Journal of Geotechnical and Geoenvironmental Engineering*. 133 (GT6), 630–640.

Ng, C.W.W., Li, J.H.M. & Yau, T.L.Y. (2001a). Behaviour of large diameter floating bored piles in saprolitic soils. *Soils and Foundations*, 41 (6), 37–52.

Ng, C.W.W., Yau, T.L.Y., Li, J.H.M. & Tang, W.H. (2001b). New failure load criterion for large diameter bored piles in weathered geomaterials. *ASCE Journal of Geotechnical and Geoenvironmental Engineering*, 127 (GT6), 488–498.

Oasys (2001).*VDISP – Vertical Displacement Analysis computer program*. London: Oasys Ltd.

O' Neill, M.W. & Poormoayed, N. (1980). Methodology for foundations on expansive clays. *Journal of the Geotechnical Emgineering Division*, ASCE, 106 (GT12).

O'Neill, M.W. & Reese, L.C. (1999). *Drilled Shafts: Construction Procedures & Design Methods*, Volumes I & II, Publication No. FHWA-IF-99-025, U.S. Dept. of Transportation. Dallas, ADSC.

Parry, R.H.G. (1978). Estimating foundation settlements in sand from plate bearing tests. *Géotechnique*. 28(1), 107–118.

Phienwej, N., Balakrishnan, E.G. & Balasubramaniam, A.S. (1994). Performance of bored piles in weathered meta-sedimentary rocks in Kuala Lumpur, Malaysia. *Proceedings of the Symposium on Geotextiles, Geomembranes and other Geosynthetics in Ground Improvement on Deep Foundations and Ground Improvement Schemes, Bangkok*. Rotterdam, Balkema. 251–260.

Poulos, H.G. (1987). From theory to practice in pile design (E.H. Davis Memorial Lecture) *Transactions of the Australian Geomechanics Society*, Sydney, 1–31.

Poulos, H.G. (1989). Pile behavior: theory and application, 29th Rankine Lecture, *Géotechnique*, 39 (3), 363–416.

Poulos, H.G. & Davis, E.H. (1980). *Pile Foundation Analysis and Design*. New York, Wiley & Sons.

Prevost, J.H. & Keane, C.M. (1990). Shear stress-strain curve generation from simple material parameters. *ASCE Journal of Geotechnical Engineering* 116 (GT8), 1255–1263.

Puppala, A.J., Enayatpour, S., Vanapalli, S., and Intharasombat, N. (2004). Review of current methods for swell characterization of subsoils for transportation infrastructure design. *ASCE, Geotechnical Special Publication No. 126 – Geotechnical Engineering for Transportation Projects*. 1, 1105–1114.

Puzrin, A.M. & Burland, J.B. (1998). Nonlinear model of small-strain behavior of soils. *Géotechnique*, 46 (1), 157–164.

Randolph, M.F. (2003). Science and empiricism in pile foundation design, 43rd Rankine Lecture, *Géotechnique*, 53 (10), 847–875.

Randolph, M.F. & Wroth, C.P. (1978). Analysis of deformation of vertically loaded piles. *ASCE Journal of the Geotechnical Engineering Division*, 104 (GT12), 1465–1488.

Randolph, M.F. & Wroth, C.P. (1979). A simple approach to pile design and the evaluation of pile tests. *Behavior of Deep Foundations*, ASTM STP 670, 484–499.

Robertson, P.K. (1990). Soil classification using the cone penetration test. *Canadian Geotechnical Journal*, 27 (1), 151–158.

Robertson, P.K. (1991). Estimation of foundation settlements in sand from CPT. In: *Vertical and Horizontal Deformation of Foundation and Embankments, ASCE Geotechnical Special Publication* No 40. 2 (27), 764–778.

Robertson, P.K., Campanella, R.G., Davies, M.P. & Sy, A. (1988). Axial capacity of driven piles in deltaic soils using CPT. *Penetration Testing 1988*. Rotterdam, Balkema. 2, 919–928.

Rocha Filho, P. (1986). Discussion on "Settlement of foundations on sand and gravel" by Burland J.B. & Burbidge, M.C. 1985, *Proceedings of Institution of Civil Engineers*. 79, 1633–1635.

Sabatini, P.J., Bachus, R.C., Mayne, P.W., Schneider, J.A. & Zettler, T.E. (2002). Evaluation of Soil and Rock Properties. *Technical Manual. FHWA-IF-02-034*. Washington, Federal Highway Administration.

Santos, J.A., Duarte, R.J.L, Viana da Fonseca, A.& Costa Esteves, E. (2005). ISC'2 experimental site – Prediction and performance of instrumented axially loaded piles. *Proceedings of the 16th International Conference on Soil Mechanics & Geotechnical Engineering, September 21 – 25 2005, Osaka, Japan*. 2, 2171–2174.

Schmertmann, J.H. (1970). Static cone to compute static settlement over sand. *ASCE Journal of the Soil Mechanics and Foundation Division*. 96(3), 1011–1043.

Schmertmann, J.H. (1978). *Guidelines for Cone Test, Performance and Design*. Federal Highway Administration, Report FHWA-TS-78209, Washington.

Schmertmann, J.H. (1986). Dilatometer to compute foundation settlement, *Proceedings of the ASCE Specialty Conference, In situ '86, VPI, Blacksburg, Virginia*, 303–321.

Schmertmann, J.H., Hartman, J.P. & Brown, P.R. (1978). Improved strain influence factor diagram. *ASCE Journal of the Geotechnical Engineering Division*. 104 (8), 1131–1135.

Schnaid, F. (2005). Geo-characterisation and properties of natural soils by *in situ* tests. *Keynote Lecture. 16th International Conference on Soil Mechanics & Geotechnical Engineering, Osaka*. Rotterdam, Millpress. 1, 3–45.

Schnaid, F., Lehane, B.M. & Fahey, M. (2004). *In situ* test characterisation of unusual geomaterials. In: Viana da Fonseca, A. & Mayne, P.W. (eds) *Proceedings of ISC-2 on Geotechnical and Geophysical Site Characterization, Porto, Portugal*. Rotterdam, Millpress. 1, 49–74.

Schneider, R.L., Muhlram, H., Tommasi, E., Medeiros, R.A., Daemon, R.F. & Nogueira, A.A. (1974). Revisão Estratigráfica da Bacia do Paraná (in Portuguese). *Congresso Brasileiro de Geologia, 1974, Porto Alegre, Sociedade Brasileira de Geologia*, 1, 41–65.

Seed, H.B. & Idriss, I.M. (1982). *Ground motion and soil liquefaction during earthquakes*, Monograph, Earthquake Engineering Research Institute, Oakland, Ca.

Seed, H.B., Woodward, R.J., Jr. & Lundgren, R. (1962). Prediction of swelling potential for compacted clays: *ASCE Journal of the Soil Mechanics and Foundation Division*. 88, (SM3), Part I, 53–87.

Singh, H., Omar, H. & Huat, B.B.K. (2006). Geological investigations for foundations. In: Huat, B.B.K., Ali, F.H., Omar, H. & Singh, H. (eds) *Foundation Engineering: Design and Construction in Tropical Soils*. London, Taylor & Francis.

Snethen, R.R. (1984). Evaluation of expedient methods for identification and classification of potentially expansive soils. *Proceedings of 5th International Conference on Expansive Soils*, Australia, 22–26.

Snethen, D.R., Johnson, L.D. & Patrick, D.M. (1977). *An Evaluation of Expedient Methodology for Identification of Potentially Expansive Soils*. Soil and Pavements Laboratory, US Army Engineering Waterway Experiment Station, Vicksburg, MS. Rep. No FHWA-RE-77-94, NTIS, PB-289-164.

Stroud, M.A. (1974). The Standard Penetration Test in insensitive clays and soft rocks. *Proceedings of the European Seminar on Penetration Testing 1*, 2 (2), 367–375.

Stroud, M.A. (1988). The Standard Penetration Test – Its application and interpretation, *Proceedings of Penetration Testing in the UK*. London, Thomas Telford. 29–48.

Tan, Y.C., Chen, C.S. & Liew, S.S. (1998). Load transfer behavior of cast-in-place bored piles in tropical residual soils of Malaysia. *Proceedings of the 13th Southeast Asian Geotechnical Conference, Taipei*, 563–571.

Tatsuoka, F. & Kohata, Y. (1995). Stiffness of hard soils and soft rocks in engineering applications. *Proceedings 1st International Conference on Pre-failure Deformation Characteristics of Geomaterials, Sapporo*. 2, 947–1063.

Tatsuoka, F. & Shibuya, S. (1992). Deformation Characteristics of Soils and Rocks from Field and Laboratory Tests. Keynote lecture in: Proceedings of the *9th Asian Regional Conference on Soil Mechanics & Foundation Engineering, Bangkok*. Rotterdam, A.A. Balkema. 2, 101–170.

Terzaghi, K. & Peck, R.B. (1967). *Soil Mechanics in Engineering Practice*. 2nd Edition, New York, John Wiley & Sons.

Toh, C.T., Ooi, T.A., Chiu, H.K. Chee, S.K. & Ting, W.H. (1989). Design parameters for bored piles in a weathered sedimentary formation. *Proceedings of the 12th International Conference on Soil Mechanics and Foundation Engineering, Rio de Janeiro*. Rotterdam, Balkema. 2, 1073–1078.

Topa Gomes, A. (2009). *Elliptical Shafts by the Sequential Method of Excavation on Vertical Direction*. The Case of Metro do Porto. PhD Thesis, Faculdade de Engenharia, University do Porto. (In Portuguese).

Touma, F.T. & Reese, L.C. (1972). *Drilled Shafts; Construction Procedures and Design Methods*. U.S. Department of Transportation, FHWA-HI-88-042, Dallas, Texas.

Tumay, M.T. & Fakhroo, M. (1981). Pile capacity in soft clays using electric QCPT data. *ASCE, Cone Penetration Testing and Experience, October 26–30, 1981, St. Louis*, 434–455.

Tuna, C. (2006). *Tests and Analysis of the Behaviour of Laterally Loaded Piles in Residual Soils from Granite*. MSc Thesis. FEUP, University of Porto (in Portuguese).

Tuna, C., Viana da Fonseca, A. & Santos, J.A. (2008). Data interpretation and analysis of the behavior of laterally loaded piles in ISC'2 experimental site by recourse of PMT and DMT based methods. In: Huang, A-B. & Mayne, P., (eds) *Proceedings of the 3rd International Conference on Site Characterization (ISC'3), 1–4 April 2008, Taiwan*. London, Taylor & Francis.

Van Der Veen, C. (1953). The bearing capacity of a pile. *Proceedings of the 3rd International Conference Soil Mechanics & Foundation Engineering, Zurich*. 2, 84–90.

Vargas, M. (1971). Geotechnics of residual soils. *Solos & Rochas. Latin-American Journal of Geotechnics*. 1, 20–41.

Vargas, M. (1985). The concept of tropical soils. *Proceedings of the 1st International Conference on Geomechanics in Tropical Lateritic and Saprolitic Soils, February 1985, Brasília, Brazil.* International Society for Soil Mechanics & Foundation Engineering. 3, 101–134.

Vaughan, P.R. (1985). Mechanical and hydraulic properties, particularly as related to their structure and mineral components. General Report. *Proceedings of the 1st International Conference on Geomechanics in Tropical Lateritic and Saprolitic Soils. February 1985, Brasília, Brazil.* International Society for Soil Mechanics & Foundation Engineering. 3, 231–336.

Vaughan, P.R., Maccarini, M. & Mokhtar, S.M. (1988). Indexing the engineering properties of residual soils. *Quarterly Journal of Engineering Geology*. 21, 69–84.

Viana da Fonseca, A. (1996). *Geomechanics in Residual Soils from Porto Granite. Criteria for the Design of Shallow Foundations*. Ph.D. Thesis, University of Porto. (In Portuguese).

Viana da Fonseca, A. (1998). Identifying the reserve of strength and stiffness characteristics due to cemented structure of a saprolitic soil from granite. *Proceedings of the 2nd International Symposium on Hard Soils – Soft Rocks, Naples*. Rotterdam, Balkema. 1, 361–372.

Viana da Fonseca, A. (1999). Surface loading tests for mechanical characterisation of a saprolitic soil from granite of Porto. *Proceedingsof the 9th Pan-American Conference on Soil Mechanics and Geotechnical Engineering, Foz do Iguassu, Brasil.* 1, 403–409.

Viana da Fonseca, A. (2001). Load Tests on residual soil and settlement prediction on shallow foundation. *ASCE Journal of Geotechnical and Geoenvironmental Engineering*. 127 (10), 869–883.

Viana da Fonseca, A. (2003). Characterizing and deriving engineering properties of a saprolitic soil from granite, in Porto. In: Tan et al. (eds) *Characterization and Engineering Properties of Natural Soils*. Lisse, Swets & Zeitlinger. 1341–1378.

Viana da Fonseca, A. (2006). *Ground Investigations and Soil Characterization by In situ Testing*. DVD – Multimedia (Films, Pictures and Technical Notes) Reg. Assoft N. 1329/D/06. Pub. Univ. of Porto (IC-FEUP).

Viana da Fonseca, A. & Almeida e Sousa, J. (2001). At rest coefficient of earth pressure in saprolitic soils from granite. *Proceedings of the 15th International Conference on Soil Mechanics & Geotechnical Engineering, Istanbul.* 1, 397–400.

Viana da Fonseca, A. & Almeida e Sousa, J. (2002). Hyperbolic model parameters for FEM analysis of a footing load test on a residual soil from granite. In: J.-P. Magnan (ed.) *PARAM 2002: International Symposium on Identification and Determination of Soil and Rock Parameters for Geotechnical Design*. Paris, Presses L'ENPC. 1, 429–443.

Viana da Fonseca, A. & Cardoso, A.S. (1999). Linearly increasing elastic analysis of surface loading tests on a saprolitic soil from granite. *Proceedings 9th Panamerican Conference on Soil Mechanics and Geotechnical Engineering, Foz do Iguassu, Brasil.* 3, 1527–1535.

Viana da Fonseca, A., Carvalho, J., Ferreira, C., Santos, J.A., Almeida, F., Pereira, E., Feliciano, J., Grade, J. & Oliveira, A. (2006) Characterization of a profile of residual soil from granite combining geological, geophysical, and mechanical testing techniques. *International Journal of Geotechnical and Geological Engineering*, 24 (5), 1307–1348.

Viana da Fonseca, A. & Coutinho R.Q. (2008). Characterization of residual soils. Keynote Lecture, in: Huang, A-B. & Mayne, P. (eds) *Proceedings of the 3rd International Conference on Site Characterization (ISC'3), 1–4 April 2008, Taiwan*. London, Taylor & Francis. 195–248.

Viana da Fonseca, A., Marques, E.A.G., Carvalho, P. & Gaspar, A. (2003). Implication of heterogeneity of Porto granite in the site investigation and classification options. Design parameters definitions for cut-and-cover stations design in Metro do Porto. *I Jornadas Luso-Espanholas de Geotecnia, CEDEX, Madrid.* (in Portuguese).

Viana da Fonseca, A., Matos Fernandes, M. & Cardoso, A.S. (1997). Interpretation of a footing load test on a saprolitic soil from granite. *Géotechnique*, 47(3), 633–651.

Viana da Fonseca, A., Matos Fernandes, M. & Cardoso, A.S. (1998). Characterization of a saprolitic soil from Porto granite by *in situ* testing. *Proceedings of the 1st International Conference on Site Characterization – ISC'98, Atlanta.* Rotterdam, Balkema. 2, 1381–1388.

Viana da Fonseca, A. & Santos, J. (2008). *International Prediction Event. Behaviour of Bored, CFA and Driven Piles in Residual Soil. ISC'2 experimental site.* University of Porto (FEUP) & Technical University of Lisbon (IST-UTL) www.fe.up.pt/sgwww/labgeo/pdf/Book-IppE-Piles-ISC2.pdf.

Viana da Fonseca, A., Santos, J.A., Massad, F. & Costa Esteves, E. (2007). Analysis of piles in residual soil from granite considering residual loads. *International Journal "Soil and Rocks"*. 30 (1), 63–80.

Viana da Fonseca, A., Vieira de Sousa, J.F. & Cruz, N. (2001). Correlations between SPT, CPT, DPL, PMT, DMT, CH, SP and PLT Tests Results on Typical Profiles of Saprolitic Soils from Granite. *Proceedingsof the International Conference on In situ Measurement of Soil Properties and Case Histories, 21–23 May 2001, Bali, Indonesia.* Parayangan Catholic University. *1, 577–584.*

Vijayvergiya, V.N. & Ghazzaly, O.I. (1973). Prediction of swell potential for natural clays. *Proceedings of the 3rd International Research and Engineering Conference on Expansive Clay Soils.*

Wahls, H.E. & Gupta, M. (1994). Settlement of shallow foundations on sand. In: *Vertical and Horizontal Deformation of Foundation and Embankments*, Geotechnical Special Publication, No 40. New York, ASCE. 1, 190–206.

Wray, W.K. (1995). So your home is built on expansible soils: a discussion of how expansive soils affect buildings. *Shallow Foundations Committee of the Geotechnical Engineering Division of the American Society of Civil Engineers, ASCE.*

Wroth, C.P. (1982). British experience with the self-boring pressuremeter. *Proceedings of the International Symposium on the Pressuremeter and its Marine Application*, Paris. 143–164.

Yogeswaran, M. (1995). Geological considerations in the development of the Kuching area. *Proceedings of the Dialogue Session on Geological and Geotechnical Considerations in Civil Works.* Geological Survey of Malaysia. Kuala Lumpur. 59–101.

STANDARDS, GOVERNMENT AND OFFICIAL PUBLICATIONS

AFNORNFP94-150 (1991). *Essai Statique de Pieu Isolé Sous Compression Axiale.* Norme Française (in French).

AFNOR DTU-13.2 (1992). *Deep Foundations for the Building Industry P11-212* (in French).

ASTM D1143-81 (1994). *Piles Under Static Axial Compressive Load. Standard Test Method.*

ASTM D420-D5779 (2004). *Standard Test Method for Prebored Pressuremeter Testing in Soils.*

ASTM D4945 (2008). *Standard Test Method for High-Strain Dynamic Testing of Deep Foundations.*

Canadian Geotechnical Society (1992) *Canadian Foundation Engineering Manual* (CFEM) 3rd Edition, Vancouver, BiTech Publishers.

EN1997-1 (2004). *Eurocode 7: Geotechnical Design – General Rules.*

FASCICULE No 62 – Titre V(1993) *Technical Rules for the Design of Foundations of Civil Engineering Structures.* Technical Manuals for Public Construction Works, Textes Officiels No 93–3 T.O. (in French).

GCO Geoguide 2 (1987). *Guide to Site Investigation.*

GEO Publication No 1/2006 (2006). *Foundation Design and Construction.*

ISSMFE (1985). Axial pile loading test – Part 1: Static loading. *Geotechnical Testing Journal.* 8(2): 79–90.

BIBLIOGRAPHY

Aoki, N. & Velloso, D.A. (1975). An approximate method to estimate the bearing capacity of piles. In: *5th Pan-American Congress of Soil Mechanics and Foundations Engineering, Buenos Aires*, 1, 367–376.

Briaud, J.L. & Tucker, L.M. (1988). Measured and predicted axial response of 98 piles. *ASCE Journal of Geotechnical Engineering,* 114 (GT9), 984–1001.

Burns, S.E. & Mayne, P.W. (1996). Small- and high-strain soil properties using the seismic piezocone. *Transportation Research Record 1548*, Washington, D.C., National Academy Press. 81–88.

Bustamante, M. & Gianeselli L. (1981). Prévision de la capacité portante des pieux isolés sous charge verticale. Règles pressiométriques et pénétrométriques. *Bulletin Liaison Laboratoire Central des Ponts et Chaussées,* 113, 83–108 (in French).

Bustamante, M. & Gianeselli, L. (1993). Design of auger displacement piles from *in situ* tests. In: *2nd International Geotechnical Seminar: Deep Foundations on Bored and Auger Piles,* Rotterdam, Balkema.

Fellenius B.H. (1989). Tangent modulus of piles determined from strain data. In: Kulhawy, F.H. (ed.) *ASCE, Geotechnical Engineering Division, The 1989 Foundation Congress.* 1, 500–510.

Fellenius, B.H., (1999). Bearing capacity – A delusion? *Proceedings of the Deep Foundation Institute Annual Meeting, October 14–16, 1999, Dearborn, Michigan.*

Fellenius, B.H. (2001a). *What capacity value to choose from the results a static loading test.* Deep Foundations Institute, Fulcrum, May 2001.

Fellenius, B.H. (2001b). Where to plot average loads from telltale measurements in piles. *Geotechnical News Magazine.* 19 (2), 32–34.

Fellenius, B.H. (2001c). From strain measurements to load in an instrumented pile. *Geotechnical News Magazine.* 19 (1), 35–38.

Fellenius, B.H. (2002a). *Basics of Foundation Design.* Electronic Edition [www.Geoforum.com].

Fellenius, B.H. (2002b). Determining the true distribution of load in piles. In: O'Neill, M.W., Townsend, F.C. (eds) *ASCE International Deep Foundation Congress: An International Perspective on Theory, Design, Construction, and Performance, February 14–16, 2002, Orlando, Florida,* Geotechnical Special Publication No. 116 (2), 1455–1470.

Fellenius, B.H. (2002c). Determining the resistance distribution in piles. Part 1: Notes on shift of no-load reading and residual load. Part 2: Method for determining the residual load. *Geotechnical News Magazine.* 20 (2), 35–38, and (3), 25–29.

Fellenius, B.H. & Altaee, A. (1995). The critical depth – How it came into being and why it does not exist. *Proceedings of the Institution of Civil Engineers, Journal of Geotechnical Engineering.* 113 (2), 107–111.

Fellenius, B.H. & Eslami, A. (2000). Soil profile interpreted from CPTu data. In: Balasubramaniam, A.S., *et al.* (eds) *Proceedings of Year 2000 Geotechnics Conference, Southeast Asian Geotechnical Society, November 27–30, 2000, Asian Institute of Technology, Bangkok, Thailand.* 1, 163–171.

Fellenius, B.H., Hussein, M., Mayne, P. & McGillivray, R. T. (2004). Murphy's law and the pile prediction event at the 2002 ASCE Geo Institute's Deep Foundation Conference. *Proceedings of the Deep Foundation Institute 29th Annual Conference on Deep Foundations, Vancouver*, 29–43.

Fellenius, B.H. & Salem, H. (2003). Prediction of response to static loading of three piles at the ISC'2 Experimental Site. *Second International Conference on Site Characterization, Porto, Portugal.*

Ferreira, C. (2009). *The Use of Seismic Wave Velocities in the Measurement of Stiffness of a Residual Soil.* Ph.D. Thesis, University of Porto, Portugal.

Ferreira, C., Viana da Fonseca, A. & Nash, D. (2011). Shear wave velocities for sample quality assessment on a residual soil. *Soils and Foundations.* 51 (4), 683–692.

Finke, K.A., Mayne, P.W. & Klopp, R.A. (1999). Characteristic piezocone response in Piedmont residual soils. *Behavioral Characteristics of Residual Soils. ASCE Geotechnical Special Publication No. 92.* Reston, Va, pp. 1–11.

Fioravante, V., Ghionna, V.N., Jamiolkowski, M. & Pedroni, S. (1995). Load carrying capacity of large diameter bored piles in sand and gravel. *Proceedings of the 10th Asian Regional Conference on Soil Mechanics and Foundation Conference, Beijing.*

Lee, J.H. & Salgado, R. (1999). Determination of pile base resistance in sands. *ASCE Journal of Geotechnical & Geoenvironmental Engineering.* 125 (GT8), 673–683.

Mayne, P.W., Brown, D.A., Vinson, J., Schneider, J.A. & Finke, K.A. (2000). Site characterization of Piedmont residual soils at the NGES, Opelika, AL. *National Geotechnical Experimentation Sites, Geotechnical Special Publication No. 93*, ASCE, Reston, Va, pp. 160–185.

Philipponnat, G. (1980). Calculus Practical Method of a single pile using SPT. *Revue Française de Géotechnique*, 10, 55–64.

Puppala, A.J., Hanchanloet, S., Jadeja, M. & Burkart, B. (1999). Evaluation of Sulfate Induced Heave by Mineralogical and Swell Tests. *Proceedings of the 11th Pan-American Conference on Soil Mechanics and Geotechnical Engineering, Foz do Iguaçu, Brazil.*

Rollins, K.M., Clayton, R.G., Mikesell, R.C., & Blaise, B.C. (2005). Drilled shaft side friction in gravelly soils. *ASCE Journal of Geotechnical and Geoenvironmental Engineering*, 131 (8), 987–1003.

Santos, J.A. (1999). *Soil Characterization by Dynamic and Cyclic Torsional Shear Tests. Application to the Study of Piles under Lateral Static and Dynamic Loadings.* Ph.D. Thesis, Technical University of Lisbon, Portugal (in Portuguese).

Sousa, C.T. (2006). *Tests and Response Analysis of Piles in Granite Residual Soil under Horizontal Actions.* M.Sc. Thesis, Faculty of Engineering of the University of Porto, Portugal (in Portuguese).

Sowers, G.F. (1994). Residual soil settlement related to the weathering profile. In: *Vertical and horizontal deformation of foundation and embankments, ASCE Geotechnical Special Publication No* 40. ASCE, New York. 2, 1689–1702.

Takesue, K., Sasao, H. & Matsumoto, T. (1998).Correlation between ultimate pile skin friction and CPT data. *Geotechnical Site Characterization* (2), Rotterdam, Balkema. 1177–1182.

Viana da Fonseca, A. (1999). Surface loading tests for mechanical characterisation of a saprolitic soil from granite of Porto. *Proceedingsof the 9th Pan-American Conference on Soil Mechanics and Geotechnical Engineering, Foz do Iguassu, Brasil.* 1, 403–409.

Viana da Fonseca, A. & Quintela, J. (2010). Modelling the behaviour of a retaining wall monitored during the excavation for deep station in Metro do Porto. Submitted to *Geomechanics and Geoengineering: An International Journal.*

Vucetic, M. & Dobry, R. (1991). Effect of soil plasticity on cyclic response. *ASCE Journal of Geotechnical Engineering* 117 (GT1), 89–107.

9

Residual soils of Hong Kong

Charles W.W. Ng & J. Xu
The Hong Kong University of Science and Technology, HKSAR, P.R. China

9.1 GENERAL DESCRIPTIONS OF DECOMPOSED ROCKS IN HONG KONG

Formation and distribution of decomposed rocks

Hong Kong is a mountainous city subject to subtropical climate conditions. Figure 9.1 shows a geological map of Hong Kong that illustrates the distributions of rock types (Irfan, 1996). The predominant rock types in Hong Kong are Mesozoic volcanic rock and granite, which constitute about 85% of rock outcrop on the island (Sewell *et al.*, 2000). The volcanic rocks include tuffs, tuffites, lavas and sedimentary rocks. They occupy about 50% of the surface area of Hong Kong. Intrusive igneous rocks, which comprise mainly granite and granodiorite, account for about 35% of the surface area of Hong Kong. A detailed classification and description of the volcanic and granitic rocks in Hong Kong is given by Sewell *et al.* (2000).

The rocks in Hong Kong are deeply weathered, typically to depths of up to 60 m in the granites and 20 m in volcanic rocks (Irfan, 1996). Weathering basically can be divided into three types: physical weathering (disintegration), chemical weathering (decomposition) and biological weathering. The two important physical processes of weathering in Hong Kong are alternate wetting and drying, and exfoliation (sheeting).

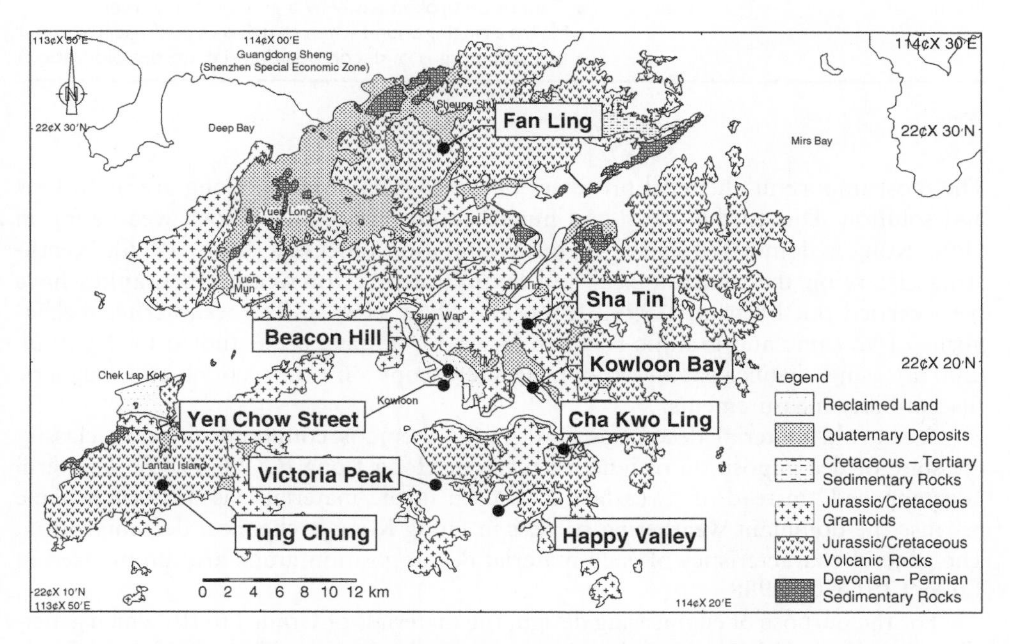

Figure 9.1 Geological map of Hong Kong together with the locations of sampling and test sites (*After* Irfan, 1996)

Table 9.1 Classification of rock material decomposition grades (GCO, 1988)

Rock Material		
Description	*Grades*	*General Characteristics*
Residual Soil	VI	• Original rock texture completely destroyed • Can be crumbled by hand and finger pressure into constituent grains
Completely Decomposed	V	• Original rock texture preserved • Can be crumbled by hand and finger pressure into constituent grains • Easily indented by the point of a geological pick • Slakes when immersed in water • Completely discoloured compared with fresh rock
Highly Decomposed	IV	• Can be broken by hand into smaller pieces • Makes a dull sound when struck by a geological hammer • Not easily indented by the point of a geological pick • Does not slake when immersed in water • Completely discoloured compared with fresh rock
Moderately Decomposed	III	• Cannot usually be broken by hand; easily broken by a geological hammer • Makes a dull or slightly ringing sound when struck by a geological hammer • Completely stained throughout
Slightly Decomposed	II	• Cannot be broken easily by a geological hammer • Makes a ringing sound when struck by a geological hammer • Fresh rock colours generally retained but stained near joint surfaces
Fresh	I	• Cannot be broken easily by a geological hammer • Makes a ringing sound when struck by a geological hammer • No visible signs of decomposition (i.e. no discolouration)

The most important chemical processes of weathering in Hong Kong are hydrolysis and solution. Due to the warm and humid subtropical climate, rock weathering in Hong Kong is dominated by the chemical alteration processes (Irfan, 1996). Weathering effects on the mineralogical and textural aspects of weathered granites have been carried out by Irfan (1996, 1999) and Shaw (1997). The weathering mechanisms of volcanic and granitic rocks in Hong Kong have been studied by Ng *et al.* (2001a) using chemical analysis, optical microscopy on thin sections and magnetic susceptibility measurements.

A six-fold material decomposition grade scheme is commonly used to classify the state of decomposition of igneous rocks in Hong Kong (GCO, 1988). The term "decomposed" instead of "weathered" is used in the material classification scheme because the dominant weathering process in Hong Kong is chemical decomposition. The general characteristics of each material decomposition grade are summarised in Table 9.1 (GCO, 1988).

For the purpose of engineering design, the materials of Grade I to III, which generally cannot be broken down by hand, are considered as rocks. The materials of Grade IV to VI, which generally can be broken down by hand into their constituent grains, are considered as soils. The soils of Grade IV and V are termed as saprolites, which preserve the original rock texture. Grade VI material is classified as residual soil and the original rock texture is completely destroyed. Because of the steep topography, residual soil is not commonly found in Hong Kong.

This state-of-the-art chapter is intended to cover the behaviour and properties of granitic and volcanic saprolites in Hong Kong (Grades IV and V). The stress-dependent soil-water characteristic curve (SDSWCC), permeability function, small strain stiffness and shear strength characteristics of some typical saprolites in Hong Kong are investigated, described and reported.

9.2 IN-SITU TEST SITES AND SAMPLING LOCATIONS

Figure 9.1 shows the locations of in-situ test sites and soil sampling locations reported in this chapter. The test site Tung Chung, located in the north of Lantau Island in Hong Kong, was selected to investigate the in-situ stress-dependent soil-water characteristic curves (SDSWCCs) and the permeability function of a saprolitic slope. The ground mainly consists of colluvium and completely decomposed tuff (CDT). A detailed ground condition of Tung Chung is described by Ng *et al.* (2011). Another three test and sampling sites were selected for investigating the small strain stiffness of saprolites in Hong Kong. Two of the sites, Kowloon Bay and Yen Chow Street, are located on the Kowloon Peninsula where granite predominates. The sampling site in the northern New Territories, Fan Ling, is underlain by coarse ash crystal tuff.

Figures 9.2a, b and c show the soil profile and results of the standard penetration tests (SPT 'N' profile) at Kowloon Bay (Ng *et al.*, 2000), Yen Chow Street (Ng and Wang, 2001) and Fan Ling (Ng and Leung, 2007b), respectively. At Kowloon Bay and Yen Chow Street, granite at varying degrees of decomposition is identified along the depth. The degree of decomposition decreases with depth, where completely decomposed granite (CDG; Grade V) overlies highly decomposed granite (HDG; Grade IV) and moderately decomposed granite (MDG; Grade III). At Fan Ling, coarse ash crystal tuff is decomposed at different degrees along the depth. Similarly, the degree of decomposition decreases with depth, where completely decomposed tuff (CDT; Grade V) overlies highly decomposed tuff (HDT; Grade IV) and moderately decomposed tuff (MDT; Grade III).

Three types of saprolites sampled included (i) completely decomposed granite (CDG) from Kowloon Bay, Yen Chow Street, Beacon Hill, Cha Kwo Ling and Happy Valley; (ii) completely decomposed volcanic (CDV) rocks from Sha Tin and Victoria Peak and (iii) completely decomposed tuff (CDT) from Tung Chung and Fan Ling. Figure 9.3 shows the particle size distribution of three saprolites and their physical properties are summarised in Table 9.2. According to the Unified Soil Classification System (ASTM, 2000), the CDG, CDV and CDT reported in this chapter may be classified as silty sand (SM), sandy elastic silt (MH) and silt (ML), respectively.

9.3 SAMPLING METHODS AND PREPARATION PROCEDURES

It is well-known that both in-situ and laboratory tests are not perfect methods to determine soil properties. Each testing method has its own advantages and disadvantages

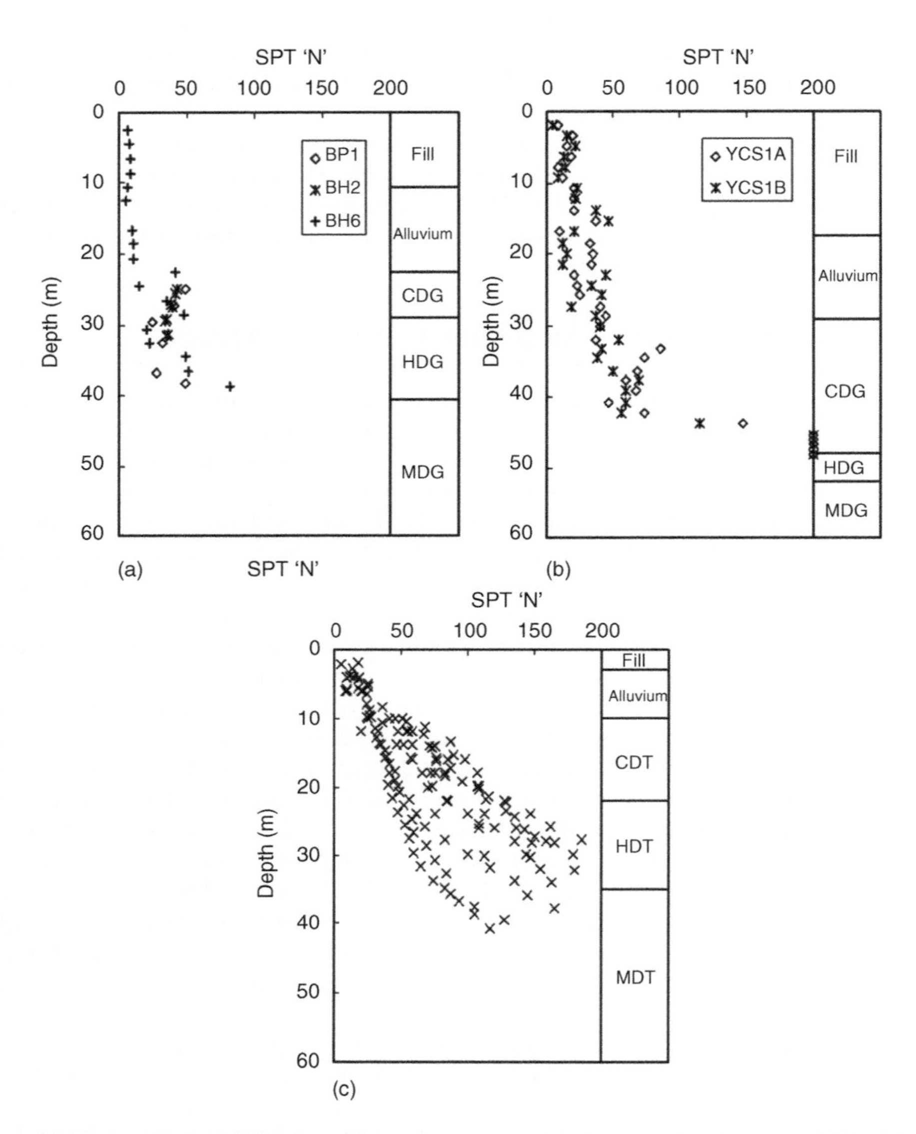

Figure 9.2 Soil profile and SPT 'N' profile at the test sites: (a) Kowloon Bay (Ng *et al.*, 2000); (b) Yen Chow Street (*After* Ng and Wang, 2001); (c) Fan Ling (*After* Ng and Leung, 2007b)

(Atkinson and Sällfors, 1991). There is no method available to test "undisturbed" samples and the soil mass in the field or in the laboratory. The "undisturbed" samples cannot be obtained in the laboratory or from the field. This is because any method of sampling and field testing will cause different degrees of sample disturbance. In Hong Kong, rotary coring and block sampling are commonly used.

Block sampling is believed to be the best available technique for soil sampling and it provides samples of fill, soils derived from in-situ rock weathering and colluvium

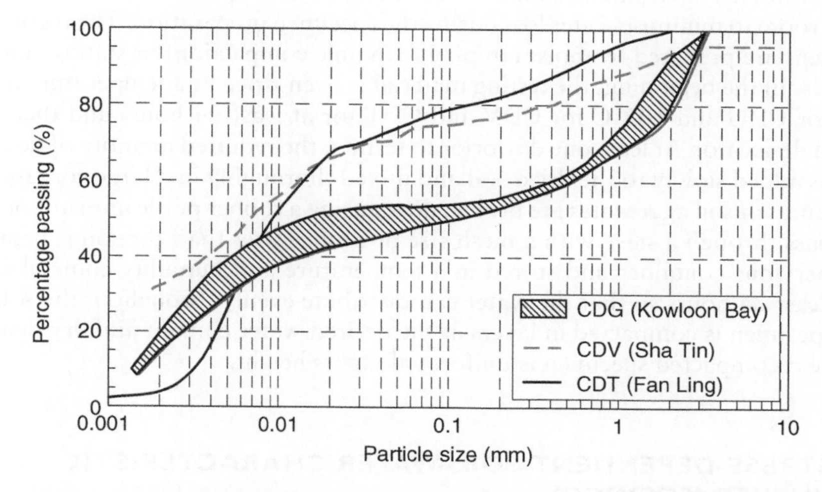

Figure 9.3 Particle size distributions of typical CDG, CDV and CDT in Hong Kong

Table 9.2 Physical properties of typical CDG, CDV and CDT in Hong Kong

	CDG (Kowloon Bay)	*CDV (Sha Tin)*	*CDT (Fan Ling)*
Classification	Silty sand (SM)	Sandy elastic silt (MH)	Silt (ML)
Liquid limit (LL) (%)	/	55.4	43
Plastic limit (PL) (%)	/	33.4	29
Plasticity index (PI) (%)	/	22	14
Specific gravity, G_s	2.63	2.62	2.73
Moisture content (%)	19–35	30	17
Dry density (kg/m^3)	1210–1580	1603 (maximum)	1730

with the least possible disturbance (GCO, 1987). The block samples are obtained by cutting exposed soil to a cubic block in trial pits or excavations. Specially oriented samples can be obtained by block sampling for the measurement of the shear strength on specific discontinuities and anisotropic soil stiffness (Ng and Leung, 2007a). On the other hand, Mazier triple-tube core-barrels, fitted with a retractable shoe, are normally used for coring soils derived from in-situ rock weathering (GCO, 1987). A Mazier core-barrel contains detachable liners within the inner barrel, which protect the core from drilling fluid and damage during extrusion. The split liners facilitate the retention of core samples during transportation to a laboratory. The core diameter is about 74 mm, which is compatible with the laboratory triaxial testing apparatus. Local experience commonly regards the Mazier sampling technique as the most suitable sampling method available for weathered granular materials at depths. The comparison of the degrees of sample disturbance associated with the block and Mazier sampling techniques and the effects of sample disturbance on the small strain stiffness of soils are given by Ng and Leung (2007a).

Intact specimens (or natural specimens) from the block or Mazier samples are carefully hand trimmed to suitable dimensions for testing the properties of natural soils in the laboratory. The trimming should be conducted in a temperature and humidity controlled room to minimise water loss during the specimen preparation. The recompacted specimens are prepared by moist tamping (dynamic compaction) or static compaction methods. In the beginning, the testing material is oven-dried at a temperature of about 50°C for CDG and 105°C for CDV and CDT for at least 24 hours and then cooled inside a desiccator. At least one day prior to testing, the required quantity of de-aerated water is added slowly to the dried soil and mixed thoroughly in a large container. The large lumps of soil aggregates are broken down using a rubber pestle until all soil aggregates pass through a sieve with a mesh size of 2 mm. After that, the soil is kept inside an impervious container and stored in a temperature and humidity controlled room for at least 24 hours so that the water can distribute evenly throughout the soil mass. Each specimen is compacted in layers at the desired water content and dry density, so that the recompacted specimen is uniform along its height.

9.4 STRESS-DEPENDENT SOIL–WATER CHARACTERISTIC CURVES (SDSWCC)

The relationship between the amount of water in a soil and soil suction (i.e. soil–water characteristics) is a key hydraulic property of an unsaturated soil. There are different terms used to describe this relationship in different disciplines. Fredlund *et al.* (2001) recommended that the term "soil-water characteristic curve (SWCC)" should be used in civil engineering-related disciplines. A SWCC represents the water retention capability of an unsaturated soil. The SWCC may be determined from both laboratory and in-situ tests by controlling soil suction while measuring the water content or measuring both of them simultaneously. A SWCC of a soil specimen in the laboratory is conventionally measured by a pressure plate extractor under zero net stress and the volume change of the soil specimen is ignored. Obviously, this approach greatly simplifies actual field conditions where the influences of stress and volume change may not be negligible. Thus, Ng and Pang (1999) developed a new one-dimensional stress-controllable volumetric pressure plate extractor (see Figure 9.4) to investigate the stress-dependent soil–water characteristic curve (SDSWCC), which considers the effects of net stress and volume change. In this section, both laboratory and field measured SDSWCC of saprolites in Hong Kong are introduced and described. The effects of stress level, stress ratio and drying and wetting cycles on SDSWCC are highlighted. Comparisons of SDSWCCs between intact and recompacted specimens and comparisons between laboratory and in-situ SDSWCCs are also discussed.

Laboratory measurements for SDSWCC

Extensive laboratory investigations on SDSWCC of saprolites have been conducted at the Hong Kong University of Science and Technology (HKUST). A SDSWCC may be influenced by many factors such as initial water content, initial dry density, particle size distribution, stress state, stress ratio and drying-wetting cycles (Ng and Pang, 2000a, b; Ng *et al.*, 2001b; Ho *et al.*, 2006; Ho *et al.*, 2007; Tse, 2007). In this section, only the effects of stress state, stress ratio and drying-wetting cycles are highlighted.

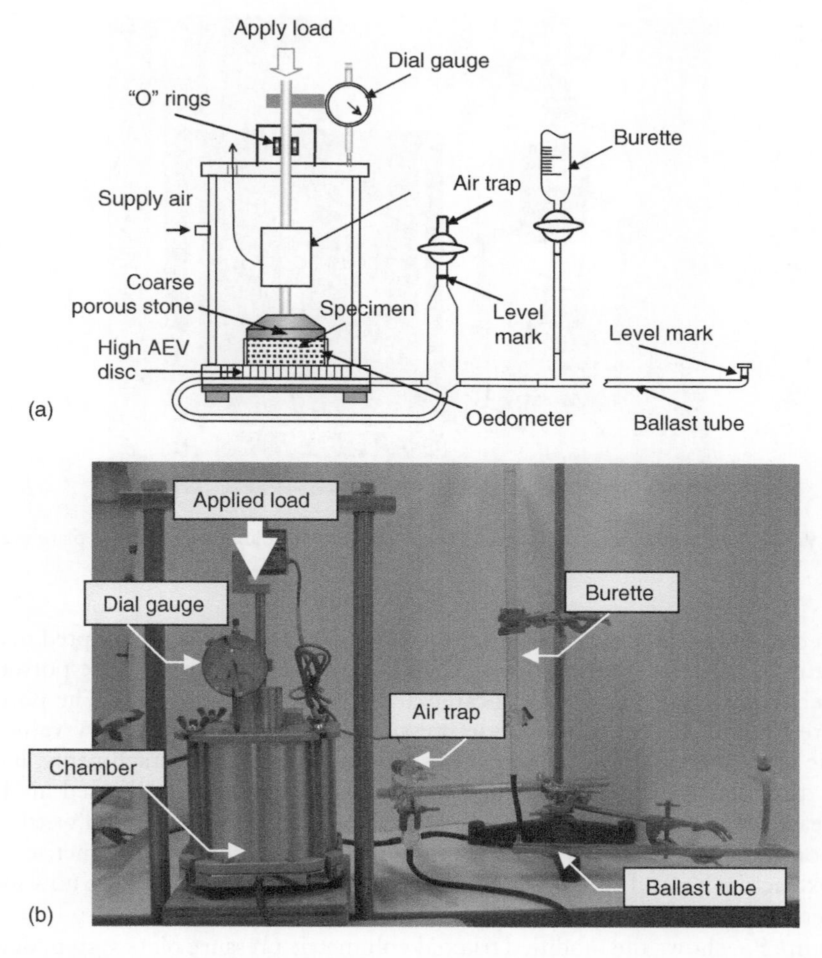

Figure 9.4 One-dimensional stress-controllable volumetric pressure plate extractor developed at HKUST (*After* Ng and Pang, 2000a): (a) schematic diagram; (b) photograph

Laboratory apparatus for measuring SDSWCC

Two types of pressure plate systems have been developed at HKUST for measuring SDSWCCs under one-dimensional (1D) (Ng and Pang, 1999, 2000a) and triaxial stress conditions (Ng *et al.*, 2001b; Ng and Menzies, 2007; Ng and Chen, 2008).

Figure 9.4 shows the one-dimensional stress-controllable volumetric pressure plate extractor for measuring SDSWCC under 1D stress conditions (Ng and Pang, 1999, 2000a). This apparatus is designed to overcome some limitations of the conventional volumetric pressure plate extractors. The vertical stress is applied through a loading frame to a soil specimen inside an oedometer ring in the airtight chamber. Because radial deformation is assumed to be zero for the K_0 stress condition, the change in total volume of the specimen is measured from the vertical displacement of the soil specimen

Figure 9.5 A commercial one-dimensional stress-controllable volumetric pressure plate extractor

using a dial gauge. As the axis-translation technique (Hilf, 1956) is adopted to control the matric suction, the pore air pressure is controlled through a coarse porous stone together with a coarse geotextile located at the top of the specimen. The pore water pressure is controlled at atmospheric pressure through a high air-entry value (AEV) ceramic disc mounted at the chamber base. A set of hysteresis attachments, including an air trap, a ballast tube and a burette, are used for flushing diffused air bubbles and measuring water volume change of a specimen during a drying and wetting cycle. Based on the concept of this one-dimensional stress-controllable volumetric pressure plate extractor (Ng and Pang, 1999), at least three similar extractors are now available commercially. One of them is shown in Figure 9.5.

Figure 9.6 shows the modified triaxial volumetric pressure plate system developed at HKUST for measuring SDSWCC under triaxial stress conditions. This system is modified from the original design by Ng *et al.* (2001b). It consists of five main components, namely a net normal stress control, a suction control, a flushing system, a measuring system for water volume change and a new measuring system for total volume change (Ng *et al.*, 2002).

An axial force, if required, is exerted on the specimen through a loading ram to apply deviator stress to the soil specimen. An internal load cell is attached to the loading ram to measure the applied axial force. A linear variable differential transformer (LVDT) is attached to the loading ram to measure the axial displacement of a soil specimen. As the axis-translation technique is adopted for controlling matric suction, the pore air pressure is controlled through a coarse porous disc placed at the top of the specimen and the pore water pressure is maintained at the atmospheric pressure through a saturated high AEV ceramic disc at the base of the specimen. Similar to the 1D stress-controllable volumetric pressure plate extractor, a set of hysteresis attachments is used to measure water volume change during a drying and wetting cycle.

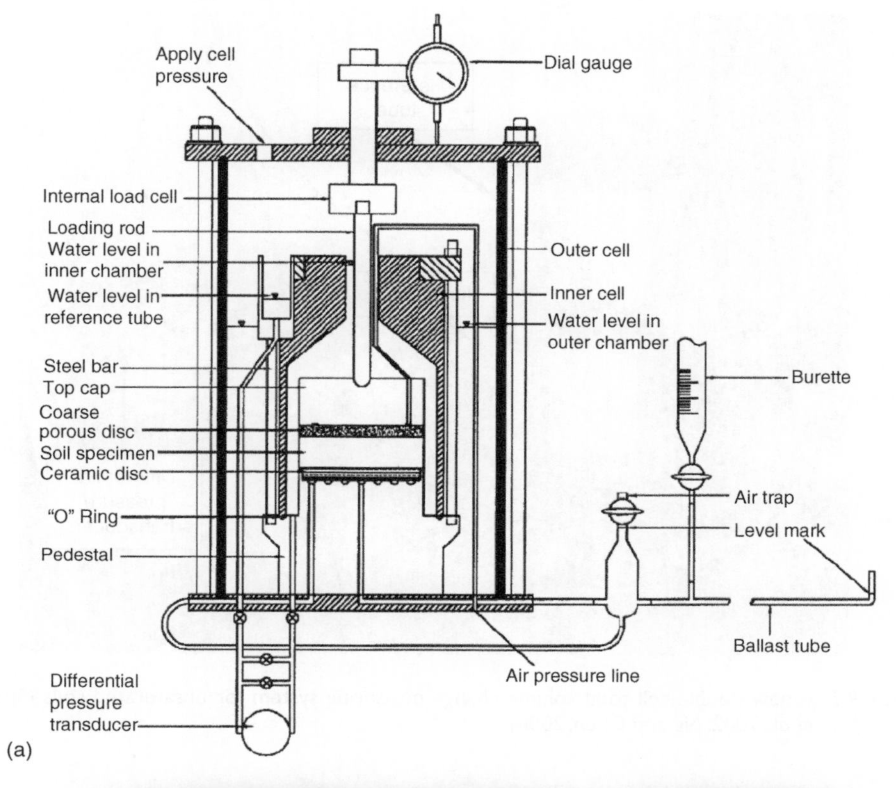

Figure 9.6 The modified triaxial volumetric pressure plate system developed at HKUST: (a) schematic diagram (*After* Ng and Menzies 2007; Ng and Chen, 2008); (b) photograph (*After* Tse, 2007)

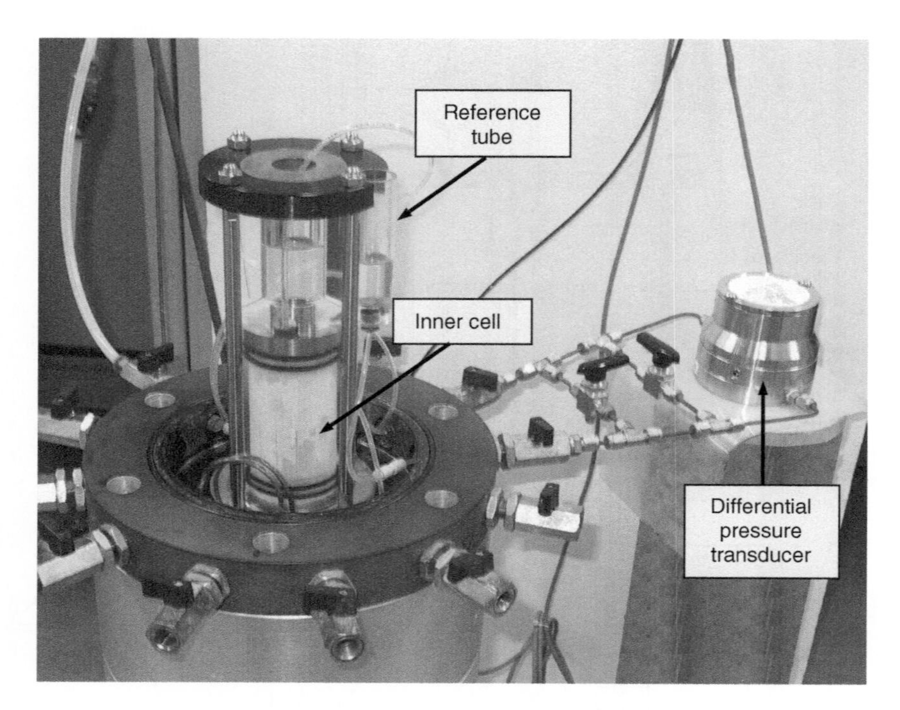

Figure 9.7 A new double-cell total volume change measuring system for unsaturated soils (*After* Ng *et al.*, 2002; Ng and Chen, 2008).

The total volume change of a soil specimen is measured by adopting a double-cell simple measuring system developed by Ng *et al.* (2002), as shown in Figure 9.7.

The basic principle of the measuring system is to record the changes in the differential pressure due to changes in the water level inside an open-ended, bottle-shaped inner cell caused by the volume change in the specimen; and inside a reference tube using an accurate differential pressure transducer (DPT). Several important steps were taken to improve the accuracy and the sensitivity of the measuring system. Detailed calibrations are carried out to account for apparent volume changes due to changes in the cell pressure, fluctuation in the ambient temperatures, creep in the inner cell wall and relative movement between the loading ram and the inner cell. The calibration results demonstrate that the measuring system is reasonably linear, reversible and repeatable. The estimated accuracy of the measuring system is in the order of 32 mm^3 (or 0.04% volumetric strain for a 38 mm diameter 76 mm height triaxial specimen), once the system is properly calibrated. Other details of this system are described by Ng *et al.* (2002).

Effects of stress on SDSWCC of natural (intact) CDV specimens

Figure 9.8 shows the measured SWCC and SDSWCCs of three natural (intact) CDV (from Sha Tin) specimens under different net normal stresses (0, 40 and 80 kPa) by both

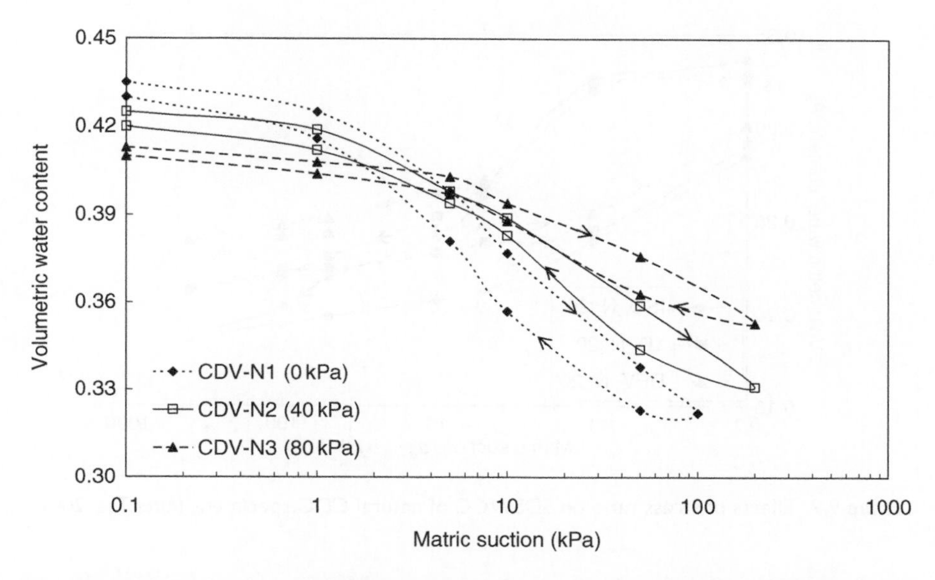

Figure 9.8 Effects of stress on SDSWCC of natural CDV specimens (*After* Ng and Pang, 2000a)

the conventional pressure plate extractor and the one-dimensional stress-controllable volumetric pressure plate extractor (see Figure 9.4). Under zero suction, soil specimens loaded to a higher net normal stress, exhibit lower initial volumetric water content. When suction changes, the soil specimen, under a higher net normal stress, shows a larger AEV and lower desorption and adsorption rates. This is probably attributable to the presence of smaller interconnected pores in the soil specimen under higher net normal stress. Besides, there is a marked hysteresis between the drying and wetting curves for all the three specimens. The size of the hysteresis loops for natural specimens seems to be independent of the range of the net normal stresses considered, except for the one that was determined by the conventional pressure plate extractor in which the volume change corrections cannot be made. The end point of the wetting curve of each specimen is lower than the corresponding starting point of the drying curve, probably due to the air trapped in the specimens. Other detailed studies of SDSWCCs of CDV (from Sha Tin) and CDG (from Happy Valley) are given by Ng and Pang (2000a, b) and Ho *et al.* (2006), respectively.

Effects of stress ratio on SDSWCC of natural CDG specimens

Figure 9.9 shows the measured SDSWCCs of three natural CDG (from Happy Valley) specimens tested at the same net mean stress of 20 kPa, but under different stress conditions i.e., 1D, isotropic and deviatoric (DEV) stress states (Tse, 2007). The test under 1D stress condition is performed in the one-dimensional stress-controllable volumetric pressure plate extractor (Figure 9.4), while the tests under isotropic and deviatoric stress states are carried out in the modified triaxial volumetric pressure plate system

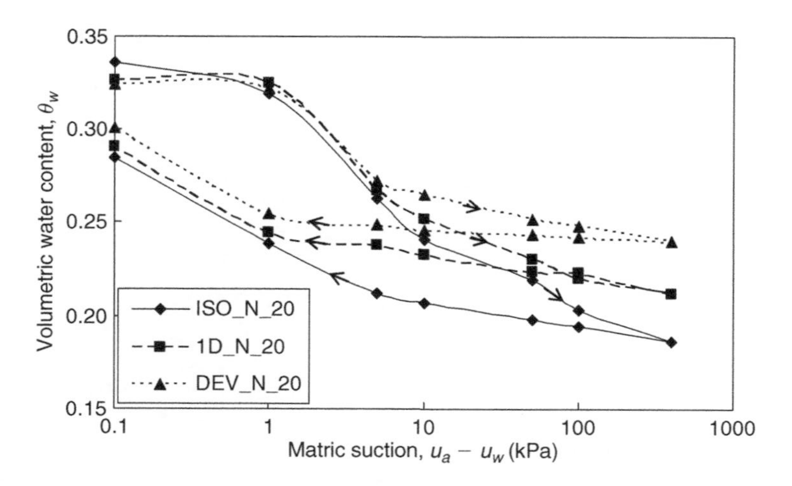

Figure 9.9 Effects of stress ratio on SDSWCC of natural CDG specimens (*After* Tse, 2007)

(Figure 9.6). The stress ratio is defined as $q/(p - u_a)$, where q and $(p - u_a)$ are the deviatoric stress and net mean stress, respectively. The stress ratios applied by the isotropic, 1D, and deviatoric stress states are 0, 0.75 and 1.2, respectively. It can be seen from the figure that the specimen under deviatoric stress state generally shows the highest water retention ability, while the specimen under isotropic stress state generally shows the lowest water retention ability. Besides, when the specimen is subjected to a higher stress ratio, a lower desorption rate, higher residual water content and smaller hysteresis loop size are observed. Detailed investigations of the stress ratio effect on SDSWCC are described by Tes (2007) and Ng and Chen (2008).

Influence of drying–wetting cycles on SDSWCC of recompacted CDV and CDG specimens

The influence of drying–wetting cycles on SDSWCC of saprolites in Hong Kong have been investigated by Ng and Pang (2000b) and Ho *et al.* (2007) using the 1D stress-controllable volumetric pressure plate extractor. Figure 9.10a shows the measured SDSWCC of a recompacted CDV (from Sha Tin) specimen subjected to three repeated drying and wetting cycles. A marked hysteresis loop between the drying and wetting paths can be observed in each cycle. The size of the hysteresis loop is the largest in the first cycle but seems to become independent in the subsequent cycles. The desorption characteristics are dependent on the drying and wetting history. The rate of desorption is relatively high during the first cycle compared with that during the second and third cycles. This may be due to the presence of relatively large voids initially. During the first wetting process, a significant volume change is likely to take place and this will result in smaller voids. Thus, a smaller desorption rate for the second and third drying and wetting cycles is observed. The adsorption characteristics of the first wetting process are also different from those of the subsequent wetting processes. The value of the matric suction, at which the soil starts to absorb water significantly, is about 50 kPa

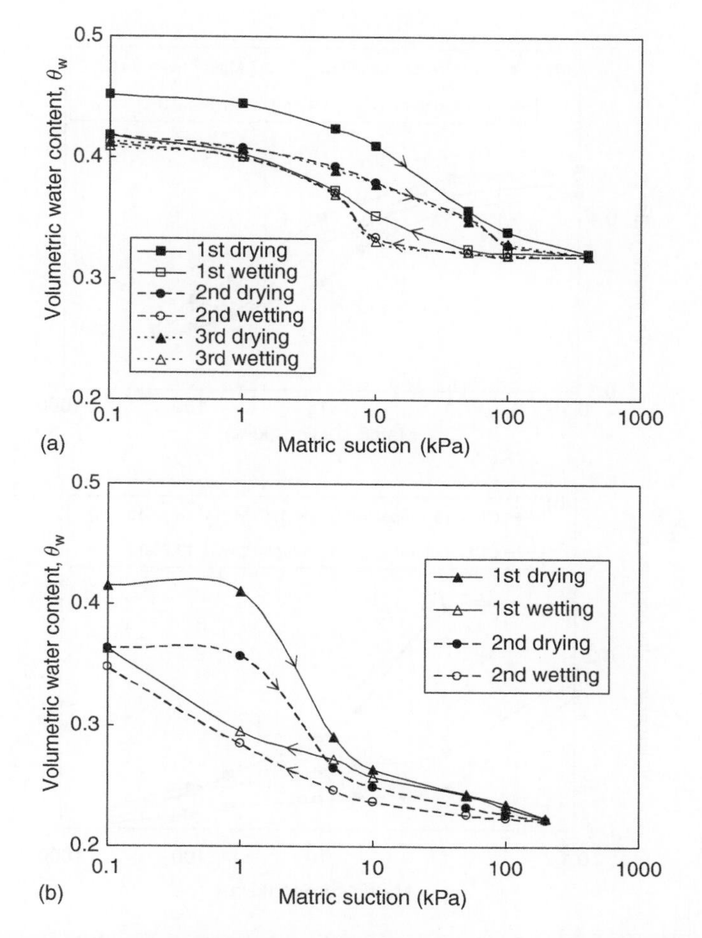

Figure 9.10 Influence of drying–wetting cycles on SDSWCCs of (a) recompacted CDV specimens (*After* Ng and Pang, 2000b); (b) recompacted CDG specimens (*After* Ho *et al.*, 2007)

during the first wetting process, higher than that during the subsequent cycles (i.e. about 10 kPa). The rates of adsorption are substantially different for the first and subsequent drying and wetting cycles at suctions ranging from 10 to 50 kPa. This might be caused by some soil structure changes after the first drying and wetting cycle.

Figure 9.10b shows the influence of drying and wetting cycles on the SDSWCC of a recompacted CDG (from Happy Valley) specimen. Two cycles of drying and wetting are applied to this specimen. Similar to the observations in SDSWCC of CDV, the size of the hysteresis loop in the first drying and wetting cycle is significantly larger than that in the second cycle. Although the shape of the two loops is surprisingly similar, the difference between the initial and final volumetric water contents of each cycle is greatly reduced in the second drying and wetting cycles. This phenomenon is also consistent with that observed from recompacted CDV specimens.

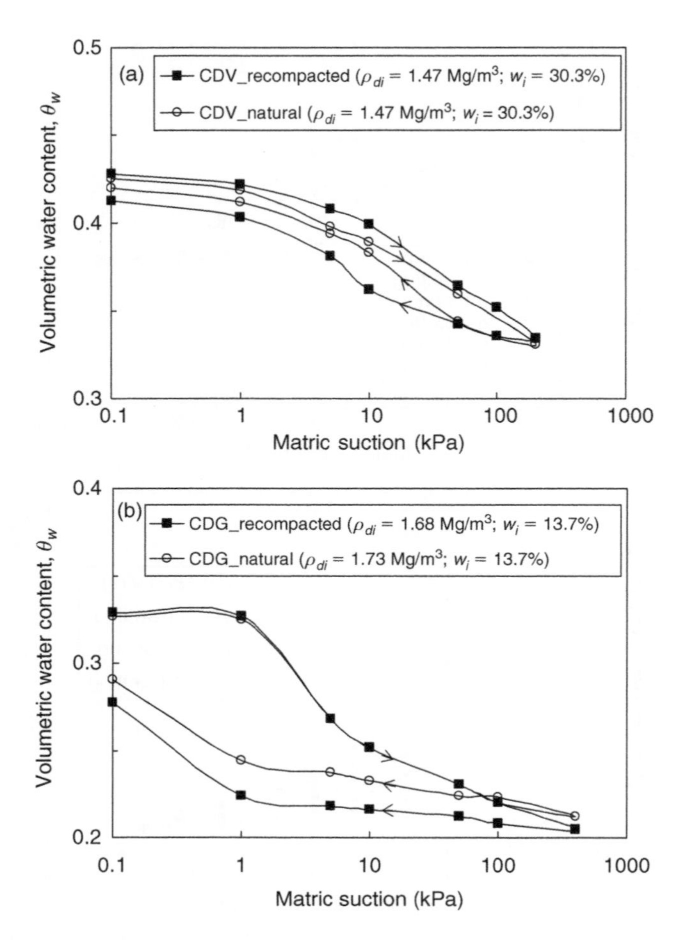

Figure 9.11 Comparison of the SDSWCCs between recompacted and natural specimens of (a) CDV (*After* Ng and Pang, 2000b); (b) CDG (*After* Tse, 2007)

Comparison between SDSWCCs of recompacted and natural CDV and CDG specimens

The SDSWCCs of a recompacted and a natural CDV (from Sha Tin) specimen are compared in Figure 9.11a. The initial water content and dry density are the same for the two specimens. Both specimens are tested by the 1D stress-controllable volumetric pressure plate extractor and loaded to the same net vertical stress of 40 kPa. The recompacted specimen seems to have a higher AEV than that of the natural one. The size of the hysteresis loop in the recompacted specimen is considerably larger than that in the natural specimen.

According to Ng and Pang (2000b), the recompacted specimen is generally expected to be more homogenous, whereas the natural specimen has relatively non-uniform pore size distribution due to various geological processes in the field. As the two specimens have the same initial density, it is reasonable to postulate that the natural specimen would have some pores larger than those in the recompacted specimen, at least statistically. Thus, the natural specimen has a slightly lower AEV and a higher desorption rate than the recompacted specimen for suctions up to 50 kPa. The desorption rates of the two soil specimens appear to be the same for suction higher than 50 kPa. On the other hand, the adsorption rates for the two specimens are considerably different. The adsorption rate of the natural specimen is substantially higher than that of the recompacted specimen. This observed behaviour may be explained by the difference in pore size distribution between the natural and recompacted specimens. As the natural soil specimen has a non-uniform pore-size distribution in which some relatively large pores exist with some relatively small pores, the presence of these small pores would facilitate the ingress of water to the specimen as the soil suction reduces. At low suctions (less than 5 kPa), the rates of adsorption of the two soil specimens appear to be the same.

Figure 9.11b shows the SDSWCCs of a natural and a recompacted CDG (from Happy Valley) specimen. The initial water contents of the two specimens are the same and the initial dry densities are similar. Both specimens are tested under the same stress level of 30 kPa. The AEV, desorption and the adsorption rates are similar for both specimens. However, the size of the hysteresis loop for the SDSWCC of the natural CDG specimen is smaller than that of the recompacted specimen. This observation is consistent with the results from CDV. This may also be due to the non-uniform pore size distribution in the natural specimen as discussed above.

In-situ measurement of SDSWCC

Sponsored by the Geotechnical Engineering Office of the Civil Engineering Development Department of HKSAR, a comprehensive field monitoring project has been undertaken on a saprolitic slope in Tung Chung, Hong Kong (see Figure 9.1) to study the hydro-mechanical responses of the slope when subjected to rainfall infiltration (Ng *et al.*, 2011; Leung *et al.*, 2011). Various instruments were installed. In order to measure in-situ permeability function (coefficient of permeability with respect to the water phase) and SDSWCC by using the Instantaneous Profile (IP) method, a flat test plot of 3.5 × 3.5 m was formed by cutting into the slope at the test location. A circular steel test ring (3 m in diameter) was embedded 100 mm into the flattened plot. Figure 9.12 shows the instrumentation scheme in the site. Ten jet-fill tensiometers (JFT_1 to JFT_10) were installed in the test ring at depths of 0.36, 0.77, 0.95, 1.17, 1.54, 1.85, 2.13, 2.43, 2.6, and 2.99 m to monitor the pore water pressure. Four time-domain reflectometry (TDR) moisture probes (TDR_1 to TDR_4) were installed at depths of 0.84, 1.85, 2.50 and 3.59 m to measure the volumetric water content in the ground. The sampling frequency of each tensiometer and TDR was 5 minutes.

To investigate the influence of wetting–drying cycles on the in-situ water permeability and SDSWCCs, the test program consisted of two wetting–drying cycles. The test program was divided into four stages: 1st wetting, 1st drying, 2nd wetting and 2nd drying. This field experiment was intentionally conducted after the wet season. According to the rainfall data obtained from Hong Kong Observatory, a negligible daily rainfall was recorded during the entire testing and monitoring period. For each wetting period, a water level of about 0.1 m was applied on the ground surface inside

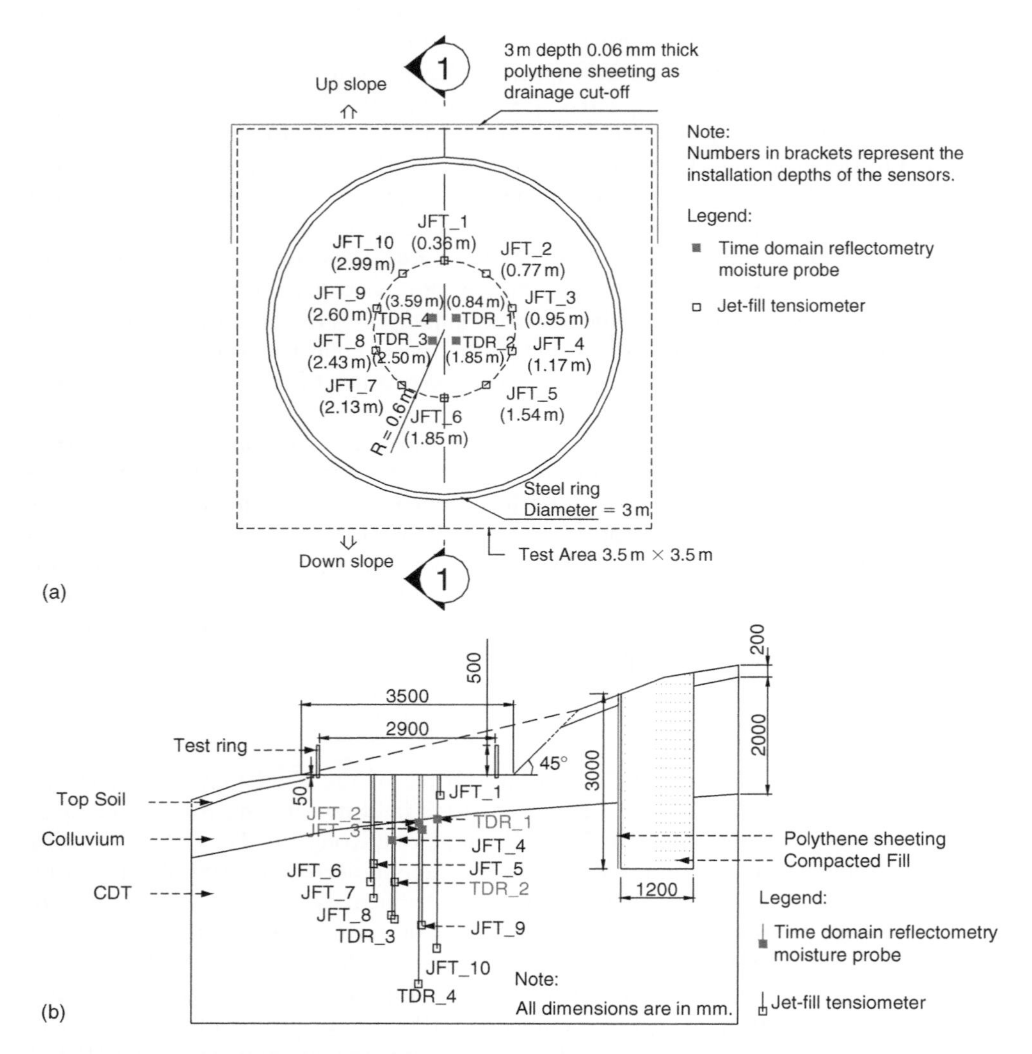

Figure 9.12 Arrangement of instrumentation (*After* Ng *et al.*, 2011): (a) instrumentation plan; (b) section 1-1 along the slope.

the ring and the water level was checked and refilled to the same level every 12 hours. For each drying period, the test plot was allowed to dry under natural evaporation.

In-situ SDSWCCs of colluvium and CDT

Under field conditions, the air phase is generally assumed to be continuous and remains at atmospheric pressure. Thus, the magnitude of the measured negative pore water pressure by the tensiometers is equivalent to the matric suction in the ground. Figure 9.13 shows the relationships between measured volumetric water

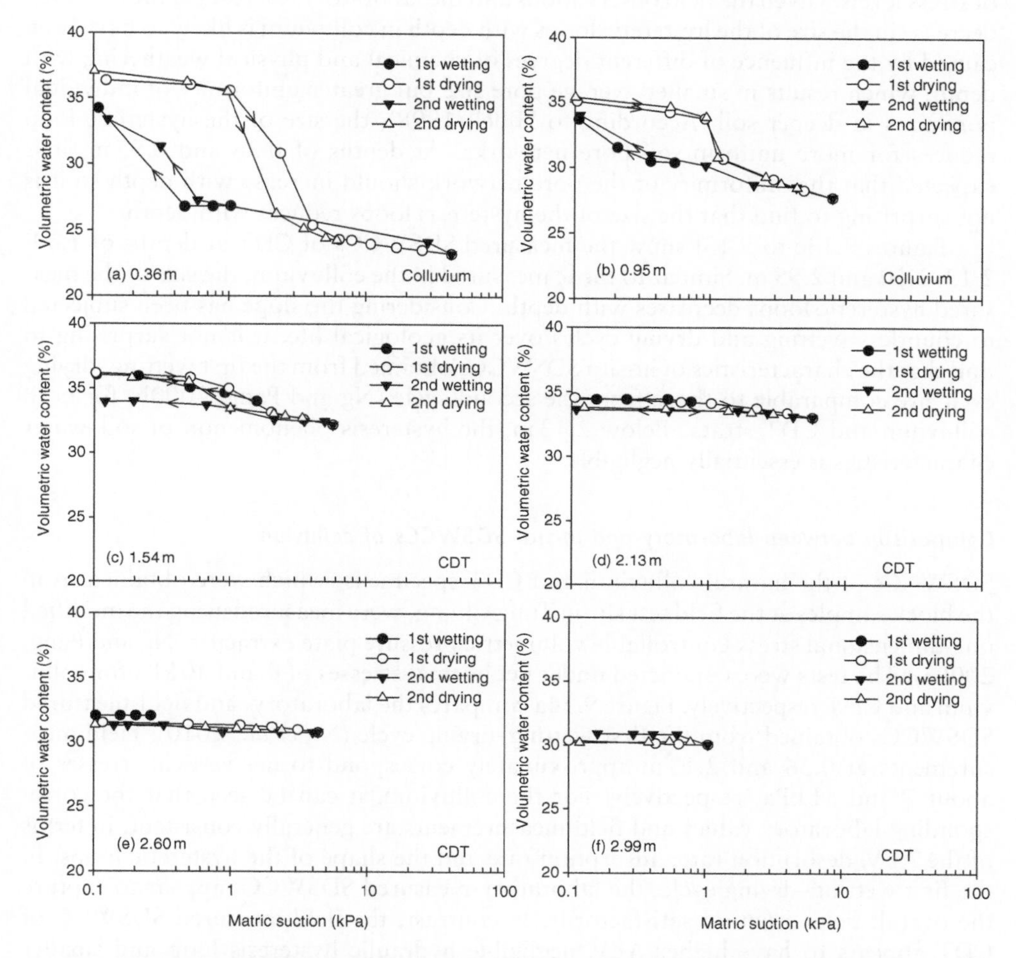

Figure 9.13 In-situ SDSWCCs of (a) colluvium at 0.36 m; (b) colluvium at 0.95 m; (c) CDT at 1.54 m; (d) CDT at 2.13 m; (e) CDT at 2.60 m; (f) CDT at 2.99 m (*After* Ng *et al.*, 2011)

content and matric suction (i.e. in-situ SDSWCCs) at six selected depths. The measured SDSWCCs are quite different along the depth, which suggests that the SDSWCCs are depth-dependent.

Within the colluvial strata (0.36 to 0.95 m), the soil appears to desaturate when the matric suction is higher than 1 kPa (i.e. air entry value ~1 kPa), as shown in Figures 9.13a and b. At a depth of 0.36 m, the reduction of volumetric water content is negligible when the matric suction is increased beyond 5.6 kPa in the first drying curve. Noticeable hydraulic hysteretic loops during wetting–drying cycles are observed, but the size of the hysteretic loop decreases with depth. This field observation is inconsistent with the previous laboratory investigation of CDV by Ng and Pang (2000a), which reported that the size of the hysteretic loops of natural CDV specimens was independent of stress levels. Given the field observations and the laboratory test results, the observed decrease in the size of the hysteretic loops with depth in colluvium is likely to have been caused by the influence of different degrees of chemical and physical weathering with depth, which results in smaller average pore size but greater uniformity of individual pore sizes in deeper soil. According to Hillel (1998), the size of the hysteretic loop reduces for more uniform soil pore networks. At depths of 0.36 and 0.95 m, it is expected that the uniformity of the pore network should increase with depth so it is not surprising to find that the size of the hysteresis loops reduces with depth.

Figures 9.13c to 9.13f show the measured SDSWCCs of CDT at depths of 1.54, 2.13, 2.60 and 2.99 m. Similar to those measured in the colluvium, the size of the measured hysteretic loops decreases with depth. Considering the slope has been subjected to countless wetting and drying cycles over its geological life, it is not surprising to find that the characteristics of in-situ SDSWCCs obtained from the first wetting–drying cycle are comparable to those from the second cycle (Ng and Pang, 2000b) for both colluvium and CDT strata. Below 2.13 m, the hysteresis phenomenon of soil–water characteristics is essentially negligible.

Comparison between laboratory and in-situ SDSWCCs of colluvium

SDSWCCs of the natural colluvium and CDT specimens, which were obtained from the block samples at the field test site in Tung Chung, were measured using the modified one-dimensional stress-controllable volumetric pressure plate extractor (Ng and Pang, 2000a). The tests were conducted under net normal stresses of 0 and 40 kPa for colluvium and CDT respectively. Figure 9.14a compares the laboratory- and field-measured SDSWCCs obtained from the first wetting–drying cycle (Ng *et al.*, 2010). Field measurements at 0.36 and 2.13 m approximately correspond to net vertical stresses of about 7 and 41 kPa, respectively. For the colluvium, it can be seen that the corresponding laboratory values and field measurements are generally consistent, in terms of the AEV, desorption rate, adsorption rate and the shape of the hysteretic loops. In the first wetting–drying cycle, the laboratory-measured SDSWCCs appear to capture the overall field situation satisfactorily. In contrast, the field-measured SDSWCC of CDT appears to have higher AEV, negligible hydraulic hysteresis loop and smaller saturated volumetric water content, when compared to those measured in the laboratory. As discussed previously, these observations may probably be attributed to the smaller and more uniform pore-size of the in-situ soils resulting from different degrees of weathering with depth. The flat shape of the in-situ SDSWCCs indicates that the CDT is within the boundary effect zone (Fredlund *et al.*, 2001), where the soil remains saturated under a given suction.

To investigate the effects of the wetting–drying cycles, laboratory- and field-measured SDSWCCs from the second wetting–drying cycles are compared in Figure 9.14b. Relatively speaking, the consistency between the laboratory and field measurements of colluvium obtained in the second cycle is not as good as those measured in the first cycle, especially at low suctions. This is mainly attributed to the smaller hysteresis loop obtained from the second cycle than that obtained in the first cycle for the laboratory specimen. Given the consistent field measurements of SDSWCCs between the first and the second cycles obtained at depths of 0.36 and 2.13 m (see Figures 9.13a and d), this seems to suggest that there was a change in

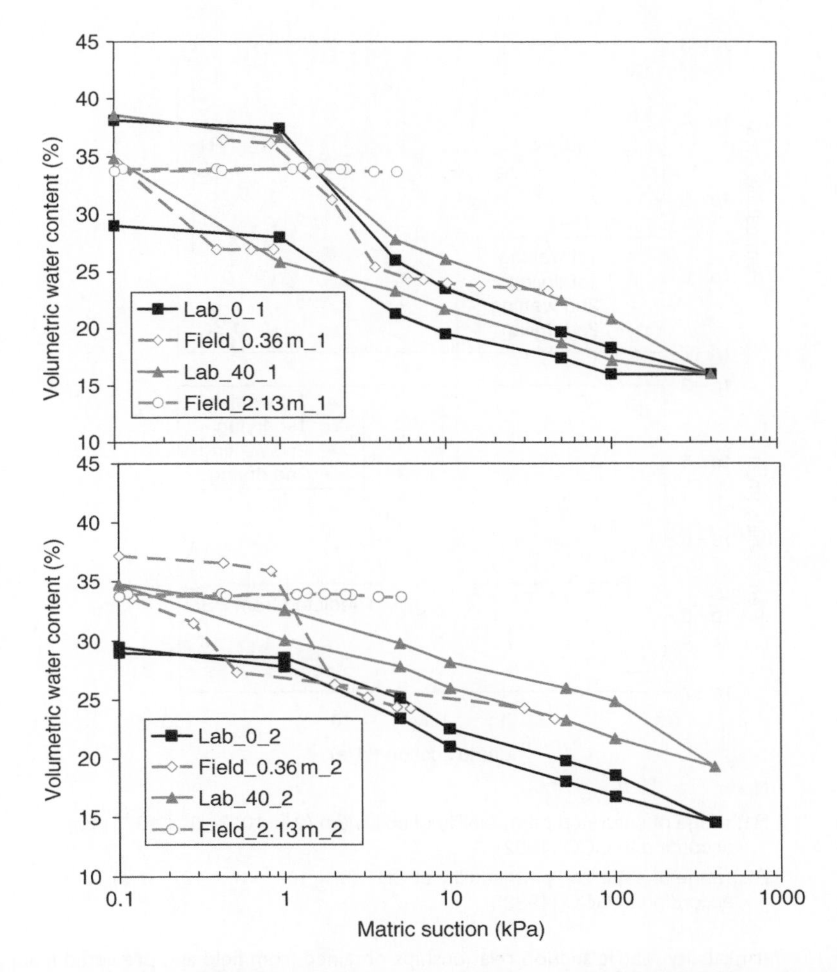

Figure 9.14 Comparisons of SDSWCCs from in-situ and laboratory measurements of colluvium and CDT from: (a) the first wetting drying cycle; (b) the second wetting-drying cycle (*After* Ng *et al.*, 2011)

the micro-structure or pore size distribution of the laboratory soil specimen after the first wetting–drying cycle, leading to different measurements being obtained during the second cycle. Detailed explanations are given by Ng *et al.* (2011).

9.5 IN-SITU PERMEABILITY FUNCTION

Field measurement technique for permeability function

The unsaturated permeability function with respect to the water phase is an important soil property required for transient seepage analysis in unsaturated soils. In the field

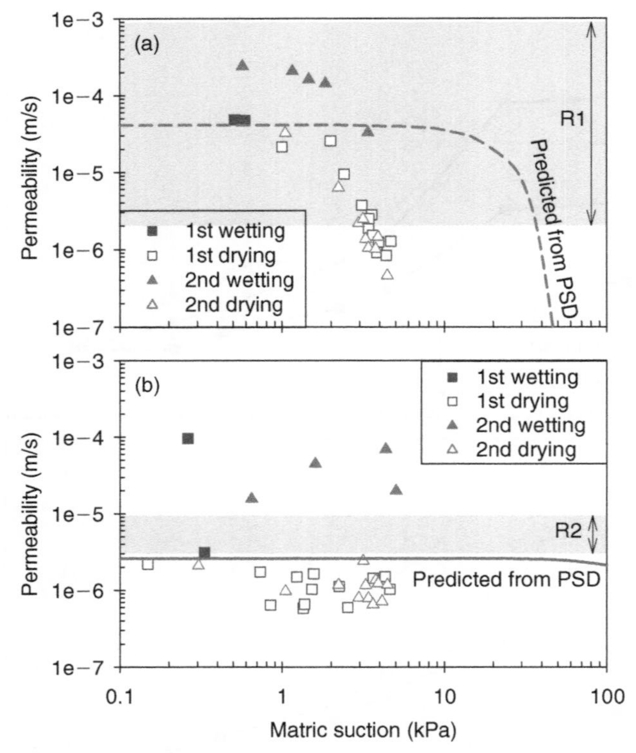

Figure 9.15 Permeability-matric suction relationships obtained from field and predicted from particle size distribution using Fredlund *et al.* (1994): (a) colluvium; (b) CDT (*After* Ng *et al.*, 2011)

monitoring project undertaken on a saprolitic slope in Tung Chung, the permeability function was measured by using the IP method. The theoretical considerations and calculations of the IP method are reported by Ng *et al.* (2011). To achieve the one-dimensional flow assumption, a 3 m deep and 1.2 m wide trench was excavated at the uphill side of the test plot to install a piece of 0.06 mm thick polythene sheeting, which acted as a cut-off sheet (see Figure 9.12). After the installation of the polythene sheeting, the trench was backfilled. Subsequently, a circular steel test ring (3 m in diameter) was embedded 100 mm into the flattened plot to retain water for the IP test.

In-situ permeability functions

Figure 9.15 shows the in-situ permeability function obtained from field measurements at depths of 0.84 and 1.85 m for the colluvium and CDT, respectively. The shaded area denotes the range of saturated permeability as reported by the GCO (1982).

As shown in Figure 9.15a, the measured unsaturated permeability in colluvium at low suction (<1 kPa) falls within the range of saturated permeability. Although the measured data are fairly scattered, the water permeability along the two wetting paths is generally higher than that of drying paths. For a given matric suction, however, the volumetric water content is higher along the drying path than that along the wetting path and thus a higher permeability may be expected along the drying path. This unusual observation might be the result of some in-situ features like cracks, fissures and relict joints in the soil or due to enhanced water connectivity in the soil structure as a result of soil structural collapse upon wetting. The measured permeability along the second wetting path is higher than that along the first wetting path by nearly one order of magnitude at a matric suction of about 0.5 kPa. This may be due to the errors introduced by the limited resolution of a tensiometer at low matric suction. At a given suction, the range of the measured permeability appears to be significant (two orders of magnitude) during the two wetting-drying cycles. This may be attributed to the heterogeneity of colluvium resulting from the presence of cracks, fissures and rootlets in the field (e.g. Basile *et al.*, 2006).

In the CDT (see Figure 9.15b), the measured data are quite scattered, mainly due to the wide spacing of the installed instruments. Nevertheless, the measured water permeability along the wetting paths is higher than that from the drying paths at a given suction. For the range of suctions measured, the permeability varies from 3×10^{-6} to 1×10^{-4} m/s along the wetting paths and from 4×10^{-7} to 3×10^{-6} m/s along the drying paths. Similar to those measured in the colluvium, the measurements from the first and second drying paths are generally consistent with each other. At a given suction, the measured permeability in colluvium is generally higher than that in CDT along all the wetting-drying paths, indicating the less permeable nature of CDT. More details about the measured permeability functions of colluvium and CDT are given by Ng *et al.* (2011).

Comparison of permeability functions obtained from field measurement and predicted from particle size distribution (PSD)

In the literature, different researchers have suggested various ways to predict permeability function from SWCCs. These methods often use a drying SWCC and saturated soil permeability at zero stress. In this section, the permeability functions obtained from field measurements are compared with those predicted from PSD. The drying SWCCs for colluvium and CDT are first predicted based on the corresponding PSDs using the method proposed by Fredlund *et al.* (1997). According to GCO (1982), the range of saturated permeability in colluvium and CDT within 5 m of ground surface is 2×10^{-6} to 9×10^{-4} m/s and 3×10^{-6} to 9×10^{-6} m/s, respectively. These reported values were obtained from constant head tests in boreholes and double-ring infiltration tests in the Midlevels area of Hong Kong Island. Using the method proposed by Fredlund *et al.* (1994), the permeability function of both colluvium and CDT may be predicted. The predicted permeability functions for both the colluvium and CDT are shown in Figure 9.15 with the field measurements by the IP method for comparisons.

Generally speaking, the predictions of permeability function using PSDs are overestimated in the drying cycles and underestimated in the wetting cycles for both the colluvium and CDT. It is obvious that the predictions are neither qualitatively nor quantitatively in agreement with the measured results, except that the predicted permeability of CDT is quantitatively in agreement with the measured drying permeability within one order of magnitude. Predicting a permeability function directly from a PSD of the two soils seems to be unsatisfactory, at least over the studied suction range. In-situ features like relict joints, fissures, cracks, rootlets and interpores may possibly be the reasons for the erroneous predictions (Ng *et al.*, 2011).

9.6 SMALL STRAIN SHEAR STIFFNESS

The deformation characteristics of many soil structures such as retaining walls, foundations and tunnels at working load are governed by the soil stiffness at small shear strains ranging from 0.01% to 1% (Mair, 1993; Ng and Lings, 1995; Atkinson, 2000). The shear modulus at very small strains (0.001% or less, Atkinson and Sällfors, 1991), G_0, is important for predicting ground deformations of soil structures subjected to static and dynamic loadings. G_0 may be determined from in-situ seismic tests including up-hole, down-hole such as suspension P-S (compression and shear wave) velocity logging, and cross-hole techniques, and laboratory tests using resonant column and bender elements. It is well known that the shear modulus decays nonlinearly and significantly with shear strain at small strain ranges (0.001% to 1%, Atkinson and Sällfors, 1991). The stiffness degradation curve may be obtained from self-boring pressuremeter tests in the field and triaxial tests with local strain measurements in the laboratory. In this section, very small strain shear modulus, stiffness anisotropy and stiffness degradation curves of typical saprolites in Hong Kong are presented and discussed.

Field measurements of very small strain shear modulus

Field measurements of shear wave velocity of decomposed granite along the depth were taken at the Kowloon Bay site (Ng *et al.*, 2000) using suspension P-S (compression and shear wave) velocity logging, and at the Yen Chow Street site using cross-hole seismic measurements (Ng and Wang, 2001). In suspension P-S velocity logging, P-wave and S-wave velocities are measured at 1 m intervals in a single uncased borehole using a down-hole probe containing a source and two receivers. By measuring the wave travel time between the two receivers, which are separated by 1 m, the average wave velocity in the region between the receivers can be determined. The induced strain is believed to be in the order of 0.001% or smaller, which is the strain level corresponding to very small strain stiffness.

For a shear wave propagating vertically with horizontal polarization, the velocity of the shear wave in the vertical plane ($v_{s(vh)}$) is measured. In cross-hole seismic velocity measurements, a hammer provides a source of horizontally propagating, vertically polarised shear waves. The waves are detected by two borehole-picks and three-component geophones located at the same horizontal level as the source in two adjacent in-line boreholes. The velocity of the horizontally propagating shear wave with vertical polarization ($v_{s(hv)}$) is calculated from the measured difference in travel times of the shear wave from the source to the two geophones and the measured path length of wave propagation along the two boreholes with geophones. With the measured

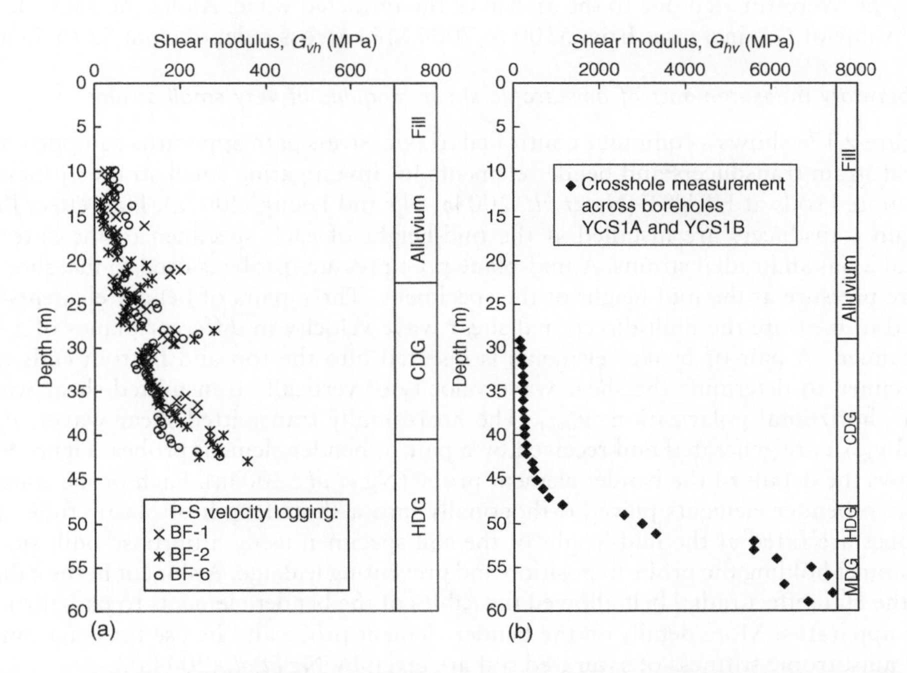

Figure 9.16 Shear modulus profiles at: (a) Kowloon Bay (*After* Ng *et al.*, 2000); (b) Yen Chow Street (*After* Ng and Wang, 2001)

shear wave velocity, the shear modulus at very small strains may be determined by the following equation:

$$G_{ij} = G_{0(ij)} = \rho v_{s(ij)}^2 \tag{9.1}$$

where i is the wave propagation direction and j is the particle motion direction, ρ is the bulk density of the soil.

In-situ measurement of very small strain shear modulus of decomposed granites

Figures 9.16a and b show the measured shear modulus profiles at Kowloon Bay (Ng *et al.*, 2000) and Yen Chow Street (Ng and Wang, 2001), respectively. The very small strain shear modulus was calculated from the measured shear wave velocity and the bulk density of the soil using Equation (9.1). The bulk density of the CDG at the two sites is about 2000 kg/m^3. The bulk densities of HDG and MDG at Yen Chow Street are 2400 and 2550 kg/m^3, respectively. As shown in the figure, the shear modulus increases gradually with depth across the CDG layer at both sites. At Kowloon Bay, the value of G_{vh} increases from about 45 MPa at a depth of 23 m to about 200 MPa at a depth of 39 m. At Yen Chow Street, the value of G_{hv} of the CDG layer increases from 50 to 280 MPa across a depth from 29 to 42 m. The values of G_{hv} at depths above the strong underlying stratum (CDG/HDG boundary, HDG/MDG boundary) may be overestimated due to the arrival of the refracted wave. Along the MDG layer, the value of G_{hv} increases from 5500 to 7000 MPa across a depth from 52 to 58 m.

Laboratory measurements of anisotropic shear modulus at very small strains

Figure 9.17a shows a computer-controlled triaxial stress path apparatus equipped with local strain transducers and bender elements for investigating small strain stiffness of saturated soils at HKUST (Ng *et al.*, 2004a, Ng and Leung, 2007a). Hall-effect local strain transducers are installed at the mid-height of each specimen to measure the local axial and radial strains. A mid-plane pore pressure probe is used to measure the pore pressure at the mid-height of the specimens. Three pairs of bender elements are used to measure the multidirectional shear wave velocity in different planes of a soil specimen. A pair of bender elements is inserted into the top and bottom ends of a specimen to determine the shear wave velocity of vertically transmitted shear waves with horizontal polarization, $v_{s(vh)}$. The horizontally transmitted shear waves, $v_{s(hv)}$ and $v_{s(hh)}$, are generated and received by a pair of bender element probes. Figure 9.18 shows the details of the bender element probe (Ng *et al.*, 2004a). Each probe consists of two bender elements placed orthogonally into a short length of plastic tube. The probes are fitted at the mid-height of the soil specimen using a purpose-built silicon grommet holding the probe in position and preventing leakage. A slot cut in the caliper of the Hall-effect radial belt allowed the cables of the bender elements to pass through the apparatus. More details on the bender element probe and its use for determining the anisotropic stiffness of saturated soil are given by Ng *et al.* (2004a).

Ng and Yung (2008) modified the triaxial apparatus for testing unsaturated soils using the axis-translation technique (Hilf, 1956). Figure 9.17b shows a schematic diagram of the modified triaxial apparatus. The air pressure is controlled through a coarse corundum disc placed on top of the soil specimen, whereas water pressure is controlled through a saturated high AEV (300 kPa) ceramic disc sealed to the pedestal of the apparatus. A base pedestal with spiral-shaped drainage groove is designed to facilitate the removal of any diffused air through the high AEV ceramic disc during long periods of unsaturated soil testing. A 16 mm-diameter recess in the center of the base pedestal is designed to house a bender element. A pre-drilled, high AEV ceramic disc is placed in the pedestal and sealed on the inner and outer circumferences of the disc with epoxy resin. Other details of the modified triaxial testing system are given by Ng and Yung (2008).

Anisotropic shear moduli of saturated CDT and CDG intact specimens at very small strains

Figure 9.19 shows the measured shear moduli ($G_{0(hh)}$, $G_{0(hv)}$ and $G_{0(vh)}$) of the saturated block and Mazier specimens of CDT under isotropic and anisotropic stress conditions (Ng and Leung, 2007a, b). The soil specimens were obtained from Fan Ling in Hong Kong. It can be seen from the figure that $G_{0(hh)}$, $G_{0(hv)}$ and $G_{0(vh)}$ all increase with the product of σ_i' and σ_j' as expected, where σ_i' and σ_j' are the effective stress in the i and j direction, respectively. Since the measured $G_{0(hh)}$ are higher than both $G_{0(hv)}$ and $G_{0(vh)}$ under the isotropic stress conditions, it is evident that the CDT is inherently an anisotropic decomposed material. This stiffness anisotropy is probably due to the horizontal layering structure of CDT resulting from various geologic processes.

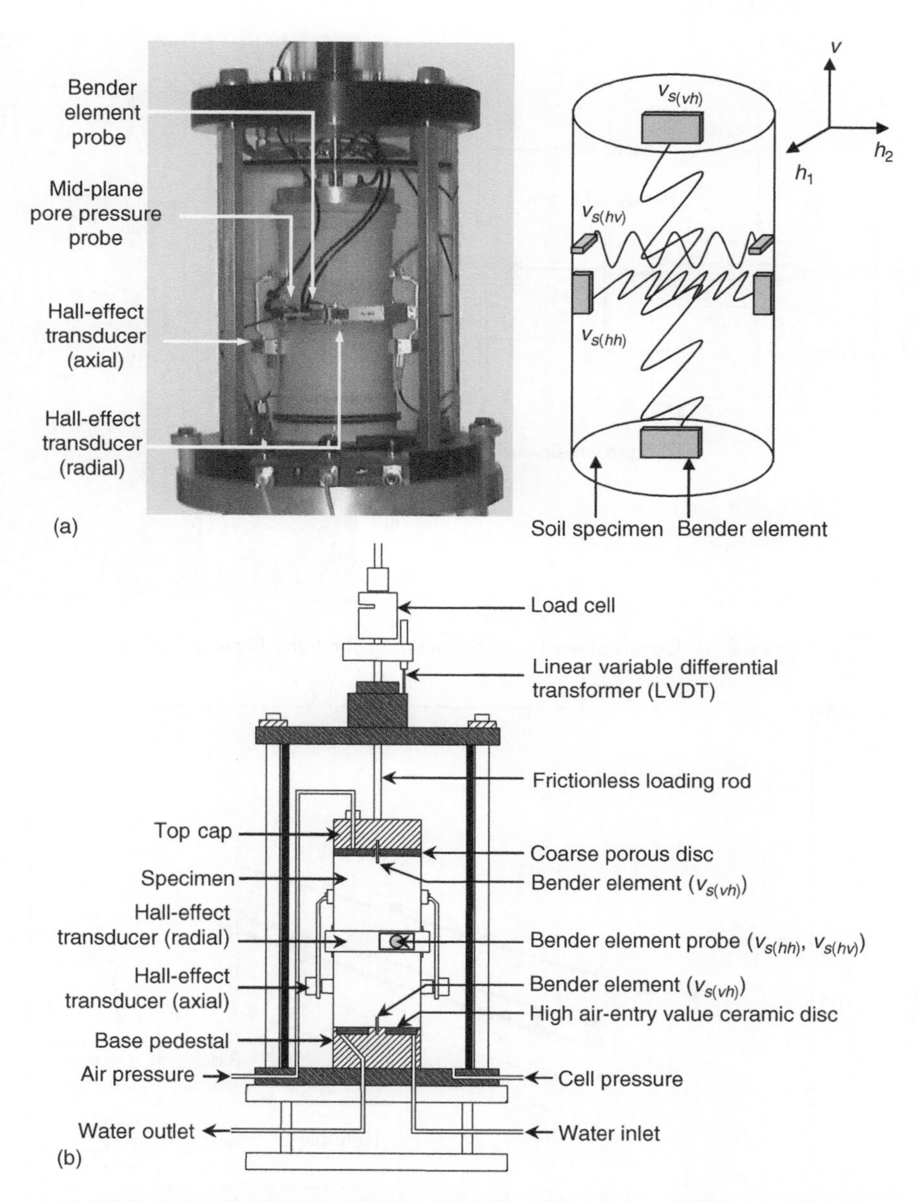

Figure 9.17 Triaxial apparatus with multidirectional shear wave velocity and local strain measurements for testing: (a) saturated soils (*After* Ng *et al.*, 2004a); (b) unsaturated soils (*After* Ng and Yung, 2008)

Besides, $G_{0(hh)}$, $G_{0(hv)}$ and $G_{0(vh)}$ of the block specimen are consistently higher than those of the Mazier specimen. This may be attributed to different degrees of sample disturbance. Sample disturbance in the Mazier specimen is considered larger than that in the block specimen (Ng and Leung, 2007a).

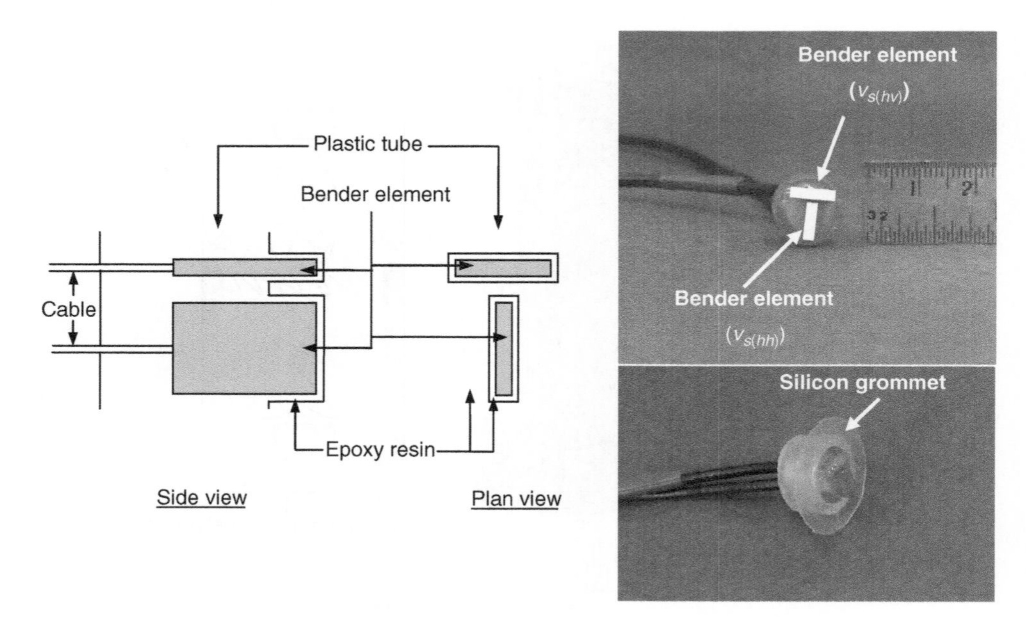

Figure 9.18 Details of the bender element probe (*After* Ng *et al.*, 2004a)

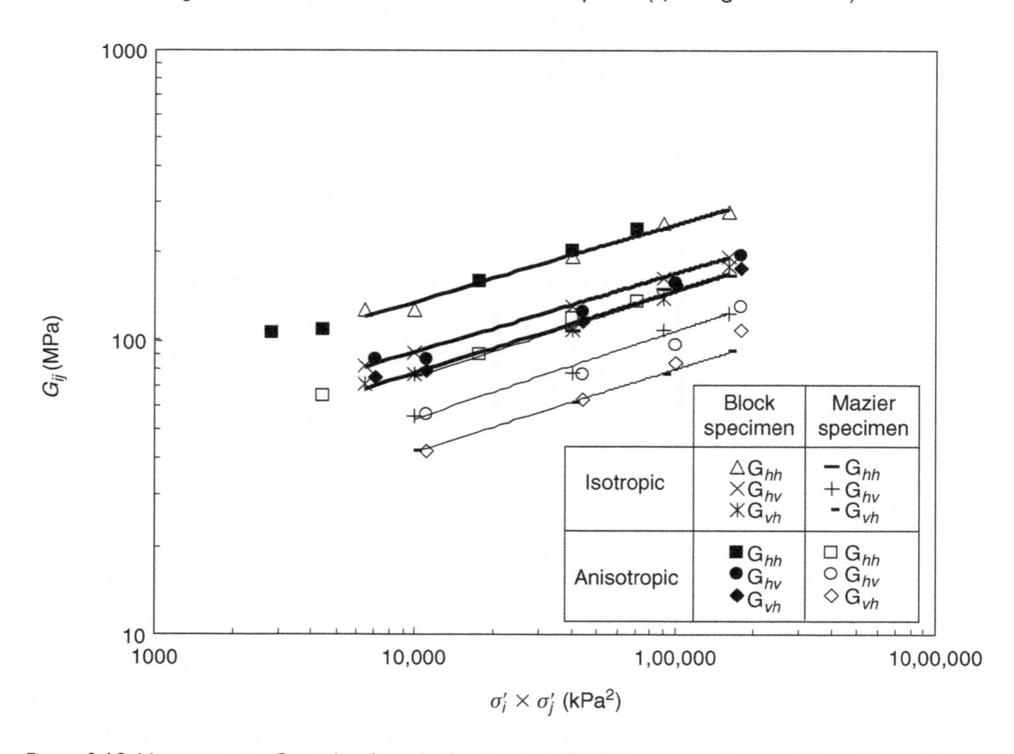

Figure 9.19 Variations in G_{ij} with $\sigma_i' \times \sigma_j'$ of saturated CDT intact specimens (*After* Ng and Leung, 2007b)

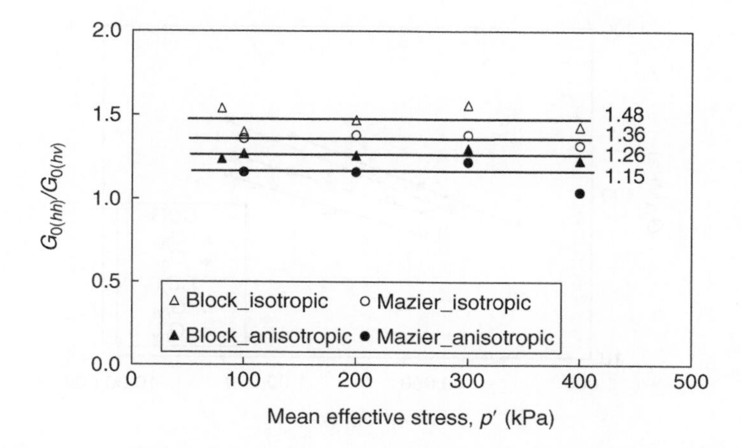

Figure 9.20 Variations in degree of stiffness anisotropy ($G_{0(hh)}/G_{0(hv)}$) with mean effective stress in saturated natural CDT specimens (*After* Ng and Leung, 2007a)

The degree of stiffness anisotropy may be expressed in terms of the ratio of the shear modulus in the horizontal plane ($G_{0(hh)}$) to that in the vertical plane ($G_{0(hv)}$) (Ng *et al.*, 2004a). To eliminate the differences in frequency and boundary effects, $G_{0(hh)}$ and $G_{0(hv)}$ are chosen for comparison since they are generated by the same bender element probe and they have the same travelling distance and boundary conditions. Figure 9.20 shows the variation in the degree of stiffness anisotropy with mean effective stress (p'). It can be seen that the average values of $G_{0(hh)}/G_{0(hv)}$ of the block and Mazier specimens under isotropic stress state are 1.48 and 1.36, respectively. Based on the theoretical derivations by Ng and Leung (2007a), this is the inherent stiffness anisotropy mainly attributed to the soil fabric. At any given p', the Mazier specimen shows a lower degree of stiffness anisotropy than those of the block specimen. This may also be due to sample disturbance (Ng and Leung, 2007a). The natural anisotropic structure of CDT may be damaged by the Mazier sampling process, resulting in the lower degree of anisotropy and smaller shear moduli in the Mazier specimen. The average values of $G_{0(hh)}/G_{0(hv)}$ for block and Mazier specimens under anisotropic stress state are 1.26 and 1.15, respectively. There is about 15% reduction in the degree of stiffness anisotropy ($G_{0(hh)}/G_{0(hv)}$) compared to the results under the isotropic stress state. The reduction in the observed degree of stiffness anisotropy was due to the stress-reduced anisotropy when the specimens were subjected to the anisotropic stress path. Other details and test results of the anisotropic stiffness of the saturated CDT are reported by Ng and Lueng (2007a).

Figure 9.21 compares measured G_{ij} of Mazier specimens of CDG (from Yen Chow Street) and CDT (from Fan Ling). The shear moduli shown in the figure are determined from Mazier specimens using bender elements in a triaxial apparatus. The CDT specimen clearly shows stiffness anisotropy, with the shear modulus in the horizontal plane higher than that in the vertical plane. The stiffness anisotropy is probably attributed to the horizontal layering structure of CDT as a result of metamorphism in the parent rock as discussed above. The stiffness anisotropy of the CDG Mazier specimen is

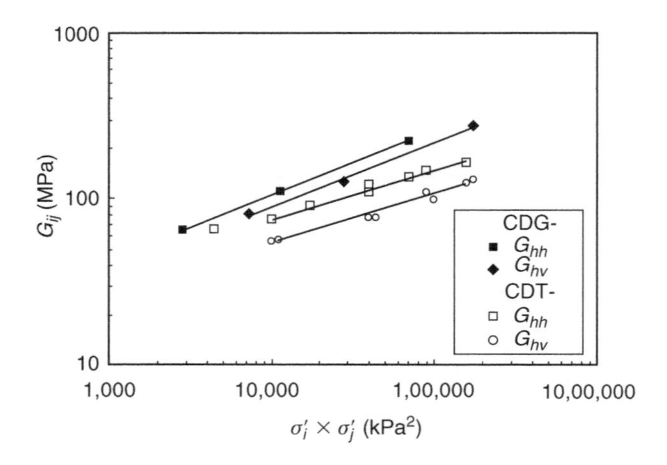

Figure 9.21 Anisotropic stiffness of CDG and CDT Mazier specimens (*After* Ng and Leung, 2007b)

obviously smaller than that of CDT as revealed in the limited set of laboratory test data. The disturbance during Mazier sampling may damage the anisotropic structure of CDG if such a soil structure exists (Ng and Leung, 2007b).

Anisotropic shear moduli of unsaturated recompacted CDT at very small strains

The anisotropic very small strain shear moduli of unsaturated recompacted CDT (from Fan Ling) specimens have been investigated theoretically and experimentally by Ng and Yung (2008) and Ng *et al.* (2009). The effects of net mean stress, matric suction, drying–wetting cycle and stress ratio on the very small strain shear stiffness and stiffness anisotropy are measured using the modified triaxial testing system shown in Figure 9.17.

Figure 9.22 shows the variations of the measured shear moduli ($G_{0(vh)}$, $G_{0(hh)}$ and $G_{0(hv)}$) with net mean stress of recompacted CDT specimens under different suction conditions (Ng and Yung, 2008). It can be seen from the figure that the measured $G_{0(vh)}$, $G_{0(hh)}$ and $G_{0(hv)}$ follow a similar trend. At any constant suction, $G_{0(vh)}$, $G_{0(hh)}$ and $G_{0(hv)}$ all increase with an increase in net mean stress but at a progressively reduced rate. The rates of increase in shear moduli with net mean stress are higher when matric suction is lower but the differences become less pronounced at high suctions. For instances, the measured $G_{0(vh)}$, $G_{0(hh)}$ and $G_{0(hv)}$ at zero suction (saturated state) increase by 143.5, 143.6 and 140.2%, respectively, as the net mean stress increases from 110 to 400 kPa. While at suction of 100 kPa, the measured $G_{0(vh)}$, $G_{0(hh)}$ and $G_{0(hv)}$ increase by 96.4, 94.1 and 89.3%, respectively, and at suction of 200 kPa, the measured $G_{0(vh)}$, $G_{0(hh)}$ and $G_{0(hv)}$ only increase by 91.2, 90.3 and 85.5%, respectively, for the same net mean stress range. The observed increase in shear moduli with suction is probably due to the meniscus water effect. The meniscus water causes an increase in the normal forces which "hold" soil particles together and hence lead to higher shear moduli. However, this beneficial holding effect could not increase infinitely, probably

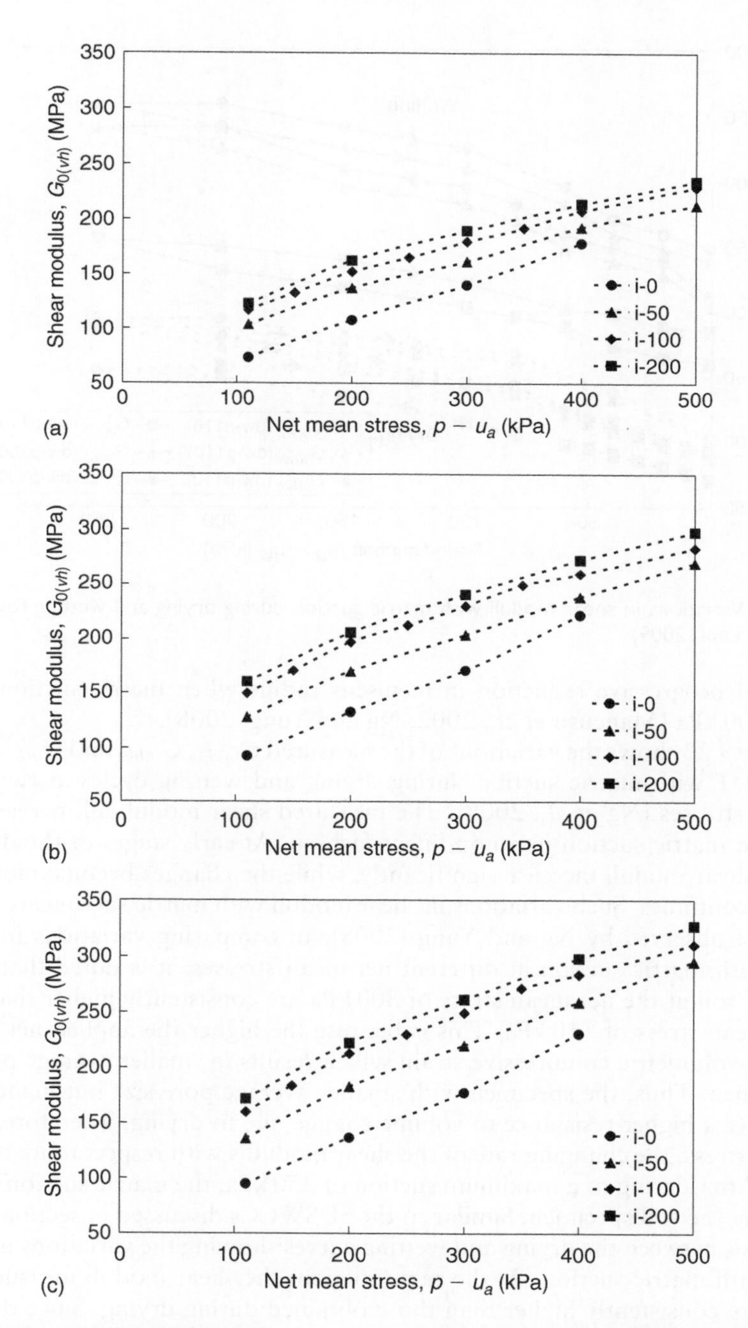

Figure 9.22 Variations of (a) $G_{0(vh)}$, (b) $G_{0(hv)}$, (c) $G_{0(hh)}$ with net mean stress of unsaturated CDT (*After* Ng and Yung, 2008)

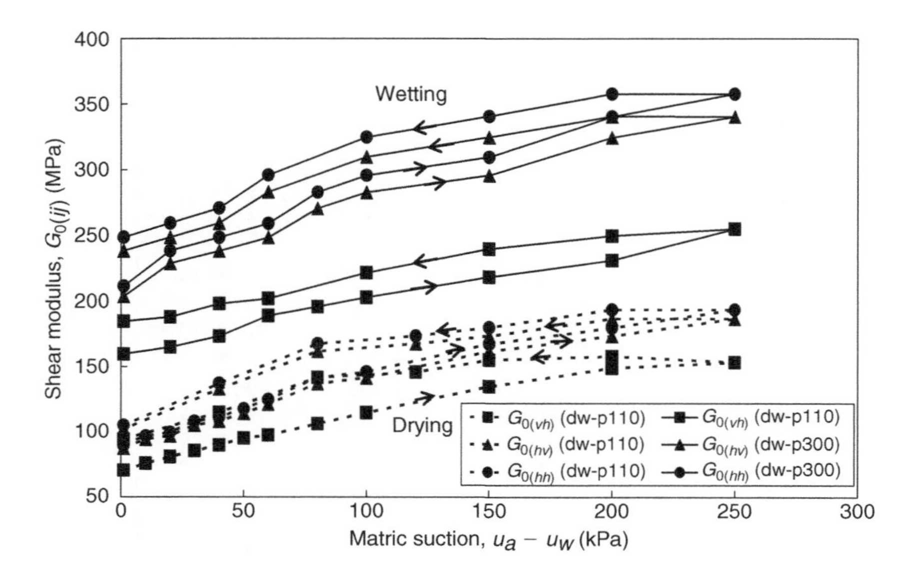

Figure 9.23 Variations in shear moduli with matric suction during drying and wetting tests (*After* Ng *et al.*, 2009)

because of progressive reduction in meniscus radius when matric suction increases beyond 100 kPa (Mancuso *et al.*, 2002; Ng and Yung, 2008).

Figure 9.23 shows the variations of the measured $G_{0(vh)}$, $G_{0(hh)}$ and $G_{0(hv)}$ of recompacted CDT with matric suction during drying and wetting cycles at two different net mean stresses (Ng *et al.*, 2009). The measured shear moduli all increase with an increase in matric suction in a non-linear fashion. At early stages of the drying process, the shear moduli increase significantly, while the changes become more gradual as drying continues. Such variations in shear moduli with matric suction are consistent with those observed by Ng and Yung (2008). In comparing variations in the shear moduli with matric suction at different net mean stresses, it is noted that the shear moduli of soil at the net mean stress of 300 kPa are consistently higher than those at the net mean stress of 110 kPa. This is because the higher the applied net stress, the larger the volumetric compressive strain which results in smaller average pore size of the specimen. Thus, the specimen with smaller average pore size but higher stiffness would have a higher resistance to volume change due to drying. Therefore, at higher net mean stress, the changing rate of the shear modulus with respect to matric suction is lower. After drying to a maximum suction of 250 kPa, the matric suction is reduced for wetting the soil specimen. Similar to the SDSWCCs discussed in section 9.4, there is hysteresis between the drying and wetting curves showing the variations in the shear moduli with matric suction. At the same suction, the shear moduli measured during wetting are consistently higher than those obtained during drying. Since the induced axial and radial strains of each soil specimen are relatively small (less than $\pm 0.3\%$), the total volume change is not significant. The shear modulus hysteresis may well be due to the fact that the water content on the adsorption curve is higher than that on the desorption curve at the same suction (Ng *et al.*, 2009).

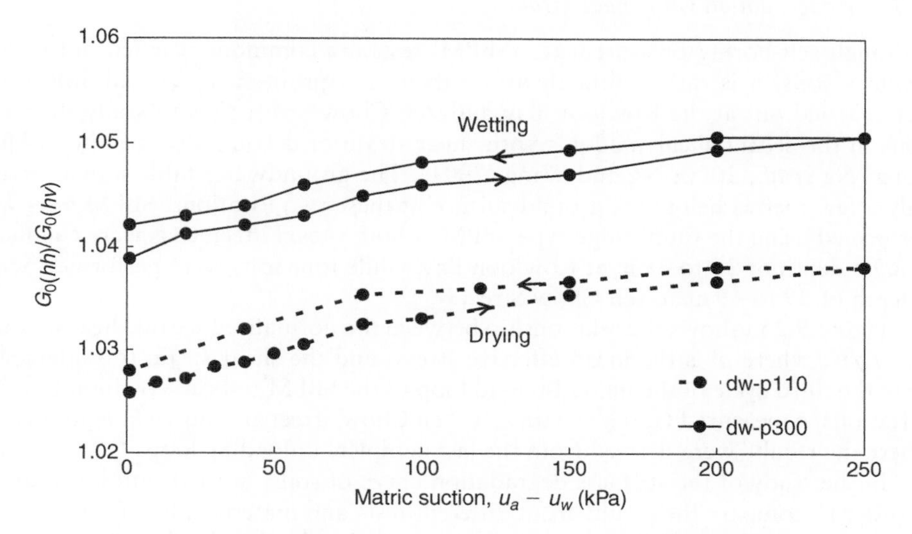

Figure 9.24 Variations in degree of stiffness anisotropy with matric suction during drying and wetting tests (*After* Ng *et al*., 2009)

Figure 9.24 shows the variations in the degree of stiffness anisotropy, $G_{0(hh)}/G_{0(hv)}$, with matric suction during the drying and wetting tests (Ng *et al.*, 2009). At different net mean stresses, the variations in the degree of stiffness anisotropy with matric suction follow a similar trend. Initially at zero suction, $G_{0(hh)}$ is about 3 and 4% higher than $G_{0(hv)}$ at net mean stresses of 110 and 300 kPa, respectively. This observed stiffness anisotropy is probably attributable to the inherent anisotropy induced in the tested material during specimen preparation (Ng and Yung, 2008). However, this observed stiffness anisotropy at zero suction may not be caused by the inherent anisotropy induced during specimen preparation only. This is because the specimens have undergone isotropic compression and wetting from the initial suction to zero suction. Therefore, $G_{0(hh)}/G_{0(hv)}$ of the specimen at the net mean stress of 300 kPa is higher than that at the net mean stress of 110 kPa. When matric suction increases, $G_{0(hh)}/G_{0(hv)}$ increases at a gradually reduced rate, but the increase is smaller than 1%. Though the magnitudes of the increase in $G_{0(hh)}/G_{0(hv)}$ with matric suction are very small, there is a clear trend illustrating that the degree of stiffness anisotropy increases with matric suction. Similar to the SDSWCCs, there is also hysteresis between the drying and wetting curves. Although the size of the hysteresis loop is very small, the trend is clear. The changing rates of $G_{0(hh)}/G_{0(hv)}$ and the sizes of the hysteresis loops at two different net mean stresses are almost the same. However, $G_{0(hh)}/G_{0(hv)}$ of the specimen at the net mean stress of 300 kPa is consistently higher than that at the net mean stress of 110 kPa. The degree of stiffness anisotropy appears not to be independent of the net stresses under isotropic stress conditions. More details of the effects of drying–wetting and stress ratio on the very small strain shear modulus of recompacted CDT are given by Ng *et al.* (2009).

Stiffness degradation with shear strain

Although self-boring pressuremeter (SBPM) tests are commonly carried out in sedimentary soils, it is rather difficult to use them in saprolites. Some trial SBPM tests were carried out at the Kowloon Bay and Yen Chow Street sites to study the variations in the secant shear modulus with shear strain of decomposed granite in Hong Kong (Ng *et al.*, 2000; Ng and Wang, 2001). The groundwater table was located at only a few metres below the ground surface at these two locations. SBPM tests were performed using the Cambridge-type SBPM at both sites. Fifteen tests were conducted along a depth of 30 to 39 m at Kowloon Bay, while four tests were performed across a depth of 39 to 49 m at Yen Chow Street.

Figure 9.25a shows the relationship between the normalised secant shear modulus ($G_{sec}/\sqrt{p'}$, where p' is the mean effective stress) and the shear strain (ε_s) interpreted from the third cycle of the unload-reload loop of the SBPM tests at Kowloon Bay. The variations in measured $G_{sec}/\sqrt{p'}$ with ε_s at Yen Chow Street are shown in Figure 9.25b, where the results were derived from the last complete unloading loop.

In the study of the stiffness degradation curve of soils, normalization is required in order to compare the results from different tests and materials. It is found that the secant shear modulus derived from the last complete unloading loop is consistent with that derived from each unload-reload loop (Ng and Wang, 2001), thus it is believed that the results shown in Figures 9.25a and b are comparable. The range of the shear modulus at very small strains determined from seismic measurements for each site is also shown in its corresponding figure. At the two sites, the relationships between $G_{sec}/\sqrt{p'}$ and ε_s for decomposed granite are generally consistent. The stiffness-strain relationship is highly non-linear. The value of $G_{sec}/\sqrt{p'}$ decreases significantly from about 300 to about 50 as ε_s increases from 0.02 to 1%. Besides, at a given shear strain, the stiffness at deeper depth is greater. This is consistent with the observations with the in-situ seismic test results as shown in Figure 9.16.

Laboratory measurements of stiffness degradation curve

Wang and Ng (2005) studied the influence of stress paths on the small strain stiffness of Mazier CDG (from Yen Chow Street) specimens from a series of constant p' triaxial compression and extension tests. The tests were performed in a triaxial apparatus equipped with Hall-effect local strain transducers as shown in Figure 9.17a. To account for the effects of the stress state and void ratio, the shear modulus determined in the triaxial stress path tests presented in this chapter is normalised by the shear modulus at very small strains, G_{vh}, obtained from the shear wave velocity measurements.

Figure 9.26a shows the variations in the normalised secant shear modulus (G_{sec}/G_{vh}) with deviatoric strain (ε_q) during constant p' compression tests at different mean effective stresses ranging from 100 to 300 kPa. The value of G_{vh} at each mean effective stress level (p') was determined from the shear wave velocity measurement using bender elements embedded in the end platens of the triaxial apparatus. As expected, the shear modulus decreases as the deviatoric strain increases. It can be seen that the test results are fairly scattered, probably due to the sample variability of the CDG specimens along the depth from 23 to 54 m. The average shear modulus degradation curve is also shown in the figure, illustrating that the average shear modulus in the compression tests drops to about 30% of the initial value (G_{vh}) at deviatoric

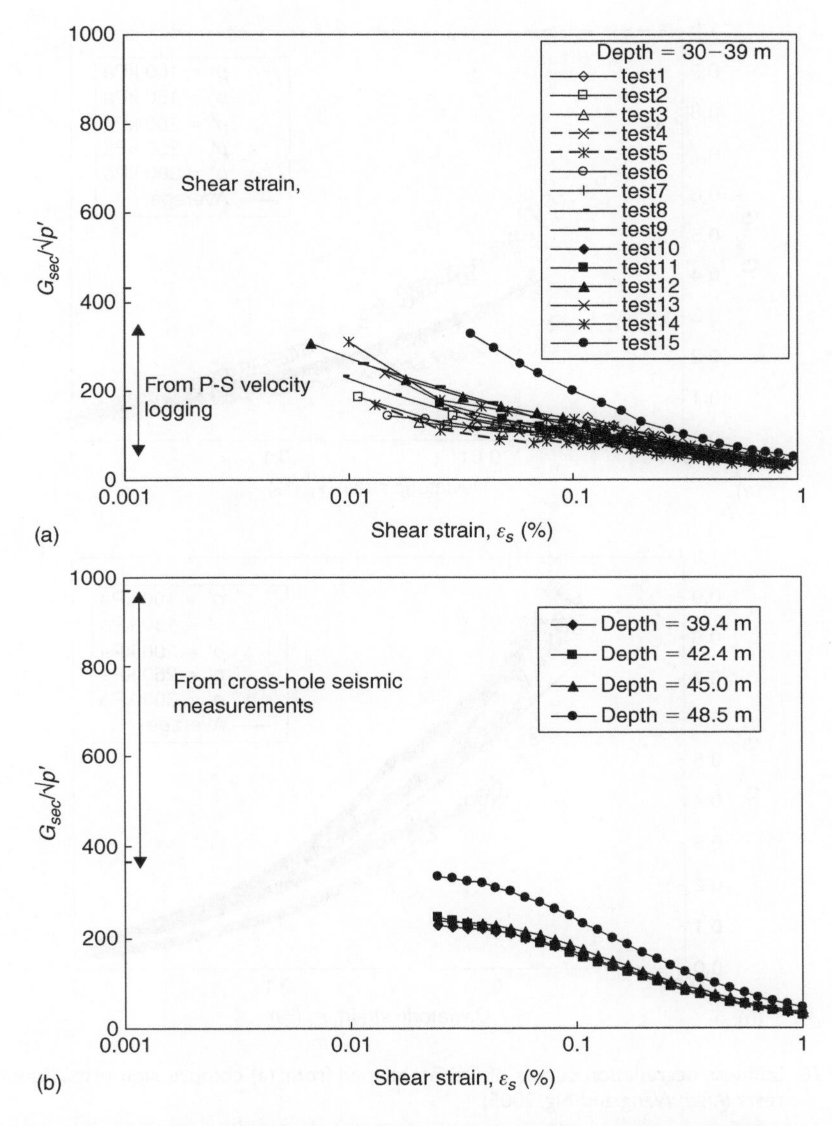

Figure 9.25 Stiffness-strain relationships of decomposed granite from self-boring pressuremeter tests at: (a) Kowloon Bay (*After* Ng *et al.*, 2000); (b) Yen Chow Street (*After* Ng and Wang, 2001)

strain of 0.01% and further reduces to 15% of the initial value (G_{vh}) at deviatoric strain of 0.1%.

A series of constant p' triaxial extension tests were also conducted on Mazier specimens of CDG taken from the same site (Wang and Ng, 2005). Figure 9.26b shows the normalised shear modulus degradation curves of the constant p' extension tests. The value of G_{vh} for normalization at each mean effective stress level (p') was the same

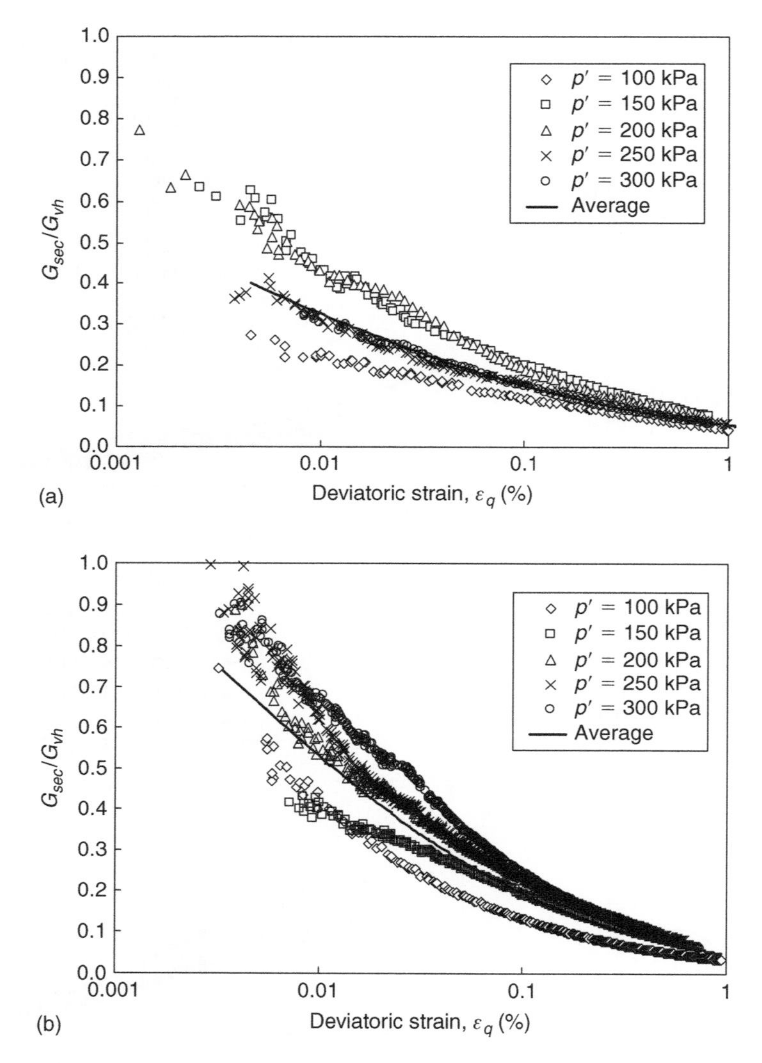

Figure 9.26 Stiffness degradation curves of CDG obtained from: (a) compression tests; (b) extension tests (*After* Wang and Ng, 2005)

as that in the compression tests. An average shear modulus degradation curve is also shown in the figure. It can be seen that the shear modulus of CDG along the extension path reduces to about 50 and 20% of the initial value (G_{vh}) at deviatoric strains of 0.01 and 0.1%, respectively. The material shows a stiffer response along the extension path, where the average G_{sec}/G_{vh} is about 61% higher than that along the compression path at a deviatoric strain of 0.01%. The difference in G_{sec}/G_{vh} between the extension and compression paths diminishes to about 20% at a deviatoric strain of 0.1% and vanishes at a deviatoric strain of 1%. The higher shear modulus along the extension

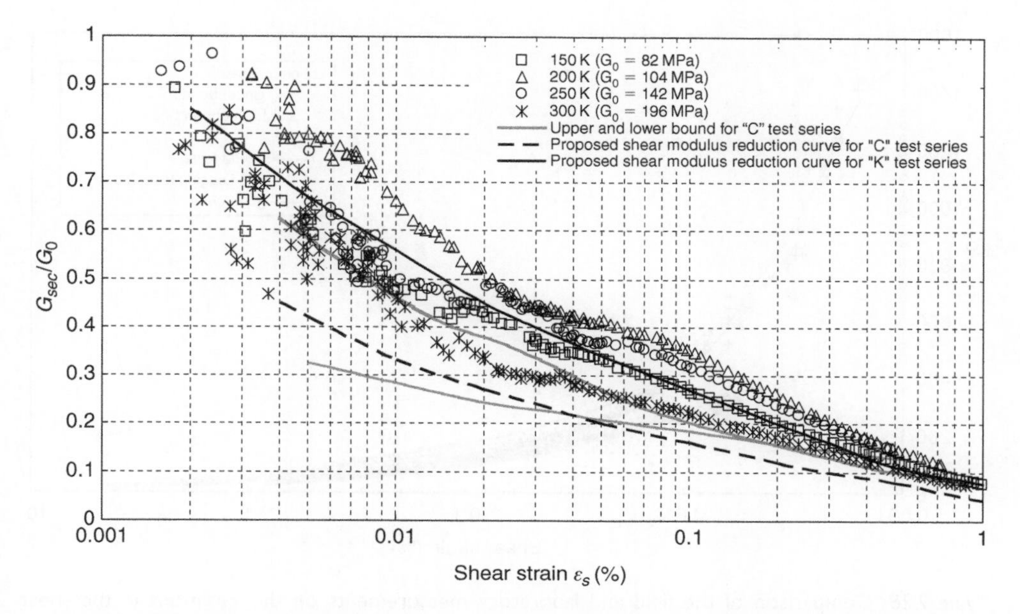

Figure 9.27 Effect of recent stress history on shear modulus degradation curves (*After* Wang and Ng, 2005)

path may be attributed to stress path reversal. During the weathering process, some of the original minerals in the granite were decomposed and leached. This weathering process is analogous to applying compression loads to soil specimens in such a way that the soils are subjected to similar downward deformation. When the intact soil specimens were loaded along the extension path, stress path reversal took place and resulted in a higher stiffness response (Atkinson *et al.*, 1990). On the other hand, the compression tests followed the continuing stress path of the natural weathering process. There was no stress path reversal, which resulted in a lower stiffness response.

Wang and Ng (2005) also investigated the effect of recent stress history on the small strain stiffness of Mazier CDG specimens. The recent stress history is defined as the rotation between the directions of the penultimate and current stress paths (Atkinson *et al.*, 1990). Figure 9.27 shows the variations of G_{sec}/G_0 with ε_s in the semi-logarithmic scale for the "K" test series; together with the lower and upper bound stiffness obtained from the "C" test series for comparison. The only difference between the "K" and "C" test series is that the "K" test series undergoes recent stress history (90° rotation of the stress path).

As expected, highly nonlinear stiffness-strain characteristics are observed in the "K" test series. The average G_{sec}/G_0 decreases from about 0.56 to 0.08 when ε_s increases from 0.01% to 1%. Although the data are quite scattered, the average shear modulus reduction curve of the "K" test series is significantly higher than that from the "C" test series, most probably due to the influence of the recent stress history. The average shear modulus reduction curve for the "K" test series is approximately 50–70% greater than that of the "C" test series. The difference in the average G_{sec}/G_0

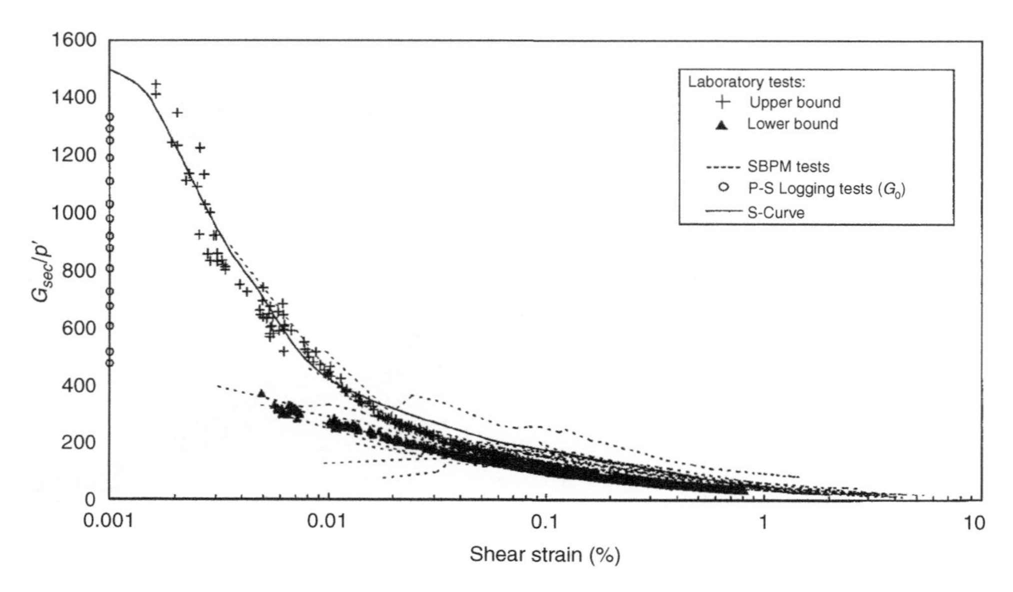

Figure 9.28 Comparison of the field and laboratory measurements on the variations in the shear modulus of CDG at Kowloon Bay with shear strain (*After* Ng *et al.*, 2000)

between the two test series decreases from 70 to 50% as ε_s increases from 0.01 to 1%. More details about the effects of recent stress history and stress paths on small strain stiffness of intact CDG are given by Wang and Ng (2005).

Comparison of the stiffness degradation curves determined in triaxial tests and in-situ SBPM tests

Figure 9.28 shows a comparison between the relationships of normalised secant shear stiffness and shear strain of CDG obtained in the field (Kowloon Bay) using the Cambridge SBPM and the suspension P-S logging method, and in the laboratory using a triaxial apparatus equipped with internal strain measuring devices (Ng *et al.*, 2000).

For plotting the data from the P-S logging method, it is assumed that the shear strain mobilised in the soil is in the order of 0.001% and that the CDG has an initial K_0 of 0.4. Despite some scatters in the measured values, reasonable consistency can be seen between the measurements obtained using the SBPM and the triaxial apparatus for shear strains greater than 0.01%. Because the recent stress history (Wang and Ng, 2005) and stress path for the soil elements around the SBPM and the soil specimens tested in the triaxial apparatus are very different, the consistency between the two sets of test results may be fortuitous. Assuming that the shear moduli obtained from the suspension P-S logging method represent the state of the soil at strains in the order of 0.001% or smaller in the figure, the determined shear moduli are in reasonable agreement with the laboratory measurements. However, significant differences in laboratory and in-situ measurements of shear stiffness were reported by Viana Da Fonseca *et al.* (1997), who compared the shear stiffness values of granitic saprolites

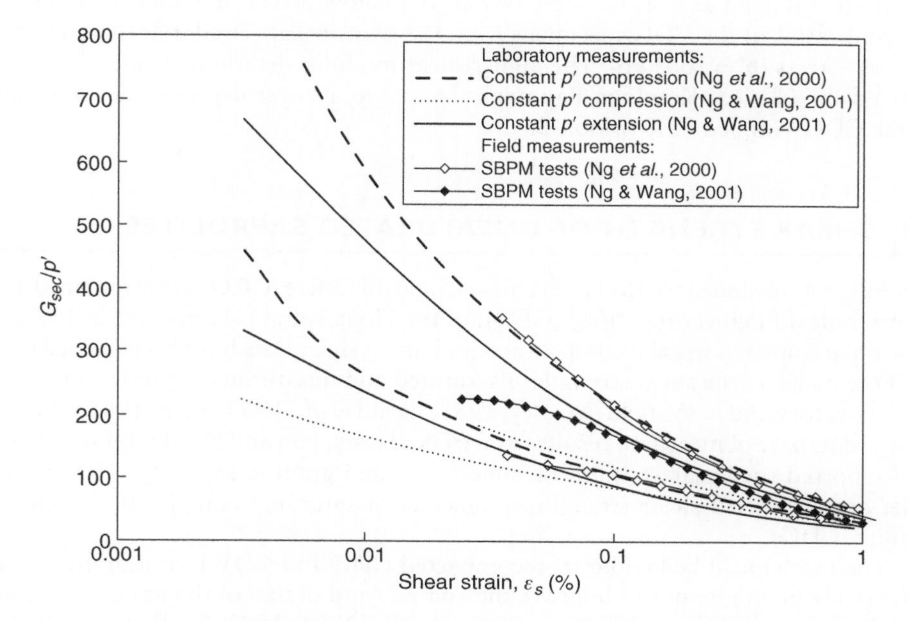

Figure 9.29 Comparison of the field and laboratory measurements on the variations in the shear modulus of CDG with shear strain (Ng and Leung, 2007b).

in Portugal obtained using the cross-hole technique and those obtained in a triaxial apparatus instrumented with local strain measuring devices. They suggested that the sample disturbance was the cause of a threefold higher measured stiffness using the cross-hole technique than in the tests in the laboratory. This does not seem to be the case in Hong Kong.

Figure 9.29 compares the variations in the stiffness degradation curves of CDG determined in triaxial tests and in-situ SBPM tests. The results of the constant p' ($p' = 200$ kPa) triaxial compression tests on the Mazier specimens of CDG at Kowloon Bay (Ng *et al.*, 2000) and Yen Chow Street (Ng and Wang, 2001) are included for comparison. The upper and lower bounds of the laboratory measurements are also shown in the figure.

The CDG at Yen Chow Street shows a higher shear modulus in the field tests than in the laboratory tests. The reasons for the discrepancy include sample disturbance and differences in stress history. Some degree of damage to the bonding and structure of natural CDG is inevitable during sampling, transportation, storage and specimen preparation. The loss of bonding and structure in soil may result in a significant reduction in soil stiffness (Cuccovillo and Coop, 1997). The shear modulus determined in the SBPM tests is derived from the unloading stress-strain curve. Unloading involves stress path reversal and thus results in a higher shear modulus. While the upper bounds of the shear modulus determined in the SBPM tests conducted at Yen Chow Street and Kowloon Bay are similar, the shear modulus determined in the laboratory tests (constant p' triaxial compression) on the CDG at Kowloon Bay is higher than that determined in similar tests on the CDG at Yen Chow Street. It should be noted that the void ratios of the CDG specimens from the two sites are similar at the same mean effective stress ($p' = 200$ kPa). The higher shear modulus determined in the laboratory tests on the CDG at Kowloon Bay might be due to a lower degree of decomposition of the CDG sampled in Kowloon Bay.

9.7 SHEAR STRENGTH OF UNSATURATED SAPROLITES

Since the establishment of the Geotechnical Control Office (GCO) (now renamed as the Geotechnical Engineering Office (GEO)) of the Hong Kong Government in 1977, significant amounts of triaxial compression and direct shear tests have been carried out by GCO to measure the shear strength of saturated and unsaturated granitic saprolites in the laboratory and in the field (Massey, 1983; Brand *et al.*, 1983; Shen, 1985). Based on a large database of measured results from triaxial tests, Pun and Ho (1996) summarised and reported the shear strength of intact saturated granitic saprolites and provided a set of generalised shear strength parameters of saturated completely decomposed granite (CDG).

The mechanical behaviour of recompacted CDG and CDV has been investigated extensively in an attempt to improve the analysis and design of the use of soil nails in initially unsaturated loose fill slopes (Ng and Chiu, 2001, 2003; Ng *et al.*, 2004b). The effects of stress state, stress path and soil suction on the stress-strain relationship, shear strength, and volumetric behaviour of the materials were assessed through a series of computer-controlled triaxial stress path tests. Specific stress path testing was carried out to mimic rainfall infiltration into an initially unsaturated soil slope by reducing soil suction but keeping other stress components constant. All test results were interpreted and modelled under a state-dependent extended critical state framework (Chiu and Ng, 2003). Moreover, Ng and Zhou (2005) investigated the effects of soil suction on dilatancy of a recompacted CDG. Some details and results of these studies are given in this section.

Laboratory apparatus for measuring shear strength of unsaturated soil

To study the shear strength characteristics of unsaturated soils, two types of apparatus, the direct shear box and triaxial apparatus, have been modified and developed at HKUST. Figure 9.30 shows the modified direct shear box developed at HKUST (Zhan, 2003; Ng and Zhou, 2005). The apparatus is modified from the one designed by Gan *et al.* (1988). Suction is applied to a soil specimen based on the axis-translation technique (Hilf, 1956). A high AEV (500 kPa) ceramic disc is installed in the lower portion of the shear box. A water chamber beneath the ceramic disc is designed to serve as a water compartment as well as a channel for flushing diffused air. The matric suction is equal to the applied pore air pressure since the applied pore water pressure is maintained at atmospheric pressure. The modified direct shear apparatus is equipped with five measuring devices. There are two LVDTs for recording horizontal and vertical displacements, a load cell for measuring the shear force, a pressure transducer for monitoring the pore water pressure, and a newly designed auto-logging water volume indicator for monitoring water volume changes in the soil specimen. All these devices

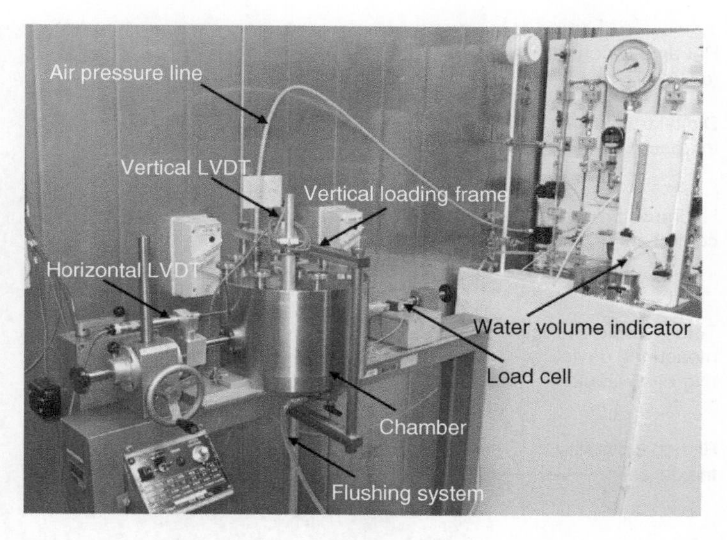

Figure 9.30 A modified direct shear box for testing unsaturated soils (*After* Zhan, 2003; Ng and Chen, 2008)

can be connected to a data logger for automatic data acquisition. Some other details of the modified direct shear box apparatus are given by Zhan (2003).

Figure 9.31 shows the computer-controlled stress-path triaxial system for unsaturated soils developed at HKUST (Zhan, 2003; Ng and Chen, 2008). This system consists of a Bishop & Wesley stress-path triaxial cell, four independent pressure controllers, five transducers, a digital transducer interface (DTI) and a computer. The four independent pressure controllers include two pneumatic controllers for controlling cell pressure and pore air pressure and two hydraulic pressure/volume controllers for controlling axial stress and pore water pressure. The water volume change of each specimen is measured by a hydraulic pressure/volume controller. The five transducers consist of an internal load cell for measurement of axial force, an LVDT (linear variable differential transformer) for measuring external axial displacement and three pressure transducers for monitoring cell pressure, pore air pressure and pore water pressure. All these five transducers are connected to the DTI for automatic data acquisition. The axis-translation technique (Hilf, 1956) is employed to control matric suction. The pore air pressure is applied through a coarse corundum disc placed on top of each soil specimen, while pore water pressure is controlled through a saturated high AEV (500 kPa) ceramic disc sealed to the base pedestal of the triaxial apparatus. The total volume change of each specimen is measured by the double-cell measuring system as shown in Figure 9.7. The details of the double-cell total volume change measuring system are given by Ng *et al.* (2002). To flush and measure the volume of diffused air, a diffused air volume indicator (DAVI) is attached to the system. This system is controlled by a closed-loop feedback scheme, which is capable of performing both stress- and strain-controlled tests in the triaxial stress space. The details of this triaxial system are given by Zhan (2003).

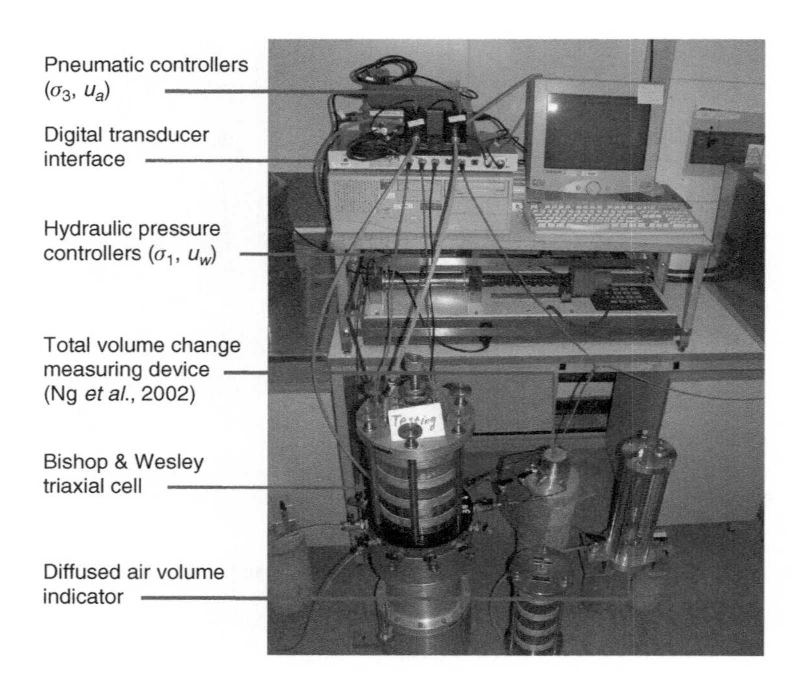

Figure 9.31 A computer-controlled triaxial stress path apparatus for testing unsaturated soils (*After* Ng and Chen, 2008)

Suction effects on the shear strength characteristics of recompacted CDV and CDG specimens

Ng and Chiu (2001, 2003) studied the suction effect on the shear strength of recompacted specimens of CDV (from Victoria Peak) and CDG (from Cha Kwo Ling) at various stress paths in a computer-controlled triaxial apparatus. Figure 9.32 shows the measured shear strength envelope of the recompacted loose CDV and CDG specimens at different suctions ranging from 0 to 150 kPa. As shown in Figure 9.32a, the shear strength envelope for unsaturated CDV specimens is approximately parallel to the one for saturated specimens. The measured angle of internal friction (ϕ') at the critical state of CDV is equal to 32.9°. Similar observations are obtained from CDG specimens as shown in Figure 9.32b. The measured angle of internal friction (ϕ') of CDG at the critical state is equal to 37.7°. All these results suggest that suction does not affect the angle of internal friction at the critical state for both CDV and CDG. Similar observations were obtained from other decomposed soils including a Korean decomposed granite (Lee and Coop, 1995) and a Hong Kong decomposed fine ash tuff and granite (Gan and Fredlund, 1996).

Although suction does not appear to affect the angle of internal friction at the critical state, it may affect the intercept of a shear strength envelope. The intercepts of the shear strength envelopes, $\mu(s)$, for recompacted CDV and CDG, can be determined from Figure 9.33. Figure 9.33 shows the determined $\mu(s)$ at different suctions. The measured results from a decomposed fine ash tuff (Gan and Fredlund, 1996) and

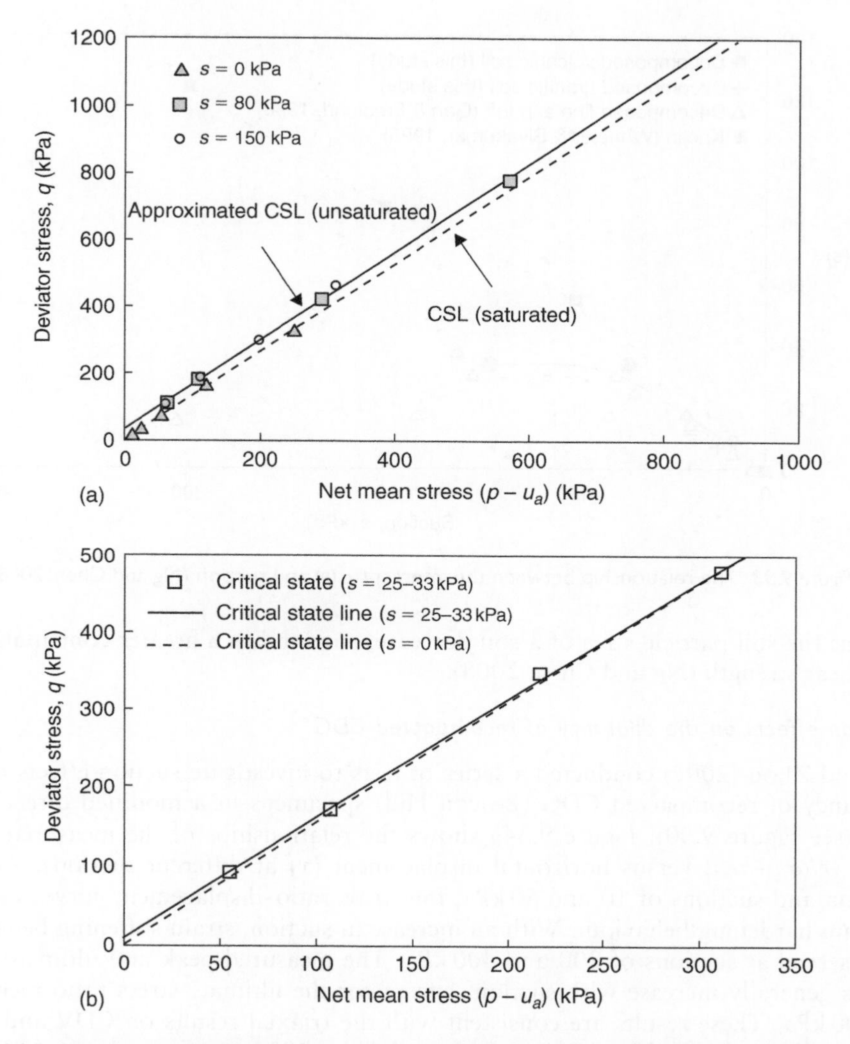

Figure 9.32 The shear strength envelopes of recompacted (a) CDV (*After* Ng and Chiu, 2001); (b) CDG (*After* Ng and Chiu, 2003) specimens at different suctions

a kaolin (Wheeler and Sivakumar, 1995) are also included in the figure for comparisons. The $\mu(s)$ of the kaolin increases with suction almost linearly. However, the CDV and the decomposed fine ash tuff exhibit a non-linear variation in the $\mu(s)$ with suction. When suction is less than about 80 kPa, the $\mu(s)$ of the CDV and the decomposed fine ash tuff increases almost linearly with suction. However, there is relatively little increase in the $\mu(s)$ of CDV and decomposed fine ash tuff when suction exceeds about 80 kPa. On the contrary, there is a reduction in the $\mu(s)$ of the decomposed fine ash tuff as the suction increases beyond around 150 kPa. As shown in the figure, the contribution of suction to the shear strength is the smallest in CDG but the largest in kaolin. This suggests

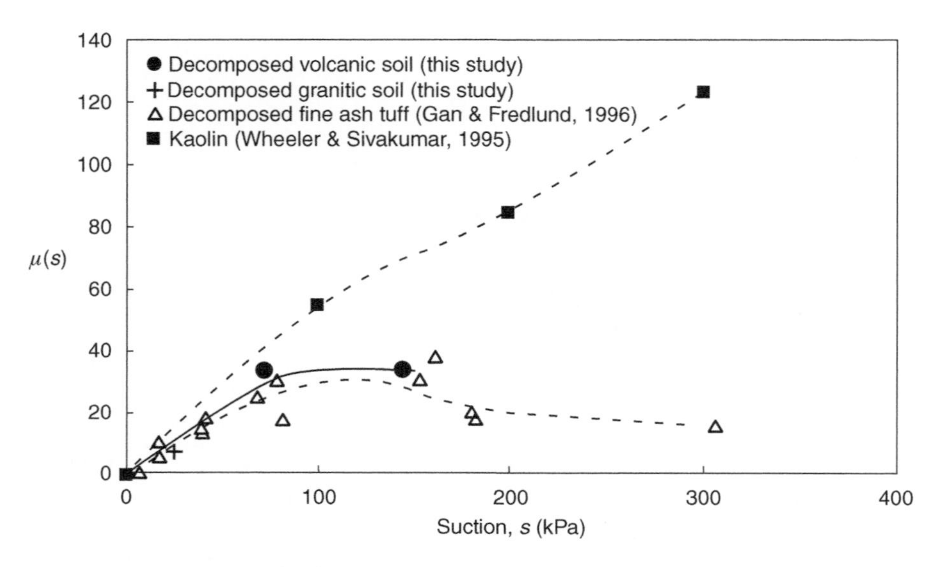

Figure 9.33 The relationship between the intercept $\mu(s)$ and suction (Ng and Chen, 2008)

that as the soil particle sizes of a soil decrease, suction has a greater contribution to the shear strength (Ng and Chen, 2008).

Suction effects on the dilatancy of recompacted CDG

Ng and Zhou (2005) conducted a series of tests to investigate suction effects on the dilatancy of recompacted CDG (Beacon Hill) specimens in a modified direct shear box (see Figure 9.30). Figure 9.34a shows the relationships of the measured stress ratio $(\tau/(\sigma_v - u_a))$ versus horizontal displacement (x) at different suctions. At zero suction and suctions of 10 and 50 kPa, the stress ratio–displacement curves indicate a strain hardening behaviour. With an increase in suction, strain softening behaviour is observed at suctions of 200 and 400 kPa. The measured peak and ultimate stress ratios generally increase with suction, except for the ultimate stress ratio measured at 200 kPa. These results are consistent with the triaxial results on CDV and CDG reported by Ng and Chiu (2001) and Ng and Chiu (2003), respectively.

Figure 9.34b shows the variations of measured dilatancy (δ_y/δ_x) of recompacted CDG versus horizontal displacement at different suctions. The dilatancy is defined as the ratio of incremental vertical displacement (δ_y) to incremental horizontal displacement (δ_x). A negative sign (or negative dilatancy) indicates expansive behaviour. Under saturated conditions, the CDG shows contractive behaviour (i.e. positive dilatancy). On the other hand, under unsaturated conditions, all the specimens display contractive behaviour initially but then change to dilative behaviour as the horizontal displacement continued to increase. The measured maximum negative dilatancy is enhanced by an increase in suction but at a reduced rate. This measured trend is consistent with the test results on a compacted silt reported by Cui and Delage (1996). The increase in maximum negative dilatancy is probably attributable to a closer particle packing (i.e. a smaller void ratio) under a higher suction. At the end of each test, the dilatancy of all specimens approached zero, indicating the attainment of the critical state.

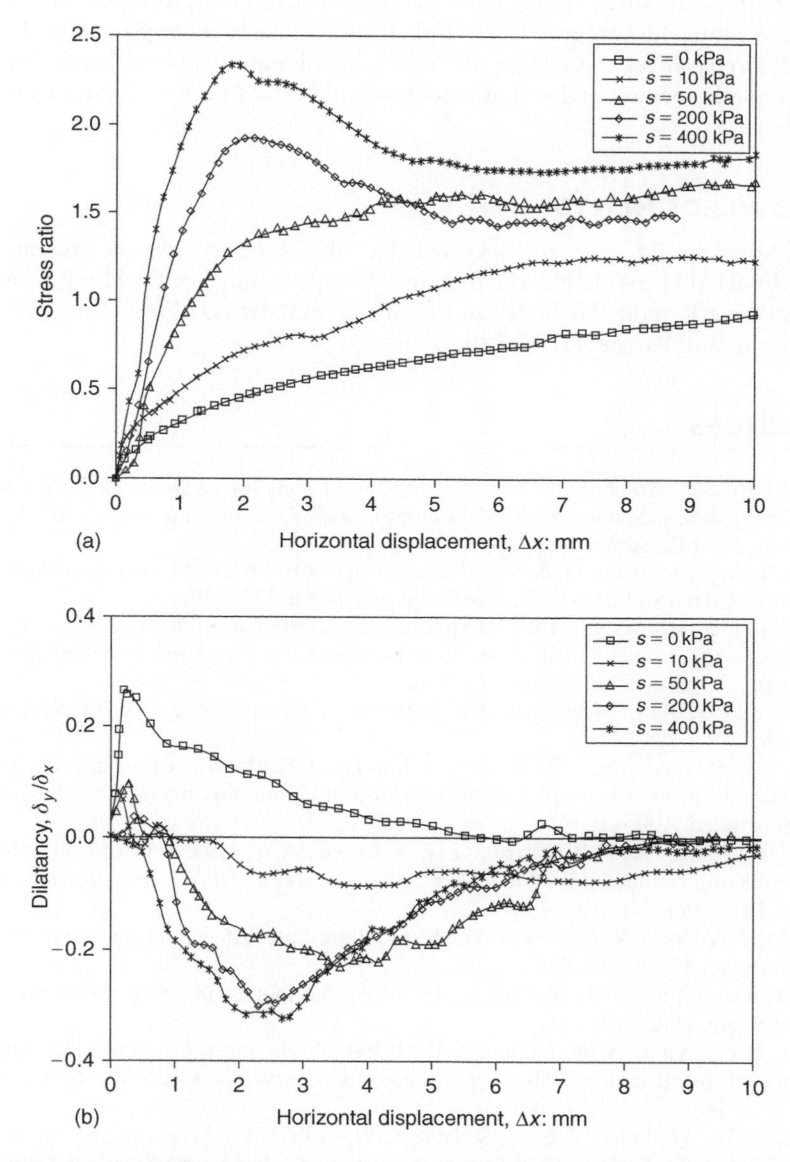

Figure 9.34 Development of the (a) stress ratio and (b) dilatancy of recompacted CDG specimens subjected to shear at different suctions (*After* Ng and Zhou, 2005)

9.8 SUMMARY

Based on field measurements and laboratory tests on intact and recompacted specimens, the hydraulic and mechanical properties of several common Hong Kong saprolites including CDV, CDG and CDT are introduced and described. These engineering properties include the stress-dependent soil-water characteristic curve (SDSWCC), permeability function, small strain shear stiffness and shear strength characteristics. Measured results from the saprolites in Hong Kong are also compared with other similar decomposed materials in other places as appropriate. For details of sample preparation and testing procedures, explanations of test results and discussion, it is highly recommended that readers should refer to the original documents and publications.

ACKNOWLEDGMENTS

The authors would like to acknowledge the support by the research grant HKUST9/CRF/09 provided by the Research Grants Council of the Hong Kong Special Administrative Region (HKSAR) and financial support (OAP06/07.EG01) provided by Ove Arup and Partners (HK) Ltd.

REFERENCES

ASTM (2000). *Standard Practice for Classification of Soils for Engineering Purposes (Unified Soil Classification System).* ASTM standard D2487. American Society of Testing and Materials, West Conshohocken, PA.

Atkinson, J.H., Richardson, D. & Stallebrass, S.E. (1990). Effect of recent stress history on the stiffness of overconsolidated soil. *Géotechnique*, 40(4): 531–540.

Atkinson, J.H. & Sällfors, G. (1991). Experimental determination of stress-strain-time characteristics in laboratory and in-situ tests. General report. *Proc. of 10th Eur. Conf. Soil Mech. & Ftn. Engng.*, Florence, Vol. 3, pp. 915–956.

Atkinson, J.H. (2000). Non-linear soil stiffness in routine design. *Géotechnique*, 50(5): 487–508.

Basile, A. Coppola, A., Mascellis, R., De., & Randazzo, L. (2006). Scaling approach to deduce field hydraulic properties and behavior from laboratory measurements on small cores. *Vadose Zone Journal*, 5: 1005–1016.

Brand, E.W., Phillipson, H.B., Borrie, G.W. & Clover, A.W. (1983). In-situ direct shear tests on Hong Kong residual soils. *Proc., Int. Symp. on Soil and Rock Investigations by In-Situ Testing*, Paris, Vol. 2, pp. 13–17.

Cuccovillo, T. & Coop, M.R. (1997). Yielding and pre-failure deformation of structured sands. *Géotechnique*, 47(3): 491–508.

Cui, Y.J. & Delage, P. (1996). Yielding and plastic behaviour of an unsaturated compacted silt. *Géotechnique*, 46(2): 291–311.

Fredlund, D.G., Xing, A. & Haung, S.Y. (1994). Predicting the permeability function for unsaturated soils using the soil-water characteristics curve. *Canadian Geotechnical Journal*, 31(4): 533–546.

Fredlund, M.D., Fredlund, D.G. & Wilson, G.W. (1997). Prediction of the soil-water characteristic curve from the grain-size distribution curve. *Proc. 3rd Brazilian Symposium on Unsaturated Soils*, Rio de Janeiro, Brazil, April 20–22, Vol. 1, pp. 13–23.

Fredlund, D.G., Rahardjo, H., Leong, E.C. & Ng, C.W.W. (2001). Suggestions and recommendations for the interpretation of soil-water characteristic curves. *Proc. 14th Southeast Asian Geotechnical Conference*, Hong Kong, Vol. 1, pp. 503–508.

Gan, J.K.M., Fredlund, D.G. & Rahardjo, H. (1988). Determination of the shear strength parameters of unsaturated soil using the direct shear test. *Canadian Geotechnical Journal*, 25(8): 500–510.

Gan, J.K.M. & Fredlund, D.G. (1996). Shear strength characteristics of two saprolitic soils. *Canadian Geotechnical Journal*, 33(4): 595–609.

Geotechnical Control Office (GCO) (1982). *Mid-level Studies: Report on Geology, Hydrology and Soil Properties*. Geotechnical Control Office, Hong Kong.

Geotechnical Control Office (GCO) (1987). *Guide to Site Investigation, Geoguide 2*. Hong Kong Government.

Geotechnical Control Office (GCO) (1988). *Guide to Rock and Soil Descriptions, Geoguide 3*. Hong Kong Government.

Hilf, J.W. (1956). *An investigation of pore-water pressure in compacted cohesive soils*. Ph.D Dissertation, Technical Memo. No. 654, U.S. Dep. of the Interior, Bureau of Reclamation, Design and Construction Div., Denver, CO.

Hillel, D. (1998). *Environmental Soil Physics*. San Diego, CA: Academic Press.

Ho, K.M.Y, Ng, C.W.W, Ho, K.K.S. & Tang, W.H. (2006). State-dependent soil-water characteristic curves (SDSWCCs) of weathered soils. *Proc. 4th Int. Conf. on Unsaturated Soils*, Arizona, USA, pp. 1302–1313.

Ho, K.M.Y., Tse, J.M.K. & Ng, C.W.W. (2007). Influence of drying and wetting history and particle size on state-dependent soil-water characteristic curves (SDSWCCs). *Proc. 3rd Asian Conf. on Unsaturated Soils*, Nanjing, China, pp. 213–218.

Irfan, T.Y. (1996). *Mineralogy and Fabric Characterization and Classification of Weathered Granitic Rocks in Hong Kong* (GEO Report No. 41). Geotechnical Engineering Office, Civil Engineering and Development Department, the Government of the Hong Kong SAR.

Irfan, T.Y. (1999). Characterization of weathered volcanic rocks in Hong Kong. *Quarterly Journal of Engineering Geology*, 32: 317–348.

Lee, I.K., and Coop, M.R. (1995). The intrinsic behaviour of a decomposed granite soil. *Géotechnique*, 45(1): 117–130.

Leung, A.K., Sun, H.W., Millis, S., Pappin, J.W., Ng, C.W.W. & Wong, H.N. (2011). Field monitoring of an unsaturated saprolitic hillslope. *Canadian Geotechnical Journal*, 48(3): 339–353.

Mair, R.J. (1993). Developments in geotechnical engineering research: application to tunnels and deep excavations. *Proc. Of the Institution of Civil Engineers: Civil Engineering*, Vol. 1, 27–41.

Mancuso, C., Vassallo, R. & d'Onofrio, A. (2002). Small strain behavior of a silty sand in controlled-suction resonant column–torsional shear tests. *Canadian Geotechnical Journal*, 39(1): 22–31.

Massey, J.B. (1983). *Shear Strength of Hong Kong Residual Soil: A Review of Work Carried Out by the Geotechnical Control Office*. Report of the Geotechnical Control Office, The Government of Hong Kong, Report 25/83.

Ng, C.W.W. & Chen, R. (2008). Invited keynote: Recompacted, natural and in-situ properties of unsaturated decomposed geomaterials. In: (eds) A.B. Huang and P.W. Mayne *Proc. 3rd Int. Conf. on Site Characteristics (TC16)*. 1–4 April 2008, Taipei pp. 117–137, London, Taylor and Francis.

Ng, C.W.W. & Chiu, C.F. (2001). Behavior of a loosely compacted unsaturated volcanic soil. *Journal of Geotechnical and Geoenvironmental Engineering*, ASCE, 127(12): 1027–1036.

Ng, C.W.W. & Chiu, C.F. (2003). Laboratory study of loose saturated and unsaturated decomposed granitic soil. *Journal of Geotechnical and Geoenvironmental Engineering*, ASCE, 129(6): 550–559.

Ng, C.W.W. & Leung, E.H.Y. (2007a). Determination of shear wave velocities and shear moduli of completely decomposed tuff. *Journal of Geotechnical and Geoenvironmental Engineering*, ASCE, 133(6): 630–640.

Ng, C.W.W. & Leung, E.H.Y. (2007b). Small-strain stiffness of granitic volcanic saprolites in Hong Kong. *Characterisation and Engineering Properties of Natural Soils*, Vol. 4, pp. 2507–2538. Invited paper for International Workshop on Natural Soil 2006, Dec. Singapore.

Ng, C.W.W. & Lings, M.L. (1995). Effects of modelling soil non-linearity and wall installation on the back-analysis of a deep excavation in stiff clay. *Journal of Geotechnical Engineering*, ASCE, 121(10): 687–695.

Ng, C.W.W. & Menzies, B. (2007). *Advanced Unsaturated Soil Mechanics and Engineering. New York:* Taylor & Francis.

Ng, C.W.W. & Pang. Y.W. (1999). Stress effects on soil-water characteristic curve and pore water pressure distributions in unsaturated soil slopes. *Proc.11th Asia Regional Conference on Soil Mechanics and Geotechnical Engineering*, Korea, Vol. 1, pp. 371–374.

Ng, C.W.W. & Pang. Y.W. (2000a). Influence of stress state on soil-water characteristics and slope stability. *Journal of Geotechnical and Geoenvironmental Engineering*, ASCE, 126(2): 157–166.

Ng, C.W.W. & Pang, Y.W. (2000b). Experimental investigations of the soil-water characteristics of a volcanic soil. *Canadian Geotechnical Journal*, 37(6): 1252–1264.

Ng, C.W.W. & Wang, Y. (2001). Field and laboratory measurements of small strain stiffness of decomposed granites. *Soils and Foundations*, 41(3): 57–71.

Ng, C.W.W. & Yung, S.Y. (2008). Determination of the anisotropic shear stiffness of an unsaturated decomposed soil. *Géotechnique*, 58(1): 23–35.

Ng, C.W.W. & Zhou, R.Z.B. (2005). Effects of soil suction on dilatancy of an unsaturated soil. *Proc. 16th ICSMGE*, Osaka, Japan: Vol. 2, pp. 559–562.

Ng, C.W.W., Pun, W.K. & Pang, R.P.L. (2000). Small strain stiffness of natural granitic saprolite in Hong Kong. *Journal of Geotechnical and Geoenvironmental Engineering, ASCE*, 126(9): 819–833.

Ng, C.W.W., Guan, P. & Shang, Y.J. (2001a). Weathering mechanisms and indices of igneous rocks of Hong Kong. *Quarterly Journal of Engineering Geology and Hydrology*, 34(2): 133–151.

Ng, C.W.W., Wang, B. & Gong, B.W. (2001b). A new apparatus for studying stress effects on soil-water characteristics of unsaturated soils. *Proc. 15th ICSMGE*, August, Istanbul, Turkey. Vol. 1, pp. 611–614.

Ng, C.W.W., Zhan, L.T. & Cui, Y.J. (2002). A new simple system for measuring volume changes in unsaturated soils. *Canadian Geotechnical Journal*, 39(2): 757–764.

Ng, C.W.W., Leung, E.H.Y. & Lau, C.K. (2004a). Inherent anisotropic stiffness of weathered geomaterial and its influence on ground deformations around deep excavations. *Canadian Geotechnical Journal*, 41(1): 12–24.

Ng, C.W.W., Fung, W.T., Cheuk, C.Y. & Zhang, L.M. (2004b). Influence of stress ratio and stress path on behaviour of loose decomposed granite. *Journal of Geotechnical and Geoenvironmental Engineering*, ASCE, 130(1): 36–44.

Ng, C.W.W., Xu, J. & Yung, S.Y. (2009). Effects of wetting-drying and stress ratio on anisotropic stiffness of an unsaturated soil at very small strains. *Canadian Geotechnical Journal*, 46(9): 1062–1076.

Ng, C.W.W., Wong, H.N., Tse, Y.M., Pappin, J.W., Sun, H.W., Millis, S.W. & Leung, A.K. (2011). A field study of stress-dependent soil-water characteristic curves and permeability of a saprolitic slope in Hong Kong. *Géotechnique*, 61(6): 511–521.

Pun, W.K. & Ho, K.K.S. (1996). *Analysis of Triaxial Tests on Granitic Saprolite Performed at Public Works Central Laboratory*. Discussion note DN 4/96, Geotechnical Engineering Office, Hong Kong Government.

Sewell, R.J., Campbell, S.D.G., Fletcher, C.J.N., Lai, K.W. & Kirk, P.A. (2000). *The Pre-Quaternary Geology of Hong Kong*. Geotechnical Engineering Office, Civil Engineering and Development Department, the Government of the Hong Kong SAR.

Shaw, R. (1997). Variations in sub-tropical deep weathering profiles over the Kowloon Granite, Hong Kong. *Journal of the Geological Society, London*, 154: 1077–1085.

Shen, J.M. (1985). *GCO Research into Unsaturated Shear Strength 1978–1982*. Geotechnical Control Office, Research Report RR 1/85, Hong Kong Government (internal report).
Tse, M.K. (2007). *Influence of Stress States on Soil-Water Characteristics, Conjunctive Surface-Subsurface Flow Modelling and Stability Analysis*. M.Phil Thesis, Hong Kong University of Science and Technology.
Viana da Fonseca, A., Matos Fermandes, M. & Silva Cardoso, A. (1997). Interpretation of a footing load test on a saprolitic soil from granite. *Géotechnique*, 47(3): 633–651.
Wang, Y. & Ng, C.W.W. (2005). Effects of stress paths on the small-strain stiffness of completely decomposed granite. *Canadian Geotechnical Journal*, 42(4): 1200–1211.
Wheeler, S.J. & Sivakumar, V. (1995). An elasto–plastic critical state framework for unsaturated soil. *Géotechnique*, 45(1): 35–53.
Zhan, L.T. (2003). *Field and Laboratory Study of an Unsaturated Expansive Soil Associated With Rain-Induced Slope Instability*. Ph.D. Thesis, Hong Kong University of Science and Technology, Hong Kong.

10

Residual soils of India

Sudhakar M. Rao
Indian Institute of Science, Bangalore, India

K.H. Venkatesh
Aptech Foundations, Bangalore, India

10.1 INTRODUCTION

India lies to the north of the equator between 8°4′ and 37°6′ North latitude and 68°7′ and 97°25′ East longitude. The geography of India is diverse and can be divided into three distinct regions, (a) the peninsular region lying to the south of the Indo-Gangetic plains, (b) the Indo-Gangetic alluvial plains stretching across northern India from Assam and Bengal on the east, through Bihar and Uttar Pradesh to Punjab on the west, and (c) the Extra-peninsula, comprising the Himalayan ranges. The peninsular region consists of basaltic rocks of various types, granites and metamorphic rocks derived from the granites. The basaltic Deccan traps are associated with the most extensive occurrence of black soil formation. In the areas occupied by Archaean gneisses and Deccan traps, extensive deposits of red soils are encountered. Large deposits of laterites are found as cappings over the Deccan traps. The geology of the soil-forming rocks is briefly considered as a prelude to soil characteristics.

10.2 THE ARCHAEAN GROUP

The Archaean rocks are mostly of igneous origin; comprising metamorphosed granitic and basaltic rocks, together with a smaller amount of sediments. The Archaean formations in the Indian shield occupy most of southern and eastern India and parts of Assam, Bihar, Madhya Pradesh and Rajasthan. The peninsular gneisses belonging to this group are the most widespread group of rocks in Southern India. They consist of a very heterogeneous mixture of different types of granites intrusive into the schistose rocks after the latter were folded, crumpled and metamorphosed. These include granites, granodiorites, gneissic granites and banded or composite gneiss. The banded gneisses consist of white bands of quartz-feldspar alternating with dark bands containing hornblende, biotite and minor accessories. The granitic group ranges in composition from granite through granodiorite to adamellite, augite-diorite, monzonite etc., and contain inclusions of hornblende rocks (Krishnan, 1982; Reddy, 1995).

Deccan traps

The Deccan traps are the most extensive geological formation of Peninsular India, with the exception of the metamorphic and igneous complex of the Archaean age. It is believed that the Deccan traps were formed as a result of sub-aerial volcanic activity during the Mesozoic era. The era was marked by the outpouring of enormous lava flows that issued through long narrow fissures or cracks in the earth's crust. The lava spread far and wide, as nearly horizontal sheets, filling up the irregularities of the pre-existing topographies. The Deccan traps cover an area of 200 000 square miles spread over Maharashtra, Gujarat, Madhya Pradesh, and parts of southern India. The Deccan traps belong to the type called plateau basalt and are uniform in composition and are composed of dolerite or basalt. The traps have been divided into three groups – Upper,

Middle and Lower. The Upper traps are 450 m thick and are spread over western India, the Middle traps are 1200 m thick and are spread over Central India, and the lower traps are 150 m thick and are spread over central and eastern India (Krishnan, 1982; Reddy, 1995).

10.3 CLIMATE

Indian climate can be subdivided into four major groups (information sourced from Climatic regions of India, http://en.wikipedia.org/wiki/Climatic_regions_of_India) as discussed below.

Tropical rainy climatic group

The regions belonging to this group experience persistent high temperatures which normally do not go below 18°C even in the coolest month. Two climatic types fall under this group.

Tropical monsoon rainforest

The west coastal lowlands, the Western Ghats and southern parts of Assam have this climate type. The region is characterised by high temperatures throughout the year, even in the hills. The rainfall here is seasonal, but heavy and is in excess of 2000 mm a year. Most of the rain is received in the period from May to November, while December to March are the dry months with scanty rainfall.

Tropical wet and dry climate

Most of the plateau of peninsular India enjoys this climate, except a semi-arid tract to the east of the Western Ghats. Winter and early summer are long dry periods with temperature above 18°C. Summer is very hot and the temperatures in the interior low level areas can go above 45°C during May. The rainy season is from June to September and the annual rainfall is between 750 to 1500 mm. Only Tamil Nadu receives rainfall during the winter months of October to December.

Dry climate group

This group consists of regions where the rate of evaporation of water is higher than the rate of moisture received through precipitation. It is subdivided into three climate types.

Tropical semi-arid steppe climate

It includes Karnataka, interior Tamil Nadu, western Andhra Pradesh and central Maharashtra. This region is a famine-prone zone with very unreliable rainfall, which varies between 400 to 750 mm annually. The coldest month is December, but even in this month the temperature remains between 20 and 24°C. The months of March to May are hot and dry with mean monthly temperatures of about 32°C.

Tropical and sub-tropical desert

Most of western Rajasthan falls under this climate type characterised by scanty rainfall. The cloud bursts are largely responsible for the all the rainfall seen in this region, which is less than 300 mm. The summer months of May and June are very hot with mean monthly temperatures in the region of 35°C and highs which can sometimes reach 50°C. During winters, the temperatures can drop below freezing in some areas due to a cold wave. There is a large diurnal range of about 14°C during summer which becomes higher by a few more degrees during winter.

Tropical and sub-tropical steppe

The region towards the east of the tropical desert running from Punjab and Haryana to Kathiawar experiences this climate type. This climate is a transitional climate falling between tropical desert and humid sub-tropical, with temperatures which are less extreme than the desert climate. The annual rainfall is between 300 to 650 mm, but is very unreliable and happens mostly during the summer monsoon season. The maximum temperatures during summer can rise to 40°C.

Humid sub-tropical climate group

The temperature during the coldest months in regions experiencing this climate falls between 18 and 0°C. It has one climatic subdivision in India.

Humid sub-tropical with dry winter

The foothills of the Himalayas, Punjab-Haryana plain adjacent to the Himalayas, Rajasthan east of the Aravalli range, Uttar Pradesh, Bihar, the northern part of West Bengal and Assam experience this climate. The rainfall is received mostly in the summer and is about 650 mm in the west and increases to 2500 mm annually to the east and near the Himalayas. The winters are mainly dry due to the land-derived winter winds which blow down the lowlands of north India towards the Bay of Bengal. The summers are hot and temperatures can reach 46°C in the lowlands. May and June are the hottest months. The winter months are mostly dry with feeble winds. Frost occurs for a few weeks in winter.

Mountain climate

In the Himalayan mountains, the temperature falls by 0.6°C for every 100 m rise in altitude and this gives rise to a variety of climates from nearly tropical in the foothills to tundra type above the snow line. The northern side of the western Himalayas, also known as the trans-Himalayan belt, is arid, cold and generally wind-swept. Most of the rainfall is in the form of snow during late winter and spring months. The places situated between 1070 and 2290 m altitudes receive the heaviest rainfall and the rainfall decreases rapidly above 2290 m. The great Himalayan range witnesses heavy snowfall during the winter months of December to February at altitudes above 1500 m. The diurnal range of temperature is also high. The states of Jammu and Kashmir, Himachal Pradesh, Uttarakhand, Sikkim and Arunachal Pradesh experience this kind of weather. The climatic regions of India are illustrated in Figure 10.1.

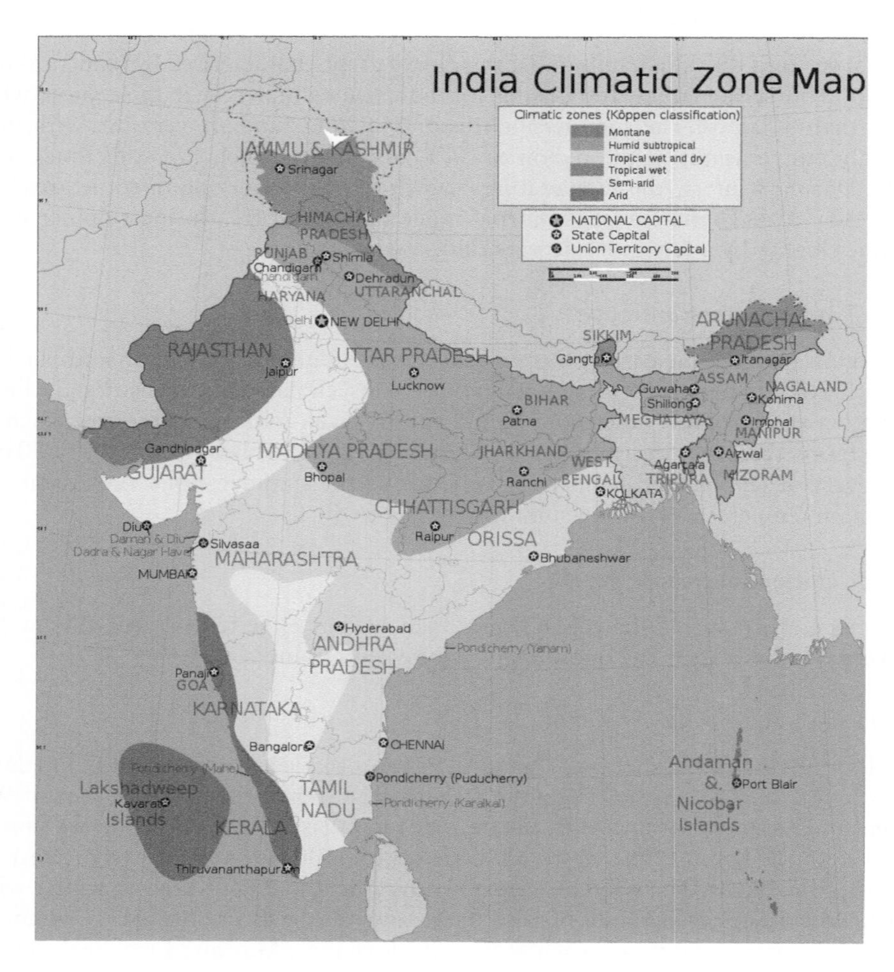

Figure 10.1 Climatic regions in India
(*source*: http://en.wikipedia.org/wiki/File:India_climatic_zone_map_en.svg)

10.4 DISTRIBUTION OF RESIDUAL SOILS

The residual soil deposits in India are of three broad types: 1. regur or black soils; 2. red soils and 3. laterites.

Black soils

These soils are black in colour and are also known as black cotton soils since they are ideal for growing cotton. These soils are mostly formed in the Deccan trap (basaltic rocks) region. Black soils are also formed in some areas of gneissic and calcareous rocks in Andhra Pradesh and southern and central Tamil Nadu. These soils cover an area of about 300,000 square km and are spread in Maharashtra, Gujarat, southern

Uttar Pradesh, eastern Rajasthan, southern and western Madhya Pradesh, northern Karnataka and parts of Andhra Pradesh and Tamil Nadu (Katti, 1975; Krishnan, 1982; Reddy, 1995). The climatic conditions under which black soils occur more frequently are those of semiarid tropical climate with low rainfall (500–800 mm/year). The landform is normally level to gently undulating and the drainage is rather poor (Kyuma and Takaya, 1969; Krishnan, 1982).

Red soils

Red soils are mostly developed on granites, gneisses, schists and ferruginous sandstones of the Archaean periods. These soils occur in regions where rainfall ranges from 500 to 1000 mm, mean annual temperature varies between 16 and 30°C, and the topography is moderately to gently sloping lands with good drainage conditions. Red soils are distributed over the states of Andhra Pradesh, Goa, Kerala, Karnataka, Tamil Nadu, Pondicherry, Orissa, West Bengal, North Eastern States and the Union territories of Andaman and Nicobar. They occupy about 700,000 square kilometres (approximately 21%) of the total geographical area in India (Bourgeon and Sehgal, 1998; Sehgal, 1998; Sehgal *et al.*, 1998; Krishnappa *et al.*, 1998).

Laterites

According to Wadia (1926), laterite is a surface formation composed essentially of a mixture of hydrated oxides of aluminium and iron with a small percentage of manganese and titanium. The percentage of oxides varies widely from alumina rich (bauxites–aluminous laterites) to deposits rich in iron oxides. The conditions favouring the formation of laterite are a warm, humid climate with plentiful and well-distributed rainfall and good drainage. They are found in the Malabar plateau of Kerala, Konkan region of Goa, south Kanara region of Karnataka, Rajahmundry region in Andhra Pradesh, Ratanagiri District in Maharashtra along the edge of plateau in the east covering small parts of Tamil Nadu, Orissa, a small part of Chhotanagpur in the north, and Meghalaya in the north-east (Nanda and Krishnamachari, 1958; Das and Mahalik, 1982; Seshagiri and Raymahashay, 1981; Basavarajaiah *et al.*, 1982; Baskar *et al.*, 2003). Laterites are formed over Precambrian charnockite, biotite gneiss, middle Miocene sandstones, Achaean khondalites, granitic gneiss and basalts. In India, large deposits of laterites are found as cappings over the Deccan Trap; the thickness of the capping being sometimes as much as 100 feet. There is usually a layer of highly ferruginous material at the surface, below which there is a layer of aluminous laterite or bauxite. These grade further below into lithomargic clay which gradually merges into unaltered rock (Krishnan, 1982). The annual rainfall ranges from 2000 to 4000 mm, while mean annual temperatures varies from 23 to 34°C.

10.5 PHYSICO-CHEMICAL PROPERTIES

Black soils

The clay fractions of black soils are dominated by the smectite group of minerals. Vertisols may be acid, neutral or alkaline in reaction. Acidic black soils may be due to

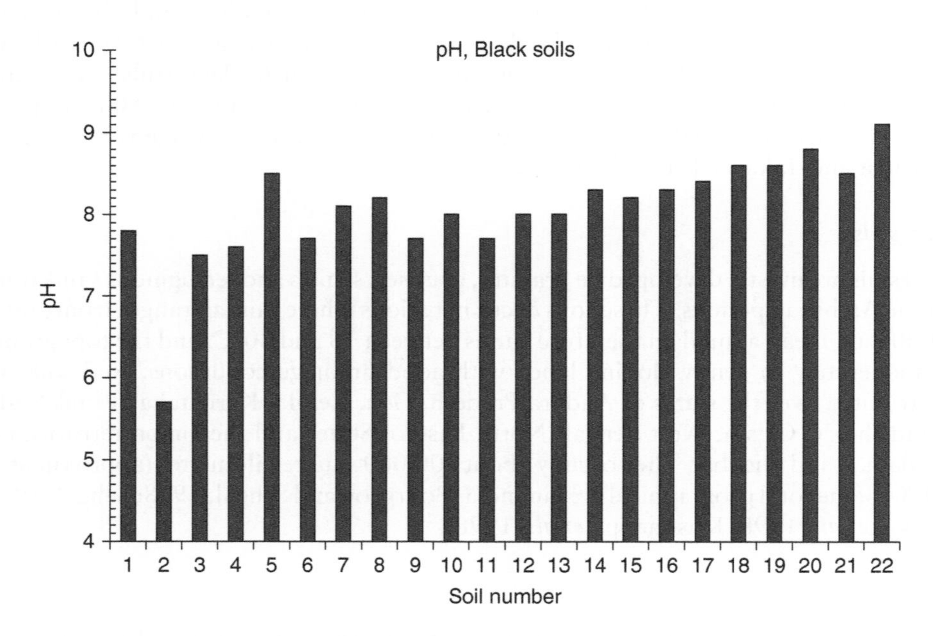

Figure 10.2 pH of black soils

acidic parent materials, dissolution of carbonates or alternating oxidation/reduction conditions. However, the majority of black soils are neutral or alkaline because they are mostly derived from calcareous or base-rich parent materials (Coulombe *et al.*, 1996). Figure 10.2 presents the pH distribution of black soils from Karnataka. The figure shows that bulk of the soils are mildly alkaline with pH values ranging from 7.5 to 8.8. According to Coulombe *et al.* (1996), the alkaline nature is indicative of the presence of sodium carbonate. Further, high pH conditions (8.5 to 9.5) promote an increase in the silicon activity and dissolution of silicate minerals.

Generally, black soils have high cation exchange capacity (CEC) ranging from 20 to 45 meq/100 g. The CEC values of the black soils are determined by the amounts of smectite and organic matter content. Calcium and magnesium are the predominant exchangeable cations, while sodium and potassium are less common. Black soils containing exchangeable sodium percentage values greater than 15 are termed sodic soils. Figure 10.3 presents the ranges of CEC values of black soils from Karnataka. The plot shows that the bulk of the samples have CEC values ranging between 40 and 60 meq/100 g.

The organic matter is an important constituent of black soils and affects their morphological, chemical and physical properties (Coulombe *et al.*, 1996). The organic matter in black soils generally ranges from 0.5 to 5% (Figure 10.4). Generally, black soils formed in a humid environment contain higher organic content than those formed in arid and semi-arid environments. Mineralogy, drainage conditions and strong clay-organic complexes of smectite and humic materials are considered to be responsible for the black colour of these soils (Singh, 1954, 1956; Skjemstad and Dalal, 1987; Coulombe *et al.*, 1996).

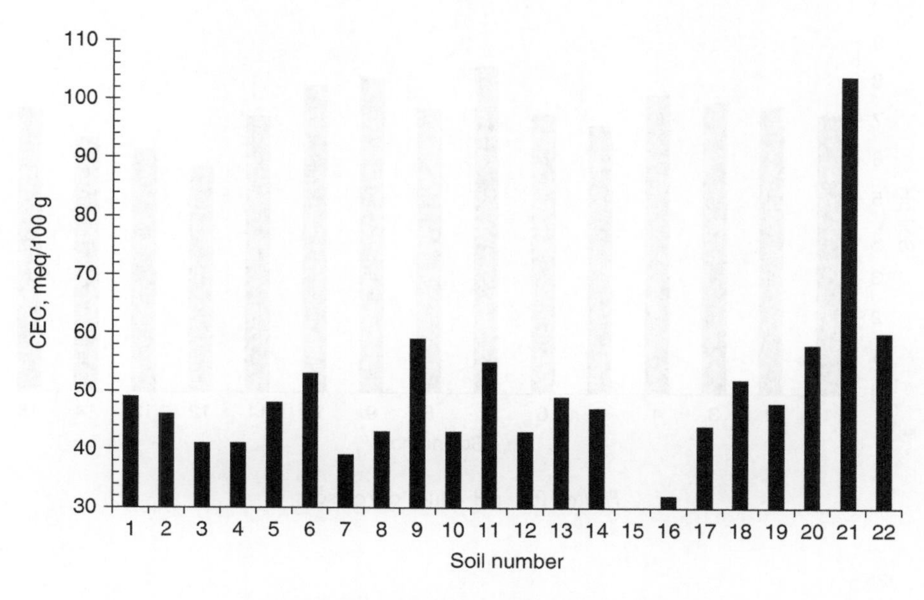

Figure 10.3 CEC of black soils

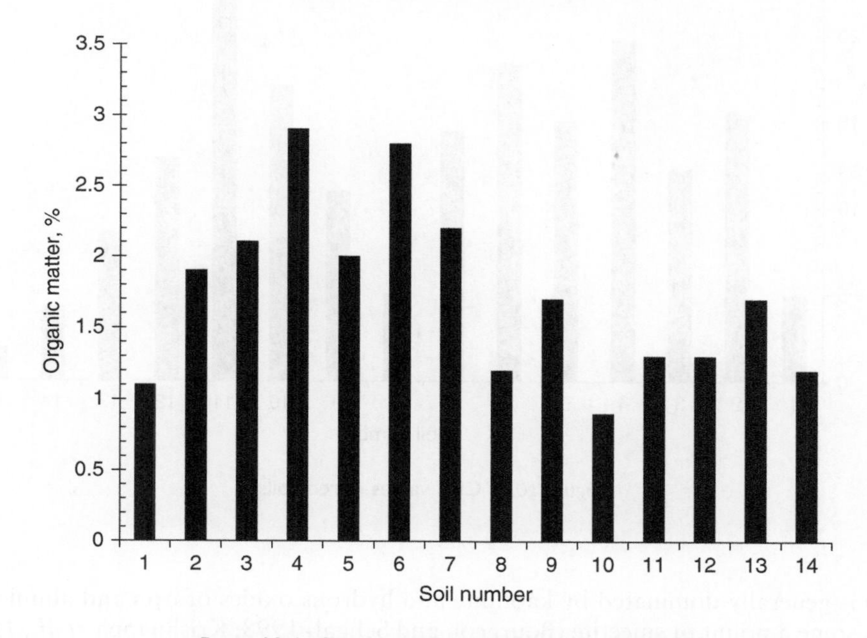

Figure 10.4 Organic matter content of black soils

Red soils

According to Sehgal (1996), red soils are considered to be those which have colours of 5 YR or redder in the soil series control section (25–150 cm). The clay fraction of these

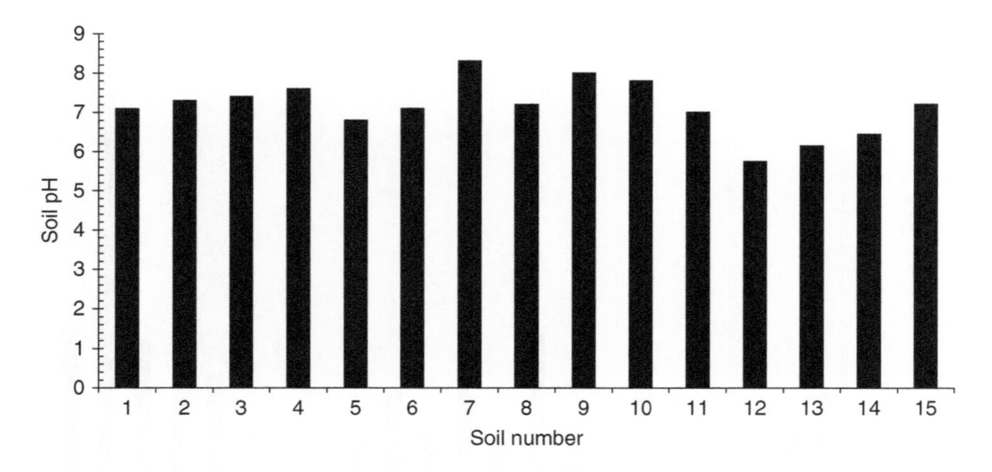

Figure 10.5 pH values of red soils

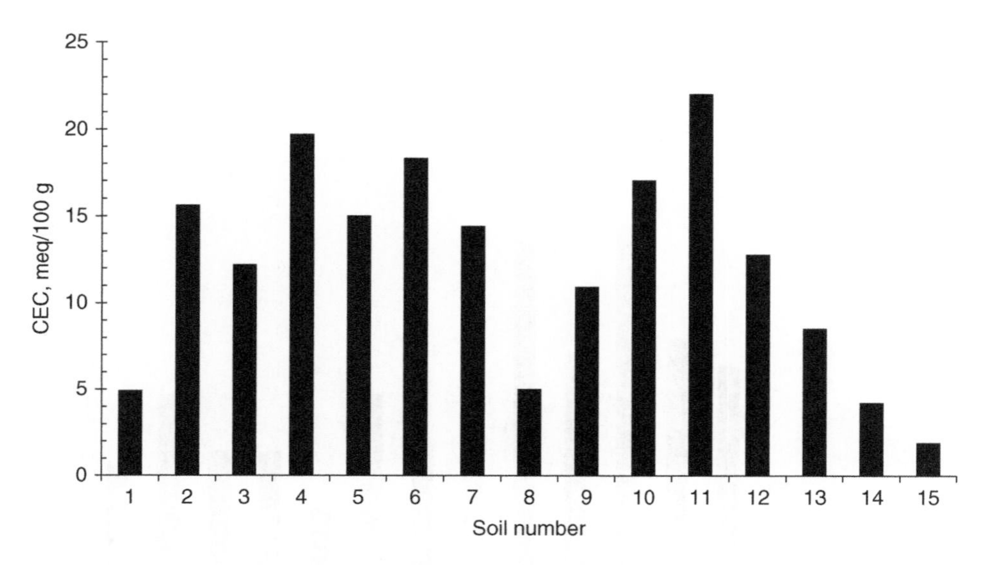

Figure 10.6 CEC values of red soils

soils is generally dominated by kaolinite and hydrous oxides of iron and aluminium, with some amount of smectite (Bourgeon and Sehgal, 1998; Krishnappa *et al.*, 1998). These soils are acidic to alkaline in reaction, with pH values ranging from 5.5 to 8.8 (Figure 10.5, red soil data from Karnataka, Tamil Nadu and Andhra Pradesh). The CEC of these soils reportedly ranges from 2 to 22 meq/100 g (Figure 10.6, red soil data from Karnataka, Tamil Nadu and Andhra Pradesh). The soils are also low in organic matter with values ranging from 0.4 to 0.8 % (Krishnappa *et al.*, 1998).

Laterites

Indian laterites from Calicut in Kerala and Rajahmundry in Andhra Pradesh have been determined to contain halloysite, kaolinite, goethite, gibbsite and quartz minerals. Acidic pH ranging from 6.1 to 6.9 has been reported for laterites from Kerala and Andhra Pradesh. Relatively high organic matter contents of 7 to 10% are reportedly present in laterites from Calicut in Kerala and Rajahmundry in Andhra Pradesh (Seshagiri and Raymahashay, 1981). However, a much lower range of organic matter content ranging from zero to 1.3% are reported for laterites from Goa (Baskar *et al.*, 2003) and central Kerala (Nambiar *et al.*, 1966). High CEC values of 29 to 31 meq/100 g have been reported for Calicut and Rajahmundry laterites that are attributed to the presence of halloysite mineral (Seshagiri and Raymahashay, 1981; Ramahashay *et al.*, 1984). However, Nambiar *et al.* (1966) reported very low CEC of 3.3 meq/100 g for laterites from central Kerala.

Laterites are differentiated from lateritic soils and non-lateritic soils based on the silica/alumina ratio. Laterites have silica/alumina ratio $\leq$1.33, lateritic soils of 1.33 to 2.00, and non-lateritic soils have silica/alumina ratio of 2.0 or more (Bawa, 1957). Nambiar *et al.* (1966) report very low silica/alumina ratio of 0.1 to 0.54 for laterites from Kerala. Nanda and Krishnamachari (1958) reported silica/alumina ratios of 0.71, 0.18, 0.12 and 1.12 for laterites from Andhra Pradesh, Assam, Maharastra and Madhya Pradesh respectively.

10.6 GEOTECHNICAL ENGINEERING DATA

Black soils

Figure 10.7 plots the distribution of clay sized fraction (<2 micron) in black soils from Karnataka. The plot shows that 70% of the examined samples had clay contents ranging between 30 and 50%. Figure 10.8 plots the Casagrande's A-line plot for the black soils and shows that the majority of the soils classify as CH-inorganic clays of high plasticity. Figure 10.9 shows that 79% of soil samples have standard Proctor MDD (maximum dry density) values ranging between 1.46 and 1.65 Mg/m^3. Likewise, 86% of compacted samples (at Standard Proctor density (MDD) and optimum moisture content (OMC)) develop unconfined compressive strengths ranging between 300 and 750 kPa (Figure 10.10), swell potentials ranging between 2.4 and 8.8% (Figure 10.11) and coefficient of permeability values ranging between 3×10^{-9} and 6×10^{-9} cm/s (Figure 10.12). The compression index values of the compacted specimens (at Standard Proctor MDD and OMC) range between 0.18 and 0.29 (Figure 10.13).

Figures 10.14 and 10.15 show the bore-hole data for black cotton soil deposits from Maharashtra and Karnataka. The data show that the residual soil profile is about 3–4 m thick and is underlain by hard strata. The SPT penetration values range from 10 to 30 for these soil deposits. The water table could not be encountered at depths of 4.5 to 6 m at both locations.

Red soils

Figure 10.16 shows that bulk of the red soils classify as CL-inorganic clays of low plasticity, while 89% of samples have Standard Proctor MDD ranging between 1.6 and

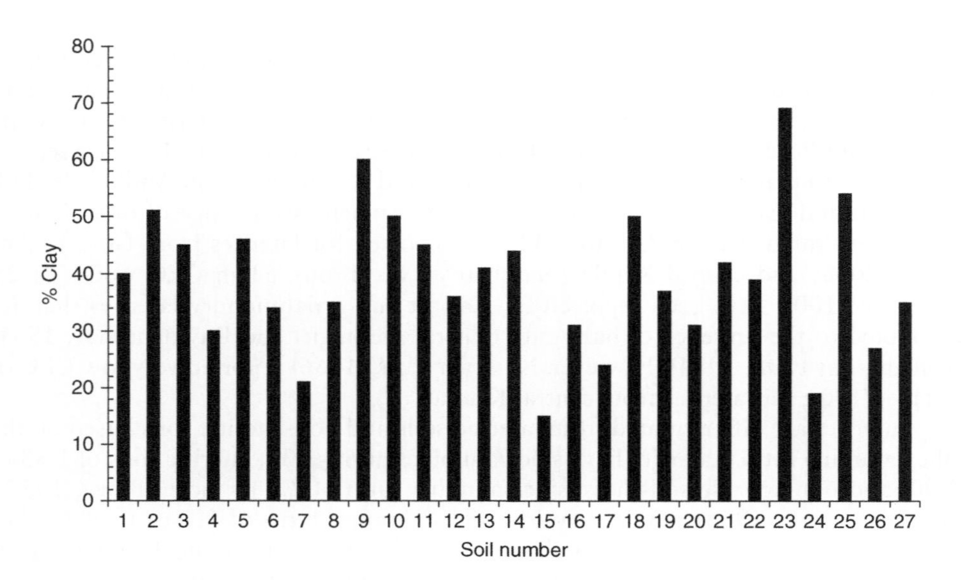

Figure 10.7 Percentage clay size distribution of black soils

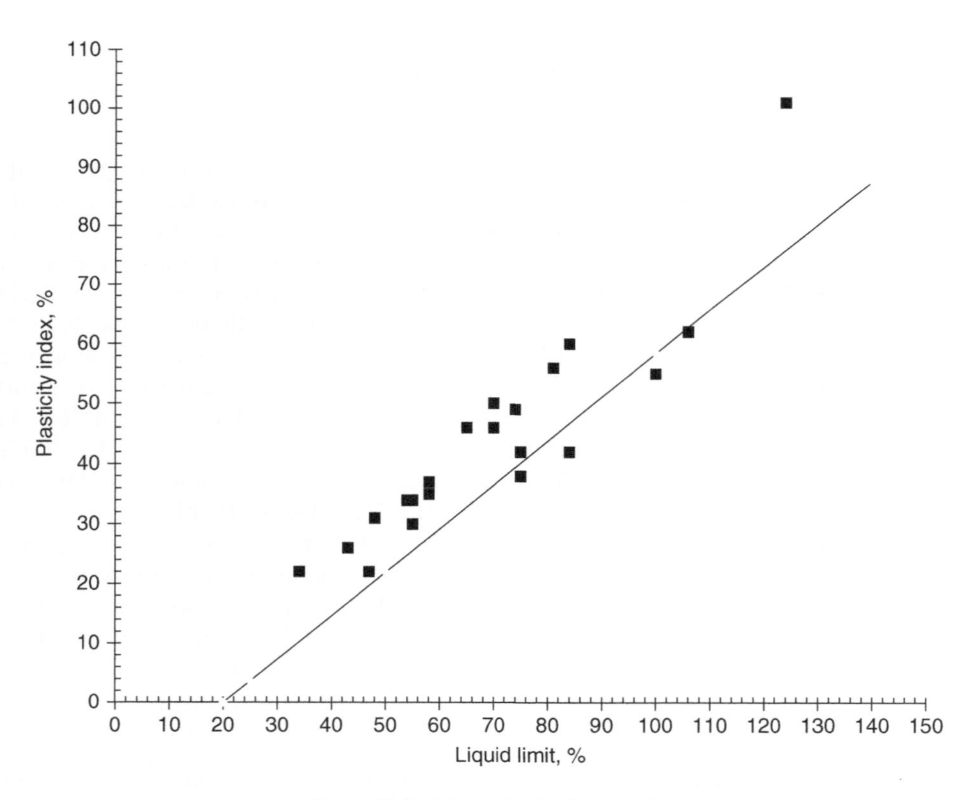

Figure 10.8 A-line plot for black soils

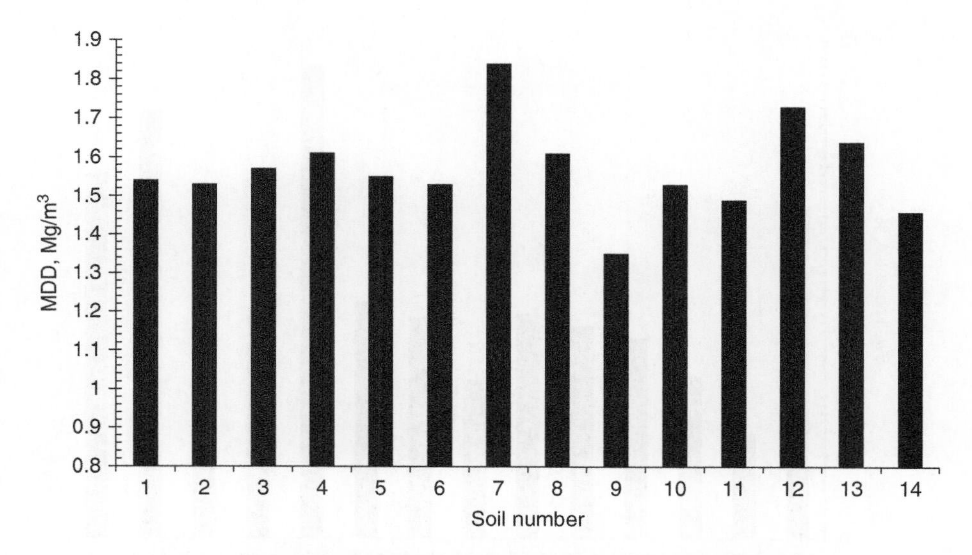

Figure 10.9 Standard Proctor, maximum dry density values of black soils

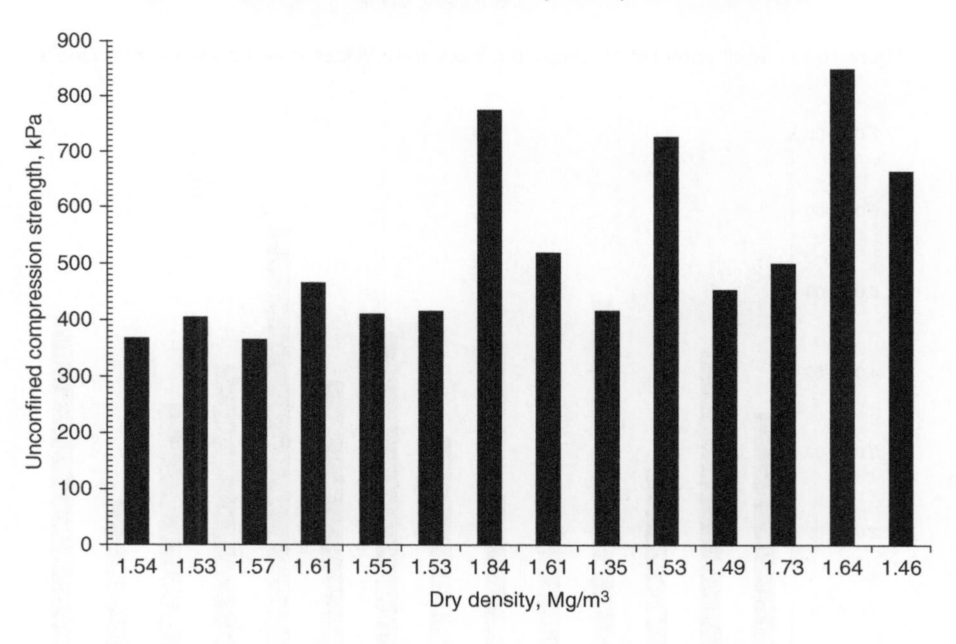

Figure 10.10 Unconfined compressive strength of compacted black soils

1.9 Mg/m^3 (Figure 10.17). Available data show that at 83% of the locations, the depth of bed-rock ranges between 3 and 10 m (Figure 10.18). 91% of the soil samples have in-situ void ratios ranging between 0.42 and 0.72 (Figure 10.19), 71% of the undisturbed samples have degree of saturation ranging between 48 and 86% (Figure 10.20), 79%

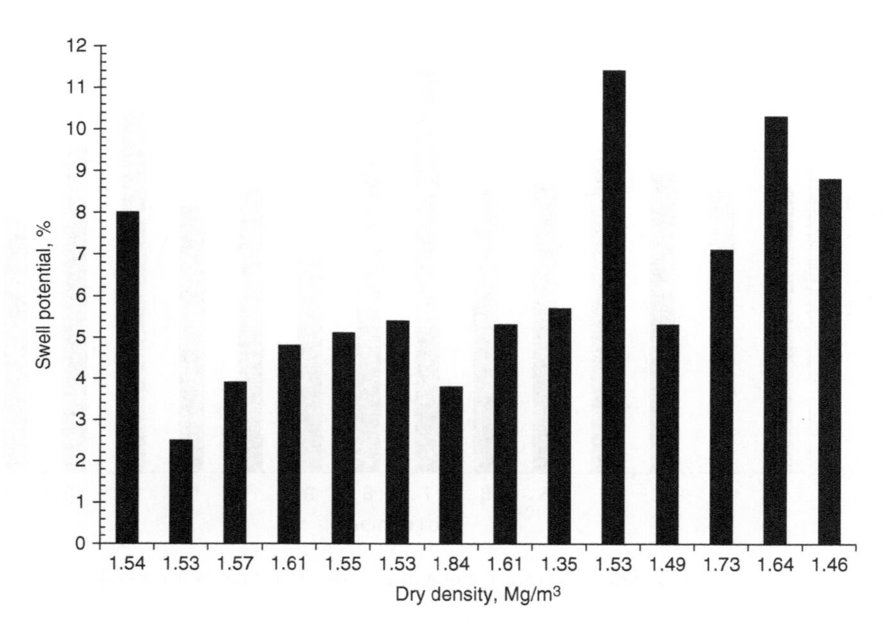

Figure 10.11 Swell potential of compacted black soils: Values at vertical stress of 6.25 kPa

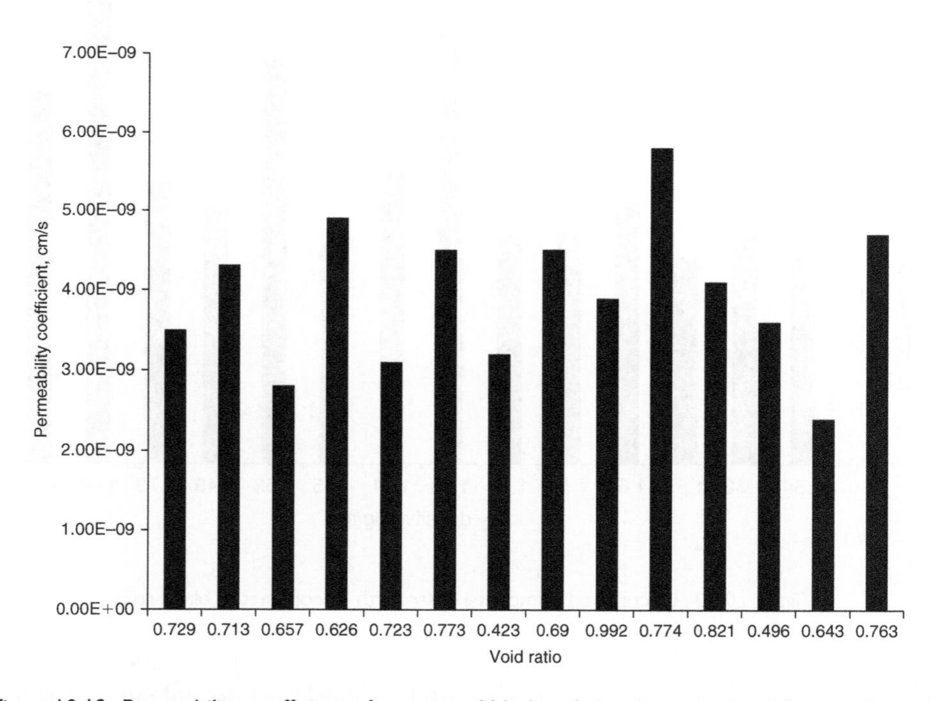

Figure 10.12 Permeability coefficient of compacted black soils: k values calculated from oedometer test data for pressure increment of 50–100 kPa

Figure 10.13 Compression index of compacted black soils: soils swollen at 6.25 kPa and consolidated to 800 kPa

Location MP – MAHARASHTRA
Borehole No. 1

Depth below ground m	Soil	Classification	SPT value	Grain size distribution: Gravel, %	Sand, %	Silt & clay, %	w_n, %	w_l, %	w_p, %
0–3	EGL	Greyish sandy clay	**SPT/1.5 m N = 20**	4	30	66	14	43.0	24
3–6		Hard rock							

Borehole terminated at 6 m depth

Figure 10.14 Bore-hole profile of black cotton soil from Maharashtra

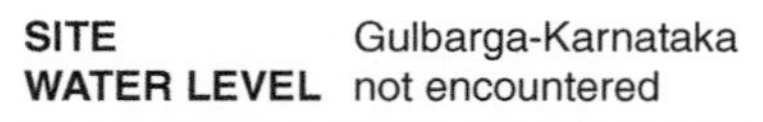

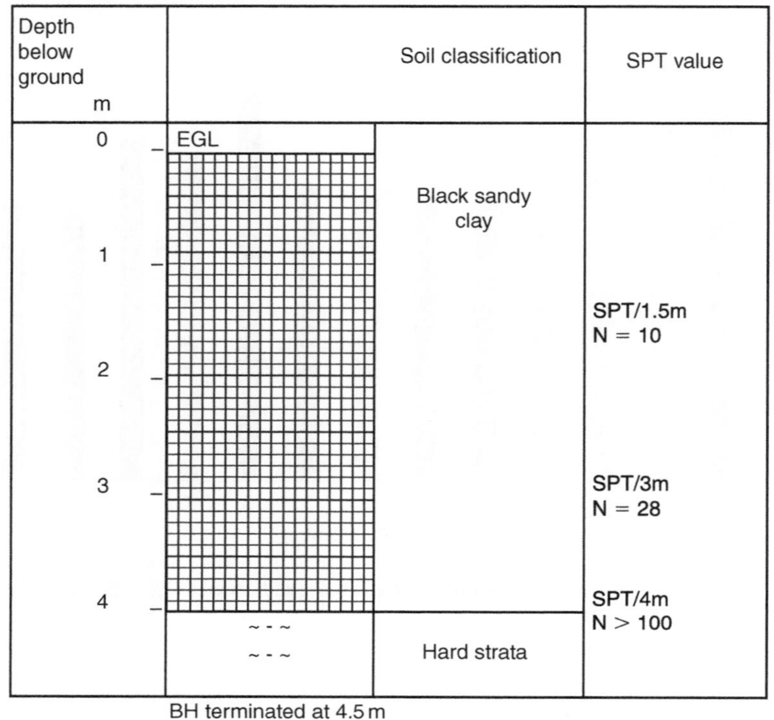

Figure 10.15 Bore-hole profile of black cotton soil from Maharashtra

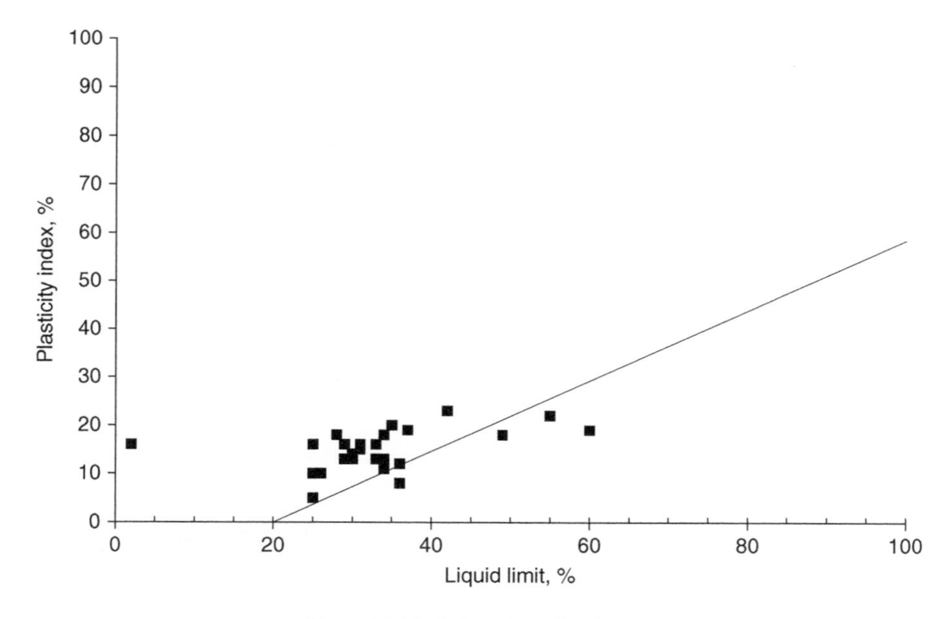

Figure 10.16 A-line plot of red soils

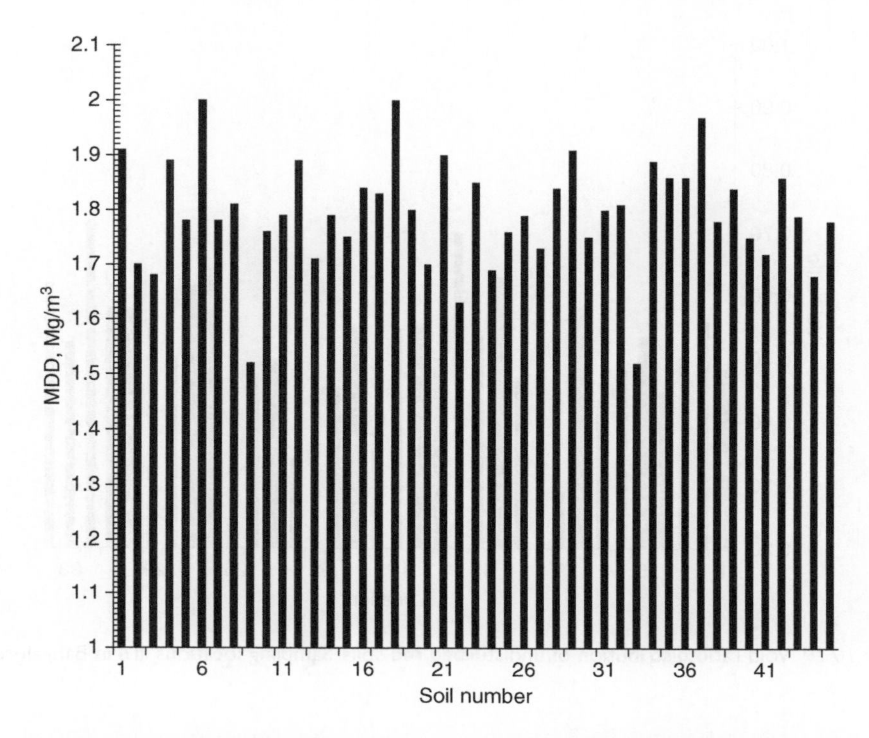

Figure 10.17 Standard Proctor MDD values of red soils: Sampling locations are in Bangalore city

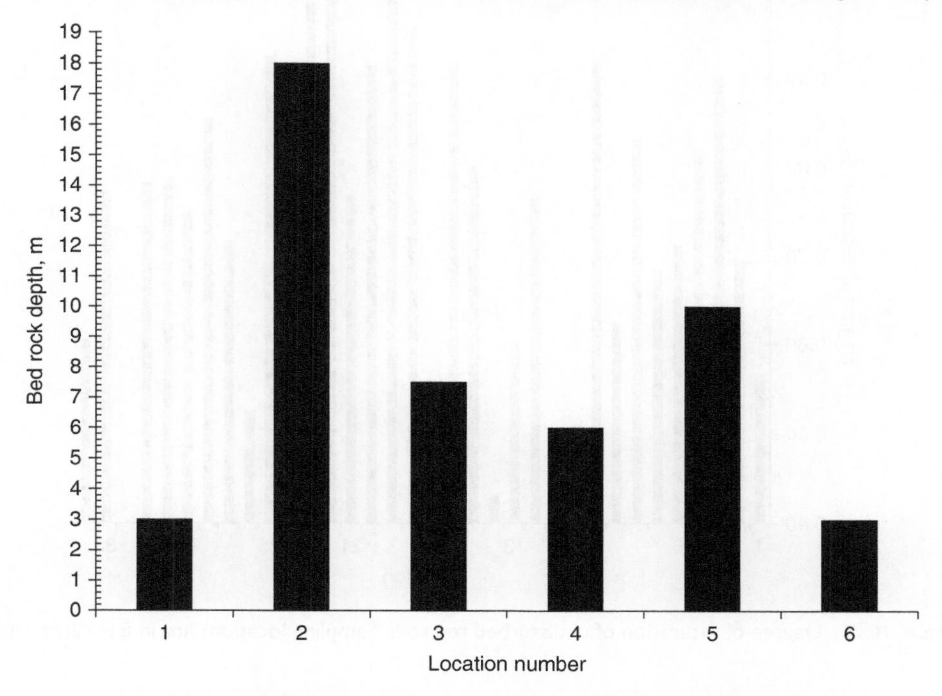

Figure 10.18 Depth of bed-rock: Sampling locations are in Bangalore city

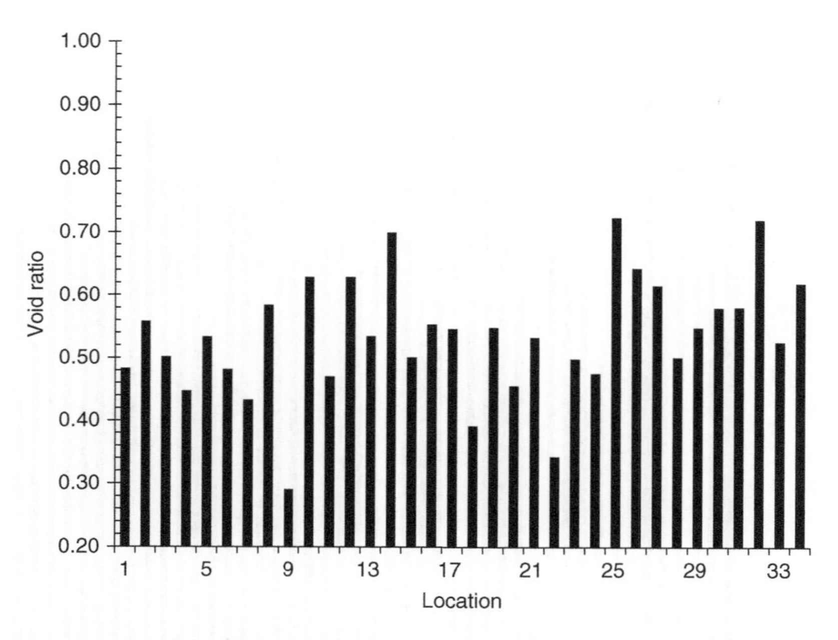

Figure 10.19 Void ratio distribution of undisturbed red soils: Sampling locations are in Bangalore city

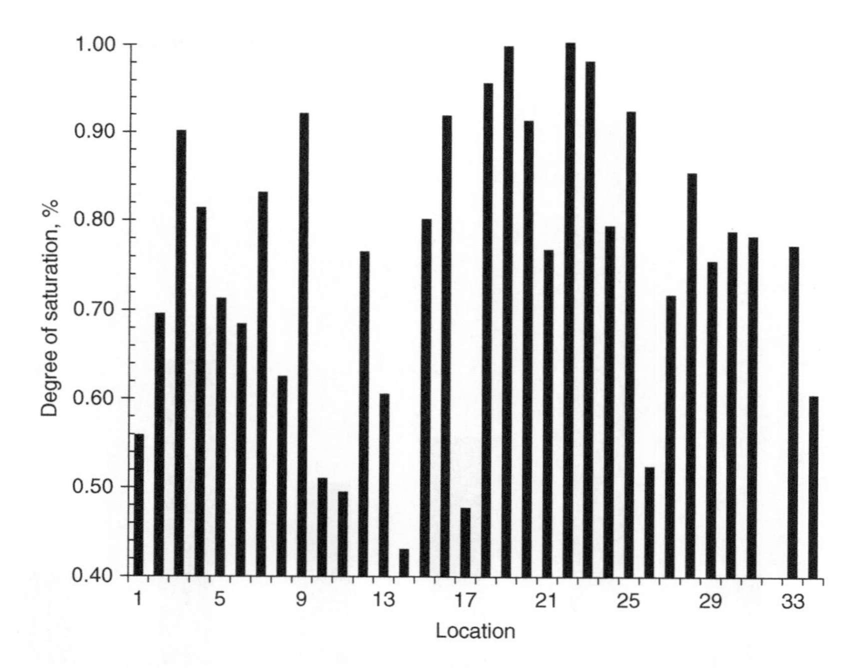

Figure 10.20 Degree of saturation of undisturbed red soils: Sampling locations are in Bangalore city

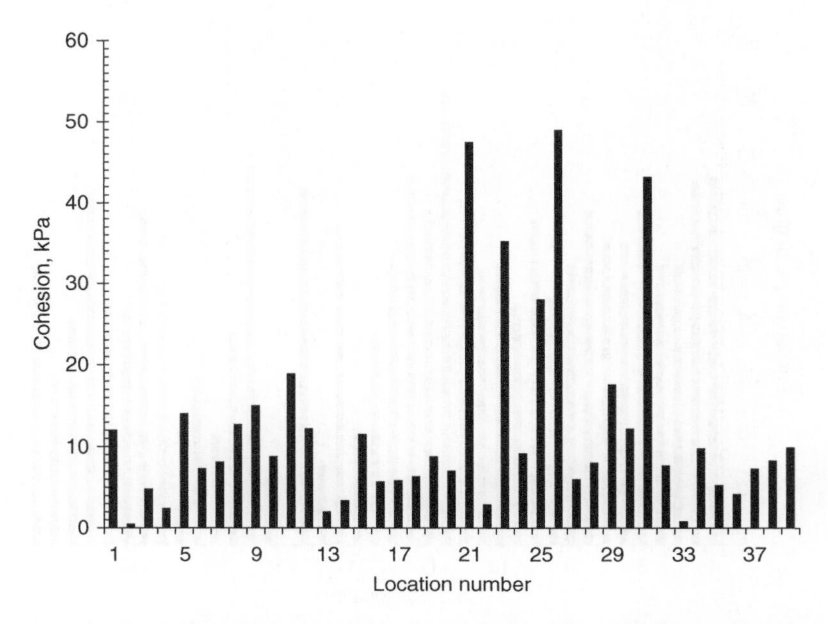

Figure 10.21 Cohesion of undisturbed red soils: Sampling locations are in Bangalore city

of the undisturbed samples develop cohesion of 2 to 20 kPa (Figure 10.21), and 64% of the samples have friction angles ranging between 29 and 39° (Figure 10.22). The red soil data reported in Figures 10.16 to 10.22 are from Ramaiah and Rao (1969).

Figures 10.23 and 10.24 present the bore-hole data for red soil deposits from Karnataka. The data show the residual soil profile is about 3–6 m thick and is underlain by hard strata. The SPT penetration values range from 10 to 32 for these soil deposits. The water table could not be encountered at depths of 3.5 to 6 m at both the locations.

Laterites

Figure 10.25 plots the distribution of clay sized fraction (<2 micron) in laterites from Kerala and South Kanara region in Karnataka (KERI, 1978; Basavarajaiah *et al.*, 1982). The plot shows that 81% of the examined samples had low clay contents ranging between 4 and 10%, about 12% of samples had clay contents below 4%, and 19% of samples had clay contents between 10 and 18%. Figure 10.26 plots the Casagrande's A-line plot for laterites from Kerala (KERI, 1978) and shows that majority of the laterites classify as CL or MH soils. Figure 10.27 plots the in-situ void ratios of laterites from Kerala and South Kanara region in Karnataka (KERI, 1978; Basavarajaiah *et al.*, 1982). The plot shows that 64% of laterites have in-situ void ratios ranging between 0.36 and 0.74; while the remaining 34% of samples had void ratios ranging between 0.74 and 1.2. Figure 10.28 shows that laterites from Kerala and South Kanara region in Karnataka (KERI, 1974; Basavarajaiah *et al.*, 1982) have standard Proctor MDD (maximum dry density) values ranging between 1.48 and 2 Mg/m^3; while 68% of laterite samples have OMC ranging between 14

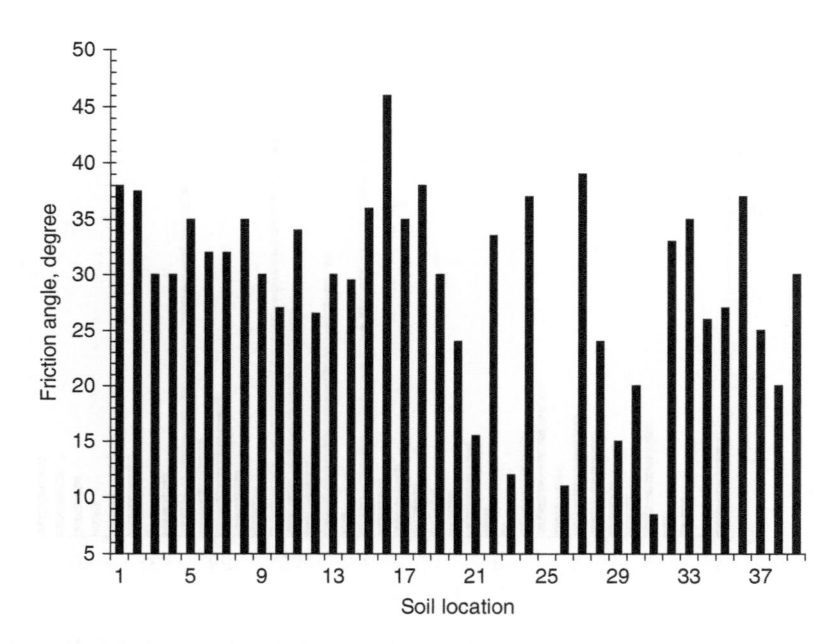

Figure 10.22 Friction angle of undisturbed red soils: Sampling locations are in Bangalore city

Depth below ground m	Soil Classification	SPT value	Grain size distribution			w_n %	w_L %	w_P, %
			Gravel %	Sand %	Silt & clay %			
0	EGL							
1		1 m/N=12						
			-	35	65	11	42.4	22.3
2	Reddish brown sandy clay							
3		3 m/N=32	-	32	68	13.6	30.8	17.6
4								
		4.5 m/N=30						
5								
6		6 m, N=31	-	31	69		30.8	17.9

Borehole terminated at 6.5 m depth

Figure 10.23 Bore-hole profile of red soil deposit from Bangalore

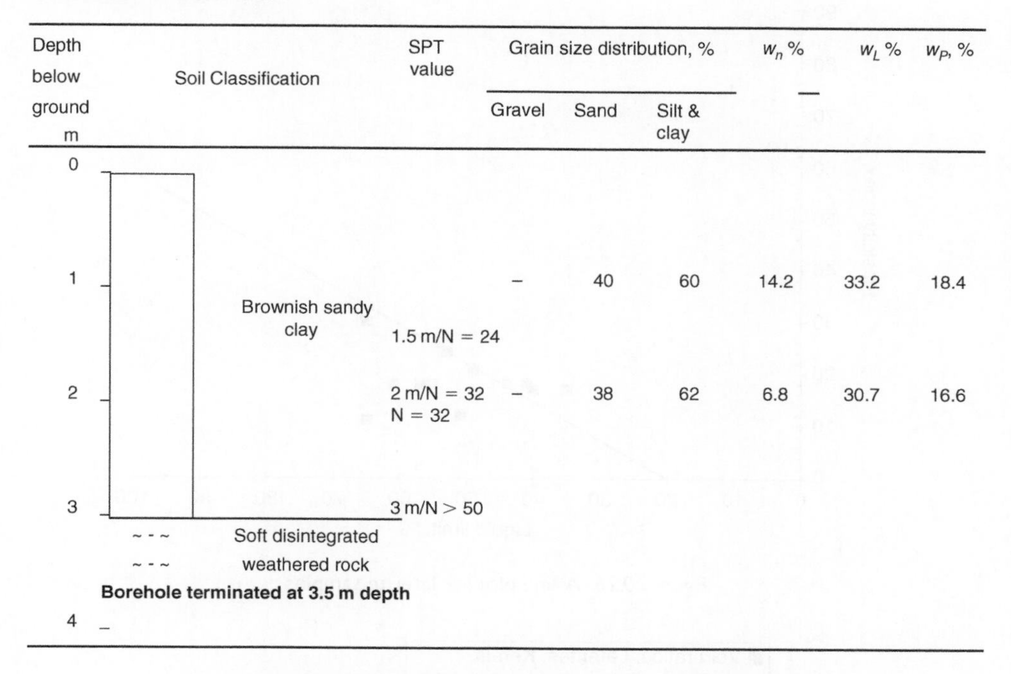

Location Bangalore, Karnataka
Water table-Not encountered
Depth below ground m
Soil Classification
SPT value
Grain size distribution, %
Gravel
Sand
Silt & clay
w_n %
w_L %
w_P, %
0
1
– 40 60 14.2 33.2 18.4
Brownish sandy clay
1.5 m/N = 24
2
2 m/N = 32
N = 32
– 38 62 6.8 30.7 16.6
3
3 m/N > 50
Soft disintegrated
weathered rock
Borehole terminated at 3.5 m depth
4

Figure 10.24 Bore-hole profile of red soil deposit from Bangalore

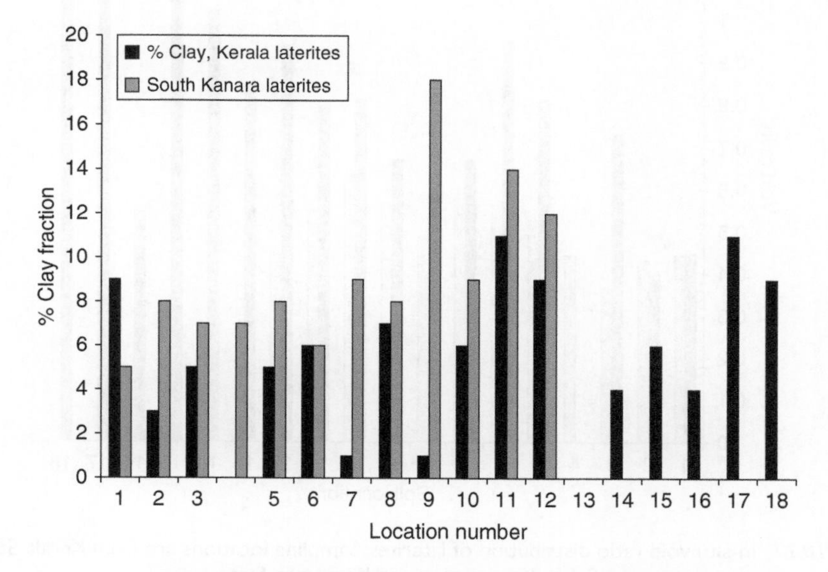

Figure 10.25 Clay sized distribution of laterite samples

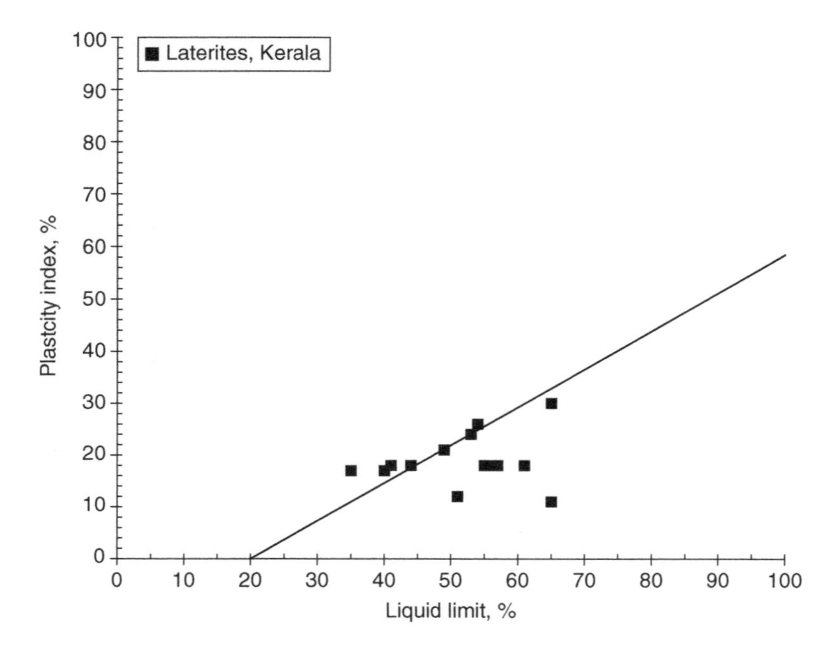

Figure 10.26 A-line plot for laterite samples

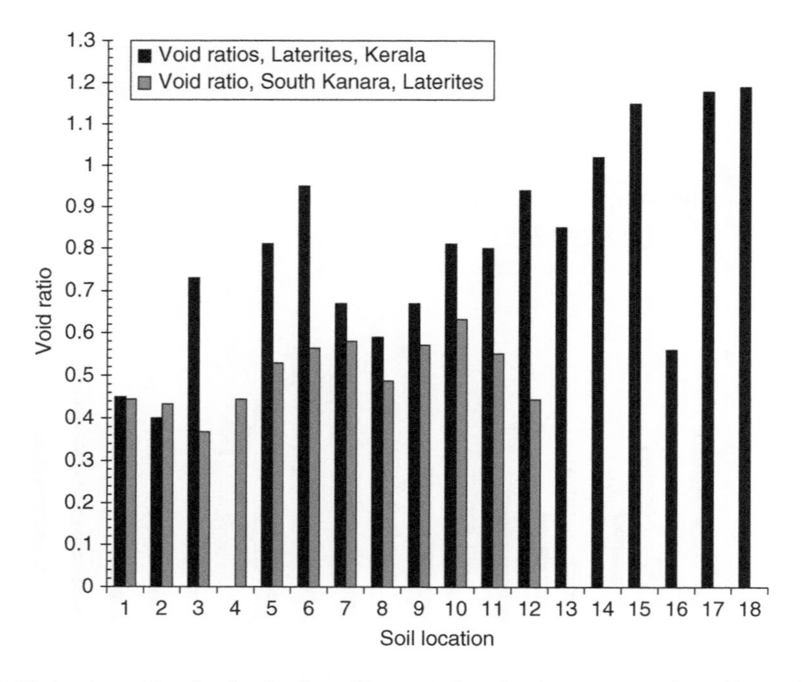

Figure 10.27 In-situ void ratio distribution of laterites: Sampling locations are from Kerala State and South Kanara region in Karnataka State

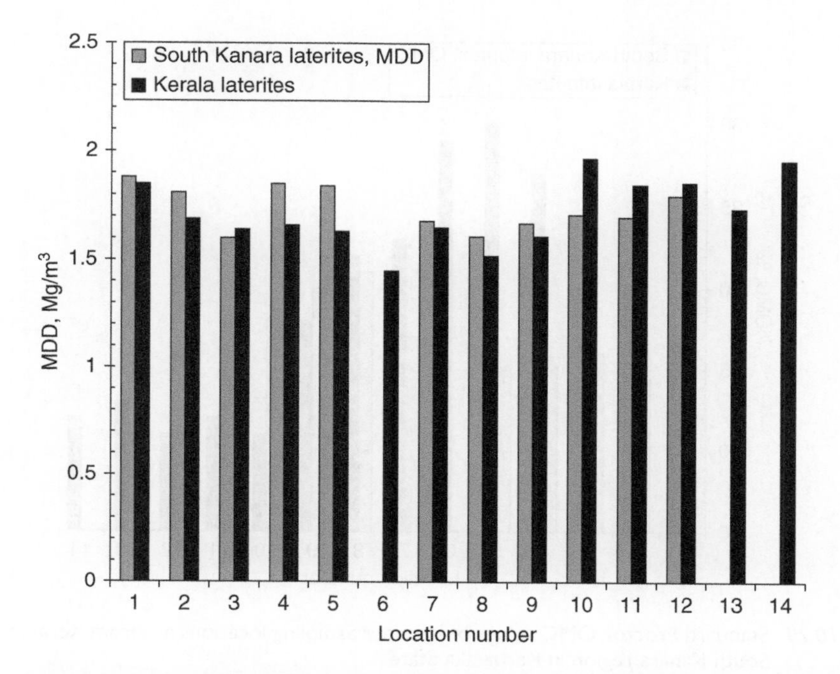

Figure 10.28 Standard Proctor MDD values for laterites: Sampling locations are from Kerala State and South Kanara region in Karnataka State

and 23%, 16% of samples have OMC values below 14%, and 12% of samples have OMC between 23 and 30% (Figure 10.29). Figure 10.30 plots the dry and wet compressive strengths of undisturbed laterites from Andhra Pradesh, Assam, Kerala, Maharashtra, Madhya Pradesh, Tamil Nadu and Rajasthan (Nanda and Krishnamachari, 1958). Data in Figure 10.30 shows that 69% of the samples exhibit dry compressive strengths between 10 and 42 MPa, 22% of samples exhibited dry compressive strengths below 10 MPa, while 9% of samples exhibited dry compressive strengths between 42 and 76 MPa. The laterite samples experienced 10 to 81% reduction in strength on wetting as illustrated by the lower wet compressive strengths of the laterites (Figure 10.30).

Figures 10.31 and 10.32 plot the friction angles and cohesion values of undisturbed laterites from South Kanara region in Karnataka (Basavarajaiah *et al.*, 1982). Data in Figure 10.32 show that 75% of samples have friction angles between 25 and 35°, 17% of samples have friction angles less than 25°, and 8% of samples have friction angles greater than 35° (40°). Data in Figure 10.33 show that 59% of samples have cohesion values ranging between 4 and 10 kPa, 33% of samples have cohesion values between 10 and 24 kPa; while 8% of samples exhibit cohesion values less than 4 kPa. The methods by which the shear strength parameters of the undisturbed laterites were determined have not been detailed in the references cited. Based on the prevailing testing practices, it is suspected that the shear strength tests were conducted on undisturbed

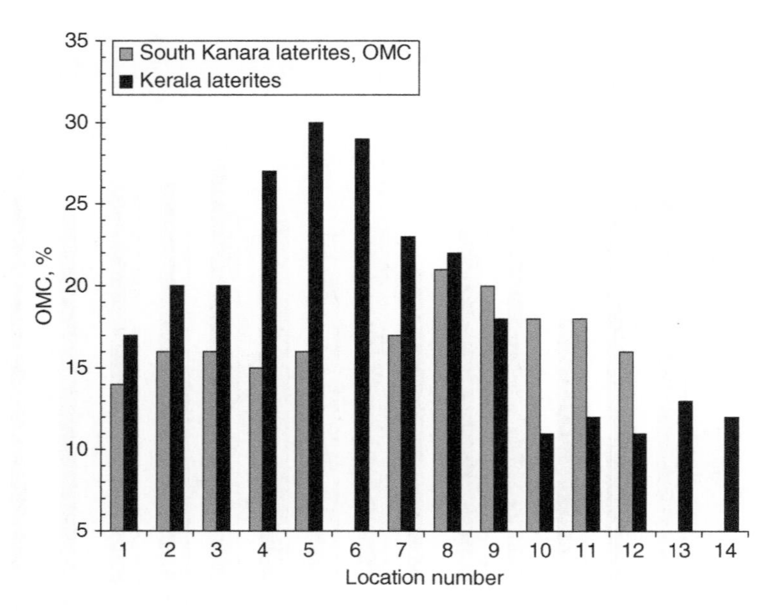

Figure 10.29 Standard Proctor OMC values of laterites: Sampling locations are from Kerala State and South Kanara region in Karnataka State

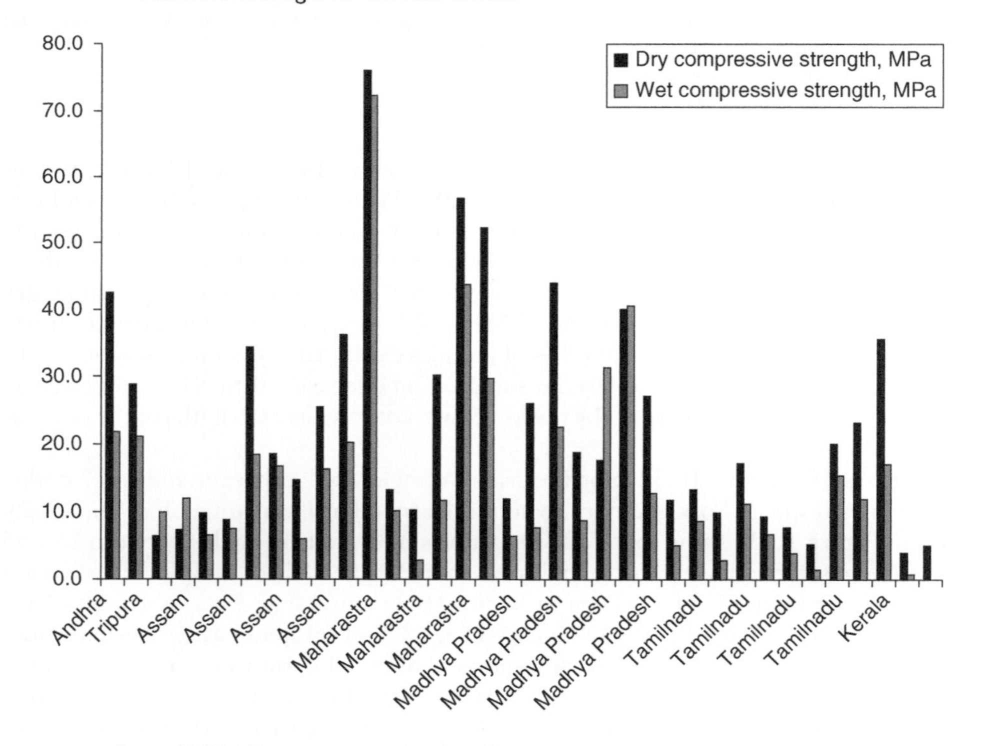

Figure 10.30 Compressive strengths of laterites in soaked and unsoaked states

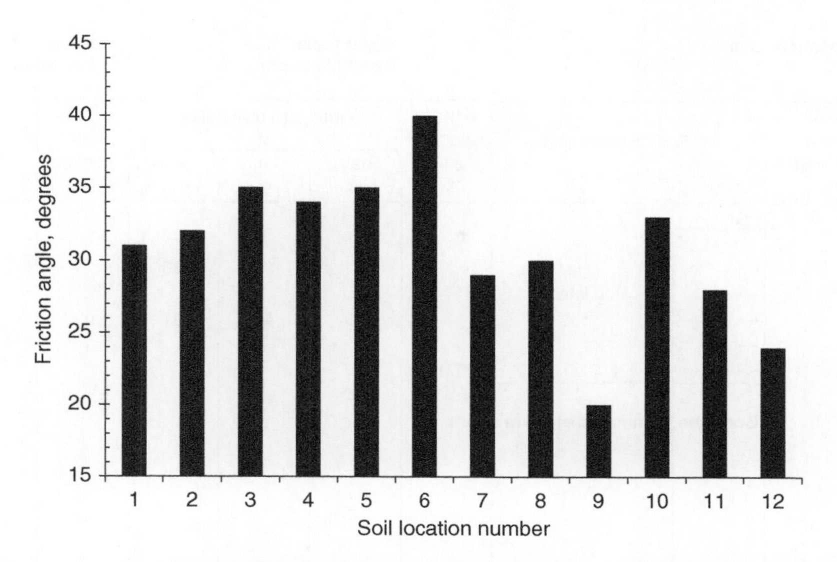

Figure 10.31 Friction angles of laterites: Sampling locations are from South Kanara region in Karnataka State

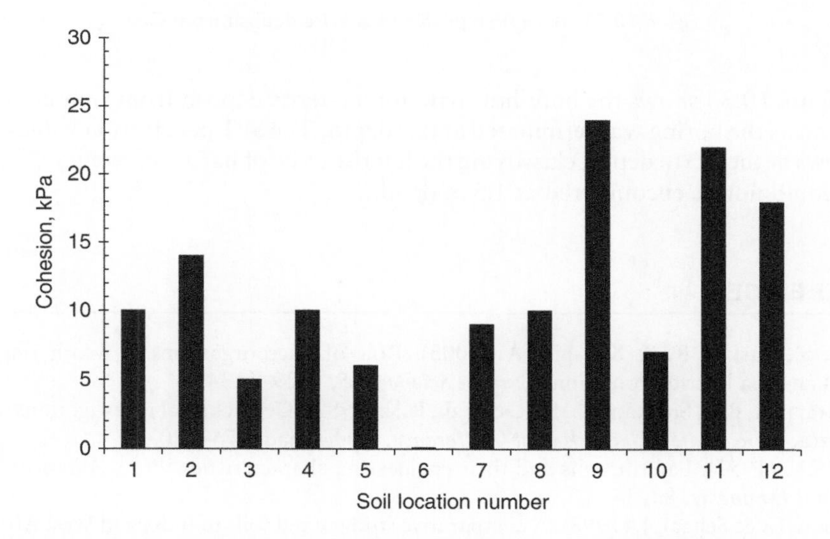

Figure 10.32 Cohesion values of laterites: Sampling locations are from South Kanara region in Karnataka State

laterite samples using the direct shear test procedure on submerged samples. Hence, at best, the friction angles and cohesion parameters reported in Figures 10.31 and 10.32 may be considered as *total* strength parameters rather than *effective* strength parameters.

Location Goa | **Water table** not encountered | **Borehole depth** 1.5 m below EGL

Depth below ground m	Soil Classification		Sample type/SPT value	Grain size distribution			Nat. water content %	
				Gravel %	Sand %	Silt & clay %		
0	EGL							
	~ - ~	Red laterite		31	33	36		
1	~ - ~			29	40	31	9.4	
	~ - ~		SPT/1.3 m N > 50					
2	**Borehole terminated at 1.5 m depth**							
3								

Figure 10.33 Bore hole profile of laterite deposit from Goa

Figure 10.33 shows the bore-hole data for a laterite deposit from Goa up to depth of 1.5 m, as the boring was terminated at this depth. The SPT penetration values exceed 50 blows at the 1.3 m depth, classifying the laterite to be of hard consistency. The water table could not be encountered at 1.5 m depth.

REFERENCES

Baskar, S., Baskar, R. & Kaushik, A. (2003). Role of microorganisms in weathering of the Konkan-Goa laterite formation. *Current Science*, 85, 1129–1134.

Basavarajaiah, B.S., Sreekantiah, H.R. & Yaji, R.K. (1982). Geotechnical investigations of South Kanara. *Proc. Indian Geotechnical Conference*, Hyderabad, 175–180.

Bawa, S.K. (1957). Laterite soils and their engineering characteristics. *Proc. American Society of Civil Engineers*, 83, 1–15.

Bourgeon, G. & Sehgal, J. (1998). A comparative study of red soils of India and West Africa, and their management for sustainable agriculture. In: J. Sehgal, W. E. Blum and K. S. Gajbhiye (eds) *Red and Lateritic Soils; Vol. 1, Managing Red and Lateritic Soils for Sustainable Agriculture*, pp. 77–92. New Delhi: Oxford and IBH Publishing Co.

Coulombe, C.E., Wilding, L.P. & Dixon, J.B. (1996). Overview of vertisols: Characteristics and impacts on society. In: D. L. Sparks (ed.), *Advances in Agronomy*, 57. New York: Academic Press.

Das, P.K. & Mahalik, N.K. (1982). Study of building failures in laterite foundation. *Proc. Conference on Construction Practices and its Implementation in Geotechnical Engineering*, Surat, India, 299–302.

Katti, R.K. (1975). Regional soil deposits of India. *Proc. Fifth Asian Regional Conference*, Vol. 2, 35–52, Bangalore, India.

KERI. (1978). Investigation of the geotechnical properties of laterites of Kerala with special reference to its origin and properties of parent rock. *Annual Review 1978 of Problem No. XXXIV (2)*. Peechi, Kerala, India: Kerala Engineering Research Institute (KERI).

Krishnan, M.S. (1982). *Geology of India and Burma*. New Delhi: CBS Publishers.

Krishnappa, A.M., Pandurangaiah, K. and Hegde, B.R. (1998). Rainfed soils of Karnataka (India): Resource development on watershed basis. In: J. Sehgal, W. E. Blum and K. S. Gajbhiye (eds) *Red and Lateritic Soils: Vol. 2, Red and Lateritic Soils of the World*, 101–113. New Delhi: Oxford and IBH Publishing Co.

Kyuma, K. & Takaya, Y. (1969). Black soils in Eastern India. *The Southeast Asian Studies*, 6, 247–256.

Nambiar, K.K.N., Balakrishnan, P.R. & Zacharias, G. (1966). Studies on laterite powder-with special reference to Kerala. *Report No. 3/66*. Kerala, India: Kerala Engineering Research Institute.

Nanda, R.L. & Krishnamachari, R. (1958). Study of soft aggregates from different parts of India with a view to their use in road construction, II-Laterites. *Journal of the Indian Roads Congress*, 22, 1–32.

Ramahashay, B.C., Bhavana, P.R., Rao, K.S. & Mehta, V.K. (1984). Ion exchange properties of lateritic soil from Calicut, Kerala. *Journal of Geological Society of India*, 25, 466–470.

Ramaiah, B.K. & Rao, S.K. (1969). Soil distribution and engineering problems in Bangalore area. *Research Bulletin, Golden Jubilee Volume*, Bangalore University, pp. 1–102.

Reddy, D.V. (1995). *Engineering Geology for Civil Engineers*. New Delhi: Oxford and IBH Publishing Co.

Sehgal, J. (1998). Red and lateritic soils: An overview, Red and Lateritic Soils. In: J. Sehgal, W. E. Blum and K. S. Gajbhiye (eds.) *Red and Lateritic Soils; Vol. 1, Managing Red and Lateritic Soils for Sustainable Agriculture*, 3–10. New Delhi: Oxford and IBH Publishing Co.

Sehgal, J., Challa, O., Thampi, C.J., Maji, A.K. & Naga Bhushana, S.R. (1998). Red and lateritic soils of India. In: J. Sehgal, W. E. Blum & K. S. Gajbhiye (eds) *Red and Lateritic Soils; Vol. 2, Red and Lateritic Soils of the World*, pp. 1–18. New Delhi: Oxford and IBH Publishing Co.

Seshagiri Rao, K. & Raymahashay, B.C. (1981). Influence of clay minerals and iron oxides on selected properties of two lateritic soils. *Indian Geotechnical Journal*, 10, 255–266.

Singh, S. (1954). A study of the black cotton soils with special reference to their coloration. *Jour. Soil Science*, 5, 289–299.

Singh, S. (1956). The formation of dark colored clay-organic complexes in Black soils. *Journal of Soil Science*, 7, 43–58.

Skjemstad, J.O. & Dalal, R.C. (1987). Spectroscopic and chemical differences in organic matter of two vertisols subjected to long periods of cultivation. *Australian Journal of Soil Research*, 25, 323–335.

Wadia, D.N. (1926). *Geology of India*. London: Macmillan Company.

11

Residual soils of Southeast Asia

Harwant Singh
Universiti Malaysia Sarawak, Kota Samarahan, Malaysia

Apiniti Jotisankasa
Kasetsart University, Bangkok, Thailand

Bujang B.K. Huat
Universiti Putra Malaysia, Serdang

11.1 INTRODUCTION

This chapter discusses the engineering practice in tropical residual soils in South-east Asia, drawing experiences from three Southeast Asian countries, namely Malaysia, Thailand and Singapore. Tropical residual soil and the factors affecting its formation in the local environment are first briefly reviewed with a discussion on the descriptive scheme used and delimitation considered as soil in engineering practice on the weathered profile. The distribution of tropical residual soils and their characteristics on different rocks follows, and then some of their physico-chemical properties are given. This is followed with a discussion on their engineering significance and how tropical residual soils affect slope stability, dam foundations, etc., drawing on local case histories as illustrations.

Tropical residual soils form in the tropical areas i.e. generally regions or climatic zones between the latitudes 20°N and S of the Equator. A thick mantle of residual soils, by definition formed *in-situ* derived from the weathering of rocks, is found in South-east Asia on account of its tropical climate. These widespread soils are a prominent feature encountered in the entire spectrum of engineering construction work spanning from hill side development, highway cut-slopes, foundation work to dam site excavations and many more. They are also widely used as construction fill materials across the board e.g. for highway embankments, earth dams, fill platforms for housing, etc.

Given the humid tropical climate that prevails, characterised by high temperatures and heavy rainfall, the formation of tropical residual soils is intense with a predominance of chemical weathering of rocks resulting in deep weathering profiles often exceeding 30 m. However, South-east Asia also spans a wide swathe of the tropics. Singapore, the nearest to the equator, followed by Malaysia to its north both experience the wettest tropical climate. Thailand, situated further north and stretching over a wider latitudinal expanse, encompasses all three types of tropical sub-climates. This has resulted in varying tropical soils according to the predominant type of tropical sub-climate. The types and properties of the tropical residual soils encountered at each individual construction site depend, of course, on the bedrock or parent material at the site, among other factors. While tropical residual soils are generally good engineering materials, some soils may be problematic.

The tropical soils and the engineering practices with respect to these soils in the aforementioned South-east Asian countries are discussed below.

11.2 RESIDUAL SOILS OF MALAYSIA

Soil formation

The entire Malaysian land mass experiences the wet type of tropical climate (Af) – with relatively abundant rainfall in every month of the year. Therefore chemical weathering of the rocks predominates, and is intense in Malaysia involving decomposition and breakdown of minerals with the rise of new, secondary minerals such as clay minerals and iron oxides/hydroxides (gibbsite, goethite, etc.). As the rock type or parent material is one of the main geological factors in the formation and properties of tropical residual soils, the residual soils derived from all three major rock classes, namely igneous, sedimentary and metamorphic rocks are present in Malaysia. The other soil forming factors are also very active.

Weathering profiles

The appraisal of the resulting soil profile is a *sine qua non* primary prerequisite for engineering characterisation of the tropical residual soils. Although various schemes for weathering profiles have been proposed for tropical residual soils, the scheme based on Little (1969) with six weathering grades classification (BS 5930 1981, British Standards Institute) developed from the working parties, namely, Core Logging (Anon., 1970), Engineering Geology Mapping (Anon., 1972) and Rock Masses (Anon., 1977). The description of the physical disintegration and morphology of the weathered profile is almost exclusively used in Malaysia. Tan (2004) states that although being derived from work done on granites (igneous rocks), its ease for use in the field and amenability for adaptation for sedimentary and meta-sedimentary rocks has, with some care or experience, made it functional and practical for engineering purposes. The tropical residual soils are considered to encompass grades IV to VI materials of the foresaid scheme by Little (1969).

Although of substantial thickness, the thickness of the various weathering profiles do differ from site to site, as well as on different parent bedrocks or rock types. Some generalisations made by Tan (2004) have granitic soil profiles often extending up to or in excess of 30 m, at times even 50 m; while residual soils of sedimentary/metasedimentary rocks, such as shales and schists are often thinner, of around ~10 m only.

Soil distribution

As residual soils are derived from the *in-situ* weathering of the parent bedrock, their distribution is therefore closely related to the distribution of the various rock types in the country. Effectively, the geologic map showing the various rock formations is a good guide to the distribution of the various types of tropical residual soils. The distribution of residual soils in Peninsular Malaysia given by Ooi (1982), of his two very broad major 'classes' of residual soils based upon their occurrence over igneous and sedimentary rock formations, is given in Figure 11.1.

This, however, at best serves as a very rough or broad guide. The first limitation, namely the very small scale of the geologic map (1:2,000,000), results in site variations or details being lost on the map. The second limitation is that sedimentary and meta-sedimentary rocks are lumped together, while in fact, they are different rock types with very contrasting characteristics which can produce very different residual soils. The residual soils have contrasting characteristics even amongst the sedimentary rocks themselves e.g. sandstone versus shale versus limestone, or metasedimentary rocks e.g. quartzite versus schist, etc. Similarly, igneous rocks spanning from granites to basalts produce entirely different residual soils, for instance, sandy versus silty/clayey soils respectively. Figure 11.2 shows a map of the extracted lithologies from the geological map (Singh and Huat, 2004).

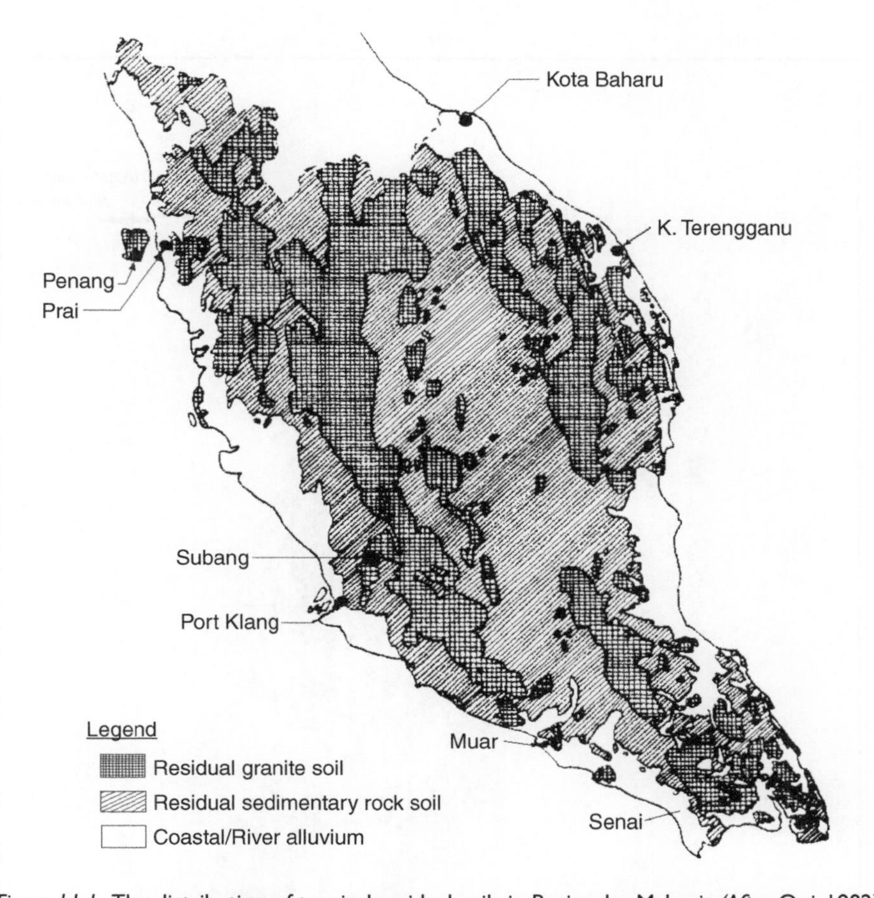

Figure 11.1 The distribution of tropical residual soils in Peninsular Malaysia (*After* Ooi, 1982)

Large tracts of areas were found that have not been lithologically demarcated although, in instances, the composite lithology is mentioned. The residual profile, therefore, cannot be assessed from the geology in areas of undifferentiated lithology. In addition, larger scale geologic maps, e.g. 1:25,000 or 1 inch = 1 mile (1:36,630), with more details are essential for engineering purposes, to be followed by detailed site mapping and investigations, including bore holes. As the geology or lithology varies laterally and vertically so do the concerned residual soils, in particular, for example, on the sedimentary and metasedimentary rock formations with their associated residual soils.

However, it must also be borne in mind that the intensive and pervasive chemical weathering in the Malaysian humid tropics has resulted in a thick layer of residual soils overlying almost all rock formations, thus shielding the bedrock from direct surface observation; in general, the rock/bedrock will only be encountered at depth. As a consequence, the geological mapping of various rock formations is itself, to some extent, based on 'soil mapping' or the ability to recognise or differentiate various types of residual soils to infer their parent bedrock.

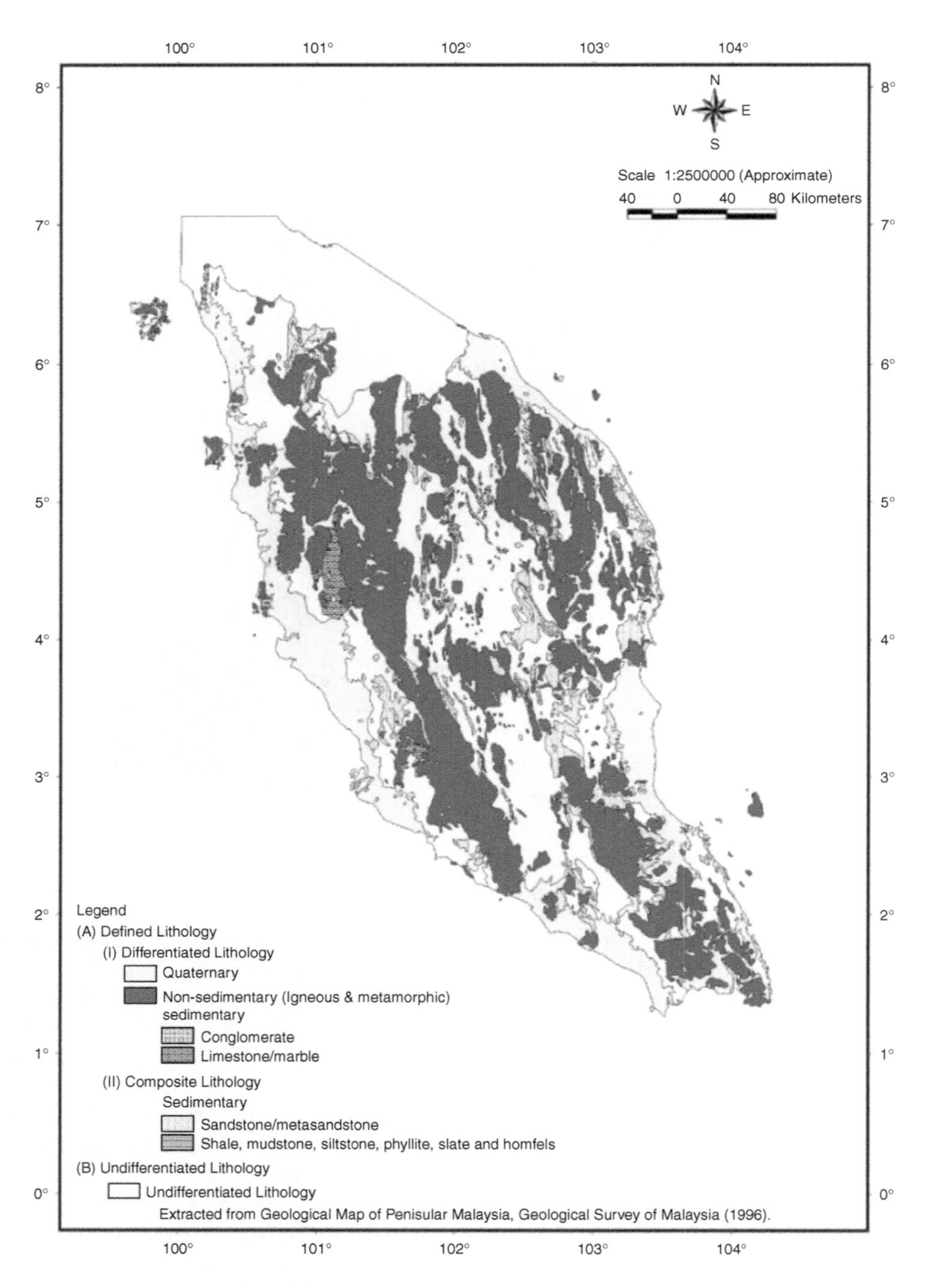

Figure 11.2 Map of Peninsular Malaysia showing the extracted reported lithology

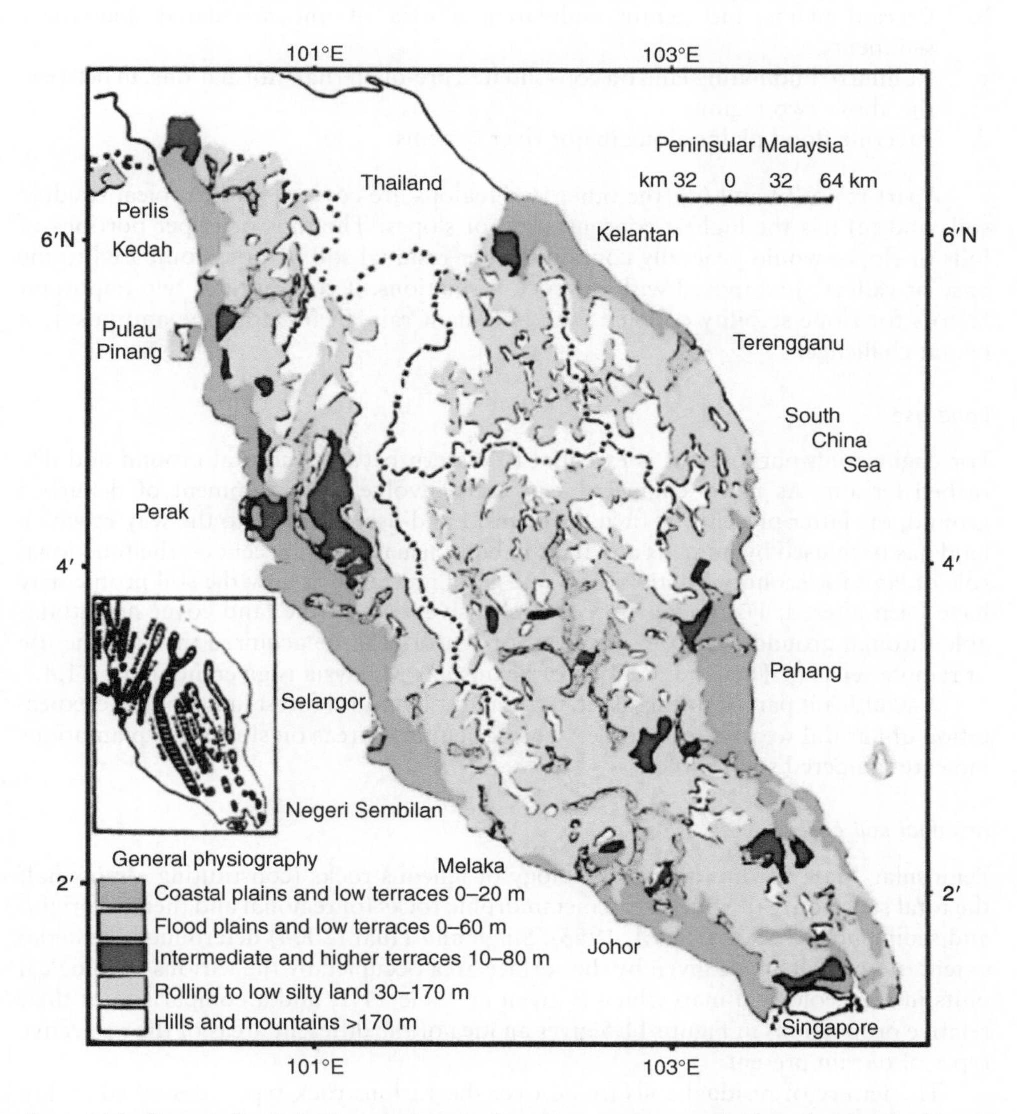

Figure 11.3 General Physical Map of Malaysia (http://www.apipnm.org/swlwpnr/reports/y_ta/z_my/mymp211.htm)

Relief and weathered profiles

The distribution of residual soils is also related to the relief. A general physical map of Malaysia is given in Figure 11.3.

For the case of Peninsular Malaysia, the physiography can be divided into the following land categories:

a. Steep mountain ranges in the central region (running from N.W. to S.E.) reaching a height of 2,400 m above mean sea level.

b. Coastal plains and gently undulating alluvia of unconsolidated quaternary sediments.
c. Rolling to undulating land on consolidated pre-quaternary formations, in between the above two regions.
d. Riverine flood plains along major river systems.

Apart from (b) and (d), the other two regions are covered with tropical residual soils and (a) has the highest concentration of slopes. The tops or upper portions of hills or slopes would generally contain thicker residual soil mantles compared to the base or valleys. Juxtaposed with the rock formations, it concentrates two important factors for slope stability coupled with abundant rainfall for drainage and poses an extant challenge.

Land use

For engineering purposes, it is essential to discern between natural ground and disturbed terrain. As many engineering projects involve re-development of disturbed ground, the latter presents its own problems. Land-use, referring to the way in which land has been used by humans and their habitat, usually with accent on the functional role of land for economic activities, is one good indicator of how the soil profile may have been altered. Fortunately, a comprehensive view of the land cover not attainable through ground-based observations or sensors can be acquired through the use of remote sensing. The land use map of Peninsular Malaysia is given in Figure 11.4.

A significant part of the highland terrain in still natural forest indicating the expectation of natural weathering profiles. The agricultural areas on slopes e.g. plantations indicate tampered soil profiles.

Residual soil cover

Peninsular Malaysia has a varied geology of igneous rocks (constituting almost half the total surface area), widespread metamorphic rocks (of regional and thermal origin) and sedimentary rocks (GSDM, 1996). Singh and Huat (2004) determined the aerial extent of each rock type given by the surface area occupied by the various lithological units in the geological map which is given in Table 11.1, and a comparison of their relative proportions in Figure 11.5 gives an idea of the different areas of the respective types of terrain present.

The nature of residual soils found over the various rock types, described in Tan (2004), is elaborated below.

Granitic Soils: Granitic soils occur widely in Peninsular Malaysia as granites form a significant part of the landmass, especially forming many of the hills and mountain ranges in the Peninsula. These soils are also used in construction, especially in hilly terrain such as in highways and dam construction. Granitic soils are generally sandy (high sand content), have lower water content and liquid limits compared to basaltic or gabbroic soils. The more well-graded nature of the granitic soils produces higher compacted densities.

Basaltic/Andesitic Soils: Basaltic and andesitic soils are not as widespread as granitic soils; occurring only in sporadic patches in Peninsular Malaysia. They are more widespread in East Malaysia, such as in the Tawau area in Sabah. There are

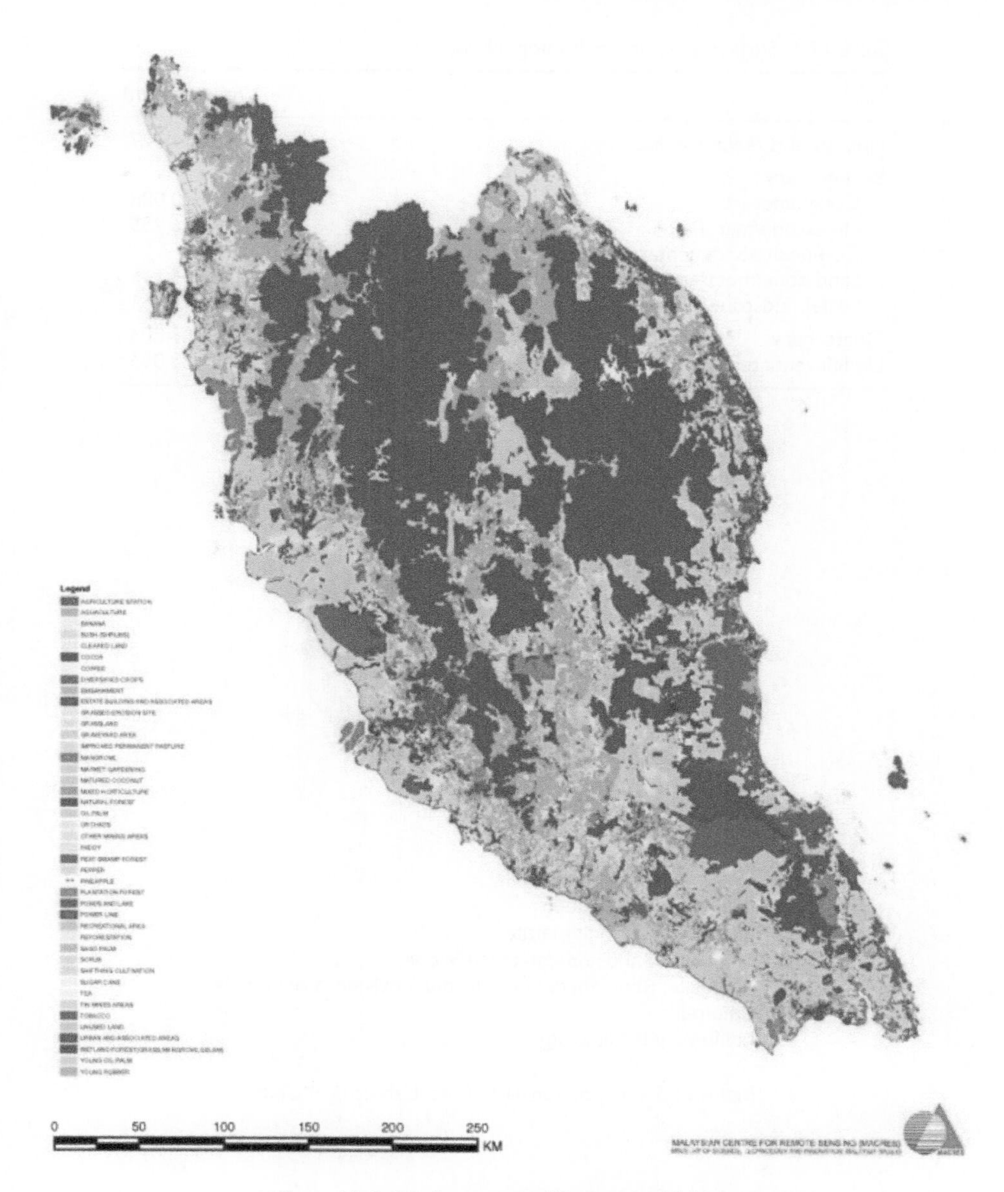

Figure 11.4 Land use map of Peninsular Malaysia
(*Source*: MACRES, MOSTI)

striking similarities between basaltic and andesitic soils as the two rock types are quite similar in texture (fine-grained volcanic rocks) and mineralogical composition. Basaltic and andesitic soils generally have higher water content due to higher fines content as compared to granitic soils. Liquid limits are thus also higher. Due to the higher fines content and not being as well-graded, they produce lower compacted densities.

Table 11.1 Surface area of the lithological units

Unit	*Area (sq. km)*
Igneous and Metamorphic	48,319,327
Sedimentary	
Conglomerate	640,088
Limestone/marble	2,126,255
Composite Sedimentary and Metamorphic	
Sandstone/metasandstone	8,834,115
Shale/mudstone/siltstone/phyllite/slate/hornfels	1,775,451
Quaternary	22,383,004
Undifferentiated Lithology	47,883,066

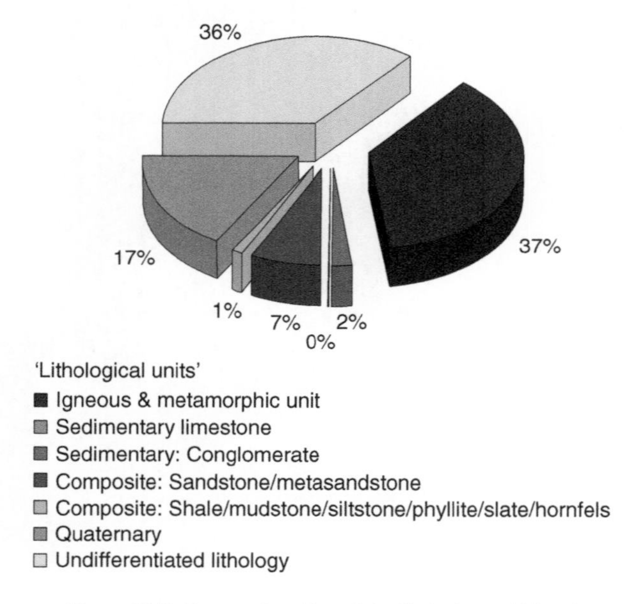

Figure 11.5 Proportionality of the 'Lithological Units'

Gabbroic Soils: Gabbroic soils occur locally in some parts of South Johor (southern state of Peninsular Malaysia) and have been studied in relation to highway cut-slope failure. The properties of gabbroic soils are similar to basaltic soils since both are basic rocks with a similar mineralogical composition, varying only in grain size, the latter being the fine-grained volcanic equivalent. Gabbroic soils have high fines content, high liquid limits and also produce low compacted densities as basaltic soils (as compared to granitic soils).

Shales and schists: Shales and schists show comparable low water content. They also have low liquid limits as they are mostly silty soils and produce low compacted densities. Shales represent soft sedimentary rocks often associated with various engineering problems, one of which is inherent slaking.

Carbonaceous Shales: Black carbonaceous shales are even more problematic since, in addition to slaking, they often contain pyrite (FeS_2) which can cause further deterioration in the shear strength of the black shales and associated residual soils.

Graphitic Schists: As in carbonaceous shales, graphitic schists also often contain pyrite which causes deterioration of the shear strength of the residual soils with time. They have been the subject of numerous studies/investigations of slope failures along major highways in Peninsular Malaysia (Tan, 1992). Highway cut-slopes in graphitic schist soils have invariably failed at slopes of 1V: 1H, 1V: 2H and even 1V: 3H. As graphitic schist soils have high carbon/graphite content, they have lower compacted densities compared to quartz-mica schist soils. The main chemical characteristic of the graphitic schist soils is the acidity (pH as low as $\sim$4) of the pore fluids due to oxidation and hydrolysis of the pyrite in the soils producing sulphuric acid as one of the by-products of the chemical reactions. The pore fluids of graphitic schist soils thus show the highest dissolved ions, with consequently high electrical conductivity values and acidity. It is the chemical properties of the graphitic schist soils that contribute to the deterioration in the shear strength (loss of cohesion, reduction in friction angle, ϕ) of these soils resulting in the frequent occurrences of cut-slope failures involving these soils.

Quartz-Mica Schists: Though not as problematic as the graphitic schists, quartz-mica schists also produce silty soils with foliations which are also subject to slope failures. Again, quartz-mica schists produce mostly silty soils, though there can be some sand due to the presence of quart layers and quartz veins.

Serpentinite: Serpentinite is an example of an ultramafic rock and occurs only in very limited places in Peninsular Malaysia. It is recognisable on site by its dark brown to red colour due to the very high Fe contents in the soils that are the products of weathering of the ultramafic rock.

Limestone: Residual soils formed by the dissolution of limestone bedrock are collectively called 'terra rossa' or residual red clays, and represent the remnants of insoluble residues or impurities in the limestone. Laterisation, or the accumulation of iron oxides in the residues, imparts a reddish colour to the soils, hence the term 'terra rossa'. However, 'terra rossa' does not appear to be common in Malaysia as not many encounters with these soils have been reported in relation to major engineering works.

Karst formation

Limestones are found in various parts of Malaysia. They are especially common in some highly populated areas like the Kelang and Kinta valleys. While they outcrop as hills with characteristic tower block topography masked by vegetation, they also form the subsurface. Limestones present distinctive challenges described further on. Figure 11.6 shows an example of the highly pinnacled limestone bedrock exposed in a previous tin mine/ex-quarry in the Sungai Way area (now Bandar Sunway), Subang Jaya, a suburb of Kuala Lumpur. Similar exposures of karstic limestone bedrock have been retained at purpose-built recreational sites like the Sunway Lagoon area as they enhance the aesthetics.

Figure 11.6 Highly irregular or pinnacled limestone bedrock at Sunway mine/quarry, Selangor, Malaysia (now Bandar Sunway, an affluent township near Kuala Lumpur)

Soil physico-chemical properties

The behaviour of a soil mass is dependent on three fundamental soil properties, namely, its physical properties, chemical properties, and composition of the soil. Systematic studies by Tan (1990a) and Tan and Anizan (1998) on the physical and physico-chemical properties of tropical residual soils in Malaysia are given in Table 11.2, which summarises the physical properties of the tropical residual soils in Peninsular Malaysia. Table 11.3 summarises the chemical properties of the residual soils. Other studies that may be referred to are of Tan (1995, 1996), Tan and Ong (1993), Tan and Tai (1999), Tan and Zulhaimi (2000), Tan and Azwari (2001), and Tan and Yew (2002).

11.2.1 Engineering applications and problem

The engineering significance and applications of tropical residual soils covering the aforementioned weathering grades IV to VI materials (soils) are discussed in the following sections. A discussion of the entire weathering profile of grade I to VI (rocks and soils) can be found in Tan (1995).

Hillside development

Perhaps the most significant engineering application in tropical residual soils and weathering profiles is in the cutting of slopes in hillside development and highway construction.

Hillside development is a major area for engineering practice in tropical terrain, with huge thicknesses of residual soil cover. Figure 11.7, showing a landslide at Bukit Antarabangsa on the outskirts of Kuala Lumpur; depicts a typical landslide in Malaysia.

The tragic incident of the collapse of the Highland Tower in Kuala Lumpur in 1993 brought forth the problems associated with hillside development in Malaysia, in addition to the problem of limestone bedrock and limestone cliffs for the Kuala Lumpur

Table 11.2 Physical properties of tropical residual soils in Peninsular Malaysia

	Granite		*Basalt*				*Schist*			
Property	*JBG*	*SL*	*SEG*	*KTN*	*Andesite*	*Gabbro*	*QS*	*GS*	*Shale*	*Serpentinite*
G_s	2.49–2.61	2.41–2.56	2.54–2.98	2.62–2.86	2.67–2.99	2.21–2.67	2.35–2.74	2.48–2.76	2.49–2.67	2.77–3.65
W_o %	12.0–25.7	27.8–41.6	11.7–40.6	17.3–54.2	16.3–50.6	37.9–57.0	1.20–29.1	4.3–35.8	9.4–41.5	26.1–69.0
LL %	55.0–91.0	58.0–87.0	50.0–84.0	31.0–85.0	57.5–99.0	72.0–96.0	26.0–64.0	26.0–79.0	30.0–81.0	37.0–96.0
PL %	33.0–50.0	31.0–45.0	29.0–44.0	26.0–52.0	42.1–85.1	36.0–52.0	21.0–44.0	18.0–45.0	22.0–42.0	30.9–70.0
PI %	22.0–44.0	26.0–42.0	21.0–40.0	4.0–36.0	8.9–45.7	36.0–45.0	4.0–22.0	4.0–37.0	8.0–39.0	6.1–47.9
G %	0	0	0–67	0–55	0–38	0	0–10	0–25	0	0–6
S %	39.0–76.0	40.0–52.0	4.0–56.0	4.0–48.0	3.0–51.0	0–9.0	3.0–84.0	1.0–49.0	26.0–72.0	1.0–65.0
M %	8.0–37.0	29.0–45.0	4.0–61.0	16.0–66.0	15.0–53.0	37.0–69.0	15.0–82.0	38.0–98.0	26.0–51.0	7.0–69.0
C %	5–37	15–23	5–64	0–63	18–79	28–60	0–42	0–47	2–25	21–72
ρd_{max} (g/cm^3)	1.45–1.58	1.41–1.55	1.26–1.80	1.22–1.60	1.15–1.96	1.27–1.39	1.39–1.76	1.33–1.62	1.42–1.76	1.24–1.71
W_{opt} %	23.4–29.0	20.0–32.0	20.0–44.0	29.0–47.5	14.5–47.5	24.0–32.0	13.3–21.2	17.1–30.2	16.0–29.0	22.0–41.0
Class (fines)	MH	MH	MH	ML-MH	MH	MH	CL-ML/MH	MH-ML	MH/CL	ML/MH-CH

JBG = Johor Baru Granite; SL = Second Link (Highway), SEG = Segamat, KTN = Kuantan, QS = Quartz-mica Schist, GS = Graphitic Schist G_s = specific gravity of particles; W_o = natural water content; LL = liquid limit; PL = plastic limit; PI = plasticity index; G = gravel; S = sand; M = silt; C = clay; ρd_{max} = maximum dry density; W_{opt} = optimum water content.

Table 11.3 Chemical properties (pore fluids chemistry) of tropical residual soils in Peninsular Malaysia

	Granite		*Basalt*				*Schist*			
Property	*JBG*	*SL*	*SEG*	*KTN*	*Andesite*	*Gabbro*	*QS*	*GS*	*Shale*	*Serpentinite*
pH	4.33–6.53	5.01–5.70	6.09–7.13	5.94–6.94	5.68–6.30	3.80–6.70	6.50–6.80	4.50–6.80	5.40–6.50	5.21–5.80
Conductivity (uS/cm)	105.8–267.0	89.7–123.0	83.0–552.0	50.0–229.0	57.0–247.0	92.9–810.0	20.0–300.0	100.00–2400.00	81.80–180.00	20.00–224.00
Na (ppm)	1.11–15.60	6.30–8.00	7.30–21.50	2.06–21.50	4.06–8.90	5.30–6.90	11.00–57.00	13.50–78.60	5.30–13.60	0.08–8.23
K (ppm)	0.80–4.60	0.50–1.10	1.23–7.39	1.25–4.31	0.05–11.17	0.36–0.89	0.20–3.30	0.30–13.20	0.70–5.00	0–7.31
Mg (ppm)	0.01–0.07	0.05–0.16	0.08–6.77	0.05–0.68	0.04–1.56	0.06–0.09	0–1.90	0–18.40	0.02–0.23	0.07–3.57
Ca (ppm)	0.20–0.66	0.40–0.60	0.40–7.23	0–3.73	0–3.68	0.10–0.80	0.60–6.60	0.20–4.50	0.40–2.47	0.67–3.97
Cl (ppm)	9.75–20.00	9.00–51.25	3.80–45.80	0	4.50–41.5	9.00–17.50	0–58.0	0–50.0	6.50–38.00	2.50–117.50
SO_4(ppm)	10.71–42.00	5.00–30.00	20.00–50.00	35.00–122.0	5.00–35.00	1.00–27.50	1.00–13.50	1.00–130.00	5.00–15.00	12.50
(Na+K)/(Mg+Ca)	2.00–30.00	11.00–15.00	1.00–19.00	3.00–17.00	3.40–113.00	7.00–82.00	2.00–17.00	4.00–28.00	7.00–19.00	0.04–3.00

Figure 11.7 Landslides at Bukit Antarabangsa on the outskirts of Kuala Lumpur in December 2008

Figure 11.8 Rockfall at Bukit Lanjan, near Kuala Lumpur, in 2003

and Ipoh areas; other urban and hillside development problems include landslides in their widest sense.

Highway construction

Highway construction is also a key area for the engineering practice in the Malaysian tropical terrain. One of the major engineering challenges is from landslides as shown in Figures 11.8 and 11.9.

Graphitic schist soils have been known to fail even at very gentle slopes of 1V: 3H. For case of the Lojing Highway, slope failure occurred at the same spot here at least three times, twice after re-grading of the slope, indicating again the weak nature of the graphitic schist soils.

The materials of different weathering grades can have different properties, e.g. shear strength, and the stable cut-slope angle may differ from grade IV to grade VI.

Figure 11.9 Slope failure in graphitic schist, Lojing Highway (East-West highway) (*After* Tan, 2004)

Different residual soils derived from different parent materials may have widely contrasting properties, e.g. granitic soils versus residual soils of graphitic schists. Cementation of soil particles, as a result of the deposition of secondary iron oxides and hydroxides from chemical weathering, enhances the shear strengths of the soils. While turfing and hydroseeding works very well in grades V and VI materials (more silty and clayey soils), problems may arise in the more sandy grade IV materials of granitic soils. Hydroseeding, for example, is only suitable for soil slopes, i.e. grades IV to VI materials only. Blindly spraying the entire cut slope without due regards to the weathering grades/profiles is wasteful, to say the least.

Foundations

Tropical residual soils are generally good bearing strata with high bearing capacities. The strength or stiffness of a tropical residual soil also generally increases with depth, e.g. SPT 'N' value increases with depth. Numerous borehole data in various tropical residual soils attest to this trend of increasing hardness/strength with depth. The exception to the rule is the slumped zone associated with limestone bedrock, which can have zero SPT 'N' values. This slumped zone can, moreover, underlie stiff to hard residual soils like those of the Kenny Hill formation in the Kuala Lumpur area; hence, a potential danger to foundation works. The slumped zone thus represents a hidden danger or soft 'bottom' that can pose problems if undetected earlier during site investigation. These weak slumped materials are characterised by very low SPT 'N' values of ~0, as shown in Figure 11.10. The slumped zone has been encountered in numerous high-rise building sites in Kuala Lumpur, including the KLCC or Petronas Twin Towers site (Hamdan and Tarique, 1995). Typically for the Kuala Lumpur area, the slumped zone comprises collapsed materials of the Kenny Hill Formation that overlie the Kuala Lumpur Limestone. The thickness of this slumped zone can vary from several meters to tens of meters. The slumped zone can also occur at great depths, say ~100 m from the ground surface. Note that overlying the slumped zone, materials representing the original, undisturbed residual soils of the Kenny Hill formation can have high SPT 'N' values of, say 30–50, or even >50, i.e. stiff to hard materials. Additional characteristics have been discussed in Tan (1988, 1990b) and Ting (1985).

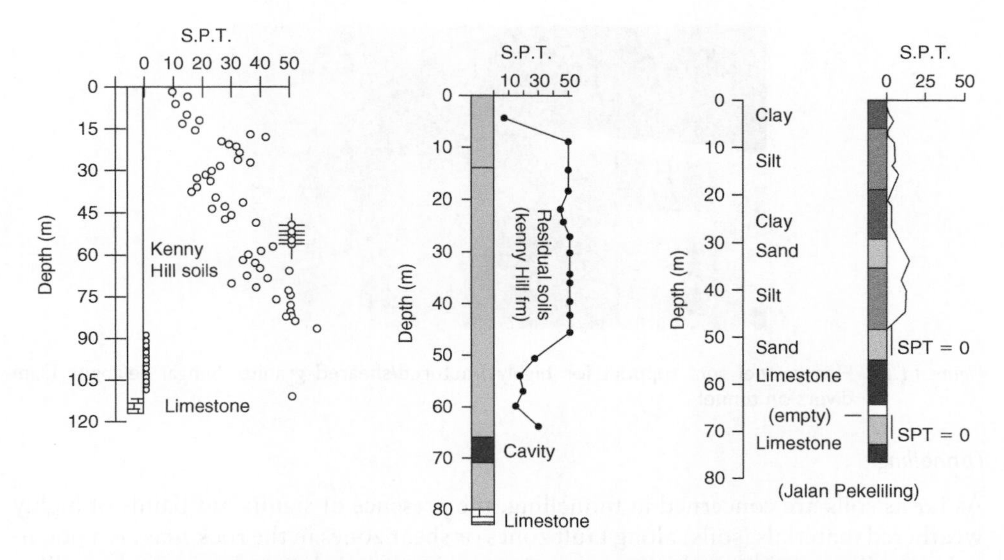

Figure 11.10 Weak collapsed soil zone (slumped zone) with SPT = 0 above limestone bedrock, Central Kuala Lumpur, Malaysia

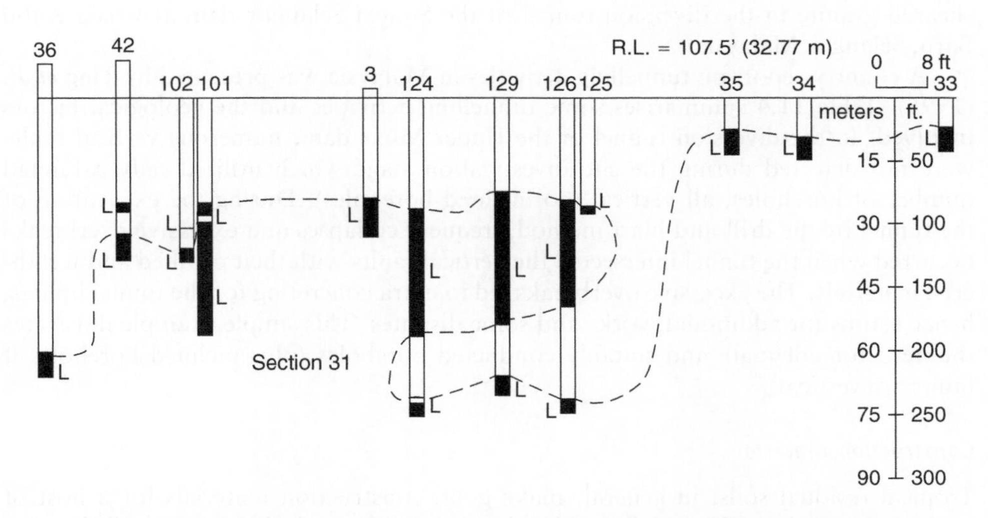

Figure 11.11 Limestone bedrock with massive overhang/sinkhole (Pan Pacific hotel, Kuala Lumpur)

In addition to the occurrence of very weak slumped materials above the limestone bedrock, irregular or highly pinnacled bedrock profile and solution cavities is yet another geological feature of great concern for foundations. Figure 11.11 shows the plan of limestone bedrock at two levels and linear trenches, troughs and overhangs.

Figure 11.12 Heavy steel sets support for highly fractured/sheared granite, Sungai Selangor Dam diversion tunnel

Tunnelling

As far as soils are concerned in tunnelling, the presence of significant bands of highly weathered materials (soils) along fault zones or shear zones in the rock mass is a potential tunnelling problem which can be catastrophic if not detected prior to tunnelling and are encountered without warning. Flow of soils and water can enter the tunnel in such a situation, leading to flooding and even collapse of the tunnel. Figure 11.12 illustrates an example of using heavy steel sets support for the highly fractured and sheared granite in the diversion tunnel of the Sungai Selangor dam at Kuala Kubu Baru, Selangor Malaysia.

A country report on tunnelling activities in Malaysia was presented by Ting *et al.* (1995). Table 11.4 summarises some tunnelling activities and the geological factors involved. In the diversion tunnel of the Upper Muar dam, numerous vertical faults were not detected during the site investigation stage which utilised only a limited number of boreholes, all vertical (no inclined boreholes). During the excavation of the tunnel by the drill-and-blast method, frequent collapses and excessive overbreaks occurred when the tunnel intersected the vertical faults with their crushed and weathered materials. The excessive overbreaks led to extra concreting for the tunnel linings, hence claims for additional works and some disputes. This simple example illustrates the need for adequate and suitably conducted boreholes (plus inclined boreholes if faults are vertical).

Construction material

Tropical residual soils, in general, make good construction materials for a host of engineering projects. Materials for the clay core of an earth dam or rockfill dam are often sourced from the grade VI, clay-rich zone of the granitic weathering profiles. More sandy soils can be obtained from the grade IV zone of the granitic weathering profiles. In engineered fills for housing platforms and highway embankments, various types of tropical residual soils ranging from granitic to shales to schist soils have been used successfully in the past. However, Toh (2003) has demonstrated that the success or otherwise of the use of compacted/engineered fills can be affected by a combination of geological factors such as lithologies and weathering grades, among other things.

Table 11.4 Some tunnelling activities in Peninsular Malaysia (*After* Ting *et al.*, 1995)

Item	*Name of project*	*Application*	*Geology*	*Observation*
(i)	Upper Muar Dam (Water supply), Negeri Sembilan	Diversion tunnel	Granite/faulting	Excessive over-breaks & collapses due to faulting.
(ii)	Kenyir Dam (Hydro-electric), Terengganu	Diversion tunnel	Granite/jointing	Sub-horizontal sheet joints required bolting.
		Pressure tunnel	Granite/jointing	Sub-vertical joints at portals required close-grid bolting.
(iii)	Batang Padang (Hydro-electic) Scheme, Pahang	Diversion tunnel Pressure tunnel Rock cavern (Power house)	Granite/faulting/ weathering	Tunnels collapsed (soil flows) due to tunnelling through weathered materials associated with fault. In-situ ground stress measurements + rock mechanic studies.
(iv)	Sg. Piah (Hydro-electric) Scheme, Pahang	Diversion tunnel Pressure tunnel	Granite/faulting+ Schist (roof pendant)	Relocation of tunnel portals away from major fault.
(v)	Pergau Dam (Hydro-electric), Kelantan	Diversion tunnel Pressure tunnel	Granite/gneiss	On going.
(vi)	Kelinchi Dam Water supply, Negeri Sembilan	Transfer tunnel	Granite/jointed/ some faults anticipated	On going. Only tunnel excavated using Tunnel Boring Machine.
(vii)	Ahning Dam (Water supply), Kedah	Tunnel for water conduit	Sandstone/shale/ conglomerate	–
(viii)	Pedu/Muda Dams (Water supply), Kedah	Transfer tunnel	Sandstone/shale/ conglomerate	–
(ix)	Karak Highway, Pahang	Highway tunnel (single)	Granite/jointing	Additional tunnel being constructed beside existing one.
(x)	Changkat Jering Highway, Perak	Highway tunnel (twin)	Granite/jointing	Massive/stable.
(xi)	Batu Arang Coal Mine, Selangor	Underground mine/tunnel/adits (abandoned)	Coal/sandstone/ shale	Sinkholes & subsidence due to tunnel collapses.
(xii)	Sg. Lembing tin mine, Pahang	Underground tin mine/numerous tunnels/adits (abandoned)	Granite/quartz dykes intruding metasedimentary rock	Seepage/pumping problems due to deep mines.
(xiii)	Kaki Bukit tin mine, Perlis	Tin mine along solution channels/ tunnels (abandoned)	Limestone hill/ bedrock	Mining alluvial tin along solution channels in limestone.
(xiv)	Hanjung Cement Plant, Perak	Access ramp/ road tunnel	Limestone hill	Quarrying for limestone. Caves/stalactites etc. along tunnel route.
(xv)	Ammunition Depot, Tg. Gelang, Kuantan, Pahang	Storage tunnels	Phylitte/Quartzite, complex folding & faults	Heavy tunnel supports steel arches + resin grouted rock bolts.

For example, the compaction of tropical residual soils derived from interbedded sandstone-shale formations can have poor results due to the presence of the harder sandstone cobbles and boulders mixed in the residual soils.

Excavation

Soil excavation would generally be non-problematic. Although secondary deposits of iron oxides/hydroxides may result in layers of iron concretions or hardpans, very hard layers of 'soil' interrupting conventional soil excavation works are normally not found in a wet type (Af) tropical climate; iron concretions or hardpans have been encountered in the Air Keroh area in Malacca, Malaysia, where numerous boulders of hardpans are left in excavation sites in schists (both graphitic and quartz-mica schists).

11.3 RESIDUAL SOILS OF THAILAND

Soil formation

The entire land mass of Thailand experiences a tropical climate. The average annual rainfall in the country ranges between 1000 and 4000 mm. The high temperature, large amount of rain as well as dense vegetation normally results in relatively deep weathering profiles, as chemical weathering of the rocks predominates and is intense. However, only the extreme south of the country experiences the Af type of tropical climate. The other part has a humid subtropical climate consisting mostly of two climatic sub-types: tropical savanna in the northern mountainous regions i.e. the Aw type, while the rest of the country is characterised by the tropical monsoonal type i.e. the Am type (Yoothong *et al.*, 1997). The climatic regions are shown in Figure 11.13.

The progressive latitudinal changes from north to south exerting a profound influence on the climate, therefore, affect the nature of the weathering profile reflected in the development of soil profiles of each sub-climatic type. Tropical residual soils derived from all three major rock types, namely igneous, sedimentary and metamorphic rocks are present.

Weathering profile

The depth of residual soils and weathering profiles are influenced by a number of factors, including parent rocks, geological structure such as fault and joint orientation, vegetation, as well as climate. An example of the weathered mantle, as in Figure 11.14, shows the ground profile in northern Thailand, in an area of mainly medium-to-coarse grain diorite parent material, with estimated residual soil thickness based on resistivity survey. The thickness of the residual soil appears irregular, ranging from 10 to 100 m with corestones being present. Steeply dipping joint sets and faults are common in some areas which perhaps also account for the irregularity of the soil thickness.

Soil distribution

The distribution of residual soils in Thailand, linked closely to the geography and geology, is described by Department of Mineral Resources of Thailand (1999) and in concise summaries by Yoothong *et al.* (1997) and Dheeradilok (1995). Briefly,

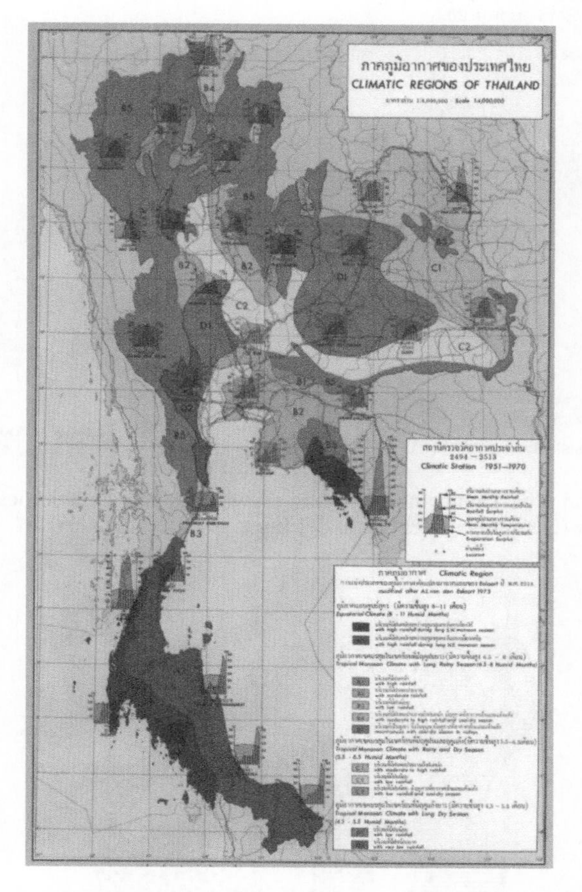

Figure 11.13 Climatic regions of Thailand (http://eusoils.jrc.ec.europa.eu/esdb_archive/EuDASM/Asia/maps/TH2000_1CL.htm)

Thailand can be divided into five main regions i.e. the central, northeastern, eastern coastal, northern and southern peninsular regions which encompass the various geological regions as shown in Figure 11.15. The geology of each region indicates the occurrence of residual soils.

Central region: The central region consists of the alluvial (extensive flood plain of the Chao Phraya river system), coastal and recent deposits bordered by hills in the east and west. While fresh water sediments (with the marine sediments more common towards the southern part) dominate this area, yet, residual soils are present. They are found in the hilly areas in the east derived from the limestone and volcanic rock (andesite, basalt, and rhyolite), and in the highlands and mountainous areas of the Tanaosi Mountain in the west formed from the weathering of various rock types, namely, metamorphic rocks (gneiss, phyllite and schist), sedimentary rocks from Cambrian up to Permian (e.g. sandstone, limestone, shale, siltstone) and igneous rocks (basalt, granite and diorite).

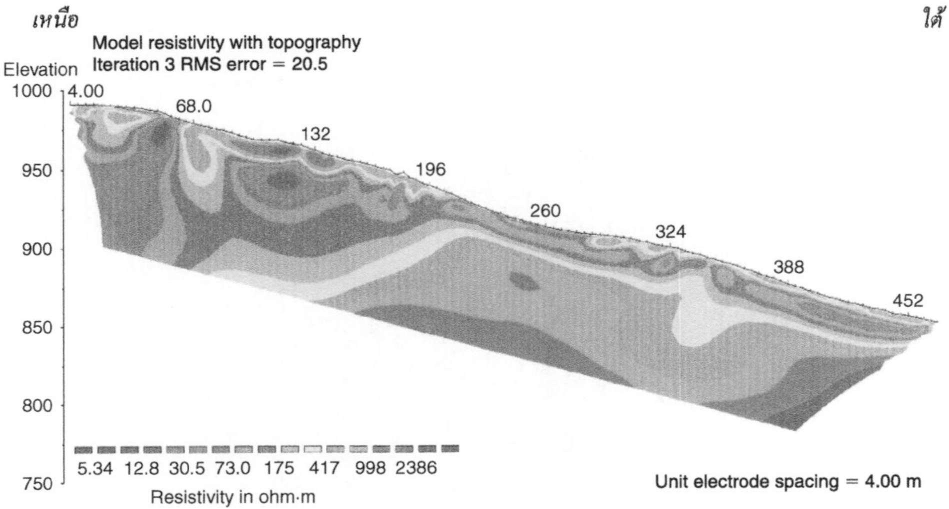

Figure 11.14 Ground profile in northern Thailand showing residual soil thickness based on resistivity survey (Department of Mineral Resource, 2007; GERD, 2008)

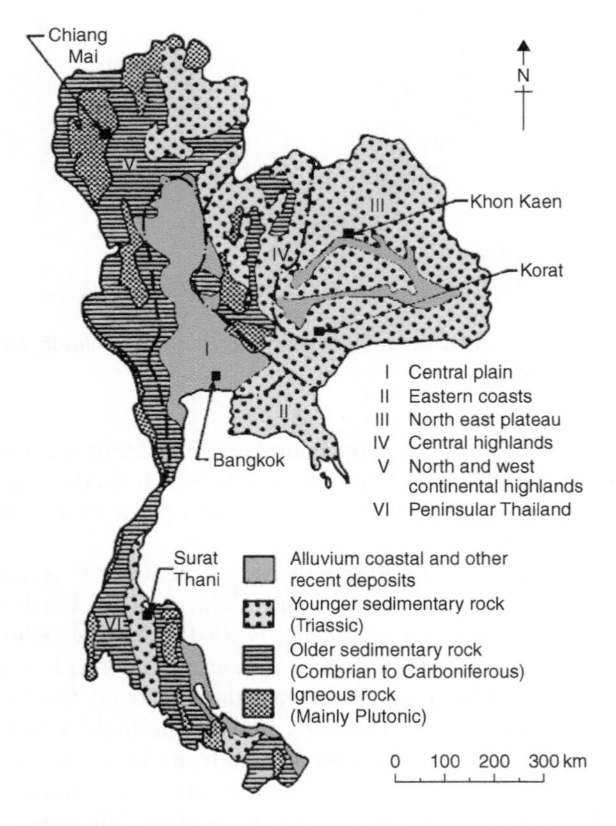

Figure 11.15 Map of Thailand showing the geology of the main geographical regions (Department of Mineral Resource, Thailand)

North-eastern region: The north-eastern region, also called the Korat plateau, is bordered by the Mekong river in the north and east and by a range of hills (mostly quartizitic sandstone) in the west and south (Phetchabun and Dong Phaya Yen Ranges in the west, and the Phanom Dongrak scarps in the south along the Thai-Cambodia border). The plateau is undulating, interspersed with low-lying hills and wide sloping valleys. Its elevations range between 250 m above sea level in the north-west and about 100 m in the south-east (Yoothong *et al.*, 1997). The geology can be described as of Jurassic–Cretaceous rocks (e.g. siltstones, sandstones, conglomerates and shales) overlain by Tertiary semi-consolidated rocks (e.g. mudstones and sandstones). Rock salt, of the Korat Group, also accumulated during the Cretaceous. Residual soils and weathered rocks can be found at shallower depths in steeper terrains. The soils of the region are mostly silty sand and sandy silt of low plasticity, though sandy clay and high-plasticity clay are not uncommon. Laterites and loess are also commonly present in the area at depths between 10–20 m (Pien-wej *et al.*, 1992).

Eastern coastal region: The eastern coastal region is bordered by hills in the north, the Gulf of Thailand in the south and west, and by the Banthat Range in the east along the Thai-Cambodia border. Numerous terrace formations and separate hills consisting of granites of different ages are common in the area. A volcanic plateau is present which consists of corundum-bearing basalt (Yoothong *et al.*, 1997).

Northern region: The northern region mainly consists of hilly and mountainous terrain. The land is flatter towards the lower part. Many wide incised plateaus and narrow inter-mountainous valleys are the common sites of habitation and major cities, such as Chaingmai. The rocks are of igneous or sedimentary origin, somewhat locally metamorphosed, folded or highly affected by past tectonic activity. Limestone ridges and volcanic plateaus are locally present. The rocks range from the Precambrian up to Quaternary. The major metamorphic rocks are gneiss and mica-schists. Outcrops of granites are of different ages and scattered volcanic rocks, such as basalt and andesites, are present

Southern region: The southern region or peninsular Thailand is characterised by a number of mountainous and hilly ranges as well as coastal terraces and plains. Granites in the region are dated from late Triassic/middle Tertiary and limestone from Ordovician/Carboniferous-Permian/Permian (Yoothong *et al.*, 1997). These rocks were locally metamorphosed into slate, schist and gneiss.

Relief and weathered profiles

The distribution of residual soils is also related to the relief. The map of the major landforms of Thailand is given in Figure 11.16.

The landforms, constituting the physiography, show the following conspicuous features of Thailand's terrain:

a. high mountains covering much of northern Thailand and extending along its entire western side down through to the peninsular,
b. plains consisting of riverine flood plain along major river systems,
c. an upland plateau in the northeast,
d. coastal plains.

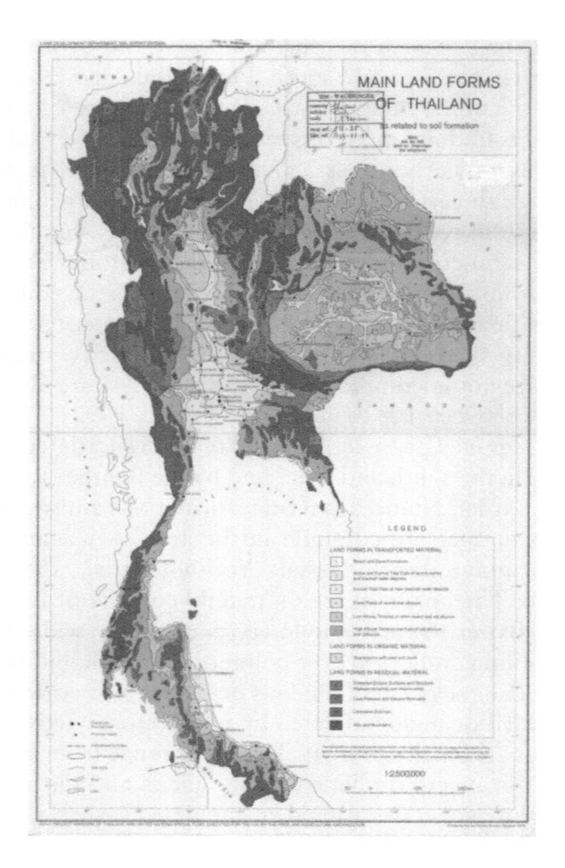

Figure 11.16 General physiography and landforms of Thailand (http://eusoils.jrc.ec.europa.eu/esdb_archive/EuDASM/Asia/maps/TH2002_SOPH.htm)

These physiographic regions are covered by either residual, organic or transported materials. Apart from (b), (d) and a substantial portion of (c), the terrain is covered with tropical residual soils.

Land use

Land-use, referring to the way in which land has been used by humans and their habitat, usually with accent on the functional role of land for economic activities, is one good indicator of natural ground and disturbed terrain. The land use map of Thailand is given in Figure 11.17.

Soil shear strength and volume change

Shear strengths of intact or compacted residual soils in Thailand have been mainly investigated in direct shear tests, both carried out *in-situ* and in the laboratory (Mairaing, 2008; Soralump and Jotisankasa, 2007). The field direct shear test has

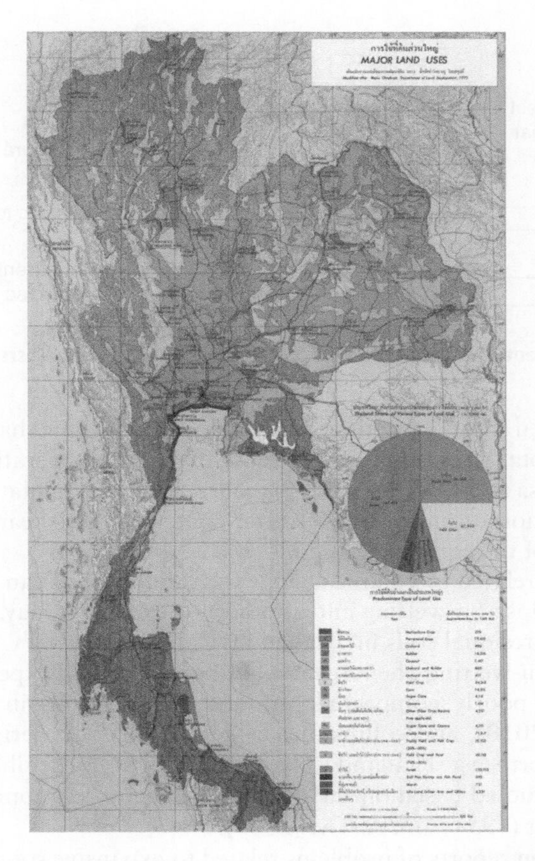

Figure 11.17 Land use map of Thailand (http://eusoils.jrc.ec.europa.eu/esdb_archive/EuDASM/Asia/maps/TH2000_3LU.htm)

been performed using the modified CBR frame (Lee *et al.*, 1985).These large *in-situ* direct shear tests have been employed whenever the influence of macrostructure such as relict joints, largest maximum particle size and the vegetation root strength need to be investigated, as well as when undisturbed samples are difficult to obtain.

A thin wall tube sampler has also been developed at the Geotechnical Engineering Research and Development (GERD) Centre of Kasetsart University, for the simple collection of soil. For sample collection, the sampler is driven into the ground within test pits and commonly into slope faces in research on slope stability.

Triaxial tests have also been carried out on re-compacted residual soils used in the construction of embankment dams, pavement materials, etc. (Cole *et al.*, 1977; Jotisankasa and Vathananukij, 2008; Jotisankasa *et al.*, 2007; Surapol *et al.*, 2008; Soralump *et al.*, 2010; Taesiri and Yuwathanon, 2005). To the author's knowledge, due to sampling difficulties, triaxial tests on intact undisturbed samples of residual soil in Thailand are still relatively scarce.

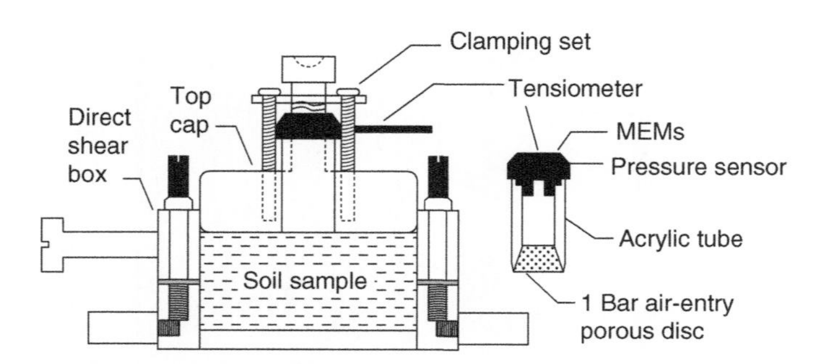

Figure 11.18 Experimental set up of the suction-monitored direct shear tests and KU-tensiometer

The unsaturated shear strength of residual soils in Thailand has mostly been investigated using the total-stress approach without explicit consideration of suction. Until recently, Jotisankasa and Mairaing (2009) incorporated the miniature tensiometers for monitoring of suction during constant-water-content direct shear tests. Figure 11.18 shows the sketch of the equipment used.

The problems related to volume change for residual soils can be divided into two types; one for collapse-on-wetting and another for expansive clay. Both these types of problems exist for residual soils in Thailand.

The collapse-on-wetting phenomenon has been observed experimentally for some intact, as well as poorly compacted, granitic residual soils in southern Thailand (Soralump *et al.*, 2010), and colluvium, as well as silty soil derived from sandstone and siltstone in northeast Thailand. The behaviour of collapsible soils in Thailand has usually been studied based on the double oedometer test approach as outlined in Jenning and Knight (1957) and ASTM (2004).

There have been reports of problems related to expansive soils including in northeast Thailand. Cracks in the superstructure of low-rise buildings are usually observed due to differential movement of the shallow foundations on these materials, which are thought to be attributed to deformation caused by changes in soil suction of these materials (e.g. Nokkaew *et al.*, 2008). Pile foundations are employed instead of the commonly used shallow footing when these problems are encountered in extreme cases (Promboon, 2008). More research is still required on this phenomenon in Thailand.

11.3.1 Engineering applications and problems

Hillside development

Slope stability is one of the most common problems related to residual soils in Thailand. The granitic hills in the north, south-east, and south of Thailand are sites where rainfall-induced landslides frequently take place during intense storms (e.g. Jotisankasa and Mairiang, 2009; Wieland, 1989). Slopes from weathered mudstone, siltstone, shale and volcanic rocks are other common terrains prone to landslides in Thailand (ADPC, 2010). These mass movements usually involve a relatively thin mantle of decomposed rocks, limited to about 0.5–3 m, which are mostly classified as low-to-medium plasticity clayey sand, silty sand (SC, SM, or CL, ML), sandy clay, silt with well-graded

Figure 11.19 Massive landslide in Krabi province triggered by persistent rainfall (photo courtesy of the Bangkok Post)

gravel, pebbles and boulders at greater depths. These rocks sometimes contain joint planes parallel to the slope surface, further inducing mass movement. A massive landslide buried hundred of houses in Krabi province (south of Thailand) in March 2011 (Figure 11.19).

As an overall disaster mitigation tool for planners, several hazard maps have been produced in Thailand based mostly on the weighted factor method, taking into account parameters such as slope gradient, soil type, rainfall intensity, past landslide records, etc. Recently, Soralump and Kulsuwan (2006) also included geotechnical factors such as the percentage of reduction in shear strength of residual soils in *in-situ* state and saturated state, geological structures, return period of rainfall etc. Figure 11.20 shows an example of such map for Phuket island off southern Thailand.

Highway construction

As for site-specific stabilisation of highway slopes in residual soils, a number of approaches have been employed by the Department of Highways, Thailand such as slope regrading, benching, berms, gabion wall, surface water and subsurface water drainage, as well as soil stabilisation using soil nailing, deep mixing, soil cement, MSE and crib walls (Taesiri and Yuwathanon, 2005). An important uncertainty in stabilisation design involves the thickness of the residual soils, and thus boreholes with geophysical methods are normally employed for site investigation, but mostly only for important sites. As for design practice, the groundwater table is one of the uncertainties, and it is sometimes estimated by visual observation of springs coming out of slopes, water levels in boreholes, observation wells, as well as the geological structure. Nevertheless, periodical observations in standpipe piezometers usually indicate no significant changes in ground water levels at depths greater than around 9–12 m (Jamnongpipatkul *et al.*, 2001) and the surges in pore water pressure perhaps only come during intense storms and are confined to shallower depths (Jotisankasa *et al.*, 2008). Due to these complications, the ground water level is usually assumed to be within 1 to 2 m below or directly at the ground surface in stability calculations (Taesiri

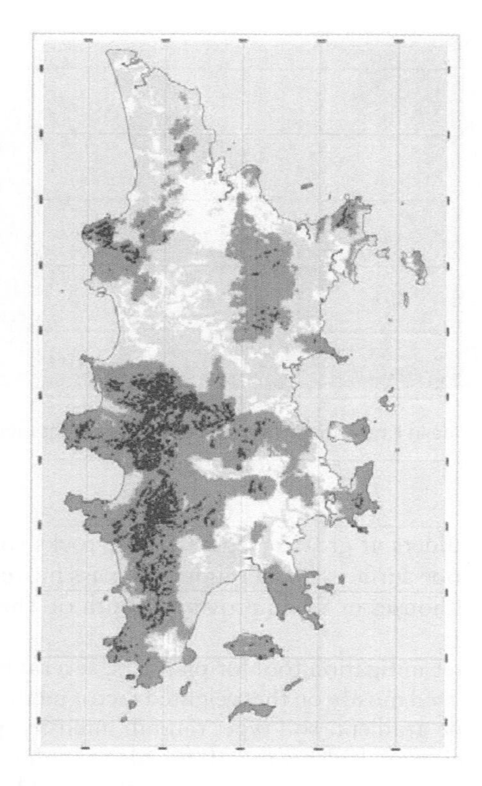

Figure 11.20 Example of landslide hazard map of Phuket considering a 1 year return period of 3 days accumulate rainfall (Soralump and Kulsuwan, 2006)

and Yuwathanon, 2005), which seems to cover the worst case scenario of slopes during intense rainstorms where pulses of positive pore water pressure have been observed (Figure 11.21).

For highway cut/filled slopes, the prevention of surface erosion and shallow mass instability is carried out mostly using shotcrete, ferro-cement, hydroseeding or vetiver grass. Since the last two decades, bio-slope stabilization using vetiver grass, in particular the *Vetiveria zizanioides*, has gained more popularity in geotechnical engineering practice in Thailand (Hengchaovanich, 2000) due to its special characters, namely; its vigorous, massive and strong subterranean root system that vertically reaches 2–5 m in depth depending on the soil type; its relatively lower cost compared to other hard engineering solutions, such as gabions and mattresses and its ability to grow very rapidly and become effective in only 4–5 months versus 2–3 years for trees or shrubs. In addition, the use of vegetation is more aesthetically pleasing to the users of highways, compared to other hard solutions.

One of the problems in using vetiver grass technology, however, is the lack of necessary nutrients in residual soils and the vigorous local weed growth. In this regard, Sanguankaeo *et al.* (2006) have investigated the proper vetiver-planting techniques

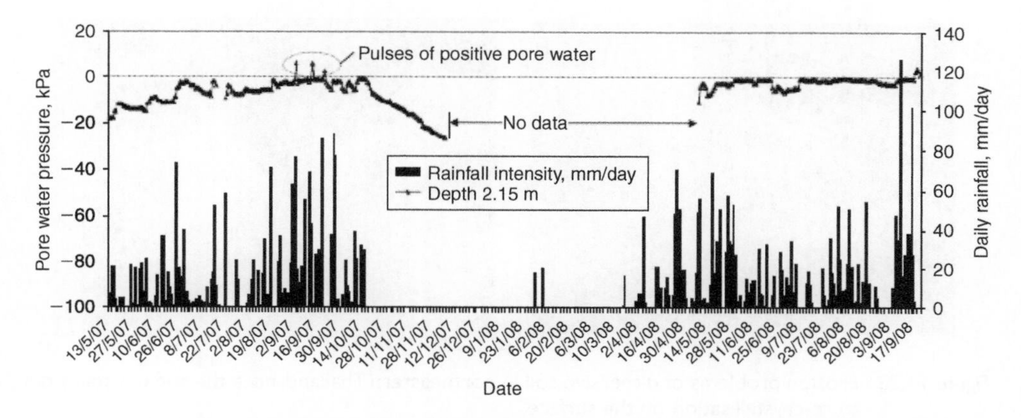

Figure 11.21 Pore water pressure and rainfall, monitored on a study slope in Nakornnayok province

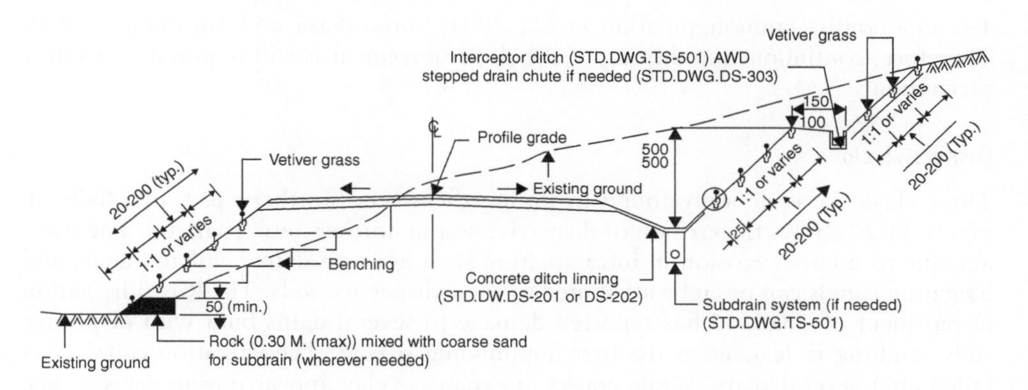

Figure 11.22 Standard drawing of 'vetiver grass planting for highway slope protection' by Royal Department of Highways, Thailand (2006)

in Thailand, including soil fertility improvement and weed control by alternating the vetiver grass with a legume species (*Arachispintoi*), optimum spacing of grasses, suitable planting periods and fertilizers. In addition, low-to-medium plasticity clayey soils with thickness of 10 to 15 cm are sometimes applied as subsoil below the vetiver grass for prevention of water infiltration to deeper ground and for improving the fertility of the subsoil. This method has been established as one of the standard methods for shallow mass stabilization of slope with gradients less than 60 degrees as shown in Figure 11.22.

Nevertheless, in many slope instability situations, a combination of different stabilisation techniques may need to be employed, depending on the specific site conditions and constraints. Figure 11.22 shows an example of the combined use of the MSE wall, gabion, vetiver grass and surface drainage, in order to stabilise a filled slope which suffered shallow mass movement. Nevertheless, in some cases, the slope movement after stabilisation has still been observed to take place at a creep rate of about

Figure 11.23 Erosion problems of dispersive soil in northeastern Thailand; note the sodium stains due to re-crystallisation on the surface

1–3 mm/month (Jamnongpipatkul *et al.*, 2001; Jotisankasa and Soralump, 2008). Therefore, continuous monitoring of the slope movement is still required as a safety precaution.

Dispersive clay

These clays are commonly found in the northeast and northern part of Thailand. Figure 11.23 shows the erosion of dispersive soil in northeastern Thailand. The damage due to internal erosion in infrastructure such as earth dams, embankments and irrigation canals can often be attributed to these dispersive soils. The Royal Irrigation Department of Thailand has reported damage to several dams built with dispersive soils resulting in leakage at the first impounding season. The irrigation canals also failed after several years. While cracks in expansive clay appear during dry seasons, these soon become preferential paths for running water in the wet seasons and the dispersive clays erodes along and through such paths. This problem can also be further exacerbated if dispersive clays are found together with expansive clays (Pornpoj *et al.*, 2006).

The clay fraction of dispersive soils erodes easily in the presence of water by the process of deflocculation. This can happen even in still water when the interparticle forces of repulsion exceed the force of attraction and the clay particles detach into suspension, making the water turbid. The main cause is the Na exchange ion in the clay lattice which is the cause of the repulsion force. No correlation has, however, been found between the plasticity index (PI) and the dispersivity of Thai soils (Nontananandh *et al.*, 2007). Standard procedures such as the Emerson Crumb test, double hydrometer, pinhole test, (e.g. Sherard *et al.*, 1976) are thus normally employed to classify dispersive clays. In many areas of northeastern Thailand, claystone and sandstone bedrock are underlying materials of these dispersive clays which somewhat vary in depth (Cole *et al.*, 1977).

When use of dispersive soils cannot be avoided, the usual treatments for them are either chemical or physical (Cole *et al.*, 1977). The chemical approach involves the use of lime treatment in order to exchange the Na ions on the clay particles with

Ca ions, hence reducing the dispersivity. The physical approach is based on the empirical evidence that the lower permeability (less than 10^{-7} m/s), achieved through adequate compaction minimises the degree of dispersivity. In order to prevent shrinkage, the cracks that could initiate erosion and evaporation should be minimised by sealing with good grass or granular cover.

11.4 RESIDUAL SOILS OF SINGAPORE

Soil formation

According to Pitts (1984), more than two-thirds of Singapore is covered by residual soils. The geology of Singapore consists of: (a) igneous rocks consisting of the *Bukit Timah granite* and the *Gombak norite*, occupying the north and central-north region; (b) sedimentary rocks of the *Jurong Formation*, occupying the west and southwest region; (c) Quaternary deposit of the *Old Alluvium* in the eastern region; and (d) recent alluvial deposits of the *Kallang Formation*, distributed throughout the island (Public Works Department, 1976; Sharma *et al.*, 1999; Leong *et al.*, 2002), as shown in Figure 11.24. Other less common geological formations in Singapore are the *Sajahat Formation*, the *Tekong Formation* and the reclaimed land. Although the *Sajahat Formation* appears only at the northeastern corner on the geological map, it has been found at several locations below the Old Alluvium and consists of well lithified quartzite, quartz sandstone and argillite. The *Tekong Formation* is a thin sequence of mainly sands and gravels.

Singapore is dominated by residual soils from two major geological formations i.e. the Bukit Timah granitic formation and the Jurong sedimentary formation (Rahardjo *et al.*, 2004).

Bukit Timah Granite: The weathering of the Bukit Timah granite has been rapid and extensive, with an average depth of weathering of 30 m and is primarily due to chemical decomposition under the humid tropical climate (Zhao *et al.*, 1994a). Its dominant granitic component is grey and medium to coarse-grained, and consists of cream or pale yellow feldspar, smoky quartz and smaller proportions of reddish-brown biotite and dark hornblende.

Gombak Norite: The Gombak Norite is an association of noritic and gabbroic rock, which outcrops in a restricted area (Bukit Panjang and Bukit Gombak) in the centre of Singapore island. These rocks are coarse-grained and plagioclase-rich, with varying amounts of orthopyroxene minerals in an intergranular structure.

Jurong Formation: The Jurong Formation covers the south, southwest and west of Singapore, with a variety of sharply folded sedimentary rocks, including conglomerate, sandstone, shale, mudstone, limestone and dolomite. It was deposited during late Triassic to early or mid-Jurassic. The Formation has been severely folded and faulted in the past as a result of tectonic movement.

Relief and weathered profiles

Slopes in Singapore, constituting of residual soils, are prone to frequent rainfall-induced slope failures (Poh *et al.*, 1985; Pitts and Cy, 1987; Tan *et al.*, 1987; Chatterjea, 1994). The relief map of Singapore, given in Figure 11.25, the surface topographical

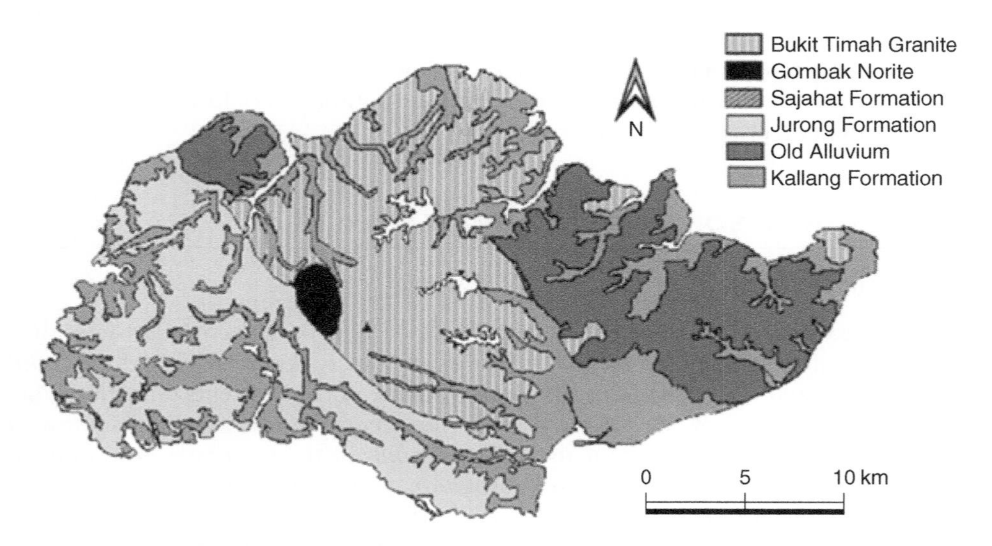

Figure 11.24 Generalised geological map of Singapore (modified *after* Rahardjo *et al.*, 2004)

Figure 11.25 Relief map of Singapore (http://www.worldofmaps.net/uploads/pics/karte_singapur.jpg)

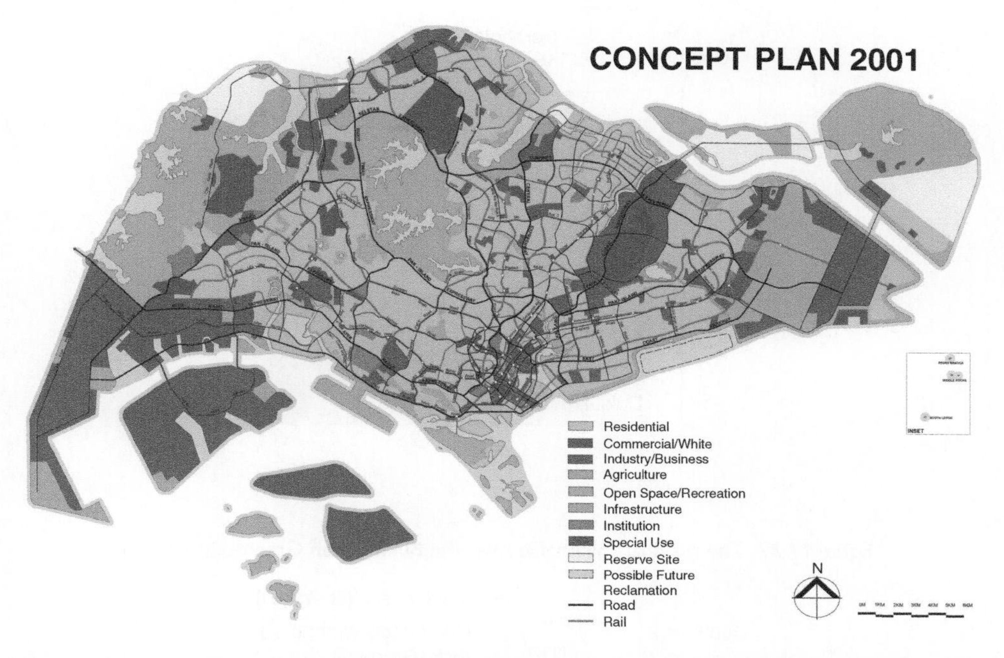

Figure 11.26 Present and future land use in Singapore (http://www.ura.gov.sg/conceptplan2001/index.html)

expression of the Bukit Timah Granite and the Jurong Formation with residual soil cover.

Land use

The land area of Singapore is 660 square kilometers and is heavily urbanised. The relief map showing the built up area and map of the Concept Plan 2001, given in Figure 11.26, which maps out the vision for the next 40 to 50 years, and shows the present and future land use. It is evident that a high proportion of the land surface is constituted from disturbed ground.

Residual soil cover

Two residual soil profiles from boreholes, over igneous and sedimentary bedrocks, described by Rahardjo *et al.* (2004) are as follows. The residual soil profile over the Bukit Timah Granite is shown in Figure 11.27. Two distinct types of residual soils were found in the boreholes of the Bukit Timah formation. The upper portion of the soil (0 to 9 m depth, borehole Y3) was completely weathered and transformed into Grade VI residual soil. The soils of this layer are clayey silt with smooth-textured soil particles, and the remnant parent rock is not observed. The soil colour ranged from reddish-brown to orange brown. At about 9 m depth, the colour gradually changes from orange-brown to yellowish-brown with white spots. From 9 to 21 m depth,

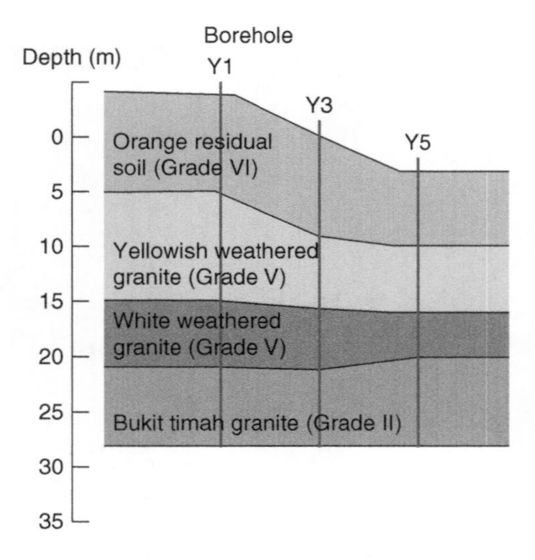

Figure 11.27 The residual soil profile over the Bukit Timah Granite (at Yishun)

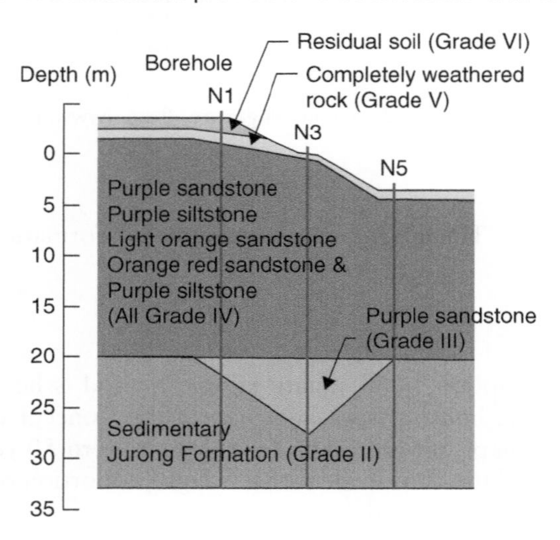

Figure 11.28 Residual soil profile over the Jurong Formation (at the Nanyang Technical University campus)

completely weathered (Grade V) granitic rock was observed. The colour changed gradually from yellowish to whitish with green and grey spots. The most apparent manifestation of residual soil at this layer was the rough texture of soil particles.

The residual soil profile over the Jurong Formation is shown in Figure 11.28. The Jurong sedimentary soil profile (borehole N3) has a purple clayey silt residual soil surface layer (Grade VI) and a completely weathered rock layer (Grade V) at 1 to 2 m depth from the ground surface. These upper layers are underlain by a mixture of highly weathered sandstone (Grade IV) that runs down to 20 m depth. From 3 to

Table 11.5 Properties of the weathered residual profile over the Bukit Timah Granite (*After* Sharma *et al.*, 1999)

Weathering Grade	ρ_{bulk} *(g/cm³)*	*SPT N-value*	*RQD (%)*	$I_{s(50)}$ *(MPa)*	δ_c *(MPa)*	V_p *(km/s)*	E_d *(GPa)*	$k(\times 10^{-9})$ *(m/s)*
Residual soil (VI)	2.09	13	–	–	–	1.2	4.5	1.0*
Completely weathered (V)	2.14	33	–	–	–	0.9	5.4	–
Highly weathered (IV)	2.32	–	45	1.8	32	0.7	5.4	5.12
Moderately weathered (III)	2.43	–	83	5.6	88	4.8	33.2	1.80
Slightly weathered	2.54	–	96	9.9	165	5.6	56.2	1.59
Fresh granite	2.66	–	99	11.1	192	5.8	60.3	0.58

Note: ρ_{bulk} = Bulk Density; SPT = Standard Penetration Test; RQD = Rock Quality Designation; $I_{s(50)}$ = Point Load Index; δ_c = Uniaxial Compressive Strength; V_p = Velocity of Compression Wave; E_d = Dynamic Modulus of Elasticity; k = Rock Mass Permeability
*In the normally consolidated range

7 m depth, a layer of weathered purple sandstone was encountered; and between 7 and 8 m depths, a purple weathered siltstone with white spot was encountered. Below 8 m depth, light orange and pink silty weathered sandstone was encountered. Between 12 and 19 m depth, porous orange and brick-red silty sandstone was encountered. Moderately weathered purple sandy siltstones (Grade III) with white spots occupy the layer between 19 and 27 m depth.

Karstic topography

The existence of limestone rock in Singapore was not detected or recorded during the publication of the "Geology of the Republic of Singapore" by PWD (1976), but the presence of limestone rock is acknowledged and greater attention is being paid during site investigations to detect its presence (Pakianathan and Jeyatharan, 2005). From boreholes drilled down to depths ranging from 20 m to 60 m below ground level in Jurong, Singapore, it is evident that limestone rock is in contact with other sedimentary rocks under reclaimed land. Limestone rock is also met in several boreholes drilled below the seabed but its seaward extent remains unknown.

Soil physico-chemical properties

Typical properties of the weathered Bukit Timah granite are summarised in Table 11.5. It should be noted that the mechanical and engineering properties change significantly at different weathering grades.

Engineering properties of residual soils of the Jurong Formation are summarised in Table 11.6.

Leong *et al.* (2003) highlighted the engineering problems of tropical residual soils in Singapore. The variable nature of the residual soils makes quantification of their engineering properties difficult. This is further compounded by the fact that the index properties of the residual soils are affected by the drying of the soil samples. Sampling residual soils for laboratory tests invariably introduces soil disturbance (Lee *et al.*, 1985; Zhao and Lo, 1994) and therefore introduces a larger scatter in the test results.

Table 11.6 Engineering properties of residual soils over the Jurong Formation (Sharma *et al.*, 1999)

Property	*Range*
Natural Water Content (w), in %	15 to 45
Bulk Density (ρ_{bulk}), in g/cm^3	1.8 to 2.2
Specific Gravity (G_s)	2.6 to 2.75
Liquid Limit (w_L), in %	28 to 60
Plastic Limit (w_p), in %	14 to 36
Permeability (k), in m/s *	10^{-6} to 10^{-9}
Compression Index (C_c)	0.1 to 0.6
Cohesion (c'), in kPa *	0 to 40
Angle of Internal Friction (ϕ'), in ° *	24 to 40

*Soils tested at both saturated and unsaturated states.

Zhao and Lo (1994) indicated that 50 mm diameter jacked-in thin-walled samplers or 100 mm diameter open samplers are suitable in firm to stiff deposits, while 63 mm diameter triple tube rotary coring is suitable in stiff to hard deposits.

11.4.1 Engineering applications and problem

Landslides

One common engineering problem encountered in Singapore residual soils is the failure of slopes due to rainfall (Pitts, 1983; Pitts and Cy, 1987; Tan *et al.*, 1987; Yang and Tang, 1997; Toll *et al.*, 1999; Toll, 2001). The factors affecting the stability of a residual soil slope due to rainfall are complex and are usually not independent of each other (Leong and Rahardjo, 1997). The quantification of the soil suction contribution to shear strength, the rate of suction loss due to rainwater infiltration and the rate of suction recovery after rainfall, are still topics of intense research in Singapore (Ramaswamy and Aziz, 1980; Pitts and Cy, 1987; Chatterjea, 1989, 1994; Lim, 1995; Gasmo, 1997; Leong *et al.*, 1998; Tsaparas *et al.*, 2003; Rezaur *et al.*, 2003; Rahardjo *et al.*, 2005; Rahardjo *et al.*, 2007; Karthikeyan *et al.*, 2008). The studies in Singapore show that, in general, matric suction in the soil increased during dry periods when evaporation was predominant; resulting in increased shear strength. Matric suction and shear strength decreased during wet periods when infiltration was predominant. The maximum changes were found to occur near the ground surface. Short duration rainfalls were shown to leave a considerable amount of matric suction in the soil, but prolonged heavy rainfalls destroyed matric suction in the soil zone near the ground surface even to the extent that a perched water table would be formed.

According to Toll (2001), minor, shallow landslides have occurred frequently on the island of Singapore. However, very few major landslides (greater than 10 m in height) have occurred. Slope failures in the sedimentary Jurong Formation and granitic Bukit Timah terrains have occurred largely on slopes with angles greater than or equal to 27°. Rainfall has been the dominant triggering event for landslides in Singapore. Observations of past landslide events suggest that a total rainfall of 100 mm within a six-day period is sufficient for minor landslides to take place. The equivalent condition for major landslides would appear to be 320 mm within 16 days but this is based on very limited data.

Excavation and tunneling

Problems related to design and construction for excavation and tunnelling in the Singapore residual soils have been described by Shirlaw *et al.* (2000), who highlighted the fact that the construction in residual soils presents difficulties in terms of a mixed ground condition where the construction process encounters several weathering grades within the same rock formation. In design, there is an uncertainty in choosing the appropriate design approach as the properties of the residual soils lie between those of a rock and a transported soil. Although residual soil is technically a soil, material features such as residual cementation and bonding, and a pore structure developed from the original rock structure is present in the residual soil. Typical design parameters relevant to transported soils must be applied with caution and engineering judgement to residual soils. Generally, the stiffness and strength of the residual soils is high when unsaturated, but may decrease rapidly with saturation. Lee *et al.* (2001) reported the effects of groundwater ingress in a soldier-piled excavation in a granitic residual soil site. The groundwater causes the soil to soften and leads to a large increase in wall deflection. However, the failure occurs only in localised soil zones.

Foundations

Broms *et al.* (1988) evaluated methods for estimating the bearing capacity and settlement of bored piles in residual soils and weathered rocks in Singapore. At working load, the bored piles carry the applied load mainly by skin friction. Uncertainties are associated with the determination of the undrained shear strength of the residual soils and weathered rocks as gradual softening takes place in the material around the borehole after the drilling and during the casting of the concrete. Pile load tests on the bored piles are difficult and costly due to the high required applied load.

REFERENCES

Ali, F. (1990). Improvement of a residual soil. *Proc. of the tenth Southeast Asian Geotechnical Conference, Taipei, Republic of China*, April 16–20, pp. 1–8.

Anon. (1970). The logging of rock cores for engineering purposes. *Quarterly Journal of Engineering Geology*, 3: 1–24.

Anon. (1972). The preparation of maps and plans in terms of engineering geology: Working Party Report. *Quarterly Journal of Engineering Geology*, 5: 193–381.

Anon. (1977). The description of rock masses for engineering purposes: Working Party Report. *Quarterly Journal of Engineering Geology*, 10: 355–388.

Asian Disaster Preparedness Center, ADPC. (2010). *Hazard Profile of Myanmar*.

ASTM Standard D2435. (2004). *Standard Test Methods for One-Dimensional Consolidation Properties of Soils Using Incremental Loading*, ASTM International, West Conshohocken, PA, US. www.astm.org.

Au Yong M.H. and Tan, B.K. (1984). Construction materials for the Sembrong and Bekok Dams, Johor. *Bulletin Geological Soc. Malaysia*, 17: 49–60.

Azwari, H.M. (2001). *Sifatfiziko-kimiadanmineralogilempungtanahbakisekitarbandar Yong Peng, Johor.* B.Sc. Hons. Thesis, Geology Programme. FST, Selangor: UKM.

Broms, B.B., Chang, M.F. and Goh, A.T.C. (1988). Bored piles in residual soil and weathered rocks in Singapore. In Van Impe (ed.), *Deep Foundations on Bored and Auger Piles*, pp. 17–34. Rotterdam: Balkema.

Buol, S.W., Southard, R.J., Graham, R.C. and McDaniel P.A. (1997). *Soil Genesis and Classification*, Iowa State Press, US, pp. 79–84.

Chatterjea, K. (1989). *Observations on the fluvial and slope processes in Singapore and their impact on the urban environment.* PhD Thesis, National University of Singapore.

Chatterjea, K. (1994). Dynamics of fluvial and slope processes in the changing geomorphic environment of Singapore, *Earth Surf. Processes Landforms*, 19: 585–607.

Cole, B.A., Chalaw, R., Pramote, M., Liggins, T.B. and Suphon, C. (1977). Dispersive clay in irrigation dams in Thailand. In: J.L. Sherard and R.S. Decker (eds.) *Dispersive Clays, Related Piping, and Erosion in Geotechnical Projects*, ASTM STP 623, American Society for Testing and Minerals, pp. 25–41.

Dames & Moore Inc. (1983). *Mass Rapid Transit System, Singapore: Detailed Geotechnical Study.* Interpretative Report, prepared for Provisional Mass Rapid Transit Authority, Singapore.

Department of Mineral Resource, Thailand (2007). *Internal Report of Geophysical Survey Using Resistivity Method in Doi Tung Development Project, Chiangrai* (in Thai).

Department of Mineral Resource (1999). *Geology of Thailand.* http://www.dmr.go.th (in Thai).

Dheeradilok, P. (1995). Quaternary coastal morphology and deposition in Thailand. *Quaternary International*, 26: 49–54.

Gasmo, J.M. (1997). *Stability of unsaturated residual soil slopes as affected by rainfall.* Masters Thesis, School of Civil & Structural Engineering, Nanyang Technological University, Singapore.

GSDM, Geological Survey of Malaysia (1996). *Annual Report*, Ministry of Primary Industries.

Hamdan, M. and Tarique, A. (1995). The Petronas Towers, the tallest building in the world. *Journal Institution of Engineers Malaysia* 56(1): 119–134.

Hengchaovanich, D. (2000). Vetiver grass technique VGT: a bioengineering and phytoremediation option for the new millennium. *Proc. of the Second International Vetiver Conference (ICV2)*, Thailand, January 2000, pp. 23–27.

Huat, B.B.K. and Muhammad, M.M. (1997). Effect of lime admixture on geotechnical properties of tropical residual soil. *Proc. International Conference on Ground Improvement Technique*, 6–8 May, Macau.

Institute of Engineers Malaysia (IEM). (1995). *Proc. of the Symposium on Hillside Development: Engineering Practice and Local By-Laws.*

Institute of Engineers Malaysia (IEM). (1996). *Proc. of the 12th S.E. Asian Geotechnical Conference, Kuala Lumpur.*

Institute of Engineers Malaysia (IEM). (1985). *Proc. of the 8th S. E. Asian Geotechnical Conference, Kuala Lumpur.*

Institute of Engineers Malaysia-Geological Society of Malaysia (IEM-GSM). (1999). *Proc. of the IEM-GSM Forum on "Karst: Geology & Engineering"*, Kuala Lumpur.

Jabatan Kerja Raya (JKR). (1985). *Pavement Design Guide.* Kuala Lumpur.

Jennings, J.E. and Knight, K. (1957). The prediction of total heave from the double oedometer test. *Proc., Symp. Expansive Clays (South Africa Institute of Civil Eng.), Johannesberg*, 7(9): 13–19.

Jotisankasa, A. and Vathananukij, H. (2008). Investigation of soil moisture characteristics of landslide-prone slopes in Thailand. *Proc.of the International Conference on Management of Landslide Hazard in the Asia-Pacific Region*, Sendai, Japan, November 11–15, p. 12.

Jotisankasa, A. and Mairaing, W. (2009). Suction-monitored direct shear testing of residual soils from landslide-prone areas. *Journal of Geotechnical and Geoenvironmental Engineering*, ASCE, 1: 119–138.

Jotisankasa, A. and Porlila, W. (2008). Development of a landslide monitoring system. *Proc. of the 2nd Technology and Innovation for Sustainable Development Conference*, January 28–29, KhonKaen, Thailand (in Thai).

Jotisankasa, A., Porlila, W., Soralump, S., Mairiang W. (2007). Development of a low cost miniature tensiometer and its applications. *Proc.3rd Asian Conference on Unsaturated Soils (Unsat-Asia 2007)*, Nanjing, China, pp. 475–480.

Jworchan, I. (2006). Minerology and chemical propertries of residual soils, Paper Number 21, *IAEG, Geological Soc. of London* (http://iaeg2006.geolsoc.org.uk/cd/PAPERS/IAEG_021.PDF).

Kannica, Y., Lek, M., Pisuth, V. and Hari, E. (1997). Clay mineralogy of Thai soils, *Applied Clay Science*, 11: 357–371.

Karthikeyan, M., Toll, D.G. and Phoon, K.K. (2008). Prediction of changes in pore-water pressure response due to rainfall events. *Proc. of the 1st European Conference on Unsaturated Soils: Advance in Geotechnical Engineering*, E-Unsat 2008, Toll *et al.* (eds), 2–4 July 2008, Durham, United Kingdom, pp. 829–834. London: Taylor & Franics.

Lee, F.H., Tan, T.S. and Wang, X.N. (2001). Groundwater effects in soldier-piled excavations in residual soils. *Proc. of Underground Singapore 2001*, Singapore, pp. 249–260.

Lee, S.L., Lo, K.W. and Leung, C.F. (1985). Sampling and testing of residual soils in Singapore. In: Brand, E.W. & Phillipson, H.B. (eds), *Sampling and Testing of Residual Soils*, pp. 153–157.

Leong, E.C. and Rahardjo, H. (1997). Factors affecting slope instability due to rainwater infiltration. *Proc. 2nd Japan National Symp. on Environmental Geotechnology*, Kyoto, Japan, pp. 163–168.

Leong, E.C. and Rahardjo, H. (1998). A review on soil classification systems. *Proc. of the International Symposium on Problematic Soils, IS-Tohoku '98, Sendai, Japan*, pp. 493–497. Rotterdam: Balkema.

Leong, E.C., H. Rahardjo, H. and S.K. Tang, S.K. (2003). Characterization and engineering properties of Singapore residual soils. *Proc. of International Workshop on Characterization and Engineering Properties of Natural Soils, Singapore*, pp. 1279–1304. Rotterdam: A. A. Balkema.

Lim, T.T. (1995). *Shear strength characteristics and rainfall induced matric suction changes in a residual soil slope*. Master Eng Thesis, School of Civil & Structural Engineering, Nanyang Technological University, Singapore.

Little, A.L. (1969). The engineering classification of residual tropical soils. *Proc. specialty session on the engineering properties of Lateritic Soil, vol. 1, 7th Int. conf. soil mechanics & foundation engineering*, Mexico City, Vol. 1, pp. 1–10.

Little, A.L. (1969). The engineering classification of residual soils. *Proc. 7th Seventh International Conference on Soil Mechanics and Foundation Engineering*, ISSMFE. Mexico, Vol. 1, pp. 1–10.

Moncharoen, L. (1983). The soil moisture map of Thailand. *Proc. 4th Int. Soil Classification Workshop, Rwanda* (Agric. Ed., 4) ASOB-AGCD, Brussels, pp. 277–295.

Newbery, J. (1971). Engineering geology in the investigation and construction of the Batang Padang hydro-electric scheme, Malaysia. *Quarterly Journal of Engineering Geology*, 3: 151–181.

Ooi, T.A. (1982). Malaysian soils and associated problems. Chapter 2 in: *Geotechnical Engineering Course*, 22 March–2 April, 1982, Univ. Malaya, Kuala Lumpur.

Pakianathan, L.J. and. Jeyatharan, K. (2005). Engineering and construction experiences in the limestones of Jurong Formation, *Proc. of The Underground Singapore Conference, Tunnelling and Underground Construction Society* (Singapore); Centre for Soft Ground Engineering,

The National University of Singapore and Geotechnical Research Centre, Nanyang Technological University, 1–2 December, Singapore.

Pitts, J. (1983). The form and causes of slope failures in an area of west Singapore Island, Singapore. *Journal of Tropical Geography*, 4(2): 162–168.

Pitts, J. and Cy, S. (1987). Insitu soil suction measurements in relation to slope stability investigations in Singapore, *Proc. 9th European Conf. on Soil Mechanics and Foundation Engineering* (eds. Hanrahan, E T, Orr, T L L and Widdis, T F), Vol. 1, pp. 79–82. Rotterdam: Balkema.

Pitts, J. (1984). A survey of engineering geology in Singapore, *J. of Southeast Asian Geotechnical Society*, 15: 1–20.

Poh, K.B., Chuah, H.L. and Tan, S.B. (1985). Residual granite soils of Singapore, *Proc. 8th SouthEast Asian Geotechnical Conf.*, Southeast Asian Geotechnical Society, Kuala Lumpur, Malaysia, 1(3), 1–9.

Public Works Department (1976). *Geology of the Rep. of Singapore*, Public Works Department, Singapore.

Rahardjo, H., Lee, T.T., Leong, E.C. and Rezaur, R.B. (2005). Response of a residual soil slope to rainfall. *Canadian Geotechnical Journal*, 42: 340–351.

Rahardjo, H., Ong, T.H., Rezaur, R.B. and Leong, E.C. (2007). Factors controlling instability of homogeneous soil slopes under rainfall. *Journal of Geotechnical and Geoenvironmental Engineering*, ASCE, 133(12): 1532–1543.

Rahardjoa, H., Aungb, K.K., Leongc, E.C. and Rezaurd, R.B. (2004). Characteristics of residual soils in Singapore as formed by weathering. *Engineering Geology*, 73: 157–169.

Raj, J.K. (1987). Clay minerals in weathered shales of red beds along the Paloh-Kluang bypass highway. *Newsletter, Geological Society of Malaysia*, 13(5): 213–219.

Raj, J.K. (1995). Clay minerals in the weathering profile of a graphitic-quartz-muscovite schist in the Kajang area, Selangor. *Newsletter, Geological Soc. of Malaysia*, 21(1): 1–8.

Ramaswamy, S.D. and Aziz, M.A. (1980). Rain induced landslides of Singapore, *Proc. Int. Symp. Landslides*, New Delhi, Vol. 1, pp. 403–306.

Ramli M. (1992). Engineering the North-South Expressway on soft ground. *Proc. Int. Conf. on Geotechnical Engineering '92 (Geotropika '92), 21–23 April, 1992, Johor Bahru*, pp. 337–356.

Rezaur, R.B., Rahardjo, H., Leong, E.C. and Lee, T.T. (2003). Hydrologic behavior of residual soil slopes. *Journal of Hydrologic Engineering*, ASCE, 8(3): 133–144.

Sharma, J.S., Chu, J. and Zhao, J. (1999). Geological and geotechnical features of Singapore: An Overview. *Tunnelling and Underground Space Technology*, 14(4): 419–431.

Sherard, J.L.F., Dunnigan, L.P. and Decker, R.S. (1976). Identification and Nature of Dispersive Soils, *Journal of the Geotechnical Engineering Division*, ASCE, 102: 287–301.

Shirlaw, J.N., Hencher, S.R. and Zhao, J. (2000). Design and construction issues for excavation and tunnelling in some tropically weathered rocks and soils. *GeoEng 2000*, Australia, pp. 1286–1329.

Singh, H. and Huat, B.B.K. (2004). Terra firma and Foundations. In: *4th Fourth International Conference on Landslides, Slope Stability, and Safety of Infrastructure*, 24–25, March, Kuala Lumpur.

Surapol, S., Seri, S., Lalit, S. and Ekawit, V. (2006). Improving the efficiency of the vetiver system in the highway slope stabilization for sustainability and saving of maintenance cost. *Proc. 4th The 4thInternational Conference on Vetiver – ICV4* Caracas, Venezuela – October 2006. Vetiver and People http://www.vetiver.org/ICV4pdfs/ICV4-PROG-IN.htm

Soralump, S. and Jotisankasa, A. (2007). Mitigation of landslide hazard in Thailand. *Proc. of Expert Symposium, Climate Change Modelling, Impacts & Adaptations and Workshop on Climate Change and Slope stability*, National University of Singapore.

Soralump, S. and Bunpoat, K. (2006). Landslide risk prioritization of Tsunami affected area in Thailand. *Proc. International Symposium on Environmental Engineering and 5th Regional Symposium on Infrastructure Development in Civil Engineering, Philippines.*

Soralump, S. Pungsuwang, D., Chantasorn, M., Inmala, N. and Alambepola, N.M.S. (2010). Landslide risk management of Patong city: Demonstration of geotechnical engineering approach, *Proc., Int. Conf. on Slope 2010: Geotechnique and Geosynthetics for Slopes*, 27–30 July, Chianmai, Thailand. pp. 37–42.

Taesiri, Y. and Yuwathanon, K. (2005). Prevention and stabilization methods for slope movements. *Internal Report of Seminar by Bureau of Materials, Analysis and Inspection*, Department of Highways, Ministry of Transport and Communication (in Thai).

Tai, T.O. (1999). *Sifatfiziko-kimiadanmineralogitanahsekitar Kota Melaka.* B.Sc. Hons. Thesis, Geology Dept., FST. Selangor: UKM.

Tan, B.K. (1983). Geotechnical aspects of the Kenyir Dam project, Trengganu, Peninsular Malaysia. *Proc. 5th Int. Congress on Rock Mechanics, ISRM*, 10–15 April 1983, Melbourne: C133–C137.

Tan, B.K. (1986). Geology and urban development of Kuala Lumpur, Malaysia. *Proc. Landplan III Symp., 15–20 Dec. 1986, Hong Kong*, pp. 127–140. Geological Society of Hong Kong Bulletin (3), Oct. 1987).

Tan, B.K. (1988). A short note on the occurrence of a soft soil zone above limestone bedrock. *Proc. Int. Conf. on Calcareous Sediments, Perth, March 15–18*, Vol. 1, pp. 35–39.

Tan, B.K. (1990a). Kajianciritanah & batuansertakestabilancerun di kawasanperbandaran Pulau Pinang (Study of soil and rock properties and slope stability in urban areas of Penang). *Final Report, Research Project No. 81/87*, July 1990, UKM.

Tan, B.K. (1990b). Subsurface geology of Ipoh area, Perak, Malaysia. *Proc. Conf. Karst Geology in Hong Kong*, 5–7 Jan. 1990, Hong Kong. Geological Society of Hong Kong Bulletin, 4: 155–166.

Tan, B.K. (1992). A survey of slope failures along the Senawang – Air Keroh Highway, Negeri Sembilan/Melaka, Malaysia. *Proc. 6th Int. Symposium on Landslides, 10–14 Feb 1992, Christchurch, Malaysia*, pp. 1423–1427.

Tan, B.K. (1994a). Investigations for the Gemencheh Dam, Negeri Sembilan. *Proc. Geotropika 94*, Malacca, 22–24 Aug 1994, Paper 3-3.

Tan, B.K. (1994b). Investigations for the Tawau Dam, Sabah, Malaysia. *Proc. 7th IAEG Congress*, Lisbon, Portugal, 5–9 Sept. 1994, V: 3707–3714.

Tan, B.K. (1995). Physico-chemical properties of granitic versus basaltic soils of Kuantan, Pahang. *Proc. Forum on soil & rock properties, Geological Society of Malaysia, Kuala Lumpur*, pp. 4.1–4.13.

Tan, B.K. (1996). Physico-chemical properties of some granitic soils from Peninsular Malaysia. *Proc. 12th SE Asian Geotech. Conf., 6–10 May, Kuala Lumpur* Vol. 1, pp. 595–600.

Tan, B.K. and Anizan, I. (1998). Sifatfiziko-kimiadanmineralogilempungbeberapajenistanah di Negeri Johor. *Final Report, Research Project No. S/7/96*, UKM, Sept. 1998.

Tan, B.K. and Azwari, H.M. (2001). Physico-chemical properties of carbonaceous shale soils in the Yong Peng area, Johor. *Proc. Geological Society of Malaysia Annual Geological Conf., PulauPangkor, Malaysia*, pp. 229–232.

Tan, B.K. and Ong, C.Y. (1993). Physico-chemical properties of granitic soils along the Ipoh-Ct. Jering Expressway, Perak, Malaysia. *Proc. 11th SE Asian Geotechnical Conf., Singapore*, pp. 217–222.

Tan, B.K. and Tai, T.O. (1999). Physico-chemical properties of graphitic schist soils from Malacca, Peninsular Malaysia. *Proc. 2nd Asian Symp. on Engineering Geology & Environment, Bangi*, Malaysia, pp. 2-8 to 2-10 (Extended Abstract).

Tan, B.K. and Yew, C.K. (2002). Physico-chemical properties of andesitic soils in the Kg. Awah area, Pahang. *Bulletin no. 45, Geological Soc. of Malaysia Annual Geological Conf.*, Kota Bharu, Malaysia, pp. 31–35.

Tan, B.K. and Zulhaimi, A.R. (2000). Physico-chemical properties of graphitic schist soils in the Rawang area, Selangor. *Proc. Geological Society of Malaysia Annual Geological Conf., Pulau Pinang*: Malaysia, pp. 283–286.

Tan, B.K. (1995). Some experiences on the weathering of rocks and its engineering significance in Malaysia, IKRAM, *Geotechnical Meeting '95, Comparative Geotechnical Engineering Practice, 7–9, June, Penang, Malaysia*, 2(6): 22.

Tan, B.K. (2004). Country case study: engineering geology of tropical residual soils in Malaysia. In: Huat, B.B.K. *et al.* (eds), *Tropical Residual Soils Engineering*. Rotterdam: A.A. Balkema.

Tan, B.K. (1994c). Investigations for the Gemencheh Dam, Negeri Sembilan, *Proc. Geotropika*, 94, Malacca, 22–24, Aug., Paper 3-3.

Tan, S.B., Tan, S.L., Lim, T.L. and Yang, K.S. (1987). Landslide problems and their control in Singapore, *Proc. 9th Southeast Asian Geotechnical Conf.*, Southeast Asian Geotechnical Soc., Bangkok, Thailand, Vol. 1, pp. 25–36.

Tarique, A. (1996). Project KLCC: Geology, soils and foundations. *Newsletter, Geol. Soc. Malaysia*, 22(2): 73–74.

Ting, W.H. (1985). Foundation in limestone areas of Malaysia. Special lecture, *Proc. 8th S.E. Asian geotechnical conf., Kuala Lumpur, Malaysia*. Vol. 2, pp. 124–136.

Ting, W.H., Ooi, T.A. and Tan, B.K. (1995). Tunnelling activities in Malaysia – Country Report. *Proc. SE Asian Symp. on Tunnelling & Underground Space Development*, 18–19 Jan. 1995, Bangkok, Thailand.

Toh, C.T. (2003). Compression of residual soils and weak rock fills. Forum on building on engineered fill. *Institution of Engineers, Malaysia*. Petaling Jaya, Malaysia.

Toll, D.G. (2001). Rainfall-induced landslides in Singapore. *Proc. Institution of Civil Engineers: Geotechnical Engineering*, 149(4): 211–216.

Toll, D.G., Rahardjo, H. and Leong, E.C. (1999). Landslides in Singapore, *Proc. 2nd International Conference on Landslides, Slope Stability and the Safety of Infrastructures, Singapore*, CI-Premier, pp. 269–276.

Tsaparas, I., Rahardjo, H., Toll, D.G. and Leong, E.C. (2003). Infiltration characteristics of two instrumented residual soil slopes. *Canadian Geotechnical Journal*, 40(5): 1012–1032.

Wieland, M. (1989). Effects of floods of November 18–23, 1988 in Southern Thailand on highway bridges and large dams. In: *Report of Two Missions of December 11–14, 1988 and of February 15–21, 1989 for Swiss Disaster Relief Unit.*

Wong, K.H. (2003). *Assessment of an acrylic polymer on the properties of soil-cement* (unpublished masters thesis). Universiti Putra Malaysia.

Yang, K.S. and Tang, S.K. (1997). Stabilising the slope of Bukit Gombak. *Proc. of 3YGEC*, Singapore. pp. 589–605.

Yew, C.K. (2002). *Sifat fiziko-kimia dan mineralogi lempung tanah baki sekitar kawasan Kg. Awah, Pahang*. B.Sc. Hons. Thesis, Geology Programme, FST. Selangor: UKM, Malaysia.

Yoothong, K., Moncharoen, L., Vijarnson, P. and Eswaran, H. (1997). Clay mineralogy of Thai soils. *Applied Clay Science*, 11: 357–371.

Zhao, J., Broms, B.B., Zhou, Y. and Choa, V. (1994). A study of the weathering of the Bukit Timah granite: Part A. Review, field observations and geophysical survey. *Bulletin of the International Association of Engineering Geology*, 49: 97–105.

Zhao, M.M. and Lo, K.W. (1994). Parameters for bearing capacity and settlement analyses of residual soils. *Proc. 7th Int. IAEG Congress*. pp. 3469–3478. Rotterdam: Balkema.

Zulhaimi, A.R. (2002). *Sifat fiziko-kimia dan mineralogi lempung tanah baki sekitar bandar Rawang, Selangor Darul Ehsan*. B.Sc. Hons. Thesis, Geology Programme, FST. Selangor: UKM, Malaysia.

Subject Index

For Product Safety Concerns and Information please contact our EU representative GPSR@taylorandfrancis.com Taylor & Francis Verlag GmbH, Kaufingerstraße 24, 80331 München, Germany